MICROBIOLOGICAL METHODS FOR ASSESSING SOIL QUALITY

MICROBIOLOGICAL METHODS FOR ASSESSING SOIL QUALITY

Edited by

Jaap Bloem
Alterra, Wageningen, The Netherlands

David W. Hopkins
University of Stirling, UK

and

Anna Benedetti
Istituto Sperimentale per la Nutrizione delle Piante, Rome, Italy

CABI Publishing

CABI Publishing is a division of CAB International

CABI Publishing
CAB International
Wallingford
Oxfordshire OX10 8DE
UK
Tel: +44 (0)1491 832111
Fax: +44 (0)1491 833508
E-mail: cabi@cabi.org
Website: www.cabi-publishing.org

CABI Publishing
875 Massachusetts Avenue
7th Floor
Cambridge, MA 02139
USA
Tel: +1 617 395 4056
Fax: +1 617 354 6875
E-mail: cabi-nao@cabi.org

© CAB International 2006. All rights reserved. No part of this publication may be reproduced in any form or by any means, electronically, mechanically, by photocopying, recording or otherwise, without the prior permission of the copyright owners.

A catalogue record for this book is available from the British Library, London, UK.

Library of Congress Cataloging-in-Publication Data

Microbiological methods for assessing soil quality / edited by Jaap Bloem, David W. Hopkins, and Anna Benedetti.
    p. cm.
  Includes index.
  ISBN 0-85199-098-3 (alk. paper)
  1. Soil microbiology. 2. Soils--Quality. 3. Soils--Analysis.
I. Bloem, Jaap, 1958- II. Hopkins, David W., Dr. III. Benedetti, Anna, Dr. IV. Title.

QR111.M39 2005
579'.1757--dc22

                                                                        2005001632

ISBN-13: 978 0 0851 99 098 9
ISBN-10: 0 85199 098 3

Typeset by Columns Design Ltd, Reading
Printed and bound in the UK by Biddles Ltd, King's Lynn

Contents

Editors		ix
Abbreviations		xi
Part I: Approaches to Defining, Monitoring, Evaluating and Managing Soil Quality		1
1	**Introduction** *Anna Benedetti and Oliver Dilly*	3
2	**Defining Soil Quality** *Richard G. Burns, Paolo Nannipieri, Anna Benedetti and David W. Hopkins*	15
3	**Monitoring and Evaluating Soil Quality** *Jaap Bloem, Anton J. Schouten, Søren J. Sørensen, Michiel Rutgers, Adri van der Werf and Anton M. Breure*	23
4	**Managing Soil Quality** *Michael Schloter, Jean Charles Munch and Fabio Tittarelli*	50
5	**Concluding Remarks** *Anna Benedetti, Philip C. Brookes and James M. Lynch*	63
Part II: Selected Methods		71
6	**Microbial Biomass and Numbers**	73
	6.1 Estimating Soil Microbial Biomass *Andreas Fließbach and Franco Widmer*	73

6.2 Microbial Biomass Measurements by Fumigation–Extraction 77
Philip C. Brookes and Rainer Georg Joergensen

6.3 Substrate-induced Respiration 84
Heinrich Höper

6.4 Enumeration and Biovolume Determination of Microbial Cells 93
Manfred Bölter, Jaap Bloem, Klaus Meiners and Rolf Möller

7 Soil Microbial Activity 114

7.1 Estimating Soil Microbial Activity 114
Oliver Dilly

7.2 Soil Respiration 117
Mikael Pell, John Stenström and Ulf Granhall

7.3 Soil Nitrogen Mineralization 127
Stefano Canali and Anna Benedetti

7.4 Nitrification in Soil 136
Annette Bollmann

7.5 Thymidine and Leucine Incorporation to Assess Bacterial Growth Rate 142
Jaap Bloem and Popko R. Bolhuis

7.6 N_2O Emissions and Denitrification from Soil 150
Ulrike Sehy, Michael Schloter, Hermann Bothe and Jean Charles Munch

7.7 Enzyme Activity Profiles and Soil Quality 158
Liz J. Shaw and Richard G. Burns

8 Soil Microbial Diversity and Community Composition 183

8.1 Estimating Soil Microbial Diversity and Community Composition 183
Jan Dirk van Elsas and Michiel Rutgers

8.2 Soil Microbial Community Fingerprinting Based on Total Community DNA or RNA 187
Jan Dirk van Elsas, Eva M. Top and Kornelia Smalla

8.3 Phospholipid Fatty Acid (PLFA) Analyses 204
Ansa Palojärvi

8.4 Substrate Utilization in Biolog™ Plates for Analysis of CLPP 212
Michiel Rutgers, Anton M. Breure and Heribert Insam

9 Plant–Microbe Interactions and Soil Quality 228

9.1 Microbial Ecology of the Rhizosphere 228
Philippe Lemanceau, Pierre Offre, Christophe Mougel, Elisa Gamalero, Yves Dessaux, Yvan Moënne-Loccoz and Graziella Berta

9.2 **Nodulating Symbiotic Bacteria and Soil Quality**   231
Alain Hartmann, Sylvie Mazurier, Dulce N. Rodríguez-Navarro, Francisco Temprano Vera, Jean-Claude Cleyet-Marel, Yves Prin, Antoine Galiana, Manuel Fernández-López, Nicolás Toro and Yvan Moënne-Loccoz

9.3 **Contribution of Arbuscular Mycorrhiza to Soil Quality and Terrestrial Ecotoxicology**   248
Silvio Gianinazzi, Emmanuelle Plumey-Jacquot, Vivienne Gianinazzi-Pearson and Corinne Leyval

9.4 **Concepts and Methods to Assess the Phytosanitary Quality of Soils**   257
Claude Alabouvette, Jos Raaijmakers, Wietse de Boer, Régina Notz, Geneviève Défago, Christian Steinberg and Philippe Lemanceau

9.5 **Free-living Plant-beneficial Microorganisms and Soil Quality**   270
Yvan Moënne-Loccoz, Sheridan L. Woo, Yaacov Okon, René Bally, Matteo Lorito, Philippe Lemanceau and Anton Hartmann

10 **Census of Microbiological Methods for Soil Quality**   296
Oliver Dilly

Index   301

Editors

Editors-in-Chief

Jaap Bloem
Department of Soil Sciences, Alterra, PO Box 47, NL-6700 AA Wageningen, The Netherlands

David W. Hopkins
School of Biological and Environmental Sciences, University of Stirling, Stirling FK9 4LA, UK

Anna Benedetti
Consiglio per la ricerca e la sperimentazione in Agricoltura, Istituto Sperimentale per la Nutrizione delle Piante, Via della Navicella, 2, 00184 Rome, Italy

Editorial Board

Richard G. Burns
School of Land and Food Sciences, The University of Queensland, Brisbane, Queensland 4072, Australia

Oliver Dilly
Lehrstuhl für Bodenschutz und Rekultivierung, Brandenburgische Technische Universität, Postfach 101344, D-03013 Cottbus, Germany

Andreas Fließbach
Research Institute of Organic Agriculture (FiBL), Ackerstrasse, CH-5070 Frick, Switzerland

Philippe Lemanceau
UMR 1229 INRA/Université de Bourgogne, 'Microbiologie et Géochimie des Sols', INRA-CMSE, BP 86510 21065, Dijon cedex, France

James M. Lynch
Forest Research, Alice Holt Lodge, Farnham GU10 4LH, UK

Yvan Moënne-Loccoz
UMR CNRS 5557 Ecologie Microbienne, Université Claude Bernard (Lyon 1), 43 bd du 11 Novembre, 69622 Villeurbanne cedex, France

Paolo Nannipieri
Dipartimento de Nutrizione delle Pianta e Scienza del Suolo, Università di Firenze, Piazzale di Cascine 28, Florence, Italy

Fabio Tittarelli
Consiglio per la ricerca e la sperimentazione in Agricoltura, Istituto Sperimentale per la Nutrizione delle Piante, Via della Navicella, 2, 00184 Rome, Italy

Jan Dirk van Elsas
Department of Microbial Ecology, Groningen University, Kerklaan 30, NL-9750 RA Haren, The Netherlands

Abbreviations

AIM	acetylene inhibition method
AM	arbuscular mycorrhiza
AO	acridine orange
AODC	acridine orange direct count
APS	ammonium persulphate
ARDRA	amplified ribosomal DNA restriction analysis
ATP	adenosine 5'-triphosphate
AUDPC	area under the disease progress curve
AWCD	average well colour development
BAS	basal respiration
(B_C/TOC)	C biomass/total organic C (ratio)
BNF	biological nitrogen fixation
CAC	citric acid cycle
CEC	cation exchange capacity
CEN	Comité Européen de Normalisation
CFE	chloroform fumigation–extraction
CFU	colony-forming units
CLPP	community-level physiological profiles
C_{mic}	microbial biomass
COST	COopération dans le domaine de la recherche Scientifique et Technique
CSLM	confocal scanning laser microscopy
CV	coefficient of variation
CWDEs	cell-wall-degrading enzymes
DAPI	4',6-diamidino-2-phenylindole-dihydrochloride
DEPC	diethyl-pyrocarbonate
DFS	differential fluorescent stain
DGGE	denaturing gradient gel electrophoresis
DMSO	dimethyl sulphoxide

dNTP	deoxynucleoside 5'-triphosphate
dpm	disintegrations per minute
dps	disintegrations per second
dsDNA	double-stranded DNA
DTAF	5-(4,6-dichlorotriazin-2-yl) aminofluorescein
dTTP	deoxythymidine triphosphate
DW	dry weight
EAP	Environmental Action Programme
EL	ester-linked
EPA	Environmental Protection Agency
EU	European Union
FAME	fatty acid methyl ester (analysis)
FAO	Food and Agriculture Organization
FB	Fluorescent Brightener
FDA	fluorescein diacetate
FID	flame ionization detector
FISH	fluorescence *in situ* hybridization
GC	gas chromatograph
GC–MS	GC coupled with a mass spectrometer
GM	genetically modified
HPLC	high-pressure liquid chromatograph
IC	ion chromatograph
INT	iodonitrotetrazolium chloride
INTF	iodonitrotetrazolium formazan
IPP	intact phospholipid profiling
IR	infrared
ISO	International Organization for Standardization
KAc	potassium acetate
MIC50	mean inhibitory concentration at 50%
MIDI or MIS	Microbial Identification System
MN_{Bas}	basal nitrogen mineralization
MPN	most probable number
MST	mean survival time
MUB	modified universal buffer
MUF	methylumbelliferyl
Nbio	N in microbial biomass
Ndfa	nitrogen derived from the atmosphere
NMDS	non-metric multidimensional scaling
Ntot	total soil nitrogen
OECD	Organization for Economic Co-operation and Development
p.a.	pro analysis (reagent purity)
PBS	phosphate-buffered saline
PCA	principal components analysis
PCR	polymerase chain reaction
PGPF	plant-growth-promoting fungi
PGPR	plant-growth-promoting rhizobacteria
PLFA	phospholipid fatty acid (analysis)

p-NPP	*p*-nitrophenyl phosphate
qCO_2	metabolic quotient
Q_N	nitrogen mineralization quotient
RAPD	random amplified polymorphic DNA
RCF	relative centrifugal force
rpm	revolutions/minute
RQ	respiratory quotient
RS	ripper subsoiling
SDS	sodium dodecyl sulphate
SEM	scanning electron microscopy
SINDI	Soil Indicators (New Zealand)
SIR	substrate-induced respiration
SOM	soil organic matter
SQI	Soil Quality Index
SSC	standard saline citrate
SSCP	single-strand conformation polymorphism
SSSA	Soil Science Society of America
TEM	transmission electron microscopy
TGGE	temperature gradient gel electrophoresis
T-RFLP	terminal restriction fragment length polymorphism
TY	tryptone-yeast extract
UV	ultraviolet
v/v	volume in volume
WHC	water-holding capacity
w/v	weight in volume

I Approaches to Defining, Monitoring, Evaluating and Managing Soil Quality

1 Introduction

ANNA BENEDETTI[1] AND OLIVER DILLY[2]

[1]*Consiglio per la ricerca e la sperimentazione in Agricoltura, Istituto Sperimentale per la Nutrizione delle Piante, Via della Navicella, 2, 00184 Rome, Italy;* [2]*Lehrstuhl für Bodenschutz und Rekultivierung, Brandenburgische Technische Universität, Postfach 101344, D-03013 Cottbus, Germany*

Introduction

Having adopted the Treaty on Biological Diversity of Rio de Janeiro (UNCED, 1992), many governments are becoming increasingly concerned about sustaining biodiversity and maintaining life support functions. In several countries, national or regional programmes have been established to monitor soil quality and/or the state of biodiversity. Most monitoring programmes include microbiological indicators, because soil microorganisms have key functions in decomposition and nutrient cycling, respond promptly to changes in the environment and reflect the sum of all factors regulating nutrient cycling (see also Chapter 3). Currently the European Union (EU) and many countries all over the world are working on legislation for the protection of soil quality and biodiversity. Policy makers, as well as land users, need indicators and monitoring systems to enable them to report on trends for the future and to evaluate the effects of soil management. This book details approaches and microbiological methods for assessing soil quality.

The European Commission has been promoting cooperation and the coordination of nationally funded research through so-called COST actions ('COopération dans le domaine de la recherche Scientifique et Technique'; http://cost.cordis.lu/src/whatiscost.cfm, accessed 27 April 2004). COST Action 831 'Biotechnology of Soil: Monitoring, Conservation and Remediation' started in October 1997 and ended in December 2002. An important aim of COST Action 831 was the development of a handbook on microbiological methods for assessing soil quality. COST Action 831 has enabled working groups of European soil microbiologists to discuss and evaluate the potential use of microbiological, biochemical and molecular tools to assess soil quality. The scientific community is constantly challenged by operative institutions, such as national and local authorities, state boards, private boards, consultants and standardization agencies, to deliver feasible methods for acquiring representative biological data on soil quality. This is extremely difficult, since soil microorganisms respond and adapt rapidly to environmental conditions. In addition, the impacts caused by human activities may be barely distinguishable from natural fluctuations, especially when changes are detected late and comparison with historical data or unaffected control sites is not possible.

Various authors have made numerous suggestions. For instance, Domsch (1980) and Domsch *et al.* (1983) proposed that any alteration, caused by either natural agents or pollutants, which returns to normal microbiological values within 30 days should be considered normal fluctuation; alterations lasting for 60 days can be regarded as tolerable, whereas those persisting for over 90 days are stress agents. Brookes (1995) suggested that no parameter should be used alone, but that related parameters should be identified and utilized together as an 'internal control', e.g. biomass carbon (C) and total soil organic C. In general, there is an approximate linear relationship between these two variables, so when soils show marked variations from what is considered to be the normal ratio between biomass C and total organic C in a particular soil management system, climate and soil type, this ratio becomes an indicator of deterioration and change in soil ecosystem functions.

Criteria for Indicators of Soil Quality

Criteria for indicators of soil quality relate mainly to: (i) their utility in defining ecosystem processes; (ii) their ability to integrate physical, chemical and biological properties; and (iii) their sensitivity to management and climatic variations (Doran, 2000). These criteria apply to soil organisms, which are thus useful indicators of sustainable land management. Ideally, soil organisms and ecological indicators should be:

1. Sensitive to variations in management;
2. Well correlated with beneficial soil functions;
3. Useful for elucidating ecosystem processes;
4. Comprehensible and useful to land managers;
5. Easy and inexpensive to measure.

Brookes (1995) proposed the following criteria for selecting a microbiological parameter as an indicator of soil pollution.

1. It should be possible to determine the property of interest accurately and precisely in a wide range of soil types and conditions.
2. Determination should be easy and of low cost, as many samples must be analysed.
3. The nature of the parameter must be such that control determinations are also possible, so that the effect of the pollutant can be assessed exactly.
4. The parameter must be sensitive enough to detect pollution, but also stable enough to avoid false alarms.

5. The parameter must have general scientific validity based on reliable scientific knowledge.
6. If the reliability of a single parameter is limited, two or more independent parameters should be selected. In this case their interrelations in unpolluted areas must also be known.

These two approaches are synergetic, as the criteria proposed by Doran (2000) focus on the sphere of interest, while Brookes' (1995) criteria identify the requisites of an indicator.

Two crucial points had to be clarified by the working groups of COST Action 831 before any choice of, or suggestion about, microbial indicators of soil quality was made:

1. Who is the handbook for?
2. How do we define soil quality?

Potential Users of this Handbook

This handbook is aimed at professionals, students and organizations working in the field of agriculture and the environment, such as:

- soil scientists, colleges, universities, libraries;
- consultants in environmental risk assessment and soil management;
- analysis laboratories, e.g. those involved in ecological monitoring;
- international (e.g. EU, Organization for Economic Cooperation and Development (OECD), Food and Agriculture Organization (FAO)), national, regional and local authorities involved in soil protection and management;
- international (e.g. ISO and CEN) and national standardization agencies.

It aims to provide clear instructions to technicians operating outside of the scientific research sector, and is meant to provide a seamless link between science and application. In contrast to earlier books on microbiological methods (for instance Alef and Nannipieri, 1995), this handbook focuses on a limited number of methods which are applicable, or already applied, in regional or national soil quality monitoring programmes. It also provides an overview of monitoring programmes implemented in several countries.

The people who create, study and assess innovative solutions using scientific methods are seldom involved directly in transferring information to end-users. This can create a knowledge gap that often leads to misinformation or poor information. The purpose of this book is to provide applicable microbiological methods for assessing soil quality. Part I provides an overview of approaches to defining, monitoring, evaluating and managing soil quality. In Part II, methods are described in sufficient detail to enable this handbook to be used as a practical guide in the laboratory. Finally, Chapter 10 gives a census of the main methods used in over 30 European soil microbiological laboratories.

Defining Soil Quality

During a COST 831 Joint Working Groups meeting on 'Defining soil quality', held in Rome in December 1998 (Benedetti *et al.*, 2000), there was broad discussion about the criteria for the definition of 'soil quality and/or qualities of soils'. This can be applied to a wide range of agricultural soils, forestry soils, grazing pastures, natural environment soils, etc., and may include different climate zones. However, the focus of our activities is in the COST domain of agriculture and biotechnology. An overview on defining soil quality is given in Chapter 2.

Evaluating Soil Quality

Once the aim and the potential users of the handbook had been defined, the next step was to establish how to evaluate soil quality, and which parameters and methods to adopt. Many questions had to be answered and were debated during a Joint Working Groups meeting on 'Evaluating soil quality', in Kiel, Germany (May 2000). The issues ranged from problems related to sampling, storage and pre-incubation of soil samples for microbiological analyses, to the choice of the most efficient methods and indicators (Bloem and Breure, 2003). An overview on evaluating soil quality is given in Chapter 3.

The methods can be divided into four groups, depending on the information they can provide:

1. Soil microbial biomass and number.
2. Soil microbial activity.
3. Soil microbial diversity and community structure.
4. Plant–microbe interactions.

Soil microbial biomass and activity are relatively easy to determine using routine methods, and are used to assess soil quality. For monitoring programmes where large amounts of samples have to be processed, often the soil is sieved, mixed and pre-incubated under standardized conditions in the laboratory to reduce variation and to facilitate comparison between different locations and different sampling dates. Direct analyses of microbial biomass and activity of field samples are also possible, and are often performed in more fundamental research. However, the higher variation found in direct analysis usually requires more replicates in space and time than with pre-incubated samples. Compared to biomass and activity, soil microbial diversity and community structure is more complicated to measure, and requires more specialized techniques, which are less easy to standardize. However, molecular techniques for their assessment are rapidly improving. The study of plant–microbe interactions is also relatively specialized and time consuming, and often requires *in situ* determinations that are rarely performed in optimal conditions. Field temperature and humidity can vary greatly and also reach extreme values which are very unfavourable for microbiological activity. Moreover, substrate concentration and pH values are seldom optimal.

Methods

Once the parameters and methods for assessing soil quality had been selected, the detailed protocol for each method was proposed and discussed during a combined meeting of working groups on 'Microbiological methods for soil quality', in Wageningen, The Netherlands (November 2001). Here, the preparation of the methods section of the handbook was initiated.

1. Soil microbial biomass and number

All the methods capable of defining the weight and number of soil microorganisms in a soil sample are included. The conventional methods for determining numbers of microbes living in soil are based on viable or direct counting procedures (Zuberer, 1994; Alef, 1995; Alef and Nannipieri, 1995; Dobereiner, 1995; Lorch *et al.*, 1995). Viable counting procedures require culturable cells and comprise two approaches: the plate count technique and the most probable number (MPN) technique. Some unculturable soil microorganisms may be potentially culturable if adequate nutritional conditions for their growth could be provided. However, many remain unculturable because they are dormant and require special resuscitation before regaining the ability to grow; or they are non-viable but still intact and detectable by microscopy (Madsen, 1996). Using specific culture media, specific functional groups of microbes can be counted. However, even with general growth media, the numbers of microbes detected are usually at least an order of magnitude lower than those obtained by direct microscopy.

Direct enumeration techniques allow the counting of total numbers of both bacteria and fungi, but usually give no indication of the composition of the respective communities. Generally, with these techniques, a known amount of homogenized soil suspension is placed on a known area of a microscope slide, the microorganisms are then stained with a fluorescent dye and are counted using a microscope (Bloem *et al.*, 1995). A disadvantage of microscopic counts is that visual counting is subjective and relatively time consuming. Therefore biochemical and physiological methods, e.g. chloroform fumigation extraction of microbial carbon and nitrogen, and substrate-induced respiration, are most commonly used (Chapter 6).

2. Soil microbial activity

Biochemical techniques are described that reveal information about the metabolic processes of microbial communities, both in their entirety (e.g.

respiration and mineralization) and according to functional groups (e.g. nitrification and denitrification).

Microbial activity can be divided into potential and actual activity. *Actual activity* means the activity microorganisms develop when conditions necessary for metabolism are less than optimal, as occurs in the open field. This activity can be determined using field sensors, but to date no serial and routine methods are available. Therefore potential activity is usually determined. *Potential activity* means metabolic activity, including enzymatic activities, that soil microorganisms are capable of developing under optimal conditions of, for example, temperature, humidity, nutrients and substrates.

Biochemical methods can be divided into two subgroups. The first includes the methods that measure active populations in their entirety, usually without adding substrates. The second contains methods that are able to define the activity and potential activity of specific organisms or metabolic groups, usually after adding specific substrates; for example, respirometric tests with specific carbon sources, and potential nitrification after addition of ammonium. A selection of commonly used methods is given in Chapter 7.

3. Soil microbial diversity and community structure

This group of methods includes the most up-to-date techniques for acquiring ecological and molecular data.

Traditionally, culturing techniques have been used for the analysis of soil microbial communities. However, only a small fraction (< 0.1%) of the soil microbial community has been determined using this approach. A number of methods are currently available for studies on soil microbial communities. The use of molecular techniques for investigating microbial diversity in soil communities continues to provide new understanding of the distribution and diversity of organisms in soil habitats. The use of RNA or DNA sequences, combined with fluorescent oligonucleotide probes, provides a powerful approach for the characterization and study of soil microbes that cannot currently be cultured. Among the most useful of these methods are those in which small subunit RNA genes are amplified from soil-extracted nucleic acids. Using these techniques, microbial RNA genes can be detected directly from soil samples and sequenced. These sequences can then be compared with those from known microorganisms. Additionally, group- and taxon-specific oligonucleotide probes can be developed from these sequences, making direct determination of microorganisms in soil habitats possible.

Phospholipid fatty acid analysis and community-level physiological profiles have also been utilized successfully by soil scientists, to access a greater proportion of the soil microbial community than can be obtained using culturing techniques. In recent years, molecular methods for soil microbial community analysis have provided new understanding of the phylogenetic diversity of microbial communities in soil (Insam *et al.*, 1997; Loczko *et al.*, 1997; Hill *et al.*, 2000); Chapter 8 describes a selection of these methods.

4. Plant–microbe interactions

The rhizosphere is recognized as the zone of influence of plant roots on the associated biota and soil (Lynch, 1998). Most studies to date have involved an ecophysiological description of this region, with emphasis on the influx of nutrients to plants, including nutrient supply mediated by symbionts (e.g. mycorrhizal fungi and nitrogen-fixing *Rhizobium* bacteria) and free-living microorganisms (e.g. plant-growth-promoting bacteria), and the efflux of photosynthetic carbon compounds, which provide essential substrates for the associated biota, from plant roots (rhizodeposition products). These qualitative and quantitative studies have been very valuable for generating energy budgets of plant and crop productivity.

Some of the methods used in the rhizosphere are the same as those used in bulk soil for determination of biomass, activity and diversity (as described in Chapters 6, 7 and 8). In addition, there are more specific techniques; for example, those for evaluating soil-nodulating potential (of nitrogen fixers), bioassays using arbuscular mycorrhizal fungi, bio-indicators for assessing phytosanitary soil quality and assessment of indigenous free-living plant-beneficial bacteria. Chapter 9 provides a selection of methods that relate soil microbial activity to plants.

Relationships Between Different Parameters and Evaluation of Results

None of the four method groups stands alone (Fig. 1.1). They can often be interfaced, and the decision to include one method in a given category rather than another is a consequence of the type of interpretation one wants to give to the results obtained. For instance, the soil adenosine triphosphate (ATP) content has been used as an indicator of both biomass (group 1) and activity (group 2). The use of ATP as an index of microbial biomass is based on the assumption that ATP is present as a relatively constant component of microbial cells, and that it is not associated with dead cells nor adsorbed to soil particles. A significant correlation was found between the ATP content and the microbial biomass of different soils (Jenkinson, 1988). However, the linear relationship between ATP and microbial biomass only holds when both are determined after soil pre-incubation at constant temperature and moisture conditions. The ATP content changes rapidly, depending on the physiological state of the cell. Therefore, it was hypothesized that ATP content measured immediately after sampling reflects microbial activity rather than biomass (Jenkinson *et al.*, 1979). The accuracy of interpretation and comparison of ATP values in different soils depends on the methods used to extract ATP from soil as well as on soil handling (Nannipieri *et al.*, 1990); for this reason, ATP determination was not included in our selection of methods.

The substrate-induced respiration (SIR) method, introduced by Anderson and Domsch (1975, 1978), depends on microbial biomass as well as activity, and reflects the metabolically active component of the microbial

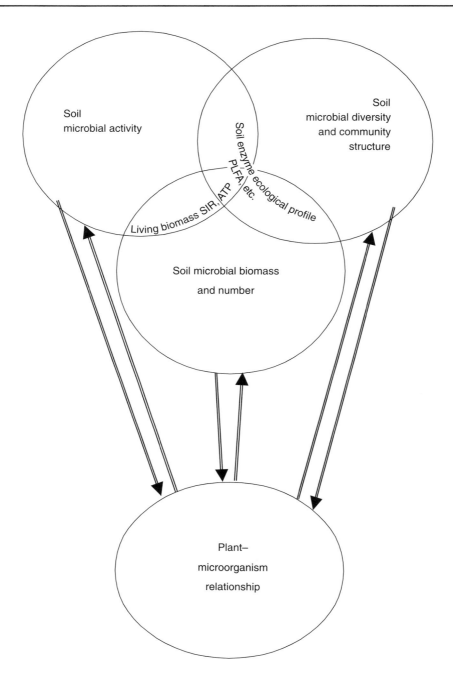

Fig. 1.1. Relationships between different soil microbiological parameters. SIR, substrate-induced respiration; ATP, adenosine triphosphate; PLFA, phospholipid fatty acid.

biomass. The microbial respiratory activity (usually determined as CO_2 evolution) of a glucose-amended soil is stimulated to a maximum within a few minutes after adding saturating amounts of substrate. The enhanced rate of respiration is usually stable for 6–8 h and is assumed to depend on the level of microbial biomass of the soil (Sparling, 1995). Thus, the initial respiratory response to glucose is taken as an index of the soil microbial biomass before the start of microbial growth (Howarth and Paul, 1994; Sparling, 1995). After about 8 h, an increase in the respiratory activity up to a plateau phase reflects microbial growth.

Similar considerations apply to the community-level physiological profile method (CLPP or Biolog^{TM}), which provides information about: (i) the structure of the microbial community (group 3); (ii) the efficiency of specific functional groups of microorganisms in metabolizing specific substrates (groups 2 and 3); or (iii) enzymatic activities (group 2). Functional diversity, as determined by CLPP, reflects both the genetic diversity and the physiological activity of organisms inhabiting the system, and is more important for the long-term stability of an ecosystem than diversity at the taxonomical level (Garland and Mills, 1991). These so-called 'multifunctional methods' may be considered as complementary to Brookes' (1995) concept of 'internal control', which describes biochemical and chemical parameters as being interrelated.

In fact, one of the most complex parts of the soil microbiologist's work is in the assessment of relationships between different parameters, as it embraces the choice of the appropriate monitoring techniques and consideration of the interpretative criteria of results obtained by the previously mentioned methods.

How should the analytically acquired results be evaluated, which information should be deduced and what strategies should be adopted? Luckily, the literature comes to our rescue and proposes approaches for the integrated processing of results, such as amoeba, star or cobweb diagrams and the use of a Soil Quality Index (SQI) to summarize large amounts of data (Chapter 3). Recently, Herrick (2000) affirmed that soil quality appears to be an ideal indicator of sustainable land management, provided that:

1. Causal relationships between soil quality and ecosystem functions are demonstrated, including biodiversity conservation, biomass production and conservation of soil and water resources.
2. The power of soil quality indicators to predict response to disturbance is increased.
3. Accessibility of monitoring systems to land managers is increased.
4. Soil quality is integrated with other biophysical and socio-economic indicators.
5. Soil quality is placed in a landscape context.

Table 1.1. Groups of microbiological, biochemical and molecular methods.

Method groups			
1. Soil microbial biomass and number	2. Soil microbial activity	3. Soil microbial diversity and community structure	4. Plant–microbe interactions
Chloroform fumigation extraction Substrate-induced respiration Direct microscopic counts	Without substrate Soil respiration N mineralization With substrate or tracer Nitrification Thymidine and leucine incorporation N_2O emission and denitrification	Molecular methods based on microbial DNA or RNA Community-level physiological profiles (BIOLOG) Phospholipid fatty acid analysis	Nodulating symbiotic bacteria Arbuscular mycorrhiza Phytosanitary soil quality Free-living plant-beneficial microorganisms

Handbook Contents

In conclusion, the first three chapters of this handbook introduce the three main topics that are decisive factors leading to the choice and subsequent use of some selected microbiological parameters as environmental indicators:

Defining soil quality → Monitoring and evaluating soil quality → Managing soil quality

Issues related to 'Managing soil quality' (Chapter 4) were presented and discussed during the final COST 831 Joint Working Groups meeting in Budapest, Hungary (September 2002).

The general section is followed by a technical section where the methods are set out into four groups, according to the classification above (Table 1.1).

A brief description is given of the potential of each method group (Chapters 6–9), with a selection of only some of the parameters available, i.e. the ones having the requisites set down in the introduction. The selected parameters are accompanied by a detailed description of the method according to the design used for ISO standardization.

The final chapter gives the results of a census of the main methods used in over 30 European laboratories which have participated in COST Action 831.

References

Alef, K. (1995) Nutrient sterilization, aerobic and anaerobic culture technique. In: Alef, K. and Nannipieri, P. (eds) *Methods in Applied Soil Microbiology and Biochemistry*. Academic Press, New York, pp. 123–133.

Alef, K. and Nannipieri, P. (1995) *Methods in Applied Soil Microbiology and Biochemistry*. Academic Press, New York.

Anderson, J.P.E. and Domsch, K.H. (1975) Measurement of bacterial and fungal contributions to soil respiration of selected agricultural and forest soil. *Canadian Journal of Microbiology* 21, 314–322.

Anderson, J.P.E. and Domsch, K.H. (1978) A physiological method for the quantitative measurement of microbial biomass in soil. *Soil Biology and Biochemistry* 10, 215–221.

Benedetti, A., Tittarelli, F., Pinzari, F. and De Bertoldi, S. (2000) *Proceedings of the Joint WGs Meeting of the Cost Action 831 Biotechnology of Soil: Monitoring, Conservation and Remediation*, 10–11 December 1998, Rome. European Communities, Luxembourg.

Bloem, J. and Breure, A.M. (2003) Microbial indicators. In: Markert, B.A., Breure A.M. and Zechmeister, H.G. (eds) *Bioindicators/Biomonitors – Principles, Assessment, Concept*. Elsevier, Amsterdam, pp. 259–282.

Bloem, J., Bolhuis, P.R., Veninga, M.R. and Wieringa, J. (1995) Microscopic methods for counting bacteria and fungi in soil. In: Alef, K. and Nannipieri, P. (eds) *Methods in Applied Soil Microbiology and Biochemistry*. Academic Press, New York, pp. 162–173.

Brookes, P.C. (1995) The use of microbial parameters in monitoring soil pollution by heavy metals. *Biology and Fertility of Soil* 19, 269–279.

Dobereiner, J. (1995) Isolation and identification of nitrogen fixing bacteria from soil and plants. In: Alef, K. and Nannipieri, P. (eds) *Methods in Applied Soil Microbiology and Biochemistry*. Academic Press, New York, pp. 134–135.

Domsch, K.H. (1980) Interpretation and evaluation of data. *Recommended Tests for Assessing the Side-effects of Pesticides on the Soil Microflora*. Weed Research Organization Technical Report No. 59, pp. 6–8.

Domsch, K.H., Jagnow, G. and Anderson, T.H. (1983) An ecological concept for the assessment of side-effects of agrochemicals on soil micro-organisms. *Residue Reviews* 86, 65–105.

Doran, J.W. (2000) Soil health and sustainability: managing the biotic component of soil quality. *Applied Soil Ecology* 15, 3–11.

Garland, J.L. and Mills, A.L. (1991) Classification and characterization of heterotrophic microbial communities on the basis of patterns of community-level sole-carbon-source utilization patterns. *Applied and Environmental Microbiology* 57, 2351–2359.

Herrick, J.E. (2000) Soil quality: an indicator of sustainable land management? *Applied Soil Ecology* 15, 75–83.

Hill, G.T., Mitkowski, N.A., Aldrich-Wolfe, L., Emele, L.R., Jurkonie, D.D., Ficke, A., Maldonado-Ramirez, S., Lynch, S.T. and Nelson, E.B. (2000) Methods for assessing the composition and diversity of soil microbial communities. *Applied Soil Ecology* 15, 25–36.

Howarth, W.R. and Paul, E.A. (1994) Microbial biomass. In: Weaver, R.W., Angle, S., Bottomley, P., Bezdicet, D., Smith, S., Tabatabai, A. and Woollen, A. (eds) *Methods of Soil Analysis. Part 2: Microbiological and Biochemical Properties*. Soil Science Society of America, Madison, Wisconsin, pp. 753–773.

Insam, H., Amor, K., Renner, M. and Crepaz, C. (1997) Changes in functional abilities of the microbial community during composting of manure. *Microbial Ecology* 31, 77–87.

Jenkinson, D.S. (1988) Determination of microbial biomass carbon and nitrogen in soil. In: Wilson, J.K. (ed.) *Advances in Nitrogen Cycling in Agricultural Ecosystems*. CAB International, Wallingford, UK, pp. 368–386.

Jenkinson, D.S., Davidson, S.A. and

Powlson, D.S. (1979) Adenosine triphosphate and microbial biomass in soil. *Soil Biology and Biochemistry* 11, 521–527.

Loczko, E., Rudaz, A. and Aragno, M. (1997) Diversity of anthropogenically influenced or disturbed soil microbial communities. In: Insam, H. and Rangger, A. (eds) *Microbial Communities Functional Versus Structural Approaches*. Springer-Verlag, Berlin, pp. 57–67.

Lorch, H.J., Benckieser, G. and Ottow, J.C.G. (1995) Basic methods for counting microorganisms in soil and water. In: Alef, K. and Nannipieri, P. (eds) *Methods in Applied Soil Microbiology and Biochemistry*. Academic Press, New York, pp. 136–161.

Lynch, J.M. (1998) What is the rhizosphere? In: Atkinsons, D. (ed.) *Proceedings of Inter Cost Actions 821, 830, 831 Meeting*. Agricultural School, 17–19 September, Edinburgh.

Madsen, E.L. (1996) A critical analysis of methods for determining the composition and biogeochemical activities of soil microbial communities *in situ*. In: Stotzky, G. and Bollag, J.M. (eds) *Soil Biochemistry*, vol. 9, Marcel Dekker, New York, pp. 287–370.

Nannipieri, P., Ceccanti, B. and Grego, S. (1990) Ecological significance of the biological activity in soil. In: Bollag, J.M. and Stotzky, G. (eds) *Soil Biochemistry*, vol. 6, Marcel Dekker, New York, pp. 293–355.

Sparling, G.P. (1995) The substrate-induced respiration method. In: Alef, K. and Nannipieri, P. (eds) *Methods in Applied Soil Microbiology and Biochemistry*. Academic Press, New York, pp. 397–404.

UNCED (United Nations Conference on Environment and Development) (1992) Agenda 21. June, Rio de Janeiro.

Zuberer, D.A. (1994) Recovery and enumeration of viable bacteria. In: Weaver, R.W., Angle, S. and Bottomley, P. (eds) *Methods of Soil Analysis. Part 2: Microbiological and Biochemical Properties*. Soil Science Society of America Book Series, No. 5, Madison, Wisconsin, pp. 119–144.

2 Defining Soil Quality

RICHARD G. BURNS,[1] PAOLO NANNIPIERI,[2] ANNA BENEDETTI[3] AND DAVID W. HOPKINS[4]

[1]*School of Land and Food Sciences, The University of Queensland, Brisbane, Queensland 4072, Australia;* [2]*Dipartimento de Nutrizione delle Pianta e Scienza del Suolo, Università di Firenze, Piazzale di Cascine 28, Florence, Italy;* [3]*Consiglio per la ricerca e la sperimentazione in Agricoltura, Istituto Sperimentale per la Nutrizione delle Piante, Via della Navicella, 2, 00184 Rome, Italy;* [4]*School of Biological and Environmental Sciences, University of Stirling, Stirling FK9 4LA, UK*

Abstract

Environmental quality is a complex concept. Defining one component of it, soil quality, is therefore usually attempted using indicators that represent, with differing levels of approximation, particular constituents, processes or conditions. In this chapter, we review briefly the ideal characteristics of a soil quality indicator and then outline some of the national frameworks for assessing soil quality that have been proposed. A recurrent theme of the existing frameworks is the use of parameters that individually give useful information, but which can be aggregated to provide an overall indicator or index of soil fertility.

Introduction

Environmental quality is a composite of the desirable properties of soil, air and water. For water and air, where relatively precise analyses can be reported, analytical data do not necessarily provide an holistic assessment of the quality of these components of the biosphere. Soils represent an even more complex environment because they are an intimate mixture of the living and non-living components and because they vary naturally in both space and time over a range of scales. Defining soil quality is, therefore, usually attempted using somewhat arbitrarily chosen chemical, biological and physical indicators which represent particular constituents, processes or conditions. A good indicator of quality must have several characteristics. It must be representative of the sites to which it is being applied; it must be accessible both in terms of the availability of the methods required to measure it and the ease with which the measurements can be interpreted by the end-user; and it must be reliable, meaning that it must be reproducible and applicable to a range of sites. Since soil quality cannot be summarized by a single component or process, its assessment must include information about several indicators. These, depending on the stated purpose, may have different scales of measurement (e.g. aggregates, horizons, profiles, catchments) and make a different proportional contribution to the evaluation of fertility. For example, in order to describe the extent, nature and likely impact of a pollutant in a particular soil, it is necessary to employ a range of indicators. These will include the concentration of the pollutants and their vertical and horizontal distribution across the site, and intrinsic soil factors such as pH, clay and organic matter content, and ion exchange capacity. All these must be considered because they will influence the bioavailability of the pollutant and, therefore, its persistence, movement and effect on selected important processes. On the other hand, if the objective is prediction of plant nutrient availability, legume nodulation or natural biological control, other factors will assume importance. From the extensive literature, it is possible to deduce several characteristics that might contribute to an ideal indicator of soil fertility, and many of these have been summarized in the literature; for example, the Organization for Economic Co-operation and Development (OECD, 1999), recognized seven categories, as follows.

1. Political relevance and user benefits; indicators should:

- provide a representative picture of the environmental conditions and of the societal pressures or reactions to the changing state of the environment;
- be simple, easy to interpret and able to indicate temporal trends;
- be reactive to environmental changes and to related human activity;
- provide a basis for international comparisons;
- have national worth and be applicable to nationally relevant regional themes;
- have threshold or reference values, such that users can evaluate the significance of the indicator values.

2. Analytical validity; analyses should:

- be well founded, both technically and scientifically;
- be based, where possible, on international standards and have international consensus in terms of validity;
- be easily applied to economic models, forecast estimates and information systems.

3. Measurability; measurements should:

- be easily available or made available at a reasonable cost:benefit ratio;
- be adequately documented and of verified quality;
- be able to be updated at regular intervals according to well-defined procedures.

4. Representativeness; indicators should:

- correlate with a specific phenomenon or characteristic;

- correlate with previously reported effects with the minimum of statistical dispersion;
- not be easily obscured by profile factors;
- have sufficient general validity to many analogous, non-identical situations.

5. Accessibility; indicators should:

- be easily measurable;
- offer the possibility of being monitored automatically;
- be easy to sample and have a threshold of analytical detection which is accessible by standard techniques.

6. Reliability; indicators should:

- have minimum systematic errors.

7. Operativeness; indicators should:

- be easily and directly utilizable for quantifying acts of intervention, costs and benefits.

The above list serves as a guide to the selection of useful indicators, but it should be recognized that no single indicator can meet all requirements. Furthermore, a major problem in the use of any indicator is the establishment of threshold or reference values. This is only possible if many data are available and, even then, is a somewhat subjective choice, based on the current and projected use of the land (see also Chapter 3).

According to the OECD (1999) the 'definitions of indicators as a concept (let alone specific indicators), vary widely' and, furthermore, different agencies and authors use their own terms and definitions. These include: variables, parameters, measures, statistical measures, proxy measures, values, measuring instruments, fractions, indices, a piece of information, empirical models of reality and signs! The OECD (1993) bravely attempted to define a few of these terms, thus:

- *parameter* – a property that is measured or observed;
- *indicator* – a value derived from parameters, which points to or provides information about, or describes, the state with a significance extending beyond that directly associated with a parameter value; and
- *index* – a set of aggregated or weighted parameters or indicators.

Any parameter that gives useful information on soil quality can be used as an indicator, and a set of indicators can be aggregated into an index. However, in order to correctly apply the terminology it is necessary to understand the meaning of each term. For example, soil organic matter content is universally recognized as an *indicator* of soil quality and, in general, the organic carbon content of an agricultural soil is a *parameter* closely correlated with the organic matter content. At temperate latitudes, the average organic matter content falls between 1% and 5%. A reduction of this quantity over a period of time is likely to be an *index* of impoverishment and could be a strong *indicator* of the loss of soil fertility and deterioration of other soil properties – such as structural stability and water retention. Similarly, the effectiveness of an amendment can be evaluated by measuring the increase in organic carbon, which then becomes an *indicator* of effectiveness of the fertilizer used. Organic matter, together with other *parameters*, can thus become an *index* of soil quality. An example of a soil quality index is given in Chapter 3. Thus, a change in the value of a single *parameter* outside a certain range can be the important *sign* of quality improvement or reduction.

International Indicators of Soil Quality

Since the beginning of the 1980s, a decrease in soil productive capacity has been observed in more than 10% of cultivated land worldwide, as evidenced by soil erosion, atmospheric pollution, amount of land in farming, excessive grazing, salinization and desertification (Francaviglia, 2004; Van-Camp *et al.*, 2004).

A definitive set of basic indicators for the evaluation of soil quality has not yet been provided, despite various international proposals, including that published by the Soil Science Society of America (SSSA). This is due mainly to the continuing difficulty in defining soil quality and how it can be assessed. Many definitions have been suggested in recent years, but one that best represents the concept was given by Doran and Parkin (1994): 'The capacity of the soil to interact with the ecosystem in order to maintain biological productivity, environmental quality and to promote animal and plant health.' This definition is similar to the three essential criteria for soil quality that were identified by the Rodale Institute (1991), namely:

- *productivity* – the soil's capacity to increase plant biological productivity;
- *environmental quality* – the soil's capacity to attenuate environmental contamination, pathogens and external damage; and
- *health of living organisms* – the interrelation between soil quality and animal, plant and human health.

The parameters for the evaluation of soil quality can be subdivided into those that are physical, chemical and biological. However, integration among them is fundamental to our understanding. Currently, definitions of soil quality standards are being discussed within international regulatory bodies. The US Environmental Protection Agency (EPA), for example, has proposed over 1800 parameters as chemical indicators of soil quality. Within the OECD, agroenvironmental indicators (there are approximately 250), including those related to soil quality, are currently being defined. So far, 58 have been proposed as soil quality indicators, but some of them are different approaches to assessment of the same indicator, for example organic matter content estimated by modelling and by analysis. The indicators and parameters in Table 2.1 have been proposed for soil and site assessment by the ISO Technical Committee 190 on 'Soil Quality' and correspond to the physical, chemical and biological parameters essential in the consideration of soil restoration.

Although approaches to assessing soil quality have developed independently in several countries, there is considerable overlap between the parameters listed in Table 2.1, those proposed by the SSSA (Table 2.2), and those used in New Zealand (Table 2.3). In the case of both the USA and

Table 2.1. Parameters proposed in a working document of ISO Technical Commitee 190 'Soil Quality', Sub-Committee 7 'Soil and Site Assessment'.

Parameters	International standards
Physical parameters	
Petrographic features	
Mineralogy	
Nature of the mother rock	
Soil profile	
Texture	
Water content	ISO 10537
Presence of roots	
Hydraulic conductivity	DIS 11275–1/DIS 11275–2
Pore water pressure	CD 15048/ISO 11276
Plasticity index	
Consistency	
Structure stability	
Degree of infiltration	
Particle size distribution	ISO 11277
Aggregation state	DIS 11273–1
Skeleton	CD 11273–2
Apparent density	FDIS 11272
Chemical parameters	
pH	ISO 10390
Redox potential	ISO 11271
Salinity	
Sodium	
Total organic carbon	ISO 10694
Carbon dioxide losses at specific temperatures	
Cation exchange capacity	ISO 11260/ISO 13526
Dry matter content	ISO 11465
Carbonates	ISO 10693
Specific electric conductivity	ISO 11265
Exchange acidity	DIS 14254
Biological indicators	
Microbial activity	ISO 14239/ISO 11266/ISO 14238/NP 15473
Harmful plant species	
Toxicity for plants	ISO 11269
Toxicity for microorganisms	
Presence of pathogens	
Microbial biomass	ISO 14240
Toxicity for macrofauna	ISO 11268

New Zealand, there is acceptance of a range of complementary parameters. The New Zealand Soil Indicators (SINDI) approach relies on a small set of indicators matched to particular national issues, so that Olsen P is, for example, prioritized as the principal indicator of soil fertility, and direct biological assessment is limited to a nitrogen mineralization assay which simultaneously provides a soil fertility indicator and acts as a surrogate for microbial biomass. Clearly, this approach reduces the demand for time-consuming and technically complex laboratory analyses. A useful feature of SINDI is that it is supported by an on-line assessment framework (http://sindi.landcare.cri.nz, accessed 25 November 2004), in which the values for the different indicators can be compared with the expected norms for particular soil types, and in which there are links to management information and advice. Although there is no minimum dataset recognized for the assessment of soil quality in Canada, the same multifaceted approach was adopted by Agriculture and Agri-Food Canada, which included assessment of the soil organic resources, structural condition, contamination and hydrological conditions (Acton and Gregorich, 1995).

Soil quality depends on several biological, chemical and physical soil properties and, theoretically, its definition should require the determination

Table 2.2. Physical, chemical and biological features proposed as basic indicators of soil quality and based on the definition of Doran and Parkin (1994).

Soil features	Methodology
Physical indicators	
Soil texture	Water-gauge method
Depth of the soil and root systems	Soil excavation and extraction
Apparent density and infiltration	Field determination with the use of infiltration rings
Water retention features	Water content at pressures of 33 kPa and 1500 kPa
Water content	Gravimetrical analysis (weight loss over 24 h at 105°C)
Soil temperature	Thermometer
Chemical indicators	
Total organic C and N	Combustion (volumetric method)
pH	Field and laboratory determinations with pH meter
Electrical conductivity	Field and laboratory determinations with a conductometer
Inorganic N (NH_4^+ and NO_3^-), P and K concentrations	Field and laboratory determinations (volumetric method)
Biological indicators	
C and N from microbial biomass	Fumigation/incubation with chloroform (volumetric method)
Potentially mineralizable N	Anaerobic incubation (volumetric method)
Soil respiration	Field determination by means of covered infiltration rings, and in the laboratory by measuring the biomass

Table 2.3. Indicators of soil quality for New Zealand used in the SINDI (Soil Indicators) scheme (http://sindi.landcare.cri.nz).

Soil property	Comments
Soil fertility indicator	
Olsen P	Plant-available phosphorus
Soil pH	Acidity or alkalinity of soil
Organic resources	
Anaerobic nitrogen mineralization	Availability of the nitrogen reserve to plants and a surrogate measure of microbial biomass
Total (organic) C	Organic matter reserves, which is also positively related to soil structure and ability to retain water
Total N	Organic N reserves
Soil physical quality	
Bulk density	Soil compaction, physical environment for roots and soil organisms
Macroporosity	Availability of water and air, retention of water, drainage properties

of these properties. Biological parameters have assumed particular importance in the assessment of soil quality because organisms respond more rapidly than most chemical and physical parameters to changes in land use, environmental condition or contamination (Doran and Parkin, 1994; Nannipieri et al., 2001; Nannipieri and Badalucco, 2002; Gil-Sotres et al., 2005). It is equally well established that soil organisms play crucial roles in many processes that underpin soil quality, such as organic matter decomposition and nutrient cycling, nitrogen fixation and aggregate formation and stabilization. For this reason, the size of the soil microbial biomass, respiration, potential nitrogen (N) mineralization, enzyme activities, abundance of fungi, nematodes and earthworms have all been used as indicators of soil quality (Lee, 1985; Doran, 1987; Dick et al., 1988; Kennedy and Papendick, 1995; Wall and Moore, 1999). The following chapters present some of the methods commonly used as indicators and critically evaluate their contribution to soil quality.

References

Acton, D.F. and Gregorich, L.J. (1995) *The Health of Our Soils – Towards Sustainable Agriculture in Canada*. Centre for Land and Biological Resources Research, Research Branch, Agriculture and Agri-Food Canada, Ottawa.

Dick, R.P., Rasmunssen, P.E. and Kerle, E.A. (1988) Influence of long term residue management on soil enzyme activities in relation to soil chemical properties of a wheat-fallow system. *Biology and Fertility of Soils* 6, 159–164.

Doran, J.W. (1987) Microbial biomass and mineralizable nitrogen distrubution in no-tillage and plowed soils. *Biology and Fertility of Soils* 5, 68–75.

Doran, J.W. and Parkin, T.B. (1994) Defining and assessing soil quality. In: *Defining Soil*

Quality for a Sustainable Environment, Soil Science Society of America Special Publication no. 35. SSSA, Madison, Wisconsin.

Francaviglia, R. (ed.) (2004) Agricultural impacts on soil erosion and soil biodiversity: developing indicators for policy analysis. *Proceedings of the OECD Expert Meeting on Soil Erosion and Soil Biodiversity Indicators, 25–28 March 2003, Rome, Italy.* OECD, Paris. Available at: http://webdomino1.oecd.org/comnet/agr/soil_ero_bio.nsf (accessed 17 December 2004).

Gil-Sotres, F., Trasar-Cepeda, C., Leiros, M.C. and Seoane, S. (2005) Different approaches to evaluating soil quality using biochemical properties. *Soil Biology & Biochemistry* 37, 877–887.

Kennedy, A.C. and Papendick, R.I. (1995) Microbial characteristics of soil quality. *Journal of Soil and Water Conservation* 50, 243–248.

Lee, K.E. (1985) *Earthworms: Their Ecology and Relationship with Soil and Land Use.* Academic Press, London.

Nannipieri, P. and Badalucco, L. (2002) Biological processes. In: Kenbi, D.K. and Nieder, R. (eds) *Handbook of Processes and Modelling in the Soil–Plant System.* The Haworth Press Inc., Binghampton, New York, pp. 57–82.

Nannipieri, P., Kandeler, E. and Ruggiero, P. (2001) Enzyme activities and microbiological and biochemical processes in soil. In: Burns, R.G. and Dick, R. (eds) *Enzymes in the Environment.* Marcel Dekker, New York, pp. 1–33.

OECD (1993) *OECD Core Set of Indicators for Environmental Performance Reviews – A synthesis report by the group on the state of the environment.* Environmental Monographs no. 83. OECD/GD(93)179. Organization for Economic Cooperation and Development, Paris.

OECD (1999) *Environmental Indicators for Agriculture*, vol. 2, *Issues and Design.* The York Workshop, Organization for Economic Cooperation and Development, Paris.

Rodale Institute (1991) *Conference Report and Abstract, International Conference on the Assessment and Monitoring of Soil Quality. Emmaus, Pennsylvania, 11–13 July 1991.* Rodale Press, Emmaus, Pennsylvania.

Van-Camp, L., Bujarrabal, B., Gentile, A.-R., Jones, R.J.A., Montanarella, L., Olazabal, C. and Selvaradjou, S.-K. (2004) Reports of the Technical Working Groups Established under the Thematic Strategy for Soil Protection. EUR 21319 EN/3, 872 pp. Office for Official Publications of the European Communities, Luxembourg. Available at: http://eusoils.jrc.it/ESDB_Archive/eusoils_docs/doc.html#OtherReports (accessed on 17 December 2004).

Wall, D.H. and Moore, J.C. (1999) Interactions underground-soil biodiversity, mutualism, and ecosystem processes. *BioScience* 49, 109–117.

3 Monitoring and Evaluating Soil Quality

JAAP BLOEM,[1] ANTON J. SCHOUTEN,[2] SØREN J. SØRENSEN,[3] MICHIEL RUTGERS,[2] ADRI VAN DER WERF[4] AND ANTON M. BREURE[2]

[1]*Department of Soil Sciences, Alterra, PO Box 47, NL-6700 AA Wageningen, The Netherlands;* [2]*Laboratory for Ecological Risk Assessment, RIVM, PO Box 1, NL-3720 BA Bilthoven, The Netherlands;* [3]*Department of General Microbiology, Institute of Molecular Biology, University of Copenhagen, Sølvgade 83H, DK 1307K Copenhagen, Denmark;* [4]*Plant Research International, PO Box 16, NL-6700 AA Wageningen, The Netherlands*

Abstract

Soil quality influences agricultural sustainability, environmental quality and, consequently, plant, animal and human health. Microorganisms are useful indicators of soil quality because they have key functions in the decomposition of organic matter, nutrient cycling and maintenance of soil structure. We summarize methods used for monitoring biomass, activity and diversity of soil organisms and show some results of the Dutch Soil Quality Network.

In contaminated soils microbial community structure was changed, but diversity was not always reduced. In contrast, microbial biomass and activity were reduced markedly. In agricultural soils there were large differences between different categories of soil type and land use. Organic management resulted in an increased role of soil organisms, as indicated by higher numbers and activity. Replacement of mineral fertilizers by farmyard manure stimulated the bacterial branch of the soil food web. Reduced availability of mineral nutrients appeared to increase fungi, presumably mycorrhizas. Bacterial DNA profiles did not indicate low genetic diversity in agricultural soils, compared with some acid and contaminated soils. Organic farms did not show higher genetic diversity than intensively farmed areas. At extensive grassland farms and organic grassland farms nitrogen mineralization was about 50% higher than on intensively farmed areas. Also, microbial biomass and activity, and different groups of soil invertebrates, tended to be higher.

Soil biodiversity cannot be monitored meaningfully with only a few simple tools. Extensive and long-term monitoring is probably the most realistic approach to obtain objective information on differences between, changes within, and human impact on, ecosystems. In most countries, microbial biomass, respiration and potential nitrogen

(N) mineralization are regarded as part of a minimum data set. Adding the main functional groups of the soil food web brings us closer to understanding biodiversity, potentially enabling us to relate the structure of the soil community to functions.

Soil Quality Monitoring and Microbiological Indicators

Following adoption of the Treaty on Biological Diversity of Rio de Janeiro (UNCED, 1992), participating governments have been concerned about the protection of endangered species, mainly plants and larger animals. Viable nature conservation areas, consisting of core areas linked by transition zones, have been developed. In addition, there is increasing concern, at both the national and the international level, about sustainable use of biodiversity and maintenance of life support functions such as decomposition and nutrient cycling (FAO, 1999; OECD, 2003; Schloter *et al.*, Chapter 4, this volume). In all soils, these vital ecosystem processes depend largely on the activities of microorganisms and small soil invertebrates that are rarely visible with the naked eye (also called 'cryptobiota').

Soil quality determines agricultural sustainability, environmental quality and, consequently, plant, animal and human health (Doran and Parkin, 1996). This chapter provides an introduction to biological approaches presently used in different countries to monitor and evaluate soil quality. Monitoring was initiated in several countries in 1992, but little information has been exchanged or published in the international literature so far. This chapter is based mainly on the experience of the Dutch Soil Quality Network (Schouten *et al.*, 2000; Bloem and Breure, 2003; Bloem *et al.*, 2004), and discussions of the working groups of EU COST Action 831 'Biotechnology of Soil, Monitoring, Conservation and Remediation'.

Soil quality has been defined as 'the capacity of a soil to function within ecosystem boundaries to sustain biological productivity, maintain environmental quality, and promote plant and animal health' (Doran and Parkin, 1994; Stenberg, 1999). The phrase 'within ecosystem boundaries' implies that each soil is different. There are no absolute quality estimates and each soil must be evaluated in relation to natural differences such as soil type, land use and climate. The term 'soil quality' is often used to describe the fitness of a soil for (agricultural) use, while the term 'soil health' is seen more as an inherent attribute regardless of land use. Often these terms are used as synonyms. There are many definitions of soil quality and soil health (Burns *et al.*, Chapter 2, this volume). Quality or health of an ecosystem is a value judgement. Although ecological health has been criticized as a nebulous concept in a scientific context, a useful consequence of that notion is that environmental monitoring programmes need to adopt a holistic ecosystem approach (Lancaster, 2000). Many different aspects need to be measured, including physical, chemical and biological characteristics. Here we focus on soil organisms and the processes they mediate.

An agricultural soil usually contains about 3000 kg (fresh weight) of soil organisms per hectare. This is equivalent to 5 cows, 60 sheep or 35 farmers living under the surface. Many thousands of species (or genotypes) contribute to a huge below-ground biodiversity. Soil invertebrates fragment dead organic matter and thus facilitate decomposition. Their direct contribution to the biochemical modification or flux of organic residue is usually small compared with the contribution of bacteria and fungi. Decomposition by bacteria and fungi causes release of mineral nutrients (mineralization) essential for plant growth. Mineralization is further performed by organisms that feed on bacteria and fungi (bacterivores and fungivores), such as protozoa and nematodes. Some small soil invertebrates (e.g. nematodes) feed directly on plant roots (herbivores). Predators eat other, usually smaller, soil invertebrates, and omnivores feed on different food sources. All these trophic interactions in the soil food web contribute to the flow of energy and nutrients through the ecosystem (Hunt et al., 1987). Models predict that the abundances of the different functional groups of organisms, i.e. the structure of the soil food web, affect the stability of the soil ecosystem (De Ruiter et al., 1995). Mycorrhizal fungi that live in symbiosis with plant roots promote the uptake of mineral nutrients by plants. Bacteria, fungi and invertebrates glue soil particles together, form stable aggregates and thus improve soil structure. Invertebrates also improve soil structure by mixing the soil (bioturbation).

The following comprise the main functional groups of the soil food web:

- Earthworms consume plant residues and soil, including (micro)organisms. Often they form the major part of the soil fauna biomass, with maximally 1000 individuals/m^2, 3000 kg fresh biomass/ha, or a few hundred kg of carbon (C) per hectare.
- Enchytraeids are relatives of earthworms with a much smaller size and a similar diet. Their population densities are between 10^2 and $10^6/m^2$, with a biomass up to 1 kg C/ha.
- Mites (fungivores, bacterivores, predators) have a size of about 1 mm, population densities of 10^4–$10^5/m^2$, and a biomass of up to 0.1 kg C/ha.
- Springtails (fungivores, omnivores) also have a size of about 1 mm. They reach population densities of 10^3–$10^5/m^2$ and a biomass of up to 1 kg C/ha.
- Nematodes (bacterivores, herbivores, fungivores, predators/omnivores) have a size of about 500 µm, population densities of 10–50/g soil, and a biomass up to 1 kg C/ha.
- Protozoa (amoebae, flagellates, ciliates) are unicellular animals with a size of 2–200 µm, population densities of about 10^6 cells/g soil, and a biomass of about 10 kg C/ha.
- Bacteria are usually smaller than 2 µm, with population densities of about 10^9 cells/g soil, and a biomass of 50–500 kg C/ha.
- Fungal hyphae usually have diameters from 2 µm to 10 µm, and reach total lengths of 10–1000 m/g soil, and a biomass of 1–500 kg C/ha.

These cryptobiota (hidden soil life) play a key role in life support functions (Bloem et al., 1997; Brussaard et al., 1997, 2003; Bloem and Breure, 2003), but are not part of any recognized list of endangered species. It is questionable whether a species-based approach is sufficient to attain a sustainable use of ecosystems inside, and especially outside, protected areas. Therefore, research networks have been initiated to monitor large areas, including agricultural soils.

Since about 1993, national or regional programmes have been established in several countries to monitor soil quality and/or the state of biodiversity (Stenberg, 1999; Nielsen and Winding, 2002). These include: Canada (23 sites), France, parts (Bundesländer) of Germany (about 350 sites; Höper, 1999; Oberholzer and Höper, 2000), parts (cantons) of Switzerland (Maurer-Troxler, 1999), the Czech Republic, the UK, Austria, the USA (21 sites; Robertson *et al.*, 1999) and New Zealand (500 sites; G. Sparling *et al.*, http://www.landcareresearch.co.nz, accessed 30 January 2004).

In The Netherlands, 200 sites are part of the Dutch Soil Quality Network, consisting of ten categories of a specific soil type with a specific land use, with 20 replicates per category (Schouten *et al.*, 2000). The replicates are mainly conventional farms. The 200 sites are representative of 70% of the surface area of The Netherlands. In addition, 50–100 sites from outside this network are sampled; for instance, organic farms or polluted areas which are supposed to be good and bad references, respectively. Each year two types of soil and land use are sampled (40 sites plus reference sites). Thus, it takes 5 years to complete one round of monitoring the whole network of 200 sites plus references. In 1993, the Dutch network started to obtain policy information on abiotic soil status. The aim was to measure changes over time and finally to evaluate the actual soil quality. A set of biological indicators has been included since 1997, consisting of microbiological indicators and several soil fauna groups, in order to take a cross-section through the soil ecosystem.

In most countries one or more microbiological indicators have been included. As part of a monitoring system, microorganisms are useful indicators of soil quality because they have key functions in decomposition of organic matter and nutrient cycling, they respond promptly to changes in the environment and they reflect the sum of all factors regulating the degradation and transformation of organic matter and nutrients (Stenberg, 1999; Bloem and Breure, 2003).

Sampling

For the application of microbiological indicators a lot of methodological choices have to be made.

How can variation in space and time be accounted for?

Mainly by taking many replicates and aiming at long-term monitoring. Samples can be taken from replicated field plots or can be pooled from larger areas. In The Netherlands, about 20 farms (replicates) spread over the country are sampled per category of soil type and land use. One mixed sample per farm (about 10–100 ha) is made up of 320 cores. These mixed samples are used for chemical, microbiological and nematode analyses. Separate soil cores or blocks (six replicates per site) are taken for analysis of mites, enchytraeids and earthworms. Some reference sites consist of smaller contaminated areas or experimental fields. Here replicated field plots (about 10 m × 10 m) are sampled.

Sampling depth is best decided by considering soil horizons and tillage depth (Stenberg, 1999). In a ploughed arable field, 0–25 cm would be appropriate; in grassland, and especially in forest, higher numbers of thinner layers would be better. However, this would result in a variable sampling depth or an increase in the number of samples by taking more than one layer. Given the large number of samples, analysing more than one depth would be too time consuming and expensive. Sampling 0–25 cm would dilute microbial activity considerably in some grassland and forest soils, where life is concentrated closer to the surface. Therefore, in the Dutch monitoring network, samples are taken from 0–10 cm depth and litter is removed before sampling. To reduce variation caused by variable weather conditions, samples are pre-incubated for 4 weeks at constant temperature (12°C) and moisture content (50% of water-holding capacity) before microbiological analyses are performed. Since each soil and land-use type in the monitoring network is analysed once every 5 years, effects of a dry summer, for instance, should be minimized.

Samples can be sieved through 2 mm or 5 mm mesh-size, or not at all. In The Netherlands and in Sweden, soil is sieved through 5 mm and 4 mm mesh sizes, respectively (it is practically impossible to pass a heavy clay soil through a 2 mm sieve). Sieving is useful to reduce variation in process rate measurements, such as respiration and mineralization, to facilitate mixing and to allow identical subsamples to be sent to different laboratories. However, sampling and sieving are major disturbances, which generally increase microbial activity and also reduce soil structure. Therefore, the results of the first week of incubation are not used for calculation of process rates.

When should samples be taken?

For microbiological parameters, early spring or late autumn is the best time. Then soil conditions are relatively mild and stable, and short-term effects of the crop are avoided. These periods are proposed in Sweden (Stenberg, 1999). In The Netherlands, for practical reasons, samples are taken from March to June. The land must be dry enough to access, and farmers prefer sampling of arable land before soil tillage and sowing new crops. Sampling of about 50 farms takes 2–3 months.

Storage and Pre-incubation of Soil

How should samples be stored: at −20°C, 1–5°C or field temperature? For how long? Obviously it is best to perform soil biological analyses soon after sampling. On the other hand, storage is inevitable when large amounts of samples from many sites have to be handled. The preferred method for storing soil samples in different countries appears to be related to the climate. In Sweden and Finland, freezing at −20°C for at most 1 year is practised. Stenberg et al. (1998) found that the effects of freezing were generally smaller than those of refrigeration. They suggested that microflora in northern soils, subjected annually to several freeze and thaw cycles, may have adapted to this stress factor. In the UK, Denmark, Germany and Switzerland soil samples are stored at 4°C, and in Italy samples are air-dried. It is generally recommended that samples for microbial analysis are stored at 2–4°C (Wollum, 1994; Nielsen and Winding, 2002). Biomass and activity usually tend to decrease during storage because available organic substrates are slowly depleted. This decrease is supposed to be slower at 4°C in a refrigerator, and may be stopped by freezing. However, in frozen samples we have found more than 50% reduction in bacterial cell numbers as counted by direct microscopy.

Using sandy soil from arable fields and grassland (Korthals et al., 1996), we investigated the effects of storage for 6 months at 12°C, 2°C and −20°C. After storage, the soil was pre-incubated for 4 weeks at 12°C and 50% water-holding capacity, and subsequently analysed. The samples were taken in May, when the moisture content in the field was between 14% and 20% (w/w), corresponding to 47–67% of the water-holding capacity of the soil. The soil was a fimic anthrosol (FAO classification) with a texture of 3% clay, 10% silt, 87% sand, an organic carbon content of 2–3% (w/w) and a pH(-KCl) of 5. The results of the stored samples were compared to results of microbiological analyses started 1 day after sampling. With all storage methods, bacterial biomass, as determined by microscopy and image analysis, did not decline in grassland soil but was strongly reduced (−70%) in arable soil (Fig. 3.1).

Bacterial growth rate (thymidine incorporation) showed the opposite: it remained high in arable soil but was strongly reduced in grassland soil during storage at 12°C and 2°C. Thus, grassland bacteria (apparently k-strategists or persisters) survived better than arable soil bacteria (apparently r-strategists or colonizers). Grassland may select for persisters because it is a more stable environment with a relatively constant food supply from grass roots, whereas arable soils may favour colonizers because substrate inputs are highly seasonal. Thus, effects of storage may be different for different microbial communities (e.g. from grassland versus arable land) and parameters (e.g. biomass versus growth rate). After freezing, growth rate had doubled in arable soil. With all storage methods, respiration (CO_2 evolution) decreased by at least 40%. N mineralization was strongly reduced after storage at −20°C and in the arable soil also at 2°C. This may have been caused by N immobilization during re-growth of bacteria when the temperature was increased. Reduction in N mineralization was less at 12°C but

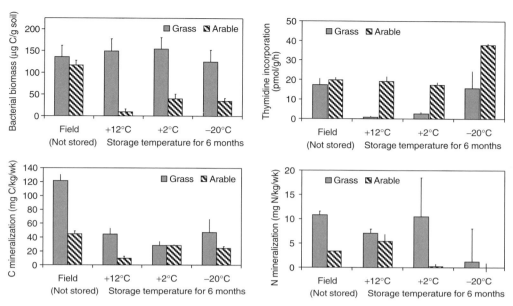

Fig. 3.1. Effect of storage of soil samples on bacterial biomass and growth rate (thymidine incorporation), soil respiration (C mineralization) and N mineralization. Error bars indicate standard error (SE), $n = 3$.

here it had decreased in grassland soil and increased in arable soil. Stenberg *et al.* (1998) reported that N mineralization capacity was greatly influenced by freezing, but that other parameters, such as basal respiration and microbial biomass, were only a little affected. After storage for 6 months, followed by 1 month pre-incubation, the number of bacterial DNA bands, as obtained by denaturing gradient gel electrophoresis, was reduced by about 20%, and some qualitative changes had occurred in the DNA banding pattern, regardless of the storage temperature. Our results support the view that soil samples for microbiological analyses should be stored for as short a time as possible (Anderson, 1987; Zelles *et al.*, 1991).

In the Dutch monitoring programme, storage of soil samples for 1–2 months is inevitable. This will cause extra variation in the results. In Germany and Switzerland, soil samples are stored for 6 months at most. In The Netherlands, a storage temperature of 12°C was chosen, which is close to the average annual soil temperature. The optimum storage method may differ for different microbiological parameters. However, using more than one storage method increases handling time and cost of monitoring. Moreover, microbiological parameters can best be related to each other when they are measured in the same portion of soil. In the Dutch soil monitoring network, after storage at 12°C, the samples are pre-incubated for 4 weeks at 12°C and 50% of the water-holding capacity. In Sweden, after freezing, samples are pre-incubated for a few days. In the UK, Germany and Italy, after storage soils are pre-incubated for 1–2 weeks. Also, the optimum time of pre-incubation may depend on the parameters to be measured. Measurement of potential microbial activities in slurries at 37°C, as in Sweden (Torstensson *et al.*, 1998), may require a shorter pre-incubation than measurements in soil under conditions similar to those in the field, as in The Netherlands.

Methods and Choice of Indicators

A range of methods is used to assess the amount (biomass), activity and diversity of soil organisms (Akkermans *et al.*, 1995; Alef and Nannipieri, 1995). No one method is best suited for all purposes and they need calibration and standardization before use (Paul *et al.*, 1999). A selection of methods is described in this book. Here, we summarize methods that are applicable to relatively large amounts of samples and which are already used in monitoring programmes.

Biomass

For methods see Chapter 6.

- Chloroform fumigation extraction (CFE): the soil is fumigated with chloroform, which permeabilizes cell membranes. The increase in extractable organic carbon (and nitrogen), compared to an unfumigated control, is a measure of the total microbial biomass (C and N).
- Substrate-induced respiration (SIR): a substrate (glucose) is added to soil at a saturating concentration and is utilized by microorganisms. The increased CO_2 evolution in the first few hours before a growth response occurs, compared to that of an unamended control, is a measure of the (responsive) microbial biomass.
- Direct microscopy can be combined with automatic image analysis (Bloem *et al.* 1995, Paul *et al.*, 1999): number and body size are determined and biomass is calculated. This can be done for different groups, e.g. fungi and bacteria. Fungi and bacteria are counted directly in soil smears after fluorescent staining. Since their numbers are much lower, soil invertebrates (nematodes, springtails, mites, etc.) are extracted from the soil before counting and identification.

Activity

For methods see Chapter 7.

- Respiration: CO_2 evolution under standardized conditions in the laboratory, without addition of substrates (basal respiration).

- Bacterial growth rate: incorporation rate of [^{3}H]thymidine and [^{14}C]leucine into bacterial DNA and proteins during a short incubation (1 h).
- Potential N mineralization: increase in mineral N under standardized moisture content and temperature in the laboratory, without addition of substrate.
- Potential nitrification: conversion of added NH_4^+ via NO_2^- to NO_3^- under optimal conditions.
- Enzyme activities, e.g. dehydrogenase, phosphatase, cellulase.

Diversity/community structure

For methods see Chapter 8.
- DNA profiles obtained by DGGE or TGGE (denaturing or temperature gradient gel electrophoresis): DNA is extracted from soil, amplified by PCR (polymerase chain reaction) and separated by gel electrophoresis. This results in a banding pattern where the number of DNA bands reflects the dominant genotypes and genetic diversity (Fig. 3.2).
- Community-level physiological profiles (CLPP): the ability to utilize a range of (31 or 95) sole-carbon-source substrates is tested in Biolog™ multiwell plates. Colour development in a well indicates utilization of a specific substrate. The pattern of colour development characterizes the functional diversity, if equal amounts of bacterial cells are added. If a fixed amount of soil is added, it reflects the number of active bacteria (Fig. 3.3).
- Phospholipid fatty acid (PLFA) analysis: PLFAs are essential membrane components of living cells. Specific PLFAs predominate in certain taxonomic groups and are relatively conservative in their concentrations within them. Measuring the concentrations of different PLFAs extracted from soils can, therefore, provide a biochemical fingerprint of the soil microbial community. The PLFA profiles reflect the community structure and show which groups are dominant. PLFAs do not, however, give any quantitative information about the number of species.
- Soil fauna: (usually microscopic) enumeration and identification of functional groups or species.

All these methods measure different aspects of microbial communities, and a combination of methods is needed for monitoring the diversity and functioning of soil microorganisms.

The choice of methods depends on the questions asked and both the expertise and budget available. Microbial biomass, respiration per unit of biomass (qCO_2) and also biodiversity are regarded as the most sensitive parameters, especially to assess the effects of soil contamination (Brookes, 1995; Giller *et al.*, 1998). The following is a list of the methods used for monitoring in several countries.

- Germany: microbial biomass (SIR), respiration, soil enzymes (Höper, 1999).

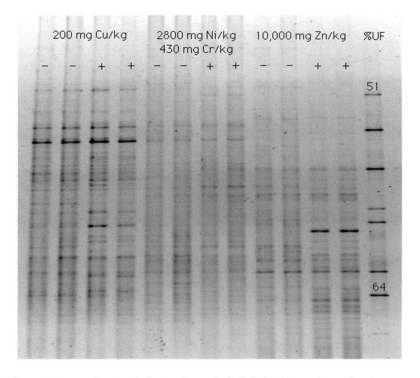

Fig. 3.2. Denaturing gradient gel electrophoresis (DGGE) DNA profiles of soils contaminated with heavy metals (+), compared to uncontaminated controls (−).

- Switzerland: earthworms, microbial biomass (CFE), respiration, N mineralization (Maurer-Troxler, 1999).
- Czech Republic: microbial biomass (SIR), respiration, N mineralization, nitrification, soil enzymes.
- United Kingdom: microbial biomass (CFE), respiration, microbial diversity (CLPP, PLFA).
- New Zealand: microbial biomass (CFE), respiration, N mineralization (G. Sparling *et al.*, available at: http://www.landcareresearch.co.nz/research/rurallanduse/soilquality/Soil_Quality_Indicators_Home.asp, accessed 30 January 2004).

We do not pretend that this list is complete. In New Zealand, the interpretation of microbial biomass and respiration measures was found to be too problematic for practical application (Carter *et al.*, 1999; Schipper and Sparling, 2000). Therefore, seven mainly abiotic soil properties were selected as core indicators of soil quality: total C, total N, mineralizable N, pH, Olsen P, bulk density and macroporosity. The only microbiological indicator used in the 500 soils project is mineralizable N as determined by anaerobic incubation under waterlogged conditions for 7 days at 40°C. The method is relatively simple (see also Canali and Benedetti, Chapter 7.3, this volume).

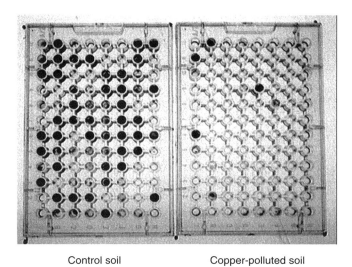

Control soil   Copper-polluted soil

Fig. 3.3. Community-level physiological profiles: utilization of different substrates in Biolog™ plates.

Mineralizable N is regarded as a measure of readily decomposed organic N, and as a measure of biological activity (Hill *et al.*, 2004). Thus, in most countries a relatively small number of variables is monitored, usually at a relative large number of sites (up to 500 in New Zealand). In the USA, in contrast, a large number of methods and variables are used at a limited number of 21 long-term ecological research sites (Robertson *et al.*, 1999).

In the Dutch Soil Quality Network, besides chemical and physical variables, nematodes have been measured since 1993. Since 1997 a wider range of biological variables has been included (Schouten *et al.*, 2000):

- bacterial biomass (microscopy and image analysis);
- bacterial growth rate (thymidine and leucine incorporation into DNA and proteins);
- bacterial functional diversity (substrate utilization profiles using Biolog™);
- bacterial genetic diversity (DNA profiles using DGGE);
- potential carbon and nitrogen mineralization (6-week laboratory incubations);
- nematodes, abundance and diversity;
- mites, abundance and diversity;
- enchytraeids, abundance and diversity;
- earthworms, abundance and diversity.

Thus, important processes and functional groups of the soil food web are included. A relatively large number of biological variables was chosen because monitoring of soil biodiversity had a high priority. Each variable is measured in a single specialized laboratory, using a single method.

Earthworms, enchytraeids, mites and nematodes are included as important functional groups of soil fauna. Microbial indicators currently used are bacterial biomass, growth rate, functional diversity and genetic diversity (Bloem and Breure, 2003). These different microbial indicators are measured in the same subsamples of soil after pre-incubation for 4 weeks at 12°C and 50% of the water-holding capacity (WHC). Potential carbon and nitrogen mineralization are determined in soil incubated for 6 weeks at 20°C and 50% WHC. C and N mineralization rates are calculated from differences in CO_2 and mineral N concentrations between week six and week one. Results of the first week are not used, to avoid disturbance effects of sample handling.

Results from Contaminated and Experimental Reference Sites

In the Dutch Soil Quality Network, large areas of about 50 ha (mainly farms) are sampled and the replicates within categories of soil type and land use are spread all over the country. Some of the reference sites were smaller long-term experimental fields and contaminated sites. Such sites are expected to show the most clear and contrasting results.

In a heavily contaminated field soil (10,000 mg zinc/kg), bacterial DGGE DNA profiles showed a significantly reduced diversity compared to remediated plots (31 versus 50 DNA bands). Also, in a slightly contaminated soil (160 mg copper/kg), the number of DNA bands was reduced from 50 to 42. However, in a soil contaminated with nickel and chromium (2800 mg/kg and 430 mg/kg, respectively) there was no reduction in the number of DNA bands. In all cases contamination caused significant qualitative changes in the DNA profiles (Fig. 3.2). Also, community-level physiological profiles (Biolog™) indicated changes in the community structure (results not shown). Thus, community structure changed, but diversity was not always reduced in seriously contaminated soils. Biomass contents for different groups of organisms, respiration and mineralization were much more reduced than diversity. Bacterial growth rate (thymidine incorporation) was the most sensitive, with decreases of more than 70% (Bloem and Breure, 2003). Thus, biological indicators showed significant reductions in contaminated soils.

Less extreme effects may be expected in agricultural soils. Sustainable agriculture aims at maintaining good crop yields with minimal impact on the environment, while at least avoiding deterioration in soil fertility and providing essential nutrients for plant growth. Further, sustainable agriculture supports a diverse and active community of soil organisms, exhibits a good soil structure and allows for undisturbed decomposition (Mäder *et al.*, 2002). Therefore, agricultural practices are adjusted to integrate organic, or more extensive, management. The main principles are to restrict stocking densities, avoid synthetic pesticides, avoid mineral fertilizers and use organic manure (Hansen *et al.*, 2001). Finally, this will result in an increased role of soil organisms, e.g. decomposers, nitrogen fixers and mycorrhizas, in plant nutrition and disease suppression.

Indeed, larger microbial biomass, diversity (Biolog™), enzyme activities, more mycorrhizas and higher earthworm abundance have been observed in organic farming during a 20-year experiment in Switzerland (Mäder et al., 2002). Application of farmyard manure, instead of mineral fertilizer, stimulates the bacterial branch of the soil food web. In fields that had received farmyard manure for 45 years (Marstorp et al., 2000), we found a fourfold greater bacterial biomass than in fields that received only mineral nitrogen (Fig. 3.4). Bacterial biomass was lower (twofold) with sewage sludge. However, a larger input of organic matter does not necessarily lead to more bacterial biomass. Arable fields under integrated farming showed greater bacterial activity but not a significantly larger bacterial biomass (Bloem et al., 1994); this was attributed to bacterivorous protozoa and nematodes, which reached 20–60% greater densities. N mineralization by the soil food web was 30% higher than in conventional fields. This compensated for a 35% smaller input of mineral fertilizer in the integrated fields and supported a crop yield of 90% of that in the conventional fields.

In grassland plots that received no (O) or incomplete (N only, or PK) mineral fertilization for 50 years, there was up to 2.5 times more fungal biomass than in plots with full mineral fertilization (NPK) (Fig. 3.5). Liming (Ca) had the same effect as fertilization because this stimulated mineralization of the high amount of organic matter (20%) in this clay soil. Part of the fungal biomass may have been mycorrhizas, which support plant nutrition at low availability of mineral nutrients. The grassland plots that received no mineral nitrogen had high plant diversity (40 species) whereas N-fertilized plots contained low plant diversity (20 species). There was no simple quantitative relationship between above-ground and below-ground diversity. The numbers of bacterial DNA bands (61–72) were not significantly different. Nevertheless, there are qualitative relationships between above- and below-ground diversity. Principal component analysis clearly separated soil bacterial DNA profiles of grassland plots from those of arable plots on the

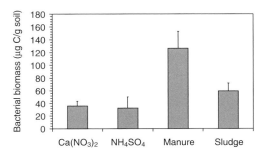

Fig. 3.4. Effect of fertilization on bacteria in arable fields, Ultuna, Sweden; experiment began 1956. Error bars indicate SE, $n = 3$.

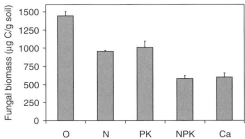

Fig. 3.5. Effect of fertilization on fungal hyphae in grassland, Ossekampen, The Netherlands; experiment began 1950. Error bars indicate SE, $n = 2$. O, no NPK.

same soil sampled on the same date. However, with samples of four different seasons, temporal variation in the DNA profiles impeded a clear separation. Spatial variation in DNA profiles was small between replicate plots on the same field, but large between different sites. The variation between 10–20 farms of the same category was as large as the differences between categories of intensive, organic and extensive grassland farms on sand. Therefore, qualitative information (community structure) is more difficult to handle than quantitative information, such as the number of species (richness) and relative abundances (evenness). Richness and evenness are combined in the Shannon diversity index (Atlas and Bartha, 1993).

The Dutch Soil Quality Network: Results of the First Year

In the first year (1997), a pilot study was performed to test whether the selected biological variables were sufficiently reproducible when applied in large-scale monitoring using mixed samples of whole farms. In that year, mites were included at only two reference farms, where the total food web structure was determined, including important functional groups such as fungi and protozoa. Thus, for these two farms, C and N mineralization were calculated using a food web model, where mineralization rates are calculated from the observed biomasses of different functional groups of soil organisms. Food web modelling was also used to predict food web stability (De Ruiter *et al.*, 1995). The budget was not sufficient to adopt this fundamental and comprehensive approach in the routine programme. Originally, potential nitrification was included in the programme, but in 1999 this indicator was replaced by potential C and N mineralization. Also in 1999, genetic diversity of bacteria was included because of increased political and scientific interest in biodiversity and because molecular techniques became available for routine use.

Results of the first year (1997) of the Dutch Soil Quality Network are summarized in Table 3.1. In this year agricultural grasslands on sea-clay and horticultural farms (bulbs and vegetables) on various soils were sampled. Sea-clay (Dutch classification) encompasses soils in loam and clay deposits of marine origin. Both types of soil are managed intensively, but frequencies of soil tillage and pesticide application are higher in horticulture. Soil pH-KCl was 6.5 in both categories. The grassland soils had a significantly higher content of clay (24% versus 7%) and organic matter (6.2% versus 2.8%) than the horticultural soils. Correspondingly, the cation exchange capacity (CEC) was significantly lower in the horticultural soils (6.5 cmol+/kg versus 22 cmol+/kg). In spite of the lower P-binding capacity, the horticultural soils had a higher phosphate content (0.77 mg P_2O_5/g versus 0.41 mg P_2O_5/g soil), which is attributed to a higher degree of fertilization. Contents of heavy metals (cadmium, copper, lead and zinc), some formerly used persistent pesticides (DDT, HCB, dieldrin and lindane) and polyaromatic hydrocarbons (phenanthrene and benzo(a)pyrene) were at acceptable concentrations and not significantly different between the categories.

Table 3.1. Indicators measured in 20 agricultural grasslands on sea-clay and 17 horticultural farms on various soils.

Soil biota	Indicators	Grassland on sea-clay ($n = 20$)	Horticulture ($n = 17$)	Statistical significance of difference Grl. – Hort.
Bacteria	Thymidine incorp. (pmol/g/h)	179	108	***
	Leucine incorp. (pmol/g/h)	847	392	***
	Bacterial biomass (mg C/g)	232.4	56	***
	Colony-forming units (10^7/g)	17.1	2.6	***
	Potential nitrification (mg NO_3-N/kg/week)	93	74	***
Biolog™	LogCFU-50 (number of CFU corresponding with 50% activity)	3.73	2.87	***
	H-coefficient (evenness of decomposition of 31 substrates)	0.39	0.60	***
	Gg50 (µg soil with 50% activity)	95	44	*
Nematodes	Abundance (number/100 g)	4,629	2,069	***
	Number of taxa	26.1	21.8	*
	Maturity index	1.77	1.47	***
	Trophic diversity index	2.12	1.51	***
	Number spec. bacterial feeding	11.4	13.3	*
	Number spec. carnivores	0.4	0.6	ns
	Number spec. hyphal feeding	2.1	2.1	ns
	Number spec. omnivores	1.0	1.2	ns
	Number spec. plant feeding	11.4	4.5	***
	Number of functional groups	3.9	4.3	ns
Enchytraeids	Abundance (number/m^2)	24,908	16,096	**
	Number of taxa	8.2	5.5	***
	Biomass (g/m^2)	5.6	1.1	***
	Number of Friderica (number/m^2)	8,654	1,300	***
Earthworms	Abundance (number/m^2)	317	40	***
	Biomass (g/m^2)	70.1	3.8	***
	Endogé-species (deeper in soil)	2.10	0.82	***
	Epigé-species (closer to surface)	1.2	0.06	***
		(n = 1)	(n = 1)	
Mites	Abundance (number/m^2)	37,900	18,100	
	Number of species	23	20	
	Number of functional groups	8	10	
Food web (model-calculations)	N-mineralization (kg N/ha/yr)	335	115	
	C-mineralization (kg C/ha/yr)	6,150	1,750	
	Stability	0.47	0.61	

CFU, colony-forming units; Grl., grassland; Hort., horticulture; ns, not significant.

In a single soil, thymidine incorporation, leucine incorporation and bacterial biomass usually show coefficients of variation (CV) of 10%, 5% and 30%, respectively. In the national monitoring programme, variation is much larger because of differences between farms. The variation between 20 replicate farms of the same category was about 30% for the growth rate measurements and about 60% for the biomass measurements. With 20 replicates, this results in standard errors of about 7% and 13% of the mean. This is sufficient to establish statistical differences between categories. Analyses of variance of (log transformed) data yielded a significance (P) < 0.001 for most indicators (Table 3.1), thus demonstrating reproducibility and discriminative power.

Thymidine and leucine incorporation indicated a 50% lower bacterial growth rate in horticulture compared to grassland. Bacterial biomass in horticulture was only 25% of the biomass in the grasslands. Thus, the specific growth rate per unit of bacterial biomass was twofold higher in the horticultural soils. Also, the results of the Biolog™ assay indicated a greater specific activity of bacteria in the horticultural soils. Using Biolog™ ECO-plates, the decomposition of 31 different carbon sources can be tested (in triplicate) in 96 wells of a microtitre plate. Of each soil suspension, four different dilutions were incubated in Biolog™ ECO-plates, and colony-forming units (CFU) were counted in parallel (for methods see Rutgers and Breure, 1999; Breure and Rutgers, 2000; Chapter 8, this volume). LogCFU-50, the number of CFU giving 50% activity, was smaller in the horticultural soils, and thus the activity per CFU greater. The activity in Biolog™ plates is measured as average colour development in the wells, reflecting respiration and growth. Potential nitrification was high in all grasslands and in most horticultural soils. On average, the rate of potential nitrification was 20% smaller in horticulture. In three (out of 17) horticultural farms, potential nitrification rate was at least 50% slower than in the other soils. Most indicators were lower in the horticultural soils. Food web modelling, applied to only two farms, indicated lower carbon and nitrogen mineralization in the soil of a horticultural farm compared to a grassland farm. Predicted soil food web stability was higher on the horticultural farm. It may be speculated that under intensive management more stress-resistant organisms are selected.

Presentation of results

The large amount of data in Table 3.1 is not suitable for presentation to either policy makers or the public. The results can be presented in a more clear and illustrative way for diagnostic purposes using the AMOEBA method (Ten Brink *et al.*, 1991), or similar cobweb or star diagrams (Stenberg, 1999; Mäder *et al.*, 2002). The former method results in an amoeba-like graphical representation of all indicator values, scaled against a historical, undisturbed or desired situation. As an example, the indicator data from one organic grassland farm were used as a reference for the intensively managed farms (Fig. 3.6).

Monitoring and Evaluating Soil Quality 39

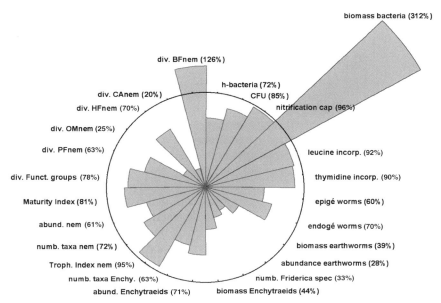

Fig. 3.6. AMOEBA presentation of indicator values from the grassland farms on sea-clay, relative to those in the reference (circle), which is set to 100%. In this case the reference is only one organic farm (from Schouten *et al.*, 2000). The graph facilitates a clear overview of the differences between categories of soil type and land use. The indicators are described in more detail in Table 3.1.

The value of each variable in the reference was scaled as 100%. This yields a circle of 100% values for the reference. In the example, mite fauna and total food web structure were omitted due to lack of data from the reference. Almost all the indicative parameters of the 20 grassland farms appeared within the 100% circle. Apparently, biodiversity within functional groups, and process rates, were lower in intensive than in organic grassland. Only bacterial biomass was much greater (312%), which was reflected in the diversity of bacterial-feeding nematodes (126%). Despite the large bacterial biomass, its activity (indicated by the number of colony-forming units, potential nitrification and [^{14}C]leucine and [^{3}H]thymidine incorporation) was approximately 90% of the reference, so the specific activity was smaller.

The indexed indicator values used to construct the AMOEBA can be further condensed into a Soil Quality Index (SQI), using the average factorial deviation from the reference value (Ten Brink *et al.*, 1991). The index is calculated as follows:

$$SQI = 10^{\log m - \frac{\sum_{i=1}^{n} |\log m - \log n_i|}{n}}$$

where m is the reference (set to 100%) and n are the measured values as percentages of the reference.

For example:

	Reference	Sample
Indicator 1	100	50
Indicator 2	100	200

log reference = 2, log sample is 1.7 and 2.3, respectively.
The (absolute) difference between reference and sample is:

For indicator 1: $|2 - 1.7| = 0.3$
For indicator 2: $|2 - 2.3| = 0.3$

The sum of the differences is 0.6 and the average difference is 0.3.
The average value of the sample is $2 - 0.3 = 1.7$ on the log scale. Back transformed this gives an SQI of $10^{1.7} = 50\%$.

Thus, in this SQI a value of 50% has the same weight as one of 200% of the reference value (both a factor two). For the AMOEBA in Fig. 3.6, this exercise resulted in an SQI value of 65% for the intensively managed farms versus the single organic farm. AMOEBA-like figures and derived indices can be used as tools for comparison (in space and time) and for relatively simple presentation of complicated results. It must be realized that they are only simplified reflections of complex ecosystems and should not be taken as absolute values. The SQI approach has been applied mainly on sites with local contamination, where an uncontaminated reference is available. It is more difficult to define a proper reference for larger areas with diffuse contamination, and for the different categories of soil type and land use of the Dutch Soil Quality Network.

The Dutch Soil Quality Network: Results of a 5-year Cycle

After the pilot study in 1997, the programme was evaluated in 1998 and continued in 1999 by sampling of grassland farms on sand with different management intensities: extensive, organic and intensive. The categories are based on the number of livestock units per hectare. One livestock unit is defined as the amount of cattle, pigs and/or poultry excreting an average of 41 kg N/ha/yr. The organic farms had been certified for at least 5 years before sampling, used compost and/or farmyard manure and no biocides, and had on average 1.6 livestock units/ha. Extensive farms used mineral fertilizer and less farmyard manure and had 2.3 livestock units/ha. Intensive farms used mineral fertilizer and farmyard manure and had 3.0 livestock units/ha. In 2000, highly intensive (Int+) farms with pigs (5.1 livestock units/ha) on sand, and forest on sand were sampled. Intensive arable farms on sand followed in 2001, when further sampling on farms was prevented by an outbreak of foot-and-mouth disease. Therefore, organic arable farms on sand and grassland farms on peat could not be sampled as planned in 2001. In 2002, sampling organic arable farms on sand was completed. Also in 2002, arable farms on sea-clay (both intensive and organic) and grassland farms on river clay (intensive) were sampled. River-clay (Dutch classification) encompasses loam and clay soils in river deposits.

Bacterial DNA profiles reflecting the dominant species or genotypes did not indicate low genetic diversity in agricultural soils. Organic farms did not show higher genetic diversity of bacteria than intensive farms. Between 48 and 69 DNA bands were found, with Shannon diversity indices from 3.57 ± 0.03 to 3.85 ± 0.02 (± SE). There may be up to 10,000 bacterial genotypes/g soil, but most species occur in very low numbers. If a species has a density below 10^6 cells/g soil (about 1/1000 of the total number), it is below the detection limit of our DGGE method (Dilly et al., 2004). Thus, this method reflects the composition of most of the bacterial biomass, but not the total number of species. Low bacterial diversity (25 DNA bands, Shannon index 2.37 ± 0.22) was only found in the usually acid forest soils, which also contained a very small bacterial biomass (Fig. 3.7).

The bacterial biomass was also very low in horticultural soils, higher in arable soils and high in grassland. Grassland contained a higher bacterial biomass on clay than on sand. The biomass appears to reflect management intensity, which is higher in arable land (tillage, pesticides, etc.) than in grassland. Similar to the biomass, potential nitrogen mineralization was very low in forest, higher in arable land and highest in grassland (Fig. 3.8).

Potential N mineralization was about 50% lower in clay than in sandy soils. At extensive and organic grassland farms on sand, N mineralization (i.e. soil fertility) was about 50% higher than at intensive farms. At the highly intensive farms, mineralization was almost as high as on the organic farms, probably because a lot of (pig) manure had been applied. Not only microbial biomass and activity, but also different groups of soil fauna, tended to be higher at organic and extensive farms (Fig. 3.9).

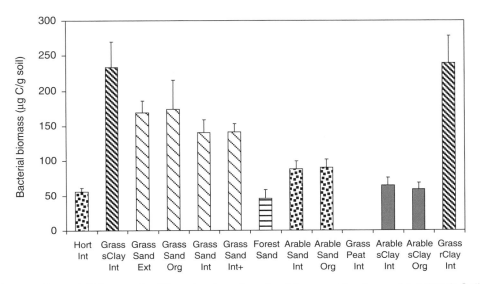

Fig. 3.7. Bacterial biomass in different categories of soil type and land use of the Dutch Soil Quality Network. Hort, horticultural; Grass, grassland; sClay, sea-clay; rClay, river-clay; Int, intensive farms; Ext, extensive farms; Org, organic farms; Int+, highly intensive farms. Error bars indicate SE, n = 20 (10 for organic farms).

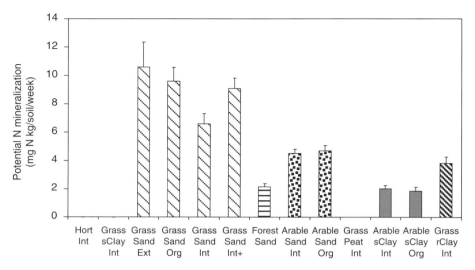

Fig. 3.8. Potential N mineralization in different categories of soil type and land use of the Dutch Soil Quality Network. Hort, horticultural; Grass, grassland; sClay, sea-clay; rClay, river-clay; Int, intensive farms; Ext, extensive farms; Org, organic farms; Int+, highly intensive farms. Error bars indicate SE, $n = 20$ (10 for organic farms).

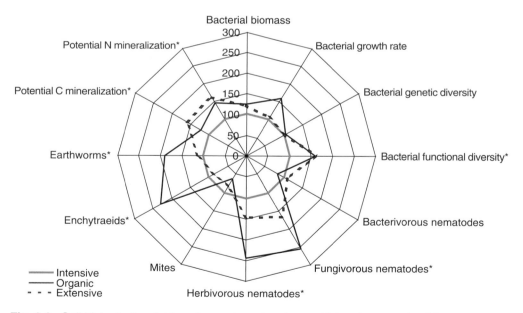

Fig. 3.9. Soil biological variables at organic, extensive and intensive grassland farms on sand. Intensive is set to 100%; * indicates a statistically significant difference ($P \leq 0.05$, analysis of variance) between categories.

Analysis of variance indicated statistically significant differences for bacterial functional diversity (colour development in Biolog™ plates/g soil; in fact, activity), fungivorous nematodes, herbivorous nematodes, enchytraeids, earthworms, potential C mineralization (respiration) and potential N mineralization. Bacterial growth rate (thymidine incorporation) was 60% higher in organic than in intensive grassland on sand (23.6 ± 4.8 versus 14.8 ± 2.9 pmol thymidine/g soil/h, ± SE, n = 10 for organic and 20 for intensive farms). Similarly, bacterial growth rate was 200% higher in organic than in intensive arable land on sand (114 ± 18 pmol/g/h versus 46.6 ± 3.9 pmol/g/h), and 33% higher in organic than in intensive arable land on clay (153 ± 26 pmol/g/h versus 115 ± 11 pmol/g/h). Both on sand and on clay there was no difference in N mineralization between intensive and organic arable farms (Fig. 3.8). The differences between categories of farms are less significant than differences between treatments in long-term experimental fields. This may be caused by several reasons. Sampling at the farm level causes greater variation than sampling in replicated experimental fields. Furthermore, it may take a long time, perhaps decades rather than 5 years, before soil organisms increase significantly after a change in management. The categories used here are rather broad, and there are relatively large differences between farms within one category. With more detailed analysis, using narrow clusters instead of broad categories of farms, trends can be detected. Bacterial biomass and bacterivorous nematodes tend to decrease with increasing stocking density on grassland farms on sand (Fig. 3.10).

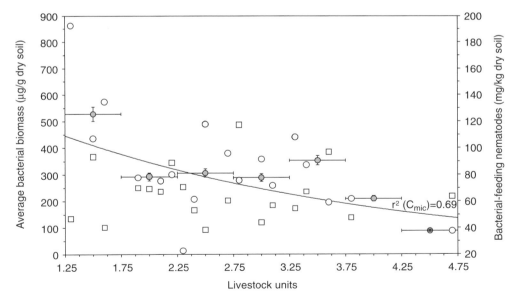

Fig. 3.10. Scatterplot of the average bacterial biomass (circles) and the bacterial-feeding nematode biomass (squares) along an increasing farming intensity gradient (livestock units/ha). The open symbols represent narrow clusters. The exponential trendline for bacterial biomass is fitted on wider clusters (closed circles) of 0.5 livestock unit width (n = 9, α = 0.05; vertical error bars for 5% bacterial biomass). (From Mulder *et al.*, 2003.)

When sufficient data (> 200 sites) become available after completing one 5-year cycle of monitoring, habitat response models can be developed to predict the value of biological variables from environmental variables such as stocking density, pH, organic matter content, clay, etc. (Oberholzer and Höper, 2000; Mulder et al., 2003).

Thus, experience with monitoring soil life shows large differences between categories of soil type and land use, and significant effects of management on most groups of organisms and potential C and N mineralization.

Measuring Soil Biodiversity: Impediments and Research Needs

Soil ecosystems are complex. Therefore, many different aspects need to be measured (Lancaster, 2000). It is important to use a set of various indicators, and not a few indicators selected *a priori*, which are supposed to be the most sensitive. Some indicators are more sensitive to contamination (e.g. bacterial growth rate); others are more sensitive to differences in soil fertility and agricultural management (e.g. N mineralization).

The complexity of biodiversity implies that time and money are major impediments for thorough monitoring. Techniques are available, but extensive monitoring is expensive. The costs of sampling and analysing one site (mixed sample of one farm) of the Dutch Soil Quality Network in 2002 were €5500. This amounts to €330,000 per year for 60 sites. Still, the important functional groups of fungi and protozoa are not (yet) included in the routine programme. Protozoa are the major bacterivores, but showed strong temporal variation, and fungi usually had a very low biomass in Dutch agricultural soils (Bloem et al., 1994; Velvis, 1997). However, recently fungal dominance was found in a few agricultural soils, especially at low mineral N fertilization (Fig. 3.5). This indicates that the role of fungi may increase when fertilization is reduced, e.g. in sustainable agriculture and set-aside land. Therefore, the fungal to bacterial ratio deserves more research as a potentially simple indicator of efficient nutrient use and sustainability (Bardgett and McAlister, 1999; Mulder et al., 2003). Based on the results obtained, in 2004 fungi were included in the indicator set. On the other hand, bacterial genetic diversity based on DGGE DNA profiles was omitted, because quantitative differences appeared to be small and qualitative results appeared difficult to interpret, especially in large-scale monitoring.

Relationships between below-ground biodiversity, in terms of species richness and function, are still not clear. Experimental reduction of biodiversity in soil did not significantly affect functioning in terms of decomposition and mineralization (Griffiths et al., 2001). Nevertheless, biodiversity may play a role in stability against perturbation and stress, in nutrient retention and in suppression of plant diseases (Griffiths et al., 2000). Fundamental research is still needed to clarify the importance of soil biodiversity. Species identity (community structure) is probably more important than species richness, but is more difficult to analyse. New techniques, such as microarrays (DNA chips), are needed to monitor large numbers of species, genes and functions, and may offer new opportunities to link diversity to function.

Using approaches as described here, biodiversity and functioning of soil ecosystems can be monitored. In pollution gradients, it is possible to use a local unpolluted control (Bloem and Breure, 2003). However, in many cases such a reference is not available. Generally, the value of an indicator is affected not only by stress factors, but also by soil type, land use and vegetation. Therefore, reference values for specific soil types have to be deduced from many observations, e.g. 20 replicates per type. The choice of a desired reference is a political rather than a scientific issue, and depends on the aims of land use. A biologically active and fertile soil is needed in (organic) farming, but a high mineralization of nutrients from organic matter may hamper conversion of agricultural land to a species-rich natural vegetation. For a specific soil and land-use type, the reference could be the current average of 20 conventional farms, or the average of 20 organic farms. Soils showing very low or very high indicator values may be suspect and need further examination. Sufficient data and experience are needed to make judgements of desirable reference values. Monitoring changes of indicators over time can reduce the importance of (subjective) reference values. Such changes may be easier to interpret than momentary values (Lancaster, 2000). Spatially extensive and long-term monitoring may be not ideal, but it is probably the most realistic approach to obtain objective information on differences between, temporal changes within, and human impact on ecosystems.

To make progress with monitoring and understanding the complexity of soil life it is necessary to start. Methods are already available. Differences between methods and laboratories occur with any variable, and can be minimized by using a limited number of specialized laboratories for each variable. For any purpose, standardization and intercalibration are necessary. Learning by doing is inevitable, thus long-term and extensive monitoring programmes need to be flexible. To begin with, monitoring programmes may be relatively simple with a limited number of variables, which should be increased when possible. Better one indicator than nothing, but the more the better. Choosing a minimum data set and ranking of indicators remains a subjective exercise. Looking at monitoring programmes established in different countries, it appears that microbial biomass, respiration and potential N mineralization are commonly regarded as part of a minimum data set. These are very useful, but more data are needed if we aim at monitoring biodiversity. Adding the main functional groups of the soil food web brings us closer to understanding biodiversity and provides the potential to relate the structure of the soil community to ecosystem functioning and environmental stress (Mulder *et al.*, 2005).

Acknowledgements

We thank many colleagues for their invaluable contribution to the results: An Vos, Popko Bolhuis, Meint Veninga, Gerard Jagers op Akkerhuis, Henk Siepel and Wim Dimmers, Alterra, Wageningen, The Netherlands; Christian Mulder, Marja Wouterse, Rob Baerselman, Niels Masselink, Margot Groot, Ruud Jeths and Hans Bronswijk, RIVM, Bilthoven, The Netherlands; Wim Didden, Sub-department of Soil Quality, Wageningen University, The Netherlands; Harm Keidel, Christel Siepman and others, Blgg bv, Oosterbeek, The Netherlands; and Ernst Witter, Department of Soil Sciences, Swedish University of Agricultural Sciences, Uppsala, Sweden, for providing samples from Ultuna (Fig. 3.4).

References

Akkermans, A.D.L., van Elsas, J.D. and De Bruijn, F.J. (eds) (1995) *Molecular Microbial Ecology Manual*. Kluwer Academic Publishers, Dordrecht, The Netherlands.

Alef, K. and Nannipieri, P. (eds) (1995) *Methods in Applied Soil Microbiology and Biochemistry*. Academic Press, London.

Anderson, J.P.E. (1987) Handling and storage of soils for pesticide experiments. In: Somerville, L. and Greaves, M.P. (eds) *Pesticide Effects on Soil Microflora*. Taylor & Francis, London, pp. 45–60.

Atlas, R.M. and Bartha, R. (1993) *Microbial Ecology. Fundamentals and Applications*. The Benjamin/Cummings Publishing Company, Redwood City, California, pp. 140–145.

Bardgett, R.D. and McAlister, E. (1999) The measurement of soil fungal:bacterial biomass ratios as an indicator of ecosystem self-regulation in temperate meadow grasslands. *Biology and Fertility of Soils* 29, 282–290.

Bloem, J. and Breure, A.M. (2003) Microbial indicators. In: Markert, B., Breure, A.M. and Zechmeister, H. (eds) *Bioindicators/Biomonitors – Principles, Assessment, Concepts*. Elsevier, Amsterdam, pp. 259–282.

Bloem, J., Lebbink, G., Zwart, K.B., Bouwman, L.A., Burgers, S.L.G.E., De Vos, J.A. and De Ruiter, P.C. (1994) Dynamics of microorganisms, microbivores and nitrogen mineralisation in winter wheat fields under conventional and integrated management. *Agriculture, Ecosystems and Environment* 51, 129–143.

Bloem, J., Veninga, M. and Shepherd, J. (1995) Fully automatic determination of soil bacterium numbers, cell volumes and frequencies of dividing cells by confocal laser scanning microscopy and image analysis. *Applied and Environmental Microbiology* 61, 926–936.

Bloem, J., De Ruiter, P.C. and Bouwman, L.A. (1997) Food webs and nutrient cycling in agro-ecosystems. In: van Elsas, J.D., Trevors, J.T. and Wellington, E. (eds) *Modern Soil Microbiology*. Marcel Dekker, New York, pp. 245–278.

Bloem, J., Schouten, A.J., Didden, W., Jagers op Akkerhuis, G., Keidel, H., Rutgers, M. and Breure, A.M. (2004) Measuring soil biodiversity: experiences, impediments and research needs. In: Francaviglia, R. (ed.) *Agricultural impacts on soil erosion and soil biodiversity: developing indicators for policy analysis*, Proceedings of an OECD expert meeting, 25–28 March 2003, Rome, Italy. OECD, Paris, p. 109–129. Available at: http://webdomino1.oecd.org/comnet/agr/soil_ero_bio.nsf (accessed 26 October 2004).

Breure, A.M. and Rutgers, M. (2000) The application of Biolog plates to characterize microbial communities. In: Benedetti, A., Tittarelli, F., De Bertoldi, S. and Pinzari, F. (eds) *Biotechnology of Soil: Monitoring, Conservation and Bioremediation*, Proceedings of the COST Action 831 joint working group meeting 10–11 December 1998, Rome, Italy. (EUR 19548), EU, Brussels, pp. 179–185.

Brookes, P.C. (1995) The use of microbial parameters in monitoring soil pollution by heavy metals. *Biology and Fertility of Soils* 19, 269–279.

Brussaard, L., Behan-Pelletier, V.M., Bignell, D.E., Brown, V.K., Didden, W., Folgarait, P., Fragoso, C., Freckman, D.W., Gupta, V.V.S.R., Hattori, T., Hawksworth, D.L., Klopatek, C., Lavelle, P., Malloch, D.W., Rusek, J., Söderström, B., Tiedje, J.M. and Virginia, R.A. (1997) Biodiversity and ecosystem functioning in soil. *Ambio* 26, 563–570.

Brussaard, L., Kuyper, T.W., Didden, W.A.M., De Goede, R.G.M. and Bloem, J. (2003) Biological soil quality from biomass to biodiversity – importance and resilience to management stress and disturbance. In: Schjønning, P., Christensen, B.T. and Elmholt, S. (eds) *Managing Soil Quality – Challenges in Modern Agriculture.* CAB International, Wallingford, UK, pp. 139–161.

Carter, M.R., Gregorich, E.G., Angers, D.A., Beare, M.H., Sparling, G.P., Wardle, D.A. and Voroney, R.P. (1999) Interpretation of microbial biomass measurements for soil quality assessment in humid temperate regions. *Canadian Journal of Soil Science* 79, 507–520.

De Ruiter, P.C., Neutel, A.-M. and Moore, J.C. (1995) Energetics, patterns of interaction strengths, and stability in real ecosystems. *Science* 269, 1257–1260.

Dilly, O., Bloem, J., Vos, A. and Munch, J.C. (2004) Bacterial diversity during litter decomposition in agricultural soils. *Applied and Environmental Microbiology* 70, 468–474.

Doran, J.W. and Parkin, T.B. (1994) Defining and assessing soil quality. In: Doran, J.W., Coleman, D.C., Bezdicek, D.F. and Stewart, B.A. (eds) *Defining Soil Quality for a Sustainable Environment*, Special Publication 35. American Society of Agronomy, Madison, Wisconsin, pp. 3–21.

Doran, J.W. and Parkin, T.B. (1996) Quantitative indicators of soil quality: a minimum data set. In: Doran, J.W. and Jones, A.J. (eds) *Methods for Assessing Soil Quality*, Special Publication 49. Soil Science Society of America, Madison, Wisconsin, pp. 25–37.

FAO (1999) *Sustaining Agricultural Biodiversity and Agro-ecosystem Functions.* Report available at: http://www.fao.org/sd/epdirect/EPre0065.htm (accessed 9 April 2004).

Giller, K.E., Witter, E. and McGrath, S.P. (1998) Toxicity of heavy metals to microorganisms and microbial processes in agricultural soils: a review. *Soil Biology and Biochemistry* 20, 1389–1414.

Griffiths, B.S., Ritz, K., Bardgett, R.D., Cook, R., Christensen, S., Ekelund, F., Sørensen, S., Bååth, E., Bloem, J., De Ruiter, P., Dolfing, J. and Nicolardot, B. (2000) Ecosystem response of pasture soil communities to fumigation-induced microbial diversity reductions: an examination of the biodiversity–ecosystem function relationship. *Oikos* 90, 279–294.

Griffiths, B.S., Ritz, K., Wheatly, R., Kuan, H.L., Boag, B., Christensen, S., Ekelund, F., Sørensen, S.J., Muller, S. and Bloem, J. (2001) An examination of the biodiversity–ecosystem function relationship in arable soil microbial communities. *Soil Biology and Biochemistry* 33, 1713–1722.

Hansen, B., Alroe, H.F. and Kristensen, E.S. (2001) Approaches to assess the environmental impact of organic farming with particular regard to Denmark. *Agriculture Ecosystems and Environment* 83, 11–26.

Hill, R., Frampton, C., Cuff, J. and Sparling, G. (2004) National soil quality review and programme design, unpublished report.

Höper, H. (1999) Bodenmikrobiologische Untersuchungen in der Bodendauerbeobachtung in Deutschland. *VBB-Bulletin* 3, 13–14. Arbeitsgruppe Vollzug Bodenbiologie, FiBL, CH-5070 Frick, Switzerland [in German].

Hunt, H.W., Coleman, D.C., Ingham, E.R., Ingham, R.E., Elliott, E.T., Moore, J.C., Rose, S.L., Reid, C.F.F. and Morley, C.R. (1987) The detrital food web in a short-grass prairie. *Biology and Fertility of Soils* 3, 57–68.

Korthals, G.W., Alexiev, A.D., Lexmond, T.M., Kammenga, J.E. and Bongers, T. (1996) Long-term effects of copper and pH on the nematode community in an agroecosystem. *Environmental Toxicology and Chemistry* 15, 979–985.

Lancaster, J. (2000) The ridiculous notion of assessing ecological health and identifying the useful concepts underneath. *Human and Ecological Risk Assessment* 6, 213–222.

Mäder, P., Fliessbach, A., Dubois, D., Gunst, L., Fried, P. and Niggli, U. (2002) Soil fertility and biodiversity in organic farming. *Science* 296, 1694–1697.

Marstorp, H., Guan, X. and Gong, P. (2000) Relationship between dsDNA, chloroform labile C and ergosterol in soils of different organic matter contents and pH. *Soil Biology and Biochemistry* 32, 879–882.

Maurer-Troxler, C. (1999) Einsatz bodenbiologischer Parameter in der langfristigen Bodenbeobachtung des Kantons Bern. *VBB-Bulletin* 3, 11–13. Arbeitsgruppe Vollzug Bodenbiologie. FiBL, CH-5070 Frick, Switzerland [in German].

Mulder, Ch., De Zwart, D., Van Wijnen, H.J., Schouten, A.J. and Breure, A.M. (2003) Observational and simulated evidence of ecological shifts within the soil nematode community of agroecosystems under conventional and organic farming. *Functional Ecology* 17, 516–525.

Mulder, Ch., Cohen, J.E., Setälä, H., Bloem, J. and Breure, A.M. (2005) Bacterial traits, organism mass, and numerical abundance in the detrital soil food web of Dutch agricultural grasslands. *Ecology Letters* 8, 80–90.

Nielsen, M.N. and Winding, A. (2002) Microorganisms as indicators of soil health. National Environmental Research Institute (NERI), Denmark, Technical Report no. 388. Available at: http://www.dmu.dk (accessed 30 January 2004).

Oberholzer, H.R. and Höper, H. (2000) Reference systems for the microbiological evaluation of soils. *Verband Deutscher Landwirtschaftlicher Untersuchungs- und Forschungsanstalten* 55, 19–34.

OECD (2003) *Agriculture and Biodiversity: Developing Indicators for Policy Analysis.* Proceedings from an OECD expert meeting Zurich, Switzerland, November 2001. OECD, Paris.

Paul, E.A., Harris, D., Klug, M. and Ruess, R. (1999) The determination of microbial biomass. In: Robertson, G.P., Coleman, D.C., Bledsoe, C.S. and Sollins, P. (eds) *Standard Soil Methods for Long-term Ecological Research.* Oxford University Press, New York, pp. 291–317.

Robertson, G.P., Coleman, D.C., Bledsoe, C.S. and Sollins, P. (eds) (1999) *Standard Soil Methods for Long-term Ecological Research.* Oxford University Press, New York.

Rutgers, M. and Breure, A.M. (1999) Risk assessment, microbial communities, and pollution induced community tolerance. *Human and Ecological Risk Assessment* 5, 661–670.

Schipper, L.A. and Sparling, G.P. (2000) Performance of soil condition indicators across taxonomic groups and land uses. *Soil Science Society of America Journal* 64, 300–311.

Schouten, A.J., Bloem, J., Didden, W.A.M., Rutgers, M., Siepel, H., Posthuma, L. and Breure, A.M. (2000) Development of a biological indicator for soil quality. *SETAC Globe* 1, 30–32.

Stenberg, B. (1999) Monitoring soil quality of arable land: microbiological indicators. *Acta Agriculturae Scandinavica, Section B, Soil and Plant Science* 49, 1–24.

Stenberg, B., Johansson, M., Pell, M., Sjödahl-Svensson, K., Stenström, J. and Torstensson, L. (1998) Microbial biomass and activities in soil as affected by frozen and cold storage. *Soil Biology and Biochemistry* 30, 393–402.

Ten Brink, B.J.E., Hosper, S.H. and Colijn, F. (1991) A quantitative method for description and assessment of ecosystems: the AMOEBA-approach. *Marine Pollution Bulletin* 23, 265–270.

Torstensson, L., Pell, M. and Stenberg, B. (1998) Need of a strategy for evaluation of arable soil quality. *Ambio* 27, 1–77.

UNCED (United Nations Conference on Environment and Development) (1992) Agenda 21. June 1992, Rio de Janeiro.

Velvis, H. (1997) Evaluation of the selective respiratory inhibition method for measuring the ratio of fungal:bacterial activity in acid agricultural soils. *Biology and Fertility of Soils* 25, 354–360.

Wollum, A.G. (1994) Soil sampling for microbiological analysis. In: Weaver, R.W., Angle, S., Bottomley, P., Bezdicek, D.F., Smith, S., Tabatabai, A. and Wollum, A.G. (eds) *Methods of Soil Analysis. Part 2: Microbiological and Biochemical Properties.* Soil Science Society of America, Madison, Wisconsin, pp. 1–14.

Zelles, L., Adrian, P., Bai, Q.Y., Stepper, K., Adrian, M.V., Fischer, K., Maier, A. and Ziegler, A. (1991) Microbial activity measured in soils stored under different temperature and humidity conditions. *Soil Biology and Biochemistry* 23, 955–962.

4 Managing Soil Quality

MICHAEL SCHLOTER,[1] JEAN CHARLES MUNCH[1] AND FABIO TITTARELLI[2]

[1]*GSF Research Centre for Environment and Health, Institute of Soil Ecology, PO Box 1129, D-85764 Neuherberg, Germany;* [2]*Consiglio per la ricerca e la sperimentazione in Agricottura, Istituto Sperimentale per la Nutrizione delle Piante, Via della Navicella, 2, 00184 Rome, Italy*

Abstract

Soil quality is defined as the 'continued capacity of soil to function as a vital living system, within ecosystem and land use boundaries, to sustain biological productivity, promote the quality of air and water environments, and maintain plant, animal and human health'. Therefore it is not surprising that the sustainable conservation of soil quality is a key issue not only in scientific discussions, but also in political controversies. However, conservation of agricultural soils is often a topic of conflict as, on the one hand, it is a major goal to produce crops with high yields and, on the other hand, soil quality should be maintained. In this chapter this clash will be described in detail and some examples of how different farming management systems affect soil quality will be given. Furthermore, how to measure soil quality easily and whether sensitive indicators are available are still open questions. Therefore, another part of this chapter will focus on questions about indicator development and use in practice.

Introduction

During recent years the European Commission has given increasing importance to the role played by soil as a strategic compartment for many of the environmental issues affecting our countries and the planet as a whole. In 2001, publishing its Sustainable Development Strategy (COM 2001, 264), the European Commission noted that soil loss and declining fertility were eroding agricultural land viability. In 2002 (Decision 1600/2002), the European Parliament and the Council laid down the Sixth Environmental Action Programme (Sixth EAP), which covers a period of 10 years, starting from 22 July 2002. The Programme addresses the key environmental objectives and priorities of the Community, which will be met through a range of measures, including legislation and strategic approaches. In Article 6,

'Objectives and priority areas for action on nature and biodiversity', the Sixth EAP foresees the development of a thematic strategy on soil protection 'addressing the prevention of, *inter alia*, pollution, erosion, desertification, land degradation, and hydrogeological risks taking into account regional diversity, including specificities of mountain and arid areas'. On the basis of this input, the European Commission published a Communication entitled 'Towards a thematic strategy for soil protection' (COM 2002, 179), which is intended as a first contribution for building on the political commitment to soil protection. The Communication listed the following as main threats to soil in the EU and candidate countries: erosion, decline in organic matter, soil contamination, soil sealing, soil compaction, decline in soil biodiversity, salinization, floods and landslides. All these threats degrade soil processes, and are considered to be driven and exacerbated by human activity. In this context, soil microbiologists have the expertise to contribute to the ongoing discussion on soil protection through the elaboration of the concept of soil quality and, in particular, of soil quality management. This is the main objective of the following chapter, which, after a short theoretical dissertation on soil quality and soil quality indicators, provides a few case studies to demonstrate the effects of different agricultural production methods on soil microbial community structure and functions.

Soil Quality and Agricultural Sustainability

Sustainable agriculture is based on the conservation of natural resources and on the concept of productivity linked closely to the maintenance of a system aimed at saving energy and resources in the mid to long term, through optimizing recycling and enhancing biodiversity, and through biological synergy. The emphasis on 'sustainable agriculture' and, more generally, on 'sustainable land use', initiated the development of the soil quality concept during the 1990s. Despite the fact that during the past 10 years many different definitions of soil quality have been proposed, the following seems to be one of the most widely accepted, and is suitable for the purpose of sustainable soil management: 'the capacity of a soil to function within ecosystem boundaries to sustain biological productivity, maintain environmental quality, and promote plant and animal health'. In a more simplistic way, this definition of soil quality, like those proposed by other authors and not reported in this chapter, refers to the capacity of a soil to function at present and in the future, for an indefinite period of time. The concept of soil quality has become a tool for assessing the sustainability of soil management systems and has been adopted by users at different educational levels, such as policy makers, land managers and farmers. According to Karlen *et al.* (2001), the use of soil quality as a tool for the assessment of human impact on natural resources is based on the distinction between inherent and dynamic characteristics of soil. Inherent characteristics are those determined by parent material, climate, vegetation and so on, which are not influenced by human activity. They contribute to an inherent soil quality, which is meaningful in determining the capacity of a soil for a specific land use. On the other hand, dynamic characteristics are those subjected to change as a consequence of a specific soil management system. So, dynamic soil quality can be measured and used to compare different agricultural practices and/or farming systems on similar soils, or in the same soil over a period of time. To succeed in soil quality assessment, the issue of which types of measurements should be made in order to evaluate the effect of management on soil function must be solved. Soils have physical, chemical and biological properties that interact in a complex way to give them their quality or capacity to function (Seybold et al., 1997). So, soil quality cannot be measured directly, but can be assessed through the measurement of changes of its attributes, or attributes of the ecosystem which are considered as indicators. Of course, we need standards for the evaluation of management systems that allow an assessment of their sustainability. Two different approaches are employed in the evaluation of sustainable management systems: the comparative assessment and the dynamic assessment.

Comparative and Dynamic Assessment of Sustainable Management Systems

Comparative and dynamic assessment are the two main approaches employed in the evaluation of sustainable management systems. In the comparative approach, the performance of a system is determined in relation to alternatives. On the other hand, in the dynamic approach a management system is assessed in terms of its performance determined over a period of time. According to the comparative approach, the characteristics, biotic and abiotic soil attributes of alternative systems are compared at time t and a decision about the relative sustainability of each system is based on the magnitude of the measured parameters. The main limit of this approach is that, if only outputs are measured, it provides little information about the process that created the measured condition. On the other hand, the main disadvantage of the dynamic approach is that it needs measurements of indicators for at least two points in time and consequently does not provide an immediate assessment of soil quality (Seybold et al., 1997). Moreover, it can be misleading in the case of a soil that functions at its highest attainable level, which it cannot improve, or when it is functioning at its lowest attainable level and cannot go lower; both these cases would show a static trend, indicating sustaining systems, but they would have completely different quality.

In our opinion, the two approaches to assessment are complementary, since they allow different scales of evaluation. While monitoring trends is more useful for evaluation at the level of the farm, the comparative assessment seems to be more suitable to a wider scale of evaluation (on a regional scale).

Indicators of Soil Quality

According to the above discussion, in order to be useful as an indicator of the sustainability of agricultural practices and of land management, indicators of soil quality should give some measure of the capacity of a soil to function in terms of plant and biological productivity and environmental quality (Seybold *et al.*, 1997). As reported in the introduction, indicators of soil quality should be:

- sensitive to long-term change in soil management and climate, but sufficiently robust not to change as a consequence of short-term changes in weather conditions;
- well correlated with beneficial soil functions;
- useful for understanding why a soil will or will not function as desired;
- comprehensible and useful to land managers;
- easy, and not expensive, to measure;
- where possible, components of existing soil databases.

Due to the complexity of the system, there cannot be a single indicator of soil quality to assess a specific soil management system, but a minimum data set of attributes regarding soil physical, chemical and biological properties must be selected. Furthermore, the suitability of soil quality indicators depends on the kind of land, land use and scale of assessment. Different land uses may require different properties of soil and, consequently, some soil quality indicators in a given situation can be more helpful than others for the purpose of the assessment (Karlen *et al.*, 2001). Another aspect, closely related to the previous one, which should always be stressed, is that the final aim of soil quality assessment is different in an agricultural soil and in a natural ecosystem. As stated by Singer and Ewing (2000), 'in an agricultural context soil quality may be managed in order to maximize production without adverse environmental effects, while in a natural ecosystem soil quality may be observed as a baseline value or set of values against which future changes in the system may be compared'. As a general consideration, each combination of soil type, land use and climate calls for a different set of practices to enhance soil quality. Whereas, within the same country, different pedoclimatic conditions require different management practices in order to reach the same sustainability goals, even more complex issues are related to soil quality management in tropical areas or, in general, in less-developed countries.

Soil Organic Carbon Pools and Processes

Since organic matter, or more specifically organic carbon and the carbon (C) cycle as a whole, can have an important effect on soil functioning, all the attributes linked to the soil C cycle are usually recommended as components in any minimum data set for soil quality evaluation.

According to a simplified scheme, soil organic matter (SOM) can be divided into two pools: non-living and living. Non-living SOM includes materials of different age and origin, which can be further divided into pools or fractions as a function of their turnover characteristics. For example, the humified fraction is more resistant to decay. The stability and longevity of this pool is a consequence of chemical structure and organo-mineral association. This pool of soil organic matter influences different aspects of soil quality, such as the fate of ionic and non-ionic compounds, the increase of soil cation exchange capacity and the long-term stability of microaggregates (Herrick and Wander, 1997). The interpretation guideline for the sake of soil quality assessment is that the higher the humified fraction of SOM, the higher is its contribution to soil quality.

Living soil organic matter represents only a small percentage of total soil organic carbon, and includes soil micro-, meso- and macroorganisms. In particular, soil microbial biomass is regarded as the most active and dynamic pool of SOM, and plays an important role in driving soil mineralization processes. Three main aspects of soil microbial biomass are usually considered for their effect on soil functions: pool size, activity and diversity. The determination of the total amount of carbon immobilized within microbial cells permits the determination of soil microbial biomass as a pool of soil organic matter. Since this pool is responsible for the decomposition of plant and animal residues and for the immobilization and mineralization of plant nutrients, it is finally responsible for the maintenance of soil fertility (Brookes, 2000). For this reason, the concept of microbial biomass has developed as an 'early warning' of changing soil conditions and as an indicator of the direction of change. Carbon mineralization activity, as a key process of the soil C cycle, determines the rapidity of the organic matter degradation process in soil. Many studies have been carried out recently in order to verify, directly or indirectly, the potential for increasing carbon storage in soil by manipulating C inputs to minimize the rate of carbon mineralization (Jans-Hammermeister et al., 1997; Fließbach and Mäder, 2000). The sensitivity of the carbon mineralization process to changes in soil management is low, because small microbial populations, in degraded soil, can mineralize organic matter to the same extent and at the same rate as large microbial populations in undegraded soils (Brookes, 1994). More sensitive to soil management changes, and more helpful as a soil quality indicator, is the combination of the two measurements, relating to both size and activity of microbial biomass. Carbon mineralization activity/unit of biomass (biomass-specific respiration) and the mineralization coefficient (respired carbon/total organic carbon) indicate efficiency in carbon utilization and energy demand. Finally, increasing importance has been given recently to soil microbial diversity measurements as indicators of community stability and of impact of stress on that community.

Interactions between the diversity of primary producers (plants) and that of decomposers (microbial communities), the two key functional groups that form the basis of all ecosystems (Loreau, 2001), have major consequences for agricultural management. Soil microorganisms control the mineralization of natural compounds and xenobiotics. Furthermore, bacteria and fungi exist at extremely high density and diversity in soils, and can react to changing environmental conditions rapidly, by adjusting: (i) activity rates; (ii) gene expression; (iii) biomass; and (iv) community structure. Some of these parameters might be perfect indicators for evaluating soil quality (Schloter *et al.*, 2003a).

Research interest in microbial biodiversity over the past 25 years has increased markedly, as microbiologists have become interested in the significance of biodiversity for ecological processes. Most of the work illustrates a dominant interest in questions concerning the effect of specific environmental factors on microbial biodiversity, the spatial and temporal heterogeneity of this biodiversity, and quantitative measures of population structure (for a review see Morris *et al.*, 2002). However, since the rise of availability of molecular genetic tools in microbial ecology in the early 1990s, it has become apparent that we know only a very small part of the diversity of the microbial world. Most of this unexplored microbial diversity seems to be hidden in yet uncultured microbes. Use of new direct methods, independent of cultivation, based on the genotype (Amann *et al.*, 1995) and phenotype (Zelles *et al.*, 1994) of the microbes, enables a deeper understanding of the composition of microbial communities. For example, using the rDNA-directed approach of dissecting bacterial communities by amplifying the 16S rDNA (*rrs*) gene from soil samples by polymerase chain reaction (PCR), and studying the diversity of the acquired *rrs* sequences, almost exclusively new sequences became apparent, which are only related to a certain degree to the well-studied bacteria in culture collections (Amann *et al.*, 1995). Based on molecular studies, it can be estimated that 1 g of soil is the habitat of more than 10^9 bacteria, belonging to about 10,000 different microbial species (Ovreas and Torsvik, 1998).

Using these new methods for studies in agricultural ecosystems, it could be clearly proven that the highest influence on microbial community structure and function, mainly in the rhizosphere, derives from used crops and applied agrochemicals. Miethling *et al.* (2000) conducted a greenhouse study with soil–plant microcosms, in order to compare the effect of the crop species lucerne (*Medicago sativa*) and rye (*Secale cereale*), soil origin (different cropping history) and a bacterial inoculant (*Sinorhizobium meliloti*) on the establishment of plant-root-colonizing microbial communities. Three community-level targeting approaches were used to characterize the variation of the extracted microbial rhizosphere consortia: (i) community-level physiological profiles (CLPP); (ii) fatty acid methyl ester (FAME) analysis; and (iii) diversity of PCR-amplified 16S rRNA target sequences from directly extracted ribosomes, determined by temperature gradient gel electrophoresis (TGGE). All approaches identified the crop species as the major determinant of microbial community characteristics. The influence of soil was consistently of minor importance, while a modification of the lucerne-associated microbial community structure after inoculation with *S. meliloti* was only consistently observed by using TGGE. In a study by Yang *et al.* (2000), the DNA sequence diversities for microbial communities in four soils affected by agricultural chemicals (mainly triadimefon and ammonium bicarbonate and their intermediates) were evaluated by random amplified polymorphic DNA (RAPD) analysis. The richness, modified richness, Shannon–Weaver index and a similarity coefficient of DNA were calculated to quantify the diversity of accessed DNA sequences. The results clearly showed that agricultural chemicals affected soil microbial community diversity at the DNA level. The four soil microbial communities were distinguishable in terms of DNA sequence richness, modified richness, Shannon–Weaver index and coefficient of DNA similarity. Analysis also showed that the amounts of organic C and microbial biomass C were low in the soil treated by pesticide (mainly triadimefon and its intermediates), but high in the soil where the chemical fertilizer (mainly ammonium bicarbonate and its intermediates) was applied. Combined, the above results may indicate that pesticide pollution caused a decrease in the soil microbial biomass but maintained a high diversity at the DNA level, compared to the control without chemical pollution. In contrast, chemical fertilizer pollution caused an increase in the soil biomass but a decrease in DNA diversity.

Managing Soil Quality: Case Studies

Managing soil quality relates to agricultural soils which are subjected to different agricultural practices as a consequence of the management system adopted. As a function of the specific objective that farmers and/or land managers and/or policy makers want to reach, relevant soil functions and/or parameters should be individualized and measured, in order to understand whether the specific soil management system adopted reaches the prefixed aims. Below are some examples of how agricultural management can influence microbial community structure and function.

Influence of precision farming and conventional agricultural management on microbial community structure and function

Precision farming summarizes cultivation practices that allow for spatial and temporal variability of soil attributes and crop parameters within an agricultural field. Distinct areas in a field are managed by applying different levels of input, depending on the yield potential of the crop in that particular area. Benefits of these actions are: reduction of the cost for crop production, thus conserving resources while maintaining high yield; and minimizing the risk of environmental pollution (Dawson, 1997).

In two studies, from Hagn *et al.* (2003a) and Schloter *et al.* (2003b), microbial community structure and function and their dynamics were investigated in relation to season, soil type and farming management practice. The research was done using soils from high- (H) and low-yield areas (L) of a field site, cultivated with winter wheat under two different farming management systems (precision farming, P; conventional farming, C) over the growing period.

It was demonstrated that the microbial biomass and the microbial community structure, measured using the phospholipid fatty acid technique and DNA-based methods in the top soils of the investigated plots, were not influenced by precision farming. Both parameters showed a typical seasonal run, which was independent of the farming management type. Microbial biomass was reduced during the summer months, due to the dry weather conditions and the hot temperatures in the top soil. The microbial community structure changed mainly after the application of fertilizers and was associated with high amounts of root exudates in late spring.

As fungal communities are essential for the reduction of soil erosion, degradation of complex organic compounds and biocontrol (Hagn et al., 2003b), fungal diversity was studied in detail using cultivation-independent methods (direct extraction of DNA from soil followed by PCR amplification of a subunit of the 18S rDNA and fingerprinting (DGGE)) as well as cultivation-dependent techniques (isolation of pure cultures). Comparison of the PCR amplicons by DGGE patterns, reflecting the *total fungal community*, showed no differences between the sampling sites and no influence of the farming management systems. Only small differences were observed over the growing period. For the identification of *active hyphae*, cultivation-dependent techniques were used. The resulting isolates were subcultured and grouped by their morphology and genotype. In contrast to the cultivation-independent approaches, clear site-specific and seasonal effects on the fungal community structure could be observed. However, no effects of the different farming management techniques were seen. These results clearly indicate that the potential fungal community (including spores) is not influenced by the investigated factors, whereas active populations show a clear response to environmental changes (soil type and season). The most abundant group, consisting of *Trichoderma* species, was investigated in more detail using strain-specific genotype-based fingerprinting techniques as well as a screening for potential biocontrol activity against the wheat pathogen *Fusarium graminearum*. The genotypic distribution, as well as the potential biocontrol activity, revealed clear site-specific patterns, reflecting the soil type and the season. A clear response of the *Trichoderma* ecotypes to different farming management techniques was not seen.

Enzyme activities in the nitrogen cycle were more affected by precision farming. Proteolytic activity was significantly increased by precision farming, especially on low-yield plots; also, nitrification and denitrification activities showed a clear response to application of fertilizers. In summer, due to the low microbial biomass, all measured activities were very low and showed no sustained reaction to farming management.

These results indicated that the structure of the entire microbial community is not influenced by precision farming compared to conventional agriculture. However, the applied farming management type had a visible influence on the induction or repression of gene transcription and expression.

Influence of changes from conventional to organic farming on microbial community structure and function

The influence of two farming systems in southern Germany on aggregate greenhouse gas emission (CO_2, CH_4 and N_2O) was investigated by Flessa et al. (2002). One system (farm A) conformed to the principles of integrated farming (recommended by the official agricultural advisory service) and the other system (farm B) followed the principles of organic farming (neither synthetic fertilizers nor pesticides were used). Farm A consisted of 30.4 ha fields (mean fertilization rate 188 kg N/ha), 1.8 ha meadows, 12.4 ha set-aside land and 28.6 adult beef steers (year-round indoor stock keeping). Farm B consisted of 31.3 ha fields, 7 ha meadows, 18.2 ha pasture, 5.5 ha set-aside land and a herd of 35.6 adult cattle (grazing period 6 months).

The integrated assessment of greenhouse gas emissions included those from fields, pasture, cattle, cattle waste management, fertilizer production and consumption of fossil fuels. Soil N_2O emissions were estimated from 25 year-round measurements on differently managed fields. Expressed per hectare farm area, the aggregate emission of greenhouse gases was 4.2 Mg CO_2 equivalents (conventional farming) and 3.0 Mg CO_2 equivalents (organic farming). Nitrous oxide emissions (mainly from soils) contributed the major part (about 60%) of total greenhouse gas emissions in both farming systems. Methane emissions (mainly from cattle and cattle waste management) were approximately 25%, and CO_2 emissions were lowest (c. 15%). Mean emissions related to crop production (emissions from fields, fertilizer production, and the consumption of fossil fuels for field management and drying of crops) were 4.4 Mg CO_2 equivalents/ha and 3.2 Mg CO_2 equivalents/ha field area for farms A and B, respectively. On average, 2.53% of total N input by synthetic N fertilizers, organic fertilizers and crop residues were emitted as N_2O-nitrogen. Total annual emissions per cattle unit (live weight of 500 kg) from enteric fermentation and storage of cattle waste were about 25% higher for farm A (1.6 Mg CO_2 equivalents) than farm B (1.3 Mg CO_2 equivalents). Taken together, these results indicated that conversion from conventional to organic farming led to reduced emissions per hectare, but yield-related emissions are not reduced.

In another study by Schloter et al. (2004), the effects on microbial community structure of conventional farming and ecological farming were compared. Plots under ecological farming for 4–40 years were compared with plots under conventional farming practice. To characterize the soil microbial community structure under both systems, a hierarchical approach was applied (phospholipid fatty acids to describe differences in superfamilies and domains, 16S rDNA pattern to describe species variability, and enrichment of *Ochrobactrum* spp. to describe microdiversity of ecotypes). To characterize functions both at the community level and at the level of the enriched *Ochrobactrum antrophi* populations, the EcoBiolog® system was used. Although no differences in the microbial biomass were observed (the two highest biomass values were found in the soil of a conventional farming plot and a plot that had been under ecological farming practice for 40 years), the results clearly indicated that there was a significant shift in microbial community structure between the conventionally farmed plots and the plots under ecological farming practice. The use of the phospholipid fatty acid (PLFA) method and the statistic evaluation of the results by principal component analysis (PCA) revealed significant changes of the microbial community on a high taxonomic level. Just 2 years after changing farming practice, shifts in the structure of the microflora were visible. The longer the ecological farming practice was applied to the plots, the greater the differences were. The main shifts were found in the Gram-positive bacteria. After extraction of total DNA from soil, amplification of the variable region V6/V8 of the 16S rDNA and fingerprinting using DGGE, three main clusters (conventional farming, 2 years' ecological farming and 40 years' ecological farming) were visible. The main differences were not found in the total diversity (Shannon index), but in the abundance of characteristic species in each soil (Simpson index). For the soils that were under ecological farming practice for 40 years, *Actinomycetes* appeared to be highly abundant. After enrichment by specific antibodies and classifying the ecotypes by the EcoBiolog® system, 24 ecotypes could be defined. Significant differences in ecotype variability were only observed in the plots that had been under ecological farming practice for 40 years. Using the EcoBiolog®system, it could be shown that differences in the microbial community structure led to a change of function: in the conventional soil specialized microbes (which can utilize only a small part of the used C sources) were dominant, whereas in the plots under ecological farming practice generalists dominated. These results could be shown at the level of the whole microbial community and were confirmed on the microdiversity level.

Sustainable Management of Soils

Sustainable management of soils is an environmental issue worldwide, but is most urgent in less-developed countries. Many of the problems faced by tropical countries concern soil degradation. As a consequence of population pressure and the lack of new land, fallow rotation is shortened, resulting in soil erosion. Since, in many tropical soils, a large proportion of available plant nutrients are in the topsoil, soil erosion causes a substantial loss in fertility and a deterioration of soil physical structure. As the cost of reclamation of eroded soil is prohibitive for many less-developed countries, soil erosion is considered to be one of the greatest threats to the sustainability of agricultural production in tropical countries (Scholes *et al.*, 1994). Other constraints to crop productivity in tropical soils are of chemical or nutrient origin: aluminium toxicity, acidity, high phosphorus fixation with iron oxides, low nutrient reserves and high secondary salinization of irrigated land (Lal, 2002). Large areas of sub-Saharan Africa and South Asia are characterized by the constraints reported above. According to Lal (2002), the goals of sustainable agriculture in these areas can be summarized as:

- food security;
- reversal of soil degradative trends;
- improvement of surface and ground water quality;
- sequestration of C in soil in order to reduce the negative effect of emission of greenhouse gases.

The identification of the goals of sustainable agriculture in tropical countries does not, in itself, solve the issue of managing soil quality, since the same objectives can be reached following different soil management systems, as a consequence of socio-economic considerations and political decisions. A typical example regarding less-developed countries is the debate about one of the most serious problems among those listed above: food security. In order to enhance food production to meet the needs of the population, during the past 30 years some areas of the tropics have experienced a dramatic increase in per capita food production. This improvement is mainly based on the introduction of new crop varieties on fertile soils heavily supplied with water, fertilizers and pesticides. While, on a regional scale and from a macroeconomic point of view, many Asian countries can now theoretically satisfy the need of their ever-growing populations for rice and wheat, in poorer rural areas the numbers of undernourished people are still extremely high. In these areas, smallholder farming systems seem to better guarantee food security. On a farm scale, traditional cropping systems based on different food crops, rather than a single high-yield cash crop, the use of indigenous crop varieties, organic fertilizers produced on-farm and minimum tillage can be more efficient for reaching self-sufficiency. In terms of sustainability of soil management for food security on a farm level, smallholder farming systems are characterized by agricultural practices that improve soil resistance to degradation and its capacity to recover rapidly after a perturbation.

Conclusion

In contrast to the high complexity of microbial communities in soil, the ideal soil microbiological and biochemical indicator to determine soil quality should be simple to measure, should work equally well in all environments and should reliably reveal which problems exist where. It is unlikely that a sole ideal indicator can be defined with a single measure, because of the multitude of microbiological components and biochemical pathways. Therefore, a minimum data set is frequently applied (Carter *et al.*, 1997). The basic indicators and the number of measures needed are still under discussion, and depend on the aims of the investigations. National and international programmes for monitoring soil quality presently include biomass and respiration measurements, but also extend to determination of nitrogen mineralization, microbial diversity and functional groups of soil fauna.

It is still unclear whether observed changes in microbial community structure are lasting, and whether naturally occurring environmental factors can influence the genotypic ability of the soil microbiota to recover after harsh conditions and become healthy again (Sparling, 1997). Research on the resilience of soil microbiota is therefore a significant task of microbial ecology.

References

Amann, R., Ludwig, W. and Schleifer, K.H. (1995) Phylogenetic identification and in situ detection of individual microbial cells without cultivation. *Microbiological Reviews* 59, 143–149.

Brookes, P.C. (1994) The use of microbial parameters in monitoring soil pollution by heavy metals. *Biology and Fertility of Soils* 19, 269–279.

Brookes, P.C. (2000) Changes in soil microbial properties as indicators of adverse effects of heavy metals. In: Benedetti, A. (ed.) *Indicatori per la Qualità del Suolo*. Fischer, Heildelberg, pp. 205–227.

Carter, M.R., Gregorich, E.G., Anderson, D.W., Doran, J.W., Janzen, H.H. and Pierce, F.J. (1997) Concepts of soil, quality and their significance. In: Gregorich, E.G. and Carter, M.R. (eds) *Soil Quality for Crop Production and Ecosystem Health*. Elsevier, Amsterdam, pp. 1–19.

Dawson, C.J. (1997) Management for spatial variability. In: Stafford, V. (ed.) *Precision Agriculture '97*. BIOS Scientific Publishers Ltd, Oxford, pp. 45–58.

Flessa, H., Ruser, R., Dörsch, P., Kamp, T., Jimenez, M.A., Munch, J.C. and Beese, F. (2002) Integrated evaluation of greenhouse gas emissions (CO_2, CH_4, N_2O) from two farming systems in southern Germany. *Agriculture, Ecosystems and Environment* 91, 175–189.

Fließbach, A. and Mäder, P. (2000) Microbial biomass and size-density fractions differ between soils of organic and conventional agricultural systems. *Soil Biology and Biochemistry* 32, 757–768.

Hagn, A., Pritsch, K., Schloter, M. and Munch, J.C. (2003a) Fungal diversity in agricultural soil under different farming management systems with special reference to biocontrol strains of *Trichoderma* spp. *Biology and Fertility of Soils* 38, 236–244.

Hagn, A., Geue, H., Pritsch, K. and Schloter, M. (2003b) Assessment of fungal diversity and community structure in agricultural used soils. *Microbiological Post* 37, 665–680.

Herrick, J.E. and Wander, M.M. (1997) Relationship between soil organic carbon and soil quality in cropped and rangeland soils: the importance of distribution, composition and biological activity. In: Lal, R., Kimble, J.M. and Stewart, B.A. (eds) *Soil Processes and the Carbon Cycle*. CRC Press, Boca Raton, Florida, pp. 405–425.

Jans-Hammermeister, D.C., McGill, W.B. and Izaurralde, R.C. (1997) Management of soil C by manipulation of microbial metabolism: daily vs. pulsed C additions. In: Lal, R., Kimble, J.M. and Stewart, B.A. (eds) *Soil Processes and the Carbon Cycle*. CRC Press, Boca Raton, Florida, pp. 321–334.

Karlen, D.L., Andrews, S.S. and Doran, J.W. (2001) Soil quality: current concepts and applications. In: Sparks, D.L. (ed.) *Advances in Agronomy*, Vol. 7. Academic Press, London, pp. 1–39.

Lal, R. (2002) The potential of soils of the tropics to sequester carbon and mitigate the greenhouse effect. In: Sparks, D.L. (ed.) *Advances in Agronomy*, Vol. 76. Academic Press, London, pp. 1–30.

Loreau, M. (2001) Microbial diversity, producer–decomposer interactions and ecosystem processes: a theoretical model. *Proceedings of the Royal Society for Biological Science* 268, 303–309.

Miethling, R., Wieland, G., Backhaus, H. and Tebbe, C.C. (2000) Variation of microbial rhizosphere communities in response to crop species, soil origin, and inoculation with *Sinorhizobium meliloti* L33. *Microbial Ecology* 40, 43–51.

Morris, C.E., Bardin, M., Berge, O., Frey-Klett, P., Fromin, N., Girardin, H., Guinebretiere, M.H., Lebaron, P., Thiery, J.M. and Troussellier, M. (2002) Microbial biodiversity: approaches to experimental design and hypothesis testing in primary scientific literature from 1975 to 1999. *Microbiological and Molecular Biology Reviews* 66, 592–616.

Ovreas, L. and Torsvik, V.V. (1998) Microbial diversity and community structure in two different agricultural soil communities. *Microbial Ecology* 36, 303–315.

Schloter, M., Dilly, O. and Munch, J.C. (2003a) Indicators for evaluating soil quality. *Agriculture, Ecosystems and Environment* 98, 255–262.

Schloter, M., Bach, H.J., Sehy, U., Metz, S. and Munch, J.C. (2003b) Influence of precision farming on the microbial community structure and selected functions in nitrogen turnover. *Agriculture, Ecosystems and Environment* 98, 295–304.

Schloter, M., Bergmüller, C., Friedel, J., Hartmann, A. and Munch, J.C. (2005) Effects of different farming practice on the microbial community structure. *Applied and Environmental Microbiology* (submitted).

Scholes, M.C., Swift, M.J., Heal, O.W., Sanchez, P.A., Ingram, J.S.I. and Dalal, R. (1994) Soil fertility in response to the demand for sustainability. In: Woomer, P.L. and Swift, M.J. *The Biological Management of Tropical Soil Fertility*. John Wiley & Sons, New York, pp. 1–14.

Seybold, C.A., Mausbach, M.J., Karlen, D.L. and Rogers, H.H. (1997) Quantification of soil quality. In: Lal, R., Kimble, J.M. and Stewart, B.A (eds) *Soil Processes and the Carbon Cycle*. CRC Press, Boca Raton, Florida, pp. 387–404.

Singer, M.J. and Ewing, S. (2000) Soil quality. In: Sumner, M.E. (ed.) *Handbook of Soil Science*. CRC Press, Boca Raton, Florida, pp. 271–298.

Sparling, G.P. (1997) Soil microbial biomass, activity and nutrient cycling as indicators of soil health. In: Pankhurst, C., Doube, B.M. and Gupta, V. (eds) *Biological Indicators of Soil Health*. CAB International, Wallingford, UK, pp. 97–119.

Yang, Y., Yao, J., Hu, S. and Qi, Y. (2000) Effects of agricultural chemicals on DNA sequence diversity of soil microbial community: a study with RAPD marker. *Microbial Ecology* 39, 72–79.

Zelles, L., Bai, Q., Ma, X., Rackwitz, R., Winter, K. and Beese, F. (1994) Microbial biomass, metabolic activity and nutritional status determined from fatty acid patterns and polyhydroxybutyrate in agriculturally managed soils. *Soil Biology and Biochemistry* 26, 439–446.

5 Concluding Remarks

ANNA BENEDETTI,[1] PHILIP C. BROOKES[2] AND JAMES M. LYNCH[3]

[1]*Consiglio per la ricerca e la sperimentazione in Agricoltura, Istituto Sperimentale per la Nutrizione delle Piante, Via della Navicella, 2, 00184 Rome, Italy;* [2]*Agriculture and Environment Division, Rothamsted Research, Harpenden AL5 2JQ, UK;* [3]*Forest Research, Alice Holt Lodge, Farnham GU10 4LH, UK*

Soil quality has been defined as the capacity of the soil to produce healthy and nutritious crops (Paperdick and Parr, 1992), but it is also important to consider the studies where indicators of soil quality have been sought for forestry (Moffat, 2003). The related concepts of soil resilience (the ability of the soil to recover after disturbance) and soil degradation (the loss of the soil's capacity to produce crops) also need to be considered alongside quality assessments (Elliott and Lynch, 1994). In this context, toxicity measurements are necessary, and recently a range of biosensors has become available for this purpose.

The aim of this handbook is to provide a practical guide on how to use biochemical, microbiological and molecular indicators to define soil quality. It must be stressed that non-experts cannot expect to resolve such a complex issue without the assistance of experts in the field; nevertheless, the topic will be addressed humbly. We are all aware that there are no absolute benchmarks for assessing biological soil fertility, while there are chemical or physical indicators that are easy to define, interpret and understand. On the contrary, biological indicators, e.g. soil microbial biomass concentration, require especially careful interpretation. There is no such thing as absolute high or low values, values that remain constant over time and in space; there are sets or families of similar values, and also other, and therefore interpretable, values. It must be kept in mind that a variability of 20% is common for biological populations, and this is one of the reasons why it is recommended that several indicators are utilized to produce an index to describe the behaviour of soil microorganisms.

An index can be a set of several indicators that are developed and deduced from parameters, as implicit in the definition provided by the OECD, 'Set of parameters or aggregate of weighed indicators' (1993). Moreover, biological indicators have to be related to physical, chemical and agronomic indices, etc. and be relevant to the nature of the type of study to be performed.

Selected biochemical, microbiological and molecular parameters available to the operator are given in this handbook (for instance in Chapter 3). Once the exact aim of the investigation has been defined, the list of parameters can usually be streamlined and reduced to answer the questions asked.

Activity carried out within the framework of EU COST Action 831 revealed that the selected factors need to be arranged in a hierarchical scale of indicators in function of the study goal. The topic of hierarchical scales of different indicators is widely debated, and it is evident that each single researcher is more skilled at using the parameters used in his or her laboratory. A guide to the hierarchical use of indicators is provided by the OECD requirements (1999), whereby indicators must:

- be clearly correlated with a certain phenomenon or a certain feature that is being investigated or monitored;
- be highly correlated with the above-mentioned effect with minimal statistical variability;
- be unobscured by much less significant responses;
- have a sufficiently generalized, albeit not identical, validity in many analogous situations.

It is clear that hierarchical levels can change depending upon whether the indicator is required for monitoring, for accurate characterization of a particular environment, for assessing or restoring previous changes, or for starting up research. If the aim is to study soil quality in terms of fertility, the hierarchical level represented in Fig. 5.1 could be applicable:

1. Biomass-carbon (C) and respiration rate.
2. Functional diversity.
3. Genetic diversity.
4. Case-by-case in-depth probes (heavy metals, genetically modified organisms, air pollution, erosion, etc.).

We should take the same approach we would all adopt if we were to have a medical check-up. The first thing a physician does is to carry out a series of routine basic examinations. The physician will probably only call for further tests if irregularities occur in these basic ones, which may indicate an underlying pathology. Then the physician may move to more sophisticated tests, indicated by the findings in the first set. These can then be followed by other, much more sophisticated and specific diagnostic investigations.

Likewise, the first step in assessing biological soil fertility, i.e. the expression of microbial turnover, is to perform simple biochemical tests. The same tests can be used effectively for environmental monitoring. The next step could be to study the functional diversity of the ecosystem, followed by genetic diversity, and then case-by-case in-depth probes. To date, some methodologies, for example the ecophysiological profile and bacterial and fungal DNA studies, are rarely utilized in nationwide, large-scale monitoring programmes.

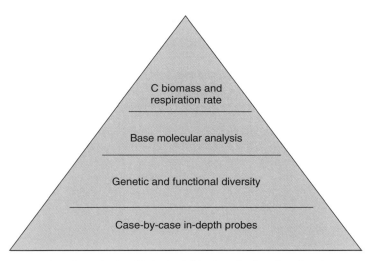

Fig. 5.1. Hierarchical scales of microbiological, biochemical and molecular parameters to define the biological fertility of a soil.

Moreover, a minimum hierarchical level must also be identified for other correlated indicators, in order to prevent false-negative and -positive results. In the case where soil fertility is related to crop yield, physical fertility is just as important as chemical and biological fertility. Obviously, however, the correct functioning of aerobic microorganisms will not occur under, for example, conditions of oxygen limitation, extremes of pH or elevated salinity. Thus, it is crucial to build other hierarchical scales, which, for chemical soil parameters, could be represented by the following:

1. Organic matter.
2. pH.
3. Available nutrients.
4. Various types of pollutants, etc.

The following parameters could be adopted for physical soil fertility indicators (Pagliai and Vignozzi, 2000; Pagliai *et al.*, 2000):

1. Porosity.
2. Aggregate stability.
3. Compactness.
4. Sealing along the profile.
5. Structure loss.
6. Superficial crusts and potential risk of their formation.
7. Fissuring.
8. Erodability.

The hierarchical scales will then be put together in an attempt to identify a minimum data set taken from the point where the different hierarchical scales overlap (Fig. 5.2).

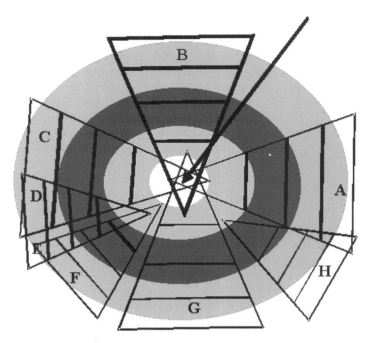

Fig. 5.2. Model showing a minimum data set for assessing soil quality and environmental sustainability. The minimum data set increases with the extent of the study (circles). A, physical parameters; B, climatic parameters; C, chemical parameters; D, microbiological parameters; E, land use; F, biological parameters; G, management; H, etc.

The extent of the study is represented by the circles in Fig. 5.2. The hierarchical scales to be used in different studies obviously differ, and range from environmental, pedological and agronomic parameters to social and economic ones. The approaches adopted in two case studies are described below.

Soil Sealing, Compaction and Erosion

Large areas of the central part of Italy are representative of the hillside environments of the Tuscan–Romagna Mountains (site 1) and of the clay hillside environments of central southern Italy (site 2). The soils, containing small amounts of organic carbon, are characterized by low structural stability and poor regeneration capacity: they must be managed correctly to minimize the potential risks of formation of surface crusts, sealed surfaces and compaction by farming machinery. The effects of such hazardous degradation in a hilly environment – the reduced rainwater infiltration rate and the creation of preferential surface runoff courses – play a role in triggering widespread and channelled erosion processes.

Soil management is crucial for the prevention and control of degradation. Different tillage systems, i.e. deep ploughing, shallow ploughing, minimum tillage and ripper subsoiling, have different effects on soil conditions. Adoption of ripper subsoiling tillage is capable of reducing structural damage caused by deep ploughing, lessening the risk of formation of surface crusts and the presence of compacted layers in the profile. This was revealed by the findings of the micromorphological analysis and quantification of the pore system in site 1. Moreover, ripper subsoiling conserves more organic carbon than does deep ploughing, especially in the topsoil layer. Also, the amounts of humified organic matter in soil managed by ripper subsoiling is greater than in soils following deep ploughing (Dell'Abate *et al.*, 2004).

Soil microbial activity has been verified by the determination of the soil microbial biomass carbon and its respiration (Table 5.1), and subsequently using the metabolic quotient (qCO_2) and the C biomass/total organic C (B_C/TOC) ratio.

Data concerning the quantity and the activity of the microbial biomass for the two different situations demonstrate that ripper subsoiling (RS) is a better management practice than deep ploughing for the maintenance of the total organic resource, and also of the living fraction. In fact, the content of microbial biomass is greater in RS for both the depths.

Results obtained for the two different management practices showed that qCO_2 was comparable in the two layers in the soil submitted to ripper subsoiling (RS), while in the soil in which the deep ploughing (DP) was practised, qCO_2 was higher at a depth of 20–40 cm.

A high value of qCO_2 in the ploughed zone (DP) highlights a situation of non-equilibrium due to the adopted practice. This conclusion can be also reached by analysing the B_C/TOC ratio: the value of microbial biomass in the deepest layer of the DP case was halved in comparison to the value found in RS to the same depth, while total organic carbon was practically constant.

In site 2, the effects of two other types of soil management (a comparison of continuous wheat and continuous lucerne) were assessed, in a field experiment established in 1994. In this case, the porosity values obtained in the 0–5 cm soil depth for the two areas on which the crops were grown show that after heavy rain, soils supporting lucerne gave a higher porosity percentage than those from the wheat-growing area. The protective action of the vegetation cover of lucerne decreased soil surface vulnerability to the

Table 5.1. Soil microbial biomass carbon (B_C), C biomass/total organic carbon ratio (B_C/TOC) and metabolic quotient (qCO_2).

Sample soil	B_C (µg/g)	B_C/TOC (%)	qCO_2 (mg CO_2-C/mg C_{mic}.h)
RS (0–20 cm)	203.8 ± 26.9	2.290 ± 0.004	0.0013 ± 0.0003
(20–40 cm)	137.7 ± 23.6	1.996 ± 0.004	0.0017 ± 0.0003
DP (0–20 cm)	121.3 ± 69.8	1.989 ± 0.012	0.0019 ± 0.0011
(20–40 cm)	73.8 ± 9.8	1.118 ± 0.01	0.0034 ± 0.0008

RS, ripper subsoiling; DP, deep ploughing.

impact of rainfall and thus lessened the risk of formation of crusts. Moreover, wheat did not seem to be the most suitable crop as it depleted the organic matter in the top horizon of the soil, while lucerne conserved the organic matter better. Removal of the finest soil particles by water erosion led to preferential loss of the most stable and most strongly sorbed organic fraction (humin), which, in contrast, accumulated in the deeper layer.

The quantity of microbial biomass in the soil cultivated with lucerne (L) was higher than that in the soil cultivated with wheat (W) (Table 5.2), and this was also confirmed by the respiration data (data not shown): compared to wheat, the lucerne crop is more effective in preserving the total soil organic carbon and also improves soil microbial life. The C biomass/total organic carbon rate had comparable values for the two crops for each depth. Also, the specific respiration (qCO_2) showed comparable values in the two cases for the two depths. It was higher under both the crops in the deepest layer, and this underlines a stress condition.

In this study, the lucerne covering showed a positive effect on the quantity of organic carbon and microbial biomass. On the other hand, microbial metabolism was not affected by the different crop (similar B_C/TOC values and metabolic quotients at the same depths). Wheat removed carbon from the soil but it did not modify microbial metabolism.

The rational answer to this environmental risk is to avoid sowing wheat in these soils, but indubitably wheat is a more profitable crop than, for example, lucerne. It is clear that the radical changes in agriculture, where the aims are to produce high yield and, at the same time, conserve 'landscaping farming' environments, must strike a balance or set a limit of degradation (Baroccio *et al.*, 2002).

Impact of Cultivation of Genetically Modified Plants

The EU Directive 18/2001 sets the general criteria on emission of genetically modified (GM) plants in Europe. For the first time, this law fixed a limit on soil, nitrogen and carbon recycling. What does this mean? Carbon and nitrogen mineralization or humification parameters should be able to detect carbon and nitrogen recycling in soil, as they are indicators of biomass activity and ecosystem functions.

Table 5.2. Soil microbial biomass carbon (B_C), C biomass/total organic carbon ratio (B_C/TOC) and metabolic quotient (qCO_2).

Sample soil	B_C (µg/g)	B_C/TOC (%)	qCO_2 (mg CO_2-C/mg C_{mic}.h)
W (0–20 cm)	153.0 ± 40.2	3.188 ± 0.004	0.0017 ± 0.0005
(20–40 cm)	89.9 ± 43.6	1.427 ± 0.008	0.0032 ± 0.0016
L (0–20 cm)	263.9 ± 144.0	3.341 ± 0.018	0.0013 ± 0.0007
(20–40 cm)	151.3 ± 63.1	1.780 ± 0.007	0.0029 ± 0.0014

W, wheat; L, lucerne.

During the past decade some authors have tested different parameters in soil growing GM and non-GM plants (Gebhard and Smalla, 1999; Lottmann *et al.*, 1999; Hopkins *et al.*, 2002; Bruinsma *et al.*, 2003; Ceccherini *et al.*, 2003). The investigation focused on soil respiration rate, biomass C, nitrification test, humification parameters, total organic carbon and the metabolic quotient to evaluate the impact of GM plants. Molecular (DGGE and PCR analysis) and ecophysiological tests (Biolog™ analysis) were also carried out. The results demonstrated that biological, chemical and ecophysiological parameters alone were not able to detect the delayed impact of GM plants on diversity and microbial biomass activity, but that molecular analysis was more efficient. Baseline biological soil fertility and also microbial population management and natural fluctuations play a crucial role in this case. Molecular analysis detects possible changes in microbial diversity, such as richness and evenness, but also reveals the persistence of transgenes in soil.

In conclusion, some useful criteria that can be adopted to select soil quality indicators are as follows:

1. Correlate soil quality with land use. Define the use of the agricultural, forestry, pasture, public nature parks, unused industrial areas, industrial areas, marginal areas, etc. The choice of indicators depends on the type of soil investigated and the indicators suitable for the type of soil. For example, it may be more useful to determine substrate-induced respiration (SIR) instead of simple CO_2 evolution in forestry or pasture soils in homeostasis, and also in soils stressed by pollution and other factors.

2. Perform a case-by-case probe on the nature of the problems to be studied. Negative/positive/unknown impact, potential pollutants, recovery actions, new growing systems, GM plants, etc. When assessing the effects of heavy metals, for example, total genetic diversity and possibly even single species diversity have to be assessed along with functional diversity and the system's resistance to perturbation. The EU Directive 18/2001 on GM plants recommends assessing nitrogen and carbon recycling when evaluating impact, while it would be constructive to specify other parameters, such as mycorrhizal infection capacity, in mycorrhizic plants.

3. Identify which hierarchical scales can be related to observations.

4. Identify a minimum data set of indicators that can be correlated with each other.

5. If available, compare the observations with historical data from the monitoring framework, or repeat the same observations during the time.

Those involved in drafting this handbook aimed to ensure that potential users address the issue positively. Serial determinations are strongly recommended in order to detect the vital cycles of soil organisms, and, in turn, to understand soil functions and better manage soil conservation. The development of a new common language can be achieved thanks to the implementation of projects for monitoring biological fertility on various scales and the creation of institutional databanks and networks. Hopefully, this handbook will achieve this goal.

References

Baroccio, F., Dell'Abate, M.T. and Benedetti, A. (2002) Effect of different soil management on organic matter conservation: chemical and microbiological indicators. In: Nardi, S., Albuzio, A., Bottacin, A., Carden, D.E., Concheri, G., Ferretti, M., Chisi, R., Malagoli, M. and Masi, A. (eds) *Proceeding of XX Convegno nazionale della Società di Chimica Agraria,* Padova, Italy, pp. 29–36.

Bruinsma, M., Kowalchuk, G.A. and van Veen, J.A. (2003) Effects of genetically modified plants on microbial communities and processes in soil. *Biology and Fertility of Soils* 37, 329–337.

Ceccherini, M.T., Potè, J., Kay, E., Van, V.T., Maréchal, J., Pietramellara, G., Nannipieri, P., Vogel, T.M. and Simonet, P. (2003) Degradation and transformability of DNA from transgenic leaves. *Applied and Environmental Microbiology* 69, 673–678.

Dell'Abate, M.T., Benedetti, A. and Baroccio, F. (2005) Example n. 1: Fagna experimental site. In: Dell'Abate, M.T., Benedetti, A., Pagliai, M. and Sequi, P. (eds) *Atlas of Soil Quality Indicators.* Agriculture and Forestry Ministry, Rome (in press).

Elliott, L.F. and Lynch, J.M. (1994) Biodiversity and soil resilience. In: Greenland, D.J. and Szabolcs, I. (eds) *Soil Resilience and Sustainable Land Use.* CAB International, Wallingford, UK, pp. 353–364.

EU Council Directive 2001/18/CE (2001) Concerning the voluntary emission on environment of genetically modified organisms.

Gebhard, F. and Smalla, K. (1999) Monitoring field releases of genetically modified sugar beet for persistence of transgenic plant DNA and horizontal gene transfer. *FEMS Microbiology Ecology* 28, 261–272.

Hopkins, D.W., Marinari, S., Webster, E.A., Tilston, E.L. and Halpin, C. (2002) Decomposition in soil of residues from plants with genetic modifications to lignin biosynthesis. *Symposium on the Impact of GMOs: Soil Microbiology and Nutrient Dynamics.* IGMO, Vienna.

Lottmann, J., Heuer, H., Smalla, K. and Berg, G. (1999) Influence of transgenic T4-lysozyme-producing potato plants on potentially beneficial plant-associated bacteria. *FEMS Microbiology Ecology* 29, 365–377.

Moffat, A.J. (2003) Indicators of soil quality for UK forestry. *Forestry* 76, 547–568.

OECD (1993) Environmental monographs n. 83. *OECD Core Set of Indicators for Environmental Performance Reviews.* OECD, Paris.

OECD (1999) Environmental Indicators for Agriculture. *Issues and Design, The York Workshop,* Vol. 2. OECD, Paris.

Pagliai, M. and Vignozzi, D. (2000) Il sistema dei pori quale indicatore delle qualità strutturali dei suoli. Accademia Nazionale delle Scienze detta dei XL. *Memorie di Scienze Fisiche e naturali* 118° XXIV, 229–238.

Pagliai, M., Pellegrini, S., Vignozzi, N., Rousseva, S. and Grasselli, O. (2000) The quantification of the effect of subsoil compaction on soil porosity and related physical properties under conventional to reduced management practices. In: Horn, R., van den Akker, J.J.H. and Arvidsson, J. (eds) *Subsoil Compaction. Distribution, Processes and Consequences. Advances in Geo Ecology* 32, 305–313.

Paperdick, R.I. and Parr, J.F. (1992) Soil quality – the key to a sustainable agriculture. *American Journal of Alternative Agriculture* 7, 2–3.

II Selected Methods

6 Microbial Biomass and Numbers

6.1 Estimating Soil Microbial Biomass

ANDREAS FLIEßBACH[1] AND FRANCO WIDMER[2]

[1]*Research Institute of Organic Agriculture (FiBL), Ackerstrasse, CH-5070 Frick, Switzerland;* [2]*Agroscope FAL Reckenholz, Reckenholzstrasse 191, CH-8046 Zürich, Switzerland*

Introduction

Soils provide the base for a rich and diverse community of microorganisms. The quality of a soil can be defined by the ability to perform functions in the ecosystem of which it is a part. Soil microbiology has been a discipline covering topics such as nitrogen and organic matter transformations and nitrogen fixation. Transformation and the effects of pollutants were topics that became important with increasing awareness of environmental damage. The potential value of the soil resource for sustainable land use has been recognized and endorses the need to understand the processes involved in organic matter cycling as well as the amount of organisms in soil and their functions. The effects of land use and management on environmental quality, or even on the global carbon balance, have been highlighted in the recent past.

Most soil functions are directly or indirectly related to the soil microbiota, which may explain the effort expended in developing techniques to determine with confidence the number, volume and diversity of organisms in soil. In ecosystem research, the cycling of most elements is driven partly or completely by the amount and activity of organisms that assimilate mineral compounds or decompose organic matter.

As early as 1909, Engberding was doubtful whether microbial numbers on agar plates would give representative cell counts for agricultural soils (Engberding, 1909). But it wasn't until 1996 that researchers were able to demonstrate experimentally that plate counts show only a very small proportion of the total soil microbial community, even though such microbes exist and are viable (Torsvik *et al.*, 1996). Therefore, techniques that are based on isolating and cultivating microorganisms are considered not to reflect the whole microbial community.

Soil fertility is a central aim of sustainable land use. Organic farming systems, especially, emphasize the role of soil microorganisms in element cycling and their value for a healthy soil (Mäder *et al.*, 2002). Reliable measurement and interpretation of biological soil quality, in a world that realizes once more the potential of the soil natural resource, is an important task.

Approaches for Estimating Soil Microbial Biomass

As for most quantitative analyses, it is necessary to account for methodological, spatial and temporal variability, which can be done by analysing an appropriate number of replicates in the laboratory and in the field. For agricultural fields, four replicate bulk samples, each consisting of at least 15 single cores, may reduce spatial variability to less than 10%. Analytical variability can be checked by a representative reference soil – stored at optimum conditions for each purpose – and being analysed with each batch of analysis.

Direct cell counting

Observing and counting microorganisms in their soil habitat is considered to reliably reflect the number and volume of microbial cells in soil. Using staining techniques, it is possible to distinguish microbes from soil particles and to determine activity and biomass. *In situ* hybridization techniques using specific RNA probes may finally serve as powerful tools for counting specific organisms or functional groups in soil. Reliable techniques in microbial biovolume estimation in soil are reviewed and described by Bölter *et al.* in Section 6.4.

The microbial biomass approach

Jenkinson and Powlson (1976a) invented an approach that overcame the isolation–cultivation bias. This chloroform fumigation incubation method was originally checked by (more time-consuming) direct microscopic measurements. They pointed out that 'soil microbes are the eye of the needle, through which all organic material that enters the soil must pass', and they defined the whole microbial community as a black box – the 'holistic approach'. This is a quantitative approach of pools and fluxes and hardly considers specific organisms or functional groups of the microbial community. The microbial biomass, as such, has been inserted as an active pool of soil organic matter in recently developed models on organic matter cycling. In Section 6.2, Brookes and Joergensen explain this approach based on chloroform fumigation.

Jenkinson and Powlson's approach served as the reference for Anderson and Domsch (1978), who invented a physiological method for estimating microbial biomass in soils that has since become very popular – the substrate-induced respiration (SIR) method. Section 6.3 by Höper deals with the advantages and shortcomings of this method as it is now used, as well as the chloroform fumigation and microscopic methods, for monitoring and survey of soils.

Molecular markers

A major step towards a more detailed analysis of the microbial community was the development of molecular markers and advances in extracting them directly from soil organisms. The advantage of these markers is their specificity to living cells or even taxa. However, differing amounts in cells of different species at different physiological states may limit their use as parameters for microbial biomass estimation.

ATP

Adenosine 5'-triphosphate (ATP) occurs in living cells and does not persist in soils. The procedure for extracting ATP from soils has been validated for a wide range of soils (Contin *et al.*, 2001) and a critical review of the method is given by Martens (2001). Provided a standardized procedure is used, Jenkinson and Ladd (1981) found, and Jenkinson (1988) and Contin *et al.* (2001) confirmed, a remarkably constant ATP content of the microbial biomass of 10–12 µmol ATP/g microbial biomass C. These authors recommend the use of a strongly acidic extractant as crucial to denature soil phosphatases, whereas Martens (2001) recommends extraction with dimethyl sulphoxide (DMSO) and Na_3PO_4, as proposed by Bai *et al.* (1988).

Membrane compounds of microbial cells

Microbial membranes contain a great number of lipid compounds that can be extracted directly from soil. The variable content of signature fatty acids in microbial cells of different taxa makes it difficult to relate them to specific biomass, but as they can be extracted quantitatively from the soil, the sum of fatty acids may serve as a biomass indicator. Zelles (1999) reviews the potential and limitations of the use of fatty acids for biomass estimation.

Nucleic acid analysis of the microbial cells

Direct nucleic acid extraction from soil has gained much attention in the past few years, since it holds great potential for describing soil microorganisms that were not accessible by use of cultivation-dependent techniques (see also Chapter 8, this volume). DNA is a biomolecule, tightly associated with living organisms, and thus might serve as a biomass indicator. However, in the past it has proven very difficult to reliably extract total DNA from soils of different textures, and it became commonly accepted that DNA quantities detected in soils are very variable and barely meaningful or useful as a biomass indicator. However, recent advances in soil DNA extraction protocols have indicated that DNA, and even RNA, can be recovered efficiently from soil (Borneman and Triplett, 1997; Bundt *et al.*, 2001; Bürgmann *et al.*, 2001), and correlation of DNA quantities and biomass has been demonstrated in a few cases (Curci *et al.*, 1997; Bundt *et al.*, 2001; Macrae *et al.*, 2001; Taylor *et al.*, 2002). A broader validation of DNA extraction protocols for quantitative extraction of DNA from soils may be beneficial in at least two ways. First, it may serve as a simple and rapidly determined biomass indicator, and, secondly, it would allow further dissection of the structure of the microbial biomass by use of molecular tools (Chapter 8, this volume).

Methods Standardized for Soil Quality Determination

ISO-certified methods are described for soil microbial biomass determination: the chloroform fumigation extraction and the substrate-induced respiration technique. For comparing quantitative results originating from different laboratories, samples have to be exchanged in order to assure reliability of the absolute values. Thresholds or minimum/maximum values for microbial biomass are often discussed. For soil quality evaluation it may be interesting to see if predicted and measured values show much of a difference. If there is, this could be used as an indication of disturbance of the system. Finally, and as pointed out by most authors in soil quality research, it is recommended to use a set of methods, in particular with respect to microbial functions, that are of ecological importance. Soil microbial biomass itself may be interpreted as the mediator of some soil functions, and apparently plays key roles in physical stability and nutrient cycling.

6.2 Microbial Biomass Measurements by Fumigation–Extraction

PHILIP C. BROOKES[1] AND RAINER GEORG JOERGENSEN[2]

[1]Soil Science Department, IACR-Rothamsted, Harpenden AL5 2JQ, UK; [2]Department of Soil Biology and Plant Nutrition, University of Kassel, Nordbahnhofstraße, D-37213 Witzenhausen, Germany

Introduction

The soil microbial biomass responds much more quickly than most other soil fractions to changing environmental conditions, such as changes in substrate inputs (e.g. Powlson *et al.*, 1987) or increases in heavy metal content (Brookes and McGrath, 1984). This, and much other similar, research supports the original idea of Powlson and Jenkinson (1976) that biomass is a much more sensitive indicator of changing soil conditions than, for example, the total soil organic matter content. Thus, the biomass can serve as an 'early warning' of such changes, long before they are detectable in other ways.

Linked parameters (e.g. biomass-specific respiration or biomass as a percentage of soil organic C) are also useful as they have their own intrinsic 'internal controls' (see Barajas *et al.*, 1999 for a discussion of this). This may permit interpretation of measurements in the natural environment, where, unlike in controlled experiments, there may not be suitable non-contaminated soil (for example) to provide good 'control' or 'background' measurements.

Here we provide experimental details of two measurements of biomass which have proved useful in environmental studies, particularly at low levels (i.e. around European Union limits) of pollution by heavy metals, namely soil microbial biomass C and biomass ninhydrin N.

Principle of the Method

Following chloroform fumigation of soil, there is an increase in the amount of various components coming from the cells of soil microorganisms which are lysed by the fumigant and made partially extractable (Jenkinson and Powlson, 1976b). Organic C (Vance *et al.*, 1987), total N and NH_4-N (Brookes *et al.*, 1985), and ninhydrin-reactive N (Amato and Ladd, 1988; Joergensen and Brookes, 1990) can be measured in the same 0.5 M K_2SO_4 extract. Further information on fumigation–extraction and other microbiological methods is given by Alef and Nannipieri (1995).

Materials and Apparatus

- A room or incubator adjustable to 25°C
- An implosion-protected desiccator
- A vacuum line (water pump or electric pump)
- A horizontal or overhead shaker
- A deep-freezer at −15°C
- Folded filter papers (e.g. Whatman 42 or Schleicher & Schuell 595 1/2)
- Glass conical flasks (250 ml)

Chemicals and Reagents

- Ethanol-free chloroform ($CHCl_3$)
- Soda lime
- 0.5 M Potassium sulphate (K_2SO_4) (87.1 g/l)

Procedure

Fumigation–extraction

A moist soil sample of 50 g is divided into two subsamples of 25 g. The non-fumigated control samples are placed in 250 ml conical flasks and then immediately extracted with 100 ml 0.5 M K_2SO_4 (ratio extractant:soil is 4:1) for 30 min by oscillating shaking at 200 rpm (or 45 min overhead shaking at 40 rpm) and then filtered through a folded filter paper. For the fumigated treatment, 50 ml glass vials containing the moist soils are placed in a desiccator containing wet tissue paper and a vial of soda lime. A beaker containing 25 ml ethanol-free $CHCl_3$ and a few boiling chips is added and the desiccator evacuated until the $CHCl_3$ has boiled vigorously for 2 min. The desiccator is then incubated in the dark at 25°C for 24 h. After fumigation, $CHCl_3$ is removed by repeated (sixfold) evacuation and the soils are transferred to 250 ml bottles for extraction with 0.5 M K_2SO_4. All treatments are replicated three times. All K_2SO_4 extracts are stored at −15°C prior to analysis.

Biomass C estimated by dichromate oxidation

Principle of the method

In the presence of strong acid and dichromate, organic matter is oxidized and Cr(+VI) reduced to Cr(+III). The amount of dichromate left is back-titrated with iron II ammonium sulphate (Kalembasa and Jenkinson, 1973) and the amount of carbon oxidized is calculated.

Additional materials and apparatus

- Liebig condenser
- 250 ml round-bottom flask
- Burette

Additional chemicals and reagents

- 66.7 mM potassium chromate ($K_2Cr_2O_7$) (19.6125 g/l)
- Concentrated phosphoric acid (H_3PO_4)
- Concentrated sulphuric acid (H_2SO_4)
- 40.0 mM iron II ammonium sulphate $[(NH_4)_2[Fe(SO_4)_2] \times 6H_2O]$
- 25 mM 1.10-phenanthroline-ferrous sulphate complex solution

All chemicals are analytical reagent grade and distilled or de-ionized water is used throughout.

- *Digestion mixture*: two parts conc. H_2SO_4 are mixed with one part conc. H_3PO_4 (v/v).
- *Titration solution*: iron II ammonium sulphate (15.69 g/l) is dissolved in distilled water, acidified with 20 ml conc. H_2SO_4 and made up to 1000 ml with distilled water.

Procedure

To 8 ml soil extract in a 250 ml round-bottom flask, 2 ml of 66.7 mM (0.4 N) $K_2Cr_2O_7$ and 15 ml of the H_2SO_4/H_3PO_4 mixture are added. The mixture is gently refluxed for 30 min, allowed to cool and diluted with 20–25 ml water, added through the condenser as a rinse. The excess dichromate is measured by back-titration with 40.0 mM iron II ammonium sulphate, using 25 mM 1.10-phenanthroline-iron II sulphate complex solution as an indicator.

Calculation of results

CALCULATION OF EXTRACTABLE ORGANIC C FOLLOWING DICHROMATE DIGESTION

C (µg/ml) = [(HB – S) / CB] × N × [VD/VS] × E × 1000

where: S = consumption of titration solution by the sample (ml); HB = consumption of titration solution by the hot (refluxed) blank (ml); CB = consumption of titration solution by the cold (unrefluxed) blank (ml); N = normality of the $K_2Cr_2O_7$ solution; VD = added volume of the $K_2Cr_2O_7$ solution (ml); VS = added volume of the sample (ml); and E = 3, conversion of Cr(+VI) to Cr(+III), assuming that, on average, all organic C is as [C(0)].

C (µg/g soil) = C (µg/ml) × (VK + SW)/DW

where: VK = volume of K_2SO_4 extractant (ml); SW = volume of soil water (ml); and DW = dry weight of sample (g).

CALCULATION OF BIOMASS C

Biomass C $(B_C) = E_C/k_{EC}$

where: E_C = (organic C extracted from fumigated soils) − (organic C extracted from non-fumigated soils) and k_{EC} = 0.38 (Vance et al., 1987).

Biomass C by UV-persulphate oxidation

Principle of the method

In the presence of potassium persulphate ($K_2S_2O_8$), extractable soil organic carbon is oxidized by ultraviolet (UV) light to CO_2, which is measured using infrared (IR) or photo-spectrometric detection.

Additional materials and apparatus

Automatic carbon analyser with IR-detection (e.g. Dohrman DC 80) or continuous-flow systems with colourimetric detection (Skalar, Perstorp).

Additional chemicals and reagents

- $K_2S_2O_8$
- Concentrated H_3PO_4
- Sodium hexametaphosphate [$(Na(PO_4)_6)n$]
- *Oxidation reagent*: 20 g $K_2S_2O_8$ are dissolved in 900 ml distilled water, acidified to pH 2 with conc. H_3PO_4 and made up to 1000 ml
- *Acidification buffer*: 50 g sodium hexametaphosphate are dissolved in 900 ml distilled water, acidified to pH 2 with conc. H_3PO_4 and made up to 1000 ml

Procedure

For the automated UV-persulphate oxidation method, 5 ml K_2SO_4 soil extract are mixed with 5 ml acidification buffer. Any precipitate of $CaSO_4$ in the soil extracts is dissolved by this procedure. The $K_2S_2O_8$ is automatically fed into the UV oxidation chamber, where the oxidation to CO_2 is activated by UV light. The resulting CO_2 is measured by IR absorption.

Calculation of results

CALCULATION OF EXTRACTABLE ORGANIC C

C (µg/g soil) = [(S × DS) − (B × DB)] × (VK + SW)/DW

where: S = C in sample extract (µg/ml); B = C in blank extract (µg/ml); DS = dilution of sample with the acidification buffer; DB = dilution of blank with the acidification buffer; VK = volume of K_2SO_4 extractant (ml); SW = volume of soil water (ml); and DW = dry weight of sample (g).

CALCULATION OF BIOMASS C

Biomass C $(B_C) = E_C/k_{EC}$

where: E_C = (organic C extracted from fumigated soils) – (organic C extracted from non-fumigated soils) and k_{EC} = 0.45 (Wu *et al.*, 1990; Joergensen, 1996a).

Biomass C by oven oxidation

Extractable soil organic C is oxidized to CO_2 at 850°C in the presence of a platinum catalyser. The CO_2 is measured by infrared absorption using an automatic analyser (Shimadzu 5050, Dimatoc 100, Analytic Jena). The new oven systems use small sample volumes – so they are able to measure C in extracts containing large amounts of salts. The procedure is similar to the automated UV-persulphate oxidation method, except that the samples are diluted with water and acidified using a few drops of HCl instead of the hexametaphosphate acidification buffer. The calculations of extractable C and biomass C are identical to those used in the automated UV-persulphate oxidation method.

Determination of ninhydrin-reactive nitrogen

Principle of the method

Ninhydrin forms a purple complex with molecules containing α-amino nitrogen and with ammonium and other compounds with free α-amino groups, such as amino acids, peptides and proteins (Moore and Stein, 1948). The presence of reduced ninhydrin (hydrindantin) is essential to obtain quantitative colour development with ammonium. According to Amato and Ladd (1988), the amount of ninhydrin-reactive compounds, released from the microbial biomass during the $CHCl_3$ fumigation and extraction by 2 M KCl, is closely correlated with the initial soil microbial biomass carbon content.

Additional apparatus

- Boiling water bath
- Photo-spectrophotometer

Additional chemicals and solutions

- Ninhydrin
- Hydrindantin
- Dimethyl sulphoxide (DMSO)
- Lithium acetate dihydrate
- Acetic acid (96%)
- Citric acid
- Sodium hydroxide (NaOH)
- Ethanol (95%)
- L-Leucine
- Ammonium sulphate (($NH_4)_2SO_4$)
- *Lithium acetate buffer*: lithium acetate (408 g) is dissolved in water (400 ml), adjusted to pH 5.2 with acetic acid and finally made up to 1 l with water
- *Ninhydrin reagent*: ninhydrin (2 g) and hydrindantin (0.3 g) are dissolved in dimethyl sulphoxide (75 ml), 25 ml of 4 M lithium acetate buffer at pH 5.2 are then added (Moore, 1968)
- *Citric acid buffer*: citric acid (42 g) and NaOH (16 g) are dissolved in water (900 ml), adjusted to pH 5 with 10 M NaOH if required, then finally made up to 1 l with water

Procedure

The procedure is described according to Joergensen and Brookes (1990) for measuring biomass C and microbial ninhydrin-reactive N in K_2SO_4 soil extracts. A 10 mM L-leucine (1.312 g/l) and a 10 mM ammonium-N [$(NH_4)_2SO_4$ 0.661 g/l] solution are prepared separately in 0.5 M K_2SO_4 and diluted within the range 0–1000 µM N. The standard solutions, K_2SO_4 soil extracts or blank (0.6 ml) and the citric acid buffer (1.4 ml) are added to 20 ml test tubes. The ninhydrin reagent (1 ml) is then added slowly, mixed thoroughly and closed with loose aluminium lids. The test tubes are then heated for 25 min in a vigorously boiling water bath. Any precipitate formed during the addition of the reagents then dissolves. After heating, an ethanol:water mixture (4 ml 1:1) is added, the solutions are thoroughly mixed again and the absorbance read at 570 nm (1 cm path length).

Calculation of results

CALCULATION OF EXTRACTED NINHYDRIN-REACTIVE N (N_{NIN})

N_{nin}(µg/g soil) = $(S - B)/L \times N \times (VK + SW)/DW$

where: S = absorbance of the sample; B = absorbance of the blank; L = millimolar absorbance coefficient of leucine; N = 14 (atomic weight of nitrogen); VK = volume of K_2SO_4 extractant (ml); SW = volume of soil water (ml); and DW = dry weight of the sample (g).

CALCULATION OF MICROBIAL NINHYDRIN-REACTIVE N

B_{nin} = (N_{nin} extracted from the fumigated soil) – (N_{nin} extracted from the non-fumigated soil)

CALCULATION OF MICROBIAL BIOMASS CARBON

Biomass C = B_{nin} × 22 (soils pH-H_2O > 5.0)
Biomass C = B_{nin} × 35 (soils pH-H_2O ≤ 5.0)

The conversion factors were obtained by correlating the microbial biomass C and B_{nin} in the same extracts of 110 arable, grassland and forest soils by the fumigation–extraction method (Joergensen, 1996b).

Discussion

Biomass measurements are certainly useful in studies of soil protection. They have the advantage that they are relatively cheap and simple, as well as being rapid. There is now a considerable amount of literature to show that these measurements are useful in determining effects of stresses on the soil ecosystem. Biomass ninhydrin measurements have two advantages over biomass C. First, a reflux digestion is not required for ninhydrin N. This makes it very suitable for situations with minimal laboratory facilities. Secondly, in both biomass C and N measurements the fraction coming from the biomass is determined following subtraction of an appropriate 'control'. With biomass C this value is often half of the total, while with biomass ninhydrin N it is commonly about 10 or less. This causes considerably less error in its determination. Both parameters are very closely correlated, however, so biomass C may be readily estimated from biomass ninhydrin N, as described above. One feature of the fumigation–extraction method frequently caused concern. Upon thawing of frozen K_2SO_4 soil extracts, a white precipitate of $CaSO_4$ occurs in near-neutral or alkaline soils. However, this causes no analytical problems in either method and may be safely ignored.

Acknowledgements

IACR receives grant-aided support from the Biotechnology and Biological Sciences Research Council of the United Kingdom.

6.3 Substrate-induced Respiration

HEINRICH HÖPER

Geological Survey of Lower Saxony, Friedrich-Missler-Straße 46/48, D-28211 Bremen, Germany

Introduction

The method of substrate-induced respiration was developed by Anderson and Domsch (1978). It is based on the principle that, under standardized conditions, the metabolism of glucose added in excess is limited by the amount of active aerobic microorganisms in the soil. During the first hours after substrate addition there is no significant growth of the microbial populations, and the respiratory response is proportional to the amount of microbial biomass in the soil. Anderson and Domsch (1978) established a conversion factor by correlating the substrate-induced respiration with the microbial biomass determined by the fumigation–incubation method: 1 ml CO_2/h corresponded to 40 mg microbial biomass carbon (C_{mic}).

For the measurements of substrate-induced respiration, Anderson and Domsch (1978) used a continuously operating CO_2-analyser (Ultragas 3, Wösthoff, Bochum, Germany); however, soil samples were not permanently aerated and were flushed with CO_2-free air only 20 min before measurement, thus leading to a temporary increase in CO_2 and depletion in O_2 in the soil sample. Later, respiration was also measured by oxygen consumption (Sapromat, Voith, Germany) in closed vessels (Beck, 1984).

Heinemeyer *et al.* (1989) developed a system where soil samples are continuously aerated with ambient air and the CO_2 production is detected by an infrared gas analyser. This very sensitive system allows the detection of respiration rates at high resolution after the addition of glucose. Concentration changes of less than 1 µl/l CO_2 are detectable. The problem of CO_2 absorption in carbonate-rich soils is overcome by the continuous flow. Kaiser *et al.* (1992) compared the soil microbial biomass as analysed by the so-called Heinemeyer device with the fumigation–extraction method according to Vance *et al.* (1987) and proposed changing the conversion factor to 30 mg C_{mic}/1 ml CO_2/h. This factor can also be used for mineral horizons of forest soils, as can be derived from the data of Anderson and Joergensen (1997) (Fig. 6.1).

The substrate-induced respiration method is based on the detection of a respiratory response of soil microorganisms on supply of glucose. Thus, only glucose-responsive and active organisms are measured. Based on this principle, the substrate-induced respiration method detects predominantly bacterial biomass. In some cases, as in peat and marsh soils, differences

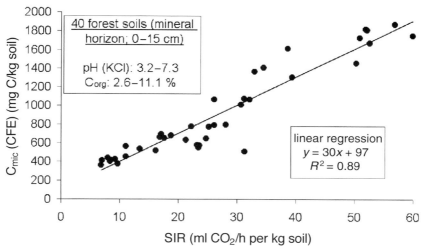

Fig. 6.1. Relation between substrate-induced respiration (SIR) and microbial biomass (C_{mic}) measured by the chloroform fumigation–extraction (CFE) method in mineral horizons of forest soils. Data were recalculated from Anderson and Joergensen (1997).

between the fumigation–extraction method and the substrate-induced respiration method were attributed to population structure, e.g. bacteria-to-fungi ratio, differing from those of common aerobic soils (Anderson and Joergensen, 1997; Brake *et al.*, 1999; Joergensen and Scheu, 1999; Chander *et al.*, 2001). Additionally, in several non-fertilized soils, nitrogen and phosphorus can become the limiting factors for microbial growth and the maximum initial respiratory response (Dilly, 1999). In this case, the amended glucose substrate should be enriched with a nitrogen, phosphorus, potassium and sulphur source (Palmborg and Nordgren, 1993).

Nevertheless, the method has been calibrated to determine the total soil microbial biomass (Anderson and Domsch, 1978) in a wide range of agricultural and forest soils. The results of the fumigation–extraction method and the substrate-induced respiration method were well correlated (Kaiser *et al.*, 1992; Anderson and Joergensen, 1997). Also, Lin and Brookes (1996) confirmed a good relationship between the fumigation–extraction method for a group of amended and unamended soils, and concluded that these soils had the same SIR response, although the activity status of the soil microorganisms and the community structure should be different.

Principle of the Method

The method of substrate-induced respiration is based on the principle that microorganisms react to the addition of glucose with an immediate respiration response that is proportional to their biomass as long as the organisms do not utilize the glucose for growth. Under standardized conditions, especially with respect to temperature and water content of the sample, and if glucose is given in excess without being inhibitory, the only factor limiting the respiratory response within the first hours is the amount of microorganisms. The optimal glucose concentration should be tested in preliminary assays. In arable soils with an expected microbial biomass below 800 mg C_{mic}/kg soil, a glucose amendment of 3000 mg/kg is often used.

A typical respiration curve is shown in Fig. 6.2, where the respiration rate is almost constant during the first 3 h after glucose amendment and increases due to microbial growth between 4 and 14 h.

The respiration rate after glucose supply can be measured by any method for measuring respiration rates in general. Nevertheless, methods that permit hourly monitoring of the respiration rates should be preferred, as the initial respiratory response of the soil sample on glucose supply can be derived from the actual curve shape (see Section 7.2). As an example of continuous measurement, the procedure using the Heinemeyer soil biomass analyser (Heinemeyer *et al.*, 1989) will be described in more detail below. Under low-budget conditions, an approach for soil respiration as developed by Jäggi (1976), based on static incubation with an alkali trap, can be used (Beck *et al.*, 1993; Anonymous, 1998; Pell *et al.*, Section 7.2).

Materials and Apparatus

General equipment

- Room or chamber at constant temperature of 22 ± 1°C
- Balance (1000 g, resolution 0.1 g)

Heinemeyer soil biomass analyser

- SIR soil biomass analyser (Heinemeyer *et al.*, 1989) with an infrared gas analyser and a flow meter for flow rates < 500 ml/min
- 24 fibreglass tubes, inner diameter 4 cm, length 25 cm. 48 sample holders of rubber foam
- Bubble meter, 100 ml, to control the flow meter
- Plastic beakers to permit homogeneous incorporation of glucose into the soil
- Hand mixer

Static incubation in airtight jars (Jäggi, 1976)

- Incubation vessels: SCHOTT-bottles with ISO thread, 250 ml
- Vessels for the soil material: flanged test tubes of polypropylene, inner diameter 29 mm × 105 mm, with lateral holes (3 cm below the fringe, 12 holes with 2 mm diameter) for gas exchange

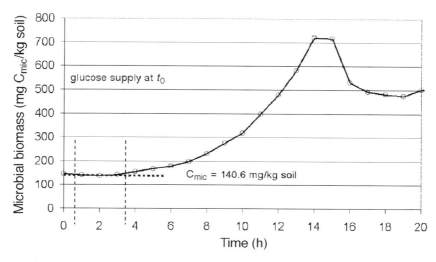

Fig. 6.2. Respiration rates converted to microbial biomass of a soil after the supply of 3000 mg glucose/kg soil.

- Rubber seal: O-ring 35 mm × 5 mm
- Burette, volume 20 ml or 10 ml, with CO_2 trap

Chemicals and solutions

General reagents

- Glucose–talcum mixture, ratio 3:7; add 150 g glucose p.a. to 350 g talcum

Heinemeyer soil biomass analyser

- Reference gases 330 µl/l CO_2 and 400 µl/l CO_2 in synthetic air
- Soda lime with indicator to produce CO_2-free air
- Demineralized water with a small amount of phosphoric acid to reduce the pH to about 5.0

Static incubation in airtight jars (Jäggi, 1976)

- Demineralized water (H_2O, electric conductivity < 5 µS/cm)
- *Sodium hydroxide solution* 0.025 M: sodium hydroxide solution 0.1 mol/l Titrisol diluted in 4 l H_2O
- *Hydrochloric acid* 0.025 M: hydrochloric acid 0.1 mol/l Titrisol diluted in 4 l H_2O

- *Phenolphthalein solution*: 0.2 g phenolphthalein ($C_{20}H_{14}O_4$, M = 318.33 g/mol) dissolved in 200 ml ethanol (60%)
- *Barium chloride solution* 0.5 M: 12.22 g barium chloride ($BaCl_2$ $2H_2O$, M = 244.28 g/mol, p.a.) dissolved in 100 ml H_2O

Procedure

General procedure

Sample preparation

Soils are partially dried or rewetted to 40–60% of the maximum water-holding capacity and passed through a 2 mm sieve. The water content of the soil should be as wet as possible but allowing the soil to be sieved and maintain the aggregate structure. Until the analysis, soil samples should be stored at 4°C. In order to reduce the effects of fresh organic matter amendments to the soil just before sampling, a pre-incubation of 1 week at about 22°C before the measurement is advised.

Determination of the optimal glucose concentration

Glucose is added in excess to the soil in order to get the maximum initial respiratory response. It has to be ascertained that the glucose does not inhibit microbial activity. At least five different concentrations should be tested. In agricultural soils, concentrations between 500 mg/kg and 6000 mg/kg soil are appropriate. Replicates are not necessary at this stage of assessment. The glucose concentration leading to the maximum initial respiratory response is optimal for measuring microbial biomass.

Measurement of microbial biomass

The measurement should last 24 h to control the exponential growth and to be sure that glucose was the only limiting factor for microbial respiration.

Heinemeyer soil biomass analyser

Glucose incorporation and measurement of respiration rate

At least three replicates should be analysed. To establish a concentration of 3000 mg glucose/kg soil, 0.5 g of the glucose–talcum mixture is added to humid soil, corresponding to 50 g on a oven-dry basis, in a plastic beaker and thoroughly mixed for 2 min using the hand mixer. Soils should be handled with care in order to avoid disaggregation and the formation of a soil paste. As quickly as possible, the samples should be transferred into the fibreglass tubes and fixed at each side by sample holders, and the measurement of the respiration rate should be started. The flow rate is adjusted for each sample to about 200 ± 200 ml/min.

Calibration

The CO_2 concentration difference has to be calibrated at least once a week, following the manual. The flow meter should be calibrated by the manufacturer every second year. An internal control using a bubble meter should be performed at least once every 6 months.

Static incubation in airtight jars (Jäggi, 1976)

Blanks

Prepare five blanks per batch without soil, handled like the soil-containing vessels, in order to be able to correct for CO_2 trapping in the alkali trap during handling and titration.

Preparation of alkali traps

Label incubation vessels and soil vessels. Add 20 ± 0.1 ml of 0.025 M sodium hydroxide solution into the incubation vessel. For one batch the same solution has to be used (to have the same relation to the blanks). The flask containing the sodium hydroxide solution has to be equipped with a CO_2 trap (soda lime) to avoid CO_2 adsorption from ambient air flowing into the flask.

First incubation time (2 h)

Glucose (60 mg) is added to the pre-incubated soil (corresponding to 20 g on an oven-dry matter basis). Soil and glucose are mixed thoroughly with a spatula and the mixture is used to fill the soil vessels. The soil vessels are transferred into a climate chamber or an incubator at 22°C for 2 h, as a starting time for glucose consumption.

Main incubation time (4 h)

After exactly 2 h the soil vessels are transferred into incubation vessels containing 20 ml 0.025 M sodium hydroxide solution, which are carefully closed and incubated for another 4 h. At least four replicates are recommended.

Titration

After 4 h, the soil vessels are removed. The sodium hydroxide solution is titrated immediately. Before titration, CO_3^{2-} is precipitated with 1 ml 0.5 M barium chloride solution. Subsequently, four drops of the phenolphthalein solution, as indicator, are added per incubation vessel and the vessel is immediately titrated with 0.025 M HCl until decoloration of the indicator. The consumed volume of the 0.025 M HCl solution is read (in ml).

Calculation

Heinemeyer soil biomass analyser

$$C_{mic} = 30\, F_s * (C_s - C_b)/SW$$

where: C_{mic} = carbon in microbial biomass (mg C/kg dry soil); F_s = flow rate of the air passing the sample (l/h); C_s = CO_2-concentration of CO_2-enriched air, coming from the sample tube (ml/l); C_b = CO_2-concentration of air coming from an empty reference tube (blank) (ml/l); SW = dry matter sample weight (kg); and 30 = constant (mg C_{mic} h/ml CO_2).

Static incubation in airtight jars (Jäggi, 1976)

The HCl readings of the five blanks are averaged (blank = BL). The amount of microbial biomass carbon (C_{mic}) for each sample (SA) is calculated as follows:

$$C_{mic} = 30(BL - SA)\frac{k \times 22 \times 1000}{1.8295 \times SW \times 4}$$

where: C_{mic} = carbon in microbial biomass (mg C_{mic}/kg dry soil); BL = mean of the HCl readings of the five blanks (ml HCl); SA = HCl readings of the samples (ml HCl); k = concentration of the HCl solution; 22 = factor (1 ml 1 M HCl corresponds to 22 mg CO_2); 1000 = conversion factor g soil into kg soil; 1.8295 = density of CO_2 at 22°C (mg/ml) (density of CO_2 at 0°C and 1013.2 hPa is 1.9768 mg/ml); SW = sample weight (g soil dry matter); and 4 = conversion factor 4 h to 1 h.

Discussion

Technical advantages of SIR

The SIR, especially when using the Heinemeyer device, has some advantages over the fumigation methods. First, it has a very low determination limit of about 5–10 mg C_{mic}/kg of soil. Thus, the method is also suitable for the assessment of subsoil samples. The standard deviation between replicates is also very low and was estimated by Höper and Kleefisch (2001), for routine analysis within the soil monitoring programme of Lower Saxony, Germany, to be in the median 2.2% and, in 90% of the cases, below 5.2% (Table 6.1).

Table 6.1. Sources of variability for substrate-induced respiration (SIR) measurements with the Heinemeyer soil biomass analyser (Heinemeyer et al., 1989) in the Soil Monitoring Programme of Lower Saxony, Germany, 1996–2000 (Höper and Kleefisch, 2001).

Sources of variability	Explanation	Coefficient of variation (median)	90% quantile
Analytical	Between 3 analytical replicates	2.2	5.2
Repeated sampling	Between 5 bulk samples of 16 sample cores each of the same plot	7.4	n.d.
Spatial	Between 4 subplots of 250 m² in the same field	9.6 12.8[a]	17.4 37.5[a]
Temporal (interannual)	Between samples of the same plot and the same annual period over 3–5 years	14.1 15.3[a]	26.1 40.9[a]

[a]Arable versus grassland soils.

Use of SIR for monitoring of soil quality

Substrate-induced respiration has been used in several soil monitoring programmes (Beck et al., 1995; Kandeler et al., 1999; Höper and Kleefisch, 2001; Machulla et al., 2001; Oberholzer and Höper, 2001; Rampazzo and Mentler, 2001). In the Soil Monitoring Programme of Lower Saxony (Germany), sources of variability were examined using a defined sampling strategy (16 cores of 4 cm diameter, 0–20 cm depth in arable land; 0–10 cm and 10–20 cm depth in grassland; sampling period between mid-February and mid-March). Variability due to space and year was rather low, and could be attributed partially to the fact that a baseline variation had to be considered, i.e. the variation between repeated samplings of the same area taken at the same time (Table 6.1). Due to the low spatial and interannual variation, it was already possible to detect significant changes of microbial biomass on some monitoring plots after 5 years. Between 1996 and 2000 a significant decrease in microbial biomass was found in some acid soils and in soils with a low organic matter input (Höper and Kleefisch, 2001).

For the evaluation of measurements, a reference system was established based on the prediction of microbial biomass from abiotic soil properties. To evaluate a given soil, the measured value was compared to the value calculated on the basis of a multiple regression equation with organic C or N content, pH and clay or sand content of the soil (Höper and Kleefisch, 2001; Oberholzer and Höper, 2001). It was possible to establish a common reference system for soils from Lower Saxony, northern Germany and Switzerland (Oberholzer and Höper, 2001). As a whole range of different land uses, climatic conditions and other soil conditions (e.g. water regime) were integrated, the standard error was rather high, ranging from 24% to 32% of the predicted value.

Use of SIR for ecotoxicological risk assessment

The SIR has been broadly used for ecotoxicological risk assessment of soils contaminated with heavy metals or organic contaminants (e.g. pesticides or TNT), for example by Beck (1981), Wilke (1988), Harden *et al.* (1993), Fließbach *et al.* (1994), Kandeler *et al.* (1996), Barajas *et al.* (1999), Chander *et al.* (2001) and Frische and Höper (2003). In contaminated soils, SIR biomass has been a more sensitive parameter than microbial biomass estimated by fumigation–extraction, and both methods were less correlated than in the above example (Chander *et al.*, 2001). Probably the glucose responsive and more active part of the microflora, determining the SIR biomass, is more sensitive to pollution than the total microbial biomass, as measured biochemically.

The use of substrate-induced respiration curves for ecotoxicological soil assessment is of growing interest (Palmborg and Nordgren, 1993; Johansson *et al.*, 1998) and has recently been standardized (ISO standard 17155). The shape of the respiration curve is changing with increasing contaminant content; especially, the time lag before the start of exponential growth and the time until maximal respiration rate increase and maximum growth rate decreases (Winkel and Wilke, 1997; Johansson *et al.*, 1998; Wilke and Winkel, 2000).

Drawbacks and limits of SIR

The SIR is a physiological method, based on the measurement of potential activity of microorganisms in the soil. It is calibrated against chloroform fumigation methods for measuring soil microbial biomass carbon in a large variety of soils. Nevertheless, differences between SIR and chloroform fumigation methods have been observed under specific situations, such as contaminated grassland soils or peatlands (Brake *et al.*, 1999; Chander *et al.*, 2001). Under these conditions, SIR, as a potential activity parameter, is obviously affected by soil conditions in a different way than microbial biomass estimated by chloroform fumigation methods, as a mass parameter.

Finally, SIR is a black-box method, not differentiating between different groups of microorganisms (e.g. bacteria versus fungi). Nevertheless, also as a black-box method, the benefit of SIR in monitoring of soil quality and ecotoxicological soil assessment has been shown in a large number of publications.

6.4 Enumeration and Biovolume Determination of Microbial Cells

MANFRED BÖLTER,[1] JAAP BLOEM,[2] KLAUS MEINERS[1] AND ROLF MÖLLER[1]

[1]*Institute for Polar Ecology, University of Kiel, Olshausenstraße 40, D-24098 Kiel, Germany;* [2]*Department of Soil Sciences, Alterra, PO Box 47, NL-6700 AA Wageningen, The Netherlands*

Introduction

Number and biomass of microorganisms are basic properties for soil ecological studies. They can be related to parameters describing microbial activity, soil health and other intrinsic soil descriptors (Kepner and Pratt, 1994). This often includes detailed analyses of the microbial communities, focusing on differentiation between individual organisms, growth forms, size classes, metabolic signatures or other specific properties.

The search for tools to count bacteria by microscopy in soil has a long tradition. The introduction of fluorescence stains and epifluorescence microscopy has found wide acceptance for aquatic and terrestrial ecological studies (Trolldenier, 1973; Zimmermann, 1975, 1977; Hobbie *et al.*, 1977). It has resulted in different applications and techniques for individual disciplines in marine microbiology, limnology and soil science (Bloem and Vos, 2004).

However, a basic restriction of total microscopic counts is the lack of discrimination between active and inactive cells. Another problem is posed by the small cell size, which is mostly less than 0.5 µm in diameter in natural samples, a fact that cannot be attributed only to dwarf or resting cells – although most of them are suspected to be non-culturable (Bakken and Olsen, 1987) and their growth rates can be very low (Bååth, 1994).

New techniques and image analysis have made microscopy an appropriate tool to get an independent insight into the soil communities. It has become a basic step for soil ecology and provides baselines for other indirect measures of biomass and activity (Bloem *et al.*, 1995a; Liu *et al.*, 2001; Bölter *et al.*, 2002). The following shall focus on basic ways of cell enumeration and related methodologies to evaluate microbial biovolume, and on microbial biomass calculated thereof. An example of results in contaminated soil is given.

Principles of Microbiological Counting

The list of appropriate dyes and staining procedures is long, and the use of fluorescent dyes has increased for various new applications. The introduction of molecular probes into microbial ecology is a main field for the application of fluorescent dyes. Although most procedures focus on bacteria, several attempts have also been made to develop techniques for studies on fungi and yeasts (e.g. Pringle *et al.*, 1989; Deere *et al.*, 1998). Some of the dyes are very sensitive and can be used to enumerate viruses (Fuhrman, 1999; Marie *et al.*, 1999). In most recent studies using fluorescent microscopy, the following dyes have been used.

DAPI (4',6-diamidino-2-phenylindole-dihydrochloride)

This is a blue fluorescent stain, which binds to double-stranded DNA (dsDNA) and RNA, but in different modes. It has become a popular stain in aquatic research for enumerations of bacteria and protozoa, and is widely used as a counterstain or parallel stain in fluorescence *in situ* hybridization (FISH) protocols. The fluorescence of DAPI, when bound to dsDNA, can best be excited at 358 nm (UV); the maximum emission is at 461 nm (Haughland, 1999). When bound to RNA, the emission maximum shifts to 500 nm. Hence, excitation can best be performed by using xenon or mercury arc lamps. DAPI is also used as a stain in laser scanning microscopy and in flow cytometry.

Acridine orange

Acridine orange (AO) has strong affinity to acidic organelles (Haughland, 1999). It can be used to stain eukaryotic and prokaryotic cells. The green and red fluorescence has been used to discriminate between live and dead cells (Strugger, 1949; Bank, 1988), viable and non-viable bacterial spores (Sharma and Prasad, 1992) or to monitor physiological activity (McFeters *et al.*, 1991). This has been attributed to the different fluorescence colours when bound to DNA and RNA (Traganos *et al.*, 1977). The dye shows bright green fluorescence when it is fixed to double-stranded nucleic acid; a red light is emitted when bound to single-stranded nucleic acid. More recently, this statement has been modified: green fluorescent bacteria are in a stationary phase and actively growing bacteria show red fluorescence (Haugland, 1999).

DTAF (5-(4,6-dichlorotriazin-2-yl) aminofluorescein)

This has been preferred for automatic image analysis of bacteria in soil smears because of the low background staining (Bloem *et al.*, 1995a,b; Paul *et al.*, 1999). It binds covalently to proteins.

Differential fluorescent stain

Differential fluorescent stain (DFS) is a nucleic acid stain, which has yielded good results for bacteria, as well as fungal hyphae, in soil. This is a mixture of europium(III) thenoyltrifluoroacetonate (europium chelate) and the disodium salt of 4,4'-bis(4-anilino-6-bis(2-hydroxyethyl)amino-S-triazin-2-ylamino)2,2'-stilbene disulphonic acid, which is also called Fluorescent Brightener (FB), Calcofluor White, Tinopal, or Fluostain I. In the DFS mixture, europium chelate stains DNA and RNA red, and FB stains cellulose and polysaccharides (cell walls) blue. Blue cells are assumed to be inactive or dead. Red cells are assumed to be active because in these cells the fluorescence is dominated by the europium chelate, indicating a higher nucleic acid content and a higher growth rate. The discrimination between active and inactive bacterial cells and fungal hyphae is an advantage of DFS (Morris *et al.*, 1997). A disadvantage is that the europium cannot be detected with a confocal laser scanning microscope for automatic image analysis (Bloem *et al.*, 1995b). Europium is a phosphorescent dye, which starts light emission microseconds after excitation. This is too slow for a confocal laser scanning microscope, but it is no problem for conventional epifluorescence microscopy. The stained cells can be identified and discriminated from other particles and thus serve as a base for their enumeration, measurement or identification.

Materials and Apparatus

Sampling and storage

To obtain samples for microbiological analyses, clean, or sometimes sterile, sampling devices are needed. Samples should be stored frozen, cool or air-dried until analysis. It should be kept in mind that any storage may influence the original community, e.g. freezing may reduce bacterial number. Pre-incubation after freezing restores bacterial number, but not necessarily to the original density (Bloem *et al.*, Chapter 3, this volume). Details on storage problems can be found in the literature for various purposes – the best strategy, however, is direct observation without long times between sampling and analyses (Ross *et al.*, 1980, West *et al.*, 1986, 1987; Turley and Hughes, 1994; Stenberg *et al.*, 1998).

Sample preparation

Various methods have been employed to separate bacteria from particles (e.g. Lindahl and Bakken, 1995). Methods using detergents or ultrasonic treatments are widely used. They can provide good results in coarse materials or organic substrates, but may cause severe problems when used in samples with high loads of silt or clay. Fine particles, just in the size of small rods or cocci, often occur after ultrasonic treatments. Before attempting such procedures, normal hand shaking can be employed as a first attempt. Bloem *et al.* (1995b) have compared different methods for preparing soil suspensions and found no positive effects of detergents and deflocculants. Too much mixing and sonication results in loss of cells. Bloem *et al.* (1995b) and Paul *et al.* (1999) recommended homogenization of soil suspensions (20 g in 190 ml) for 1 min at maximum speed (20,000 rev/min) in a (Waring) blender, for counting both bacteria and fungi. Soil samples with high contents of coarse material can be allowed to settle for a short time (approximately 1 min) before subsamples are used. Soil suspensions fixed with formaldehyde (3.7% final concentration) can be stored for 1 week at 2°C. It is safest to make the preparations as soon as possible after fixation and store the stained slides.

Staining and filtration

Preparation of filters

Identification and discrimination of small objects need a well-defined background. The use of black (polycarbonate) membrane filters is recommended. A pore size of 0.2 µm should be used for counting bacteria. The staining can take place on the filter, or in solution and then be filtered by low vacuum (0.2 kP/cm^2). The filter is then mounted on a microscope slide and the stained bacteria are viewed with an epifluorescence microscope. Unbound dyes can interfere with the filter material and cause background staining, so a washing step may be necessary.

Preparation of soil smears

Bacteria can also be counted directly on a glass slide in a soil smear (Babiuk and Paul, 1970). A smear is prepared by drying 10 µl of a homogenized soil suspension on a printed microscopic slide. Bloem *et al.* (1995b) reported less background staining and less fading of fluorochromes in smears than on filters. Soil films in smears are completely flat, which is a great advantage for automatic image analysis.

Microscopy

The correct optical presentation of objects below the micrometer range may raise several problems, due to the limited depth of focus of normal lenses. The use of confocal laser scanning microscopy is thus a good tool because images of different focal planes are combined to one image with extended focus. In addition, halo effects are avoided. Illumination and halo effects may lead to overestimation of sizes. Kato (1996) makes the point for careful consideration of halo effects and low visibility of cell protoplasm with

DAPI, which may result in underestimation of biovolumes. DAPI has been reported to yield about 40% lower estimates of cell volumes than AO (Suzuki *et al.*, 1993).

Discrimination between small bacteria and unspecific particles needs to be performed carefully. Human subjectivity is another important factor when comparing and analysing data from various laboratories (Domsch *et al.*, 1979; Nagata *et al.*, 1989), beside problems with accurate size measurement (Suzuki *et al.*, 1993), which has important effects on biovolume calculations.

Equipment

Epifluorescence microscope fitted with filters for excitation of cells with blue light (wavelength $c.$ 470 nm, for AO and DTAF) and UV ($c.$ 365 nm, for DAPI and DFS), and equipped with a 100× oil-immersion lens for bacteria and a 40× (50×) lens(es) for fungi. For direct counting and sizing, an eyepiece graticule can be used (May, 1965), e.g. G12 New Porton Grid (Graticules Ltd, Tonbridge, Kent, UK). For digital image analysis, videocameras can be connected to personal computers equipped with image analysis software.

Chemicals and Solutions

General supplies

- Particle-free water
- Prestained polycarbonate filters (diameter: 25 mm; pore size: 0.2 µm); it is also possible to use normal polycarbonate filters stained in a solution of 2 mg/l Irgalan-Black in 2% acetic acid fixed with 0.2% formaline. After staining for about 12 h the filters are washed with particle-free distilled water until no black stain is visible in the washing water
- Cellulose acetate filters (diameter: 25 mm; pore size: 0.4–0.6 µm) must be used as backing filters to support homogeneous distribution of bacteria on the polycarbonate filter
- Clean microscope slides and cover slips
- Non-fluorescent immersion oil (e.g. Cargill Type A, Cargill Ltd, Cedar Grove, New Jersey, USA)
- Tips for micro-pipettes (200 µl and 1–10 ml)
- Box for storage of microscope slides
- For soil smears: printed microscope slides (e.g. Cel-Line (Erie Scientific, Portsmouth, New Hampshire, USA) or Bellco (Vineland, New Jersey, USA)) with a hole of 12 mm diameter in the centre. Wipe the slides with 70% ethanol and, finally, with a little undiluted liquid soap to promote even spreading of soil suspension

Solutions

Filter all the following solutions through a 0.2 µm membrane before use.

- Particle-free formalin (37%) (two times 0.2 µm filtered)
- For bacteria on filters:
  – acridine orange (AO) (e.g. Sigma Chemical Co., St Louis, Missouri, USA), 1 mg/ml (can be stored at 4°C for some days). If it is not fixed with formalin, the solution needs to be filtered (0.2 µm) before its next use. Contamination occurs frequently!
  – as an alternative to AO, the dye DAPI (e.g. Sigma Chemical Co.) can be used; prepare stock solution: 1 mg/ml (can be stored frozen in the dark for several weeks); and working solution: 10 µg/ml
- For soil smears stained with DTAF (bacteria) or DFS (bacteria and fungi):
  – buffer solution consisting of 0.05 M Na_2HPO_4 (7.8 g/l) and 0.85% NaCl, adjusted to pH 9
  – stain solution consisting of 2 mg DTAF dissolved in 10 ml of the buffer (should not be stored for longer than a day)
  – DFS solution is prepared by dissolving 3.5 g/l europium chelate (Kodak, Eastman Fine Chemicals, Rochester, New York, USA) and 50 mg/l fluorescent brightener, $C_{40}H_{42}N_{12}O_{10}S_2$ Na_2 (FW 960.9, Fluostain I, Sigma Chemical Co.), in 96% ethanol. Fluorescent brightener $C_{40}H_{42}N_{12}O_{10}S_2$ without Na_2 needs addition of NaOH (1 drop/ml) to get it into solution (Serita Frey, personal communication). After a few minutes, when the powder has dissolved completely, dilute to 50% ethanol with an equal volume particle-free water. In order to avoid high counts in blanks (preparations without soil added) due to precipitation of europium chelate, it is better to prepare the DFS solution 1 day before use and to filter the stain through a 0.2 µm pore-size membrane immediately before use. With all stains, blanks should be checked regularly.

Procedure

The protocol for the method is simple: the objects are stained with an appropriate fluorochrome, filtered on to a polycarbonate membrane and counted by epifluorescence microscopy. The individual steps are easy and can even be performed in field labs.

Preparation of the soil suspension

Weigh approximately 1 g fresh soil into a clean glass vial, add 10 ml of particle-free distilled water fixed with particle-free buffered formalin (final concentration: 1%) and shake vigorously for 1 min.

High dilution of the soil is also important to minimize masking of bacteria by soil particles. Masking is likely to occur when more than 1 mg of soil is added per cm² on the microscope slide (Bloem *et al.*, 1995b). Coarse particles are removed by settling for 1 min before subsamples are taken for the filtration (see below). Many laboratories homogenize the soil suspensions to disperse bacteria, e.g by using a blender for 1 min at maximum speed (Bloem *et al.*, 1995b, Paul *et al.*, 1999).

Staining

Equip the filtration unit with a cellulose-acetate backing filter (0.4–0.6 µm pore size) to spread the vacuum evenly. Put the black-stained polycarbonate filter (shiny side up) on top of it. The polycarbonate-filter must fit close and flat to the backing filter, without any air bubbles or folds.

1. Staining with AO

Pipette 5 ml of particle-free water into the funnel, add 100 µl of the soil suspension, add 500 µl of the AO solution, mix the water/sample/dye solution carefully and stain for about 3 min.

2. Staining with DAPI

Pipette 5 ml of particle-free water into the funnel, add 100 µl of the soil suspension, mix carefully and filter the samples down at low vacuum (maximum: –150 mbar) until approximately 1 ml remains. Stop filtration by releasing the vacuum from the filtration unit, add 700 µl of the DAPI working solution and stain for 8 min.

After staining (methods 1 or 2), suck the solution softly with low vacuum (maximum: –200 mbar), until all the water has gone. Transfer the dry polycarbonate filter from the vacuum device, and mount it (bacteria on top) on a microscope slide, using a thin smear of non-fluorescent immersion oil (e.g. Cargill Type A). Wynn-Williams (1985) recommends Citifluor® as a photofading retardant, which also accentuates different colour contrasts between organisms. Take care that no air is under the filter. Add a small drop of the immersion oil on top of the filter and mount a cover slip. Press the cover slip down carefully until the oil moves out from the edges of the cover slip and the filter. Lateral movements of the cover slip must be avoided; they can result in unequal distribution of the bacteria. To avoid cross-contamination, wash the funnel of the filtration unit thoroughly with particle-free distilled water between samples. The cellulose-acetate backing filter can be used for several filtrations.

3. Staining bacteria in soil smears with DTAF

Smear 10 µl soil suspension evenly in the hole on a glass slide. The water-repellent coating keeps the suspension in a well-defined area. Allow the smears to air-dry completely; this fixes the organisms to the slide. The slides are placed flat on paper tissue in a plastic tray. Flood the spots of dried soil with drops of stain for 30 min at room temperature. To prevent drying, the tissue is moistened before, and the trays are covered during staining. Rinse the slides three times for 20 min with buffer and finally for a few seconds with water, by putting them in slide holders and passing them through four baths. After air-drying, mount a cover slip with a small drop of immersion oil. The edges of the cover slip can be sealed with nail varnish. The slides can be stored at 2°C for at least a year (Bloem and Vos, 2004).

4. Staining fungal hyphae in soil smears with DFS

Prepare smear as above. After air-drying, stain the slides for 1 h in a bath with DFS solution. Flooding with drops of DFS is also possible if drying is effectively prevented during staining. Evaporation of ethanol may lead to precipitation of europium chelate, resulting in fluorescent spots, which may be confused with bacteria. Rinse the slides three times for 5 min in baths with 50% ethanol. After air-drying, mount a cover slip with a small drop of immersion oil. The slides can be stored at 2°C for at least a year (Morris *et al.*, 1997; Bloem and Vos, 2004).

Examination of the slide

Check the slide for one focal plane. A large drop of immersion oil can result in floating cells, i.e. a slide with two focal plains. Check for equal distribution of cells on the filter. Bacterial number per counting area should range between 20 and 50 cells. If the number of cells is too low or too high, different aliquots of the sample dilution (10–500 µl) or different sample inputs (0.1–3 g) can be used. Filters can be frozen immediately and stored at –20°C in the dark for later analysis.

Counting and calculations

Counting proceeds by use of an epifluorescence microscope at randomly chosen fields following a cross-pattern on the slide. To minimize subjectivity, ensure that the filter is not observed while the field of view is changed. A minimum of 400 cells or a minimum of 20 counting squares can be recommended as a general rule. The optimum number of fields depends on the average number of bacteria per field. If the bacteria are randomly distributed, and there are at least 25 cells per field, ten fields are usually sufficient. When there are only a few cells per field, it is better to increase the number of counted fields. With a free computer program, the random distribution of bacteria on the slide and the optimum number of fields can be checked during counting (Bloem et al., 1992).

Total numbers of bacteria are calculated from the mean count of bacteria per counting area, the effective filtration area, the dilution factor of the soil suspension, the amount of soil used and the filtration volume as follows:

$$TBC = (D \times B \times M)/W$$

where: TBC = total bacterial count per g soil (cells/g); D = dilution caused by suspension and subsequent subsampling of the soil; B = mean count of bacteria per counting area; M = microscope factor (filtration area/area of counting field); and W = weight of oven-dry soil sample (g).

Sizing of the bacteria can be performed with different methods, e.g. with eyepiece graticules or digital image analysis systems. Fungi are counted at about 400-fold magnification. Check for unstained (brown) hyphae using transmitted light instead of epifluorescence. In one or two transects over the filter (about 50–100 fields), hyphal lengths are estimated by counting the number of intersections of hyphae with (all) the lines of the counting grid (Bloem et al., 1995b, Paul et al., 1999). Usually, many fields are needed because most fields contain no hyphae.

The hyphal length, H (µm), is calculated as:

$$H = (n \times \pi \times a)/(2 \times l)$$

where: n = number of intersections per grid; a = grid area (µm^2); l = total length of lines in the counting grid (µm).

The total length of fungal hyphae F (per mg soil) is calculated as:

$$F = H \times 10^{-6} \times (A/B) \times (1/S)$$

where: H = hyphal length (µm/grid); 10^{-6} = conversion of µm to m; A = area of the slide covered by sample; B = area of the grid; and S = amount of soil on the slide.

Calculation of biovolume

Several methods have been used for the estimation of bacterial biovolume, including electronic sizing, flow cytometry and different microscopic techniques (Bratbak, 1985). The latter are scanning electron microscopy (SEM), transmission electron microscopy (TEM), confocal laser scanning microscopy, normal light microscopy and widely used epifluorescence microscopic techniques (Bratbak and Dundas, 1984; Bratbak, 1993; Bloem et al., 1995a). The microscopic estimation of cell volumes is based on measurement of the linear dimensions (length, width, perimeter) of individual bacterial cells. These parameters may be obtained with an eyepiece graticule, using photomicrographs or videocamera-equipped microscopes and digital image analysis (Bloem et al., 1995a; Posch et al., 1997).

Linear dimensions are converted to cell volumes using stereometric formulas. A widely used formula, applicable for cocci and rods as well as filaments, has been given by Krambeck et al. (1981):

$$V = (\pi/4) \times W^2 \times (L - W/3)$$

where: V = volume (μm^3); L = length (μm); and W = width (μm); for cocci, length = width. This equation can also be used for fungal hyphae when the diameter has been estimated.

Other calculations distinguish between distinct bacterial morphotypes and use various geometrical bodies to approximate cell shapes. Additional formulas are based on multiple measurement features derived by computer-assisted microscopy (Posch *et al.*, 1997; Blackburn *et al.*, 1998). Assuming a linear relationship between length and width for bacteria, an approximation for increasing cell width with cell length can be applied (Zimmermann, 1975; Bölter *et al.*, 1993). Special demands for volume calculations need to be taken into consideration when cyanobacteria, fungi or algae are under inspection, and special conversion factors at species level or for morphological groups need to be taken into consideration (Bölter, 1997).

Calculation of bacterial biomass

The bacterial biomass is calculated by multiplication of the product of cell number and average cell volume with a conversion factor. The easiest conversion can be performed by the assumption that 80% of the biovolume consists of water, and the remaining dry weight (20%) is considered to be 50% carbon. This rough estimate has found several refinements due to specific habitats, populations, or other considerations. Problems arise since microbial communities are natural consortia of bacteria and fungi. Van Veen and Paul (1979) describe significantly different ratios for bacteria, yeasts and filamentous fungi, depending on actual water content. The spans for specific weights of bacteria and fungi range from 0.11 g/cm³ to 1.4 g/cm³.

Three different models have been used to convert bacterial numbers and cell volumes into biomass values (Norland, 1993).

- The constant ratio model assumes a constant, size-independent carbon:biovolume ratio and neglects the potential condensation of carbon in smaller cells.
- The constant biomass model assumes a constant carbon content of cells with different biovolumes, e.g. a standard carbon unit per cell.
- The allometric model assumes that the dry-weight:volume ratio depends on cell volume, and that smaller bacteria have a higher dry-weight:volume ratio than larger ones. This model considers that the cell quota of different cell constituents may vary with cell size. The allometric models are expressed as power functions:

$$m = C \times V^a$$

where: m = carbon content; C = conversion factor (carbon per unit volume) V = cell volume; and a = scaling factor.

All models neglect the fact that cell quotas of different major constituents, including water, vary not only with cell size but also with bacterial species, and depend on the physiological condition of the organisms (Bratbak, 1985; Fagerbakke *et al.*, 1996; Troussellier *et al.*, 1997). Most of the experimentally derived conversion factors are based on single species or distinct assemblages (mainly planktonic bacteria). Transferring these factors to other environments is problematic.

As a rough estimate, fungal biomass can be calculated from the biovolume using a specific carbon content of 1.3×10^{-13} g C/μm^3. Because smaller cells tend to have a higher density, a higher specific carbon content of 3.1×10^{-13} g C/μm^3 is used for bacteria (Bloem *et al.*, 1995b).

Methodological Remarks

Clean containers and working solutions are a must for reliable results of bacterial numbers. Controls (blank filters) should be performed routinely to check for bacterial contamination of staining solutions and particle-free water used in the protocol.

A potential source of error is the subjectivity of the method, e.g. differences between the operators' judgement of what is a bacterial cell and what is detritus, and particle or size classifications. It is important to compare individual results and discuss criteria of cell identification in order to get consistency between researchers. Determination of bacterial counts of relatively small sampling sets should be performed by one person.

The effective filter area is not the total area of the filter, but the area of the filter through which the soil suspension has been filtered. The effective filter area varies with different filtration units.

Error propagation and statistical considerations

Many working steps are involved from first sample preparation to final results on counts and biomass. Each of these carries a specific error that influences the final result. Therefore, one of the most important questions is: how exact is the final result?

Errors are unavoidable, but they should be reduced to a minimum. Typical errors during biomass calculation (Table 6.2) are due to:

- *Sample collection*: this error source is not predictable. It is necessary to keep in mind that samples and subsamples have to be representative for the site of analysis.
- *Sample preparation*: weighing with lab balances (sample mass, water content) may cause an error of approximately 0.1% to maximally 1%. Dilutions and aliquots, handled with pipettes, can have an error of approximately 1%. Effective filtration or smear area can be estimated with a precision of *c.* 1% (blunt diameter).
- *Image analysis*: calibration of the length measurement is part of image analysis; its resolution is restricted to pixel size. The resulting counting field area provides another error of approximately 2%.

Table 6.2. Error Propagation – example.

	Typical value	Typical error	Relative error (%)
Sample preparation			
Sample weight	1 g	0.001 g	0.1
Water content	10 %	0.1 %	1.0
Dilution	10 ml	0.1 ml	1.0
Aliquot	100 µl	1 µl	1.0
Diameter filter	20 mm	0.1 mm	0.5
Filter area	2.9 cm^2	3.0 mm^2	1.0
Calibration			
Pixel length	0.1 µm	1 nm	1.0
Counting field	2340 µm^2	48 µm^2	2.1
Counting			
Objects per counting field due to Poisson distribution	e.g. 356/26 = 13.7	0.2	1.4
Error propagation			
Mean cell volume			+7.15 / –6.68 %
Total bacterial number			+6.64 % / –6.75 %
Bacterial biovolume			+13.79 % / –13.90 %

Counting and calculations

Not all objects on a filter or smear can be counted. Thus, statistical methods need to be used to assess the variation and statistical significance of the counts. The first question to be answered is how many organisms need to be counted? It is recommended to analyse at least 20 counting fields, each of which should contain at least 20 objects, up to a maximum of 50. The optimal number of objects per counting field must be obtained by diluting the original sample.

All counts per counting field must fit a Poisson distribution in order to be summarized by a significant mean value. The test for Poisson distribution can be performed by the chi-square test. The error of the mean is calculated by square root of (B/n), where B is the mean count of bacteria per counting area, and n is the total count. This mean value for the counting fields can be used to calculate a total count valid for the entire sample. It is obvious that the error due to the counting procedure is much greater than other errors, e.g. the effective filter area. Hence, much effort must be put into its minimization by aiming for great accuracy in sample preparation and data evaluation. Counts of bacteria are often used for determinations of biomass, using a uniform mean cell volume or a mean carbon content per cell, or via the determination of a sample-specific biovolume. The first approach is a rough estimate, neglecting individual properties of the communities, and should be used only when details of the community cannot be obtained.

The other way offers a more precise data set and provides data for comparisons between individual communities and samples. The classical way for this procedure is the use of size classes of objects (Bölter *et al.*, 1993). The boundaries of such classes can be preset classes according to empirical knowledge about ecologically reasonable categories, which may provide a best fit to the community from known data or the size classes, or they can be calculated in order to obtain optimal fits, for example:

$$b = (x_{max} - x_{min})/[1 + 3.32\log(n)]$$

where: b = width of size class; x_{max} = maximal x value; x_{min} = minimal x value; and n = number of observations.

The use of preset size classes is recommended since comparisons between numbers of objects of individual size classes can be performed. Descriptions of communities can be performed using histograms. Contents of size classes can be compared by appropriate statistical tests. The number of organisms per size class can be aggregated if individual classes do not contain enough objects to perform such tests. It is necessary to mention that descriptors of size histograms, e.g. mean values or medians, need special attention in the case of open classes (e.g. all cells < 0.25 µm). For such classes, only medians are allowed as descriptors (Sachs, 1984; Lozan and Kausch, 1998). Extremes can be omitted by the '4 sigma-rule' (Sachs, 1984).

Comparisons between total numbers or numbers per size class become possible by various measures, their skewness and kurtosis or other statistical properties. Often, bimodal distributions or log-normal distributions can be observed. The latter are most typical for data obtained in natural environments. They are characterized by high standard deviations or high variation coefficients (Sachs, 1984). It is necessary to discriminate between rods and cocci for size classifications.

Before further use of calculated means, etc., the distribution has to be checked at least for symmetry. Normal distribution can be assessed by the David test (two-sided test, 5%) (Lozan and Kausch, 1998). As stated above, non-normal distributions are most typical for data obtained in natural environments. In this case, data have to be transformed into a normal distribution first. A logarithmic transformation is mostly successful. All relevant calculations and tests can be performed on a log-normal distribution as on a normal distribution. For further use in biomass calculations, results must be retransformed.

Estimating a maximum error of the final result

To gain a statement of the precision of the final results of mean cell volume, total bacterial number or bacterial biovolume, all individual errors have to be considered together as they appear in the equations. Following error propagation, the absolute values sum up to a maximum overall error of the final result (Table 6.2). This can be used for worst-case scenarios. An easy way to evaluate maximum error is to add up the percentages of individual errors.

Discussion

The use of direct counts and estimates of bacterial volume, biomass or other parameters derived thereof is an important tool to understand the microbial habitat and microbial processes. As an example, results from a site polluted with heavy metals are given in Fig. 6.3. At the site of a galvanizing company, soil samples were taken at the most polluted spot around a former basin, and at distances of 10 m and 50 m (unpolluted control). Nickel (Ni) and chromium (Cr) contents were 2800/430, 930/1300 and < 5.0/< 10 mg/kg dry soil, respectively. In the sandy soil under grass, the following characteristics were determined: pH-KCl 6.0, 5.7 and 5.9; organic matter 3.9%, 4.5% and 2.9% (w/w); and clay 8.0%, 2.5% and 3.7%. Bacteria were measured by image analysis in DTAF-stained soil smears (Bloem et al., 1995a,b). In the most polluted soil, bacterial number was 50% lower than in the unpolluted control (Fig. 6.3a). The average cell volume was 30% smaller (Fig 6.3b). Thus, the bacterial biomass (calculated from number and volume) was only 35% of the level in the unpolluted soil (Fig. 6.3c). The results of the microscopic measurements were confirmed by independent measures of bacterial growth rate by thymidine incorporation (see Section 7.5, this volume) – the bacterial growth rate was greatly reduced (Fig. 6.3d). Similar levels of heavy metals in clay soil showed no significant ecological effects. The bioavailability of contaminants is reduced by clay, organic matter and a higher pH. Thus, actual ecological effects of contamination can be demonstrated by measuring soil microorganisms.

This view into the 'home' of the organisms under evaluation, and the use of proper descriptors, provides insights in the community structure and thus allows better and more relevant interpretations of results from biochemical or physiological approaches. Shifts in populations can be followed and their inherent changes become visible. This holds especially true for speculations on the 'active' biomass or 'active' community involved in metabolic processes. New techniques of differential staining procedures can be applied, associated with enormous progress in digital image analysis. Nevertheless, care must be taken while interpreting individual local aspects and projecting data from microscopic views into the full scale of the environment. This problem, mostly neglected, holds true not only for this method, but also needs to be respected in all calculations in environmental research.

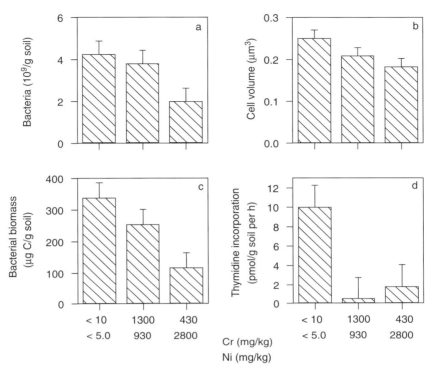

Fig. 6.3. Reduced bacterial number (a), mean cell volume (b), biomass (c) and growth rate (d) in heavy metal (chromium and nickel) contaminated soil. Error bars indicate the least significant difference at $P = 0.05$.

Normative References for Chapter 6

http://www.iso.org (accessed 27 October 2004)

ISO 14240–1 (1997) Soil quality – Determination of soil microbial biomass – Part 1: Substrate-induced respiration method.

ISO 14240–2 (1997) Soil quality – Determination of soil microbial biomass – Part 2: Fumigation–extraction method.

ISO 17155 (2002) Soil quality – Determination of abundance and activity of soil microflora using respiration curves.

References for Chapter 6

Alef, K. and Nannipieri, P. (1995) *Methods in Applied Soil Microbiology and Biochemistry*. Academic Press, London.

Amato, M. and Ladd, J.N. (1988) Assay for microbial biomass based on ninhydrin-reactive nitrogen in extracts of fumigated soils. *Soil Biology and Biochemistry* 20, 107–114.

Anderson, J.P.E. and Domsch, K.H. (1978) A physiological method for the quantitative measurement of microbial biomass in soils. *Soil Biology and Biochemistry* 10, 215–221.

Anderson, T.H. and Joergensen, R.G. (1997) Relationship between SIR and FE estimates of microbial biomass C in deciduous forest soils at different pH. *Soil Biology and Biochemistry* 29, 1033–1042.

Anonymous. (1998) Bestimmung der mikrobiellen Biomasse (Substratinduzierte Respiration; Isermeyer-Ansatz), Methode B-BM-IS. In: Agroscope FAL (ed.) *Referenzmethoden der Eidgenössischen landwirtschaftlichen Forschungsanstalten*, Vol. 2: *Bodenuntersuchungen zur Standort-Charakter-isierung*. Agroscope FAL Reckenholz, Eidgenössische Forschungsanstalt für Agrarökologie und Landbau, Zürich-Reckenholz.

Bååth, E. (1994) Thymidine and leucine incorporation in soil bacteria with different cell size. *Microbial Ecology* 27, 267–278.

Babiuk, L.A. and Paul, E.A. (1970) The use of fluorescein isothiocyanate in the determination of the bacterial biomass of grassland soil. *Canadian Journal of Microbiology* 16, 57–62.

Bai, Q.Y., Zelles, L., Scheunert, I. and Korte, F. (1988) A simple effective procedure for the determination of adenosine triphosphate in soils. *Chemosphere* 17, 2461–2470.

Bakken, L.R. and Olsen, R.A. (1987) The relationship between cell size and viability of soil bacteria. *Microbial Ecology* 13, 103–114.

Bank, H.L. (1988) Rapid assessment of islet viability with acridine orange and propidium iodide. *In Vitro Cellular and Developmental Biology* 24, 266–273.

Barajas, A.M., Grace, C., Ansorena, J., Dendooven, L. and Brookes, P.C. (1999) Soil microbial biomass and organic C in a gradient of zinc concentrations around a spoil tip mine. *Soil Biology and Biochemistry* 31, 867–876.

Beck, T. (1981) Untersuchungen über die toxische Wirkung der in Siedlungsabfällen häufigen Schwermetalle auf die Bodenmikroflora. *Zeitschrift für Pflanzenernährung und Bodenkunde* 144, 613–627.

Beck, T. (1984) Mikrobiologische und biochemische Charakterisierung landwirtschaftlich genutzter Böden. I. Mitteilung: die Ermittlung einer bodenmikrobiologischen Kennzahl. *Zeitschrift für Pflanzenernährung und Bodenkunde* 147, 456–466.

Beck, T., Öhlinger, R. and Baumgarten, A. (1993) Bestimmung der Biomasse mittels Substrat-Induzierter Respiration (SIR). In: Schinner, F., Öhlinger, R., Kandeler, E. and Margesin, R. (eds) *Bodenbiologische Arbeitsmethoden*. Springer Verlag, Berlin, pp. 68–72.

Beck, T., Capriel, P., Borchert, H. and Brandhuber, R. (1995) The microbial biomass in agricultural soils, 2nd communication. The relationships between microbial biomass and chemical and physical soil properties. *Agribiological Research* 48, 74–82.

Blackburn, N., Hagström, A., Wikner, J., Cuadros-Hansson, R. and Bjørnsen, P.K. (1998) Rapid determination of bacterial abundance, biovolume, morphology, and growth by neural network-based image analysis. *Applied and Environmental Microbiology* 64, 3246–3255.

Bloem, J. and Vos, A. (2004) Fluorescent staining of microbes for total direct counts. In: Kowalchuk, G.A., De Bruijn, F.J., Head, I.M., Akkermans, A.D.L. and van Elsas, J.D. (eds) *Molecular Microbial Ecology Manual*, 2nd edn. Kluwer Academic Publishers, Dordrecht, The Netherlands, pp. 861–874.

Bloem, J., Van Mullem, D.K. and Bolhuis, P.R. (1992) Microscopic counting and calculation of species abundances and statistics in real time with an MS-DOS personal computer, applied to bacteria in soil smears. *Journal of Microbiological Methods* 16, 203–213.

Bloem, J., Veninga, M. and Shepherd, J. (1995a) Fully automatic determination of soil bacterial numbers, cell volumes and frequencies of dividing cells by confocal laser scanning microscopy and image analysis. *Applied and Environmental Microbiology* 61, 926–936.

Bloem, J., Bolhuis, P.R., Veninga, M.R. and Wieringa, J. (1995b) Microscopic methods for counting bacteria and fungi in soil. In: Alef, K. and Nannipieri, P. (eds) *Methods in Applied Soil Microbiology and Biochemistry*. Academic Press, London, pp. 162–173.

Bölter, M. (1997) Microbial communities in

soils and on plants from King George Island (Arctowski Station, Maritime Antarctica). In: Battaglia, B., Valencia, J. and Walton, D.W.H. (eds) *Antarctic Communities: Species, Structure and Survival*. Cambridge University Press, Cambridge, pp. 162–169.

Bölter, M., Möller, R. and Dzomla, W. (1993) Determination of bacterial biovolume with epifluorescence microscopy: comparison of size distributions from image analysis and size classifications. *Micron* 24, 31–40.

Bölter, M., Bloem, J., Meiners, K. and Möller, R. (2002) Enumeration and biovolume determination of microbial cells – a methodological review and recommendations for applications in ecological research. *Biology and Fertility of Soils* 36, 249–259.

Borneman, J. and Triplett, E.W. (1997) Rapid and direct method for extraction of RNA from soil. *Soil Biology and Biochemistry* 29, 1621–1624.

Brake, M., Höper, H. and Joergensen, R.G. (1999) Land use-induced changes in activity and biomass of microorganisms in raised bog peats at different depths. *Soil Biology and Biochemistry* 31, 1489–1497.

Bratbak, G. (1985) Bacterial biovolume and biomass estimations. *Applied and Environmental Microbiology* 49, 1488–1493.

Bratbak, G. (1993) Microscope methods for measuring bacterial biovolume: epifluorescence microscopy, scanning electron microscopy, and transmission electron microscopy. In: Kemp, P.F., Sherr, B.F., Sherr, E.B. and Cole, J.J. (eds) *Handbook of Methods in Aquatic Microbial Ecology*. Lewis Publishers, Boca Raton, Florida, pp. 309–317.

Bratbak, G. and Dundas, I. (1984) Bacterial dry matter content and biomass estimations. *Applied and Environmental Microbiology* 48, 755–757.

Brookes, P.C. and McGrath, S.P. (1984) The effects of metal toxicity on the soil microbial biomass. *Journal of Soil Science* 35, 341–346.

Brookes, P.C., Landman, A., Pruden, G. and Jenkinson, D.S. (1985) Chloroform fumigation and the release of soil nitrogen: a rapid direct extraction method for measuring microbial biomass nitrogen in soil. *Soil Biology and Biochemistry* 17, 837–842.

Bundt, M., Widmer, F., Pesaro, M., Zeyer, J. and Blaser, P. (2001) Preferential flow paths: biological 'hot spots' in soils. *Soil Biology and Biochemistry* 33, 729–738.

Bürgmann, H., Pesaro, M., Widmer, F. and Zeyer, J. (2001) A strategy for optimizing quality and quantity of DNA extracted from soil. *Journal of Microbiological Methods* 45, 7–20.

Chander, K., Dyckmans, J., Höper, H. and Jörgensen, R.G. (2001) Long-term effects on soil microbial properties of heavy metals from industrial exhaust deposition. *Journal of Plant Nutrition and Soil Science* 164, 657–663.

Contin, M., Todd, A. and Brookes, P.C. (2001) The ATP concentration in the soil microbial biomass. *Soil Biology and Biochemistry* 33, 701–704.

Curci, M., Pizzigallo, M.D.R., Crecchio, C., Mininni, R. and Ruggiero, P. (1997) Effects of conventional tillage on biochemical properties of soils. *Biology and Fertility of Soils* 25, 1–6.

Deere, D., Shen, J., Vesey, G., Bell, P., Bissinger, P. and Veal, D. (1998) Flow cytometry and cell sorting for yeast viability assessment and cell selection. *Yeast* 30, 147–160.

Dilly, O. (1999) Nitrogen and phosphorus requirement of the microbiota in soils of the Bornhoeved Lake district. *Plant and Soil* 212, 175–183.

Domsch, K.H., Beck, T., Anderson, J.P.E., Söderström, B., Parkinson, D. and Trolldenier, G. (1979) A comparison of methods for soil microbial population and biomass studies. *Zeitschrift für Pflanzenernährung und Bodenkunde* 142, 520–522.

Engberding, D. (1909) Vergleichende Untersuchungen über die Bakterienzahl im Ackerboden in ihrer Abhängigkeit von äusseren Einflüssen. *Centralblatt für Bakteriologie Abt II* 23, 569–642.

Fagerbakke, K.M., Heldal, M. and Norland, S. (1996) Content of carbon, nitrogen, oxygen, sulfur and phosphorus in native aquatic and cultured bacteria. *Aquatic Microbial Ecology* 10, 15–27.

Fließbach, A., Martens, R. and Reber, H.H. (1994) Soil microbial biomass and activity in soils treated with heavy metal contaminated sewage sludge. *Soil Biology and Biochemistry* 26, 1201–1205.

Frische, T. and Höper, H. (2003) Soil microbial parameters and luminescent bacteria assays as indicators for *in-situ* bioremediation of TNT-contaminated soils. *Chemosphere* 50, 415–427.

Fuhrman, J.A. (1999) Marine viruses and their biogeochemical and ecological effects. *Nature* 399, 541–548.

Harden, T., Joergensen, R.G., Meyer, B. and Wolters, V. (1993) Soil microbial biomass estimated by fumigation–extraction and substrate-induced respiration in two pesticide-treated soils. *Soil Biology and Biochemistry* 25, 679–683.

Haughland, R.P. (ed.) (1999) *Molecular Probes: Handbook of Fluorescent Probes and Research Chemicals*. Molecular Probes Inc., 7th edn, Eugene, Oregon (CD-rom edn). Available at: http://www.probes.com (accessed 27 October 2004).

Heinemeyer, O., Insam, H., Kaiser, E.-A. and Walenzik, G. (1989) Soil microbial biomass and respiration measurements: an automated technique based on infrared gas analysis. *Plant and Soil* 116, 191–195.

Hobbie, J.E., Daley, R.J. and Jasper, S. (1977) Use of nucleopore filters for counting bacteria by fluorescence microscopy. *Applied and Environmental Microbiology* 33, 1225–1228.

Höper, H. and Kleefisch, B. (2001) Untersuchung bodenbiologischer Parameter im Rahmen der Boden-Dauerbeobachtung in Niedersachsen. Bodenbiologische Referenzwerte und Zeitreihen. *Arbeitshefte Boden* 2001/4, 1–94.

Jäggi, W. (1976) Die Bestimmung der CO_2-Bildung als Mass der bodenbiologischen Aktivität. *Schweizerische landwirtschaftliche Forschung* 15, 371–380.

Jenkinson, D.S. (1988) Determination of microbial biomass carbon and nitrogen in soil. In: Wilson J.R. (ed.) *Advances in Nitrogen Cycling in Agricultural Systems*. CAB International, Wallingford, UK, pp. 368–386.

Jenkinson, D.S. and Ladd, J.N. (1981) Microbial biomass in soil: measurement and turnover. In: Paul, E.A. and Ladd, J.N. (eds) *Soil Biochemistry*, Vol. 5. Marcel Decker, New York, pp. 415–471.

Jenkinson, D.S. and Powlson, D.S. (1976a) The effects of biocidal treatments on metabolism in soil – V. A method for measuring soil biomass. *Soil Biology and Biochemistry* 8, 209–213.

Jenkinson, D.S. and Powlson, D.S. (1976b) The effects of biocidal treatments on metabolism in soil – I. Fumigation with chloroform. *Soil Biology and Biochemistry* 8, 167–177.

Joergensen, R.G. (1996a) The fumigation–extraction method to estimate soil microbial biomass: calibration of the k_{EC} value. *Soil Biology and Biochemistry* 28, 25–31.

Joergensen, R.G. (1996b) Quantification of the microbial biomass by determining ninhydrin-reactive N. *Soil Biology and Biochemistry* 28, 301–306.

Joergensen, R.G. and Brookes, P.C. (1990) Ninhydrin-reactive nitrogen measurements of microbial biomass in 0.5 M K_2SO_4 soil extracts. *Soil Biology and Biochemistry* 22, 1023–1027.

Joergensen, R.G. and Scheu, S. (1999) Response of soil microorganisms to the addition of carbon, nitrogen and phosphorus in a forest Rendzina. *Soil Biology and Biochemistry* 31, 859–866.

Johansson, M., Pell, M. and Stenström, J. (1998) Kinetics of substrate-induced respiration (SIR) and denitrification: applications to a soil amended with silver. *Ambio* 27, 40–44.

Kaiser, E.-A., Mueller, T., Joergensen, R.G., Insam, H. and Heinemeyer, O. (1992) Evaluation of methods to estimate the soil microbial biomass and the relationship with soil texture and organic matter. *Soil Biology and Biochemistry*, 24, 675–683.

Kalembasa, S.J. and Jenkinson, D.S. (1973) A comparative study of titrimetric and gravimetric methods for the determination of organic carbon in soil. *Journal of the Science of Food and Agriculture* 24, 1085–1090.

Kandeler, E., Kampichler, C. and Horak, O. (1996) Influence of heavy metals on the functional diversity of soil microbial communities. *Biology and Fertility of Soils* 23, 299–306.

Kandeler, E., Tscherko, D. and Spiegel, H. (1999) Long-term monitoring of microbial biomass, N mineralisation and enzyme activities of a Chernozem under different tillage management. *Biology and Fertility of Soils* 28, 343–351.

Kato, K. (1996) Image analysis of bacterial cell size and diversity. In: Colwell, R.R., Simidu, U. and Ohwada, K. (eds) *Microbial Diversity in Time and Space.* Plenum Press, New York, pp. 141–147.

Kepner, R.L. and Pratt, J.R. (1994) Use of fluorochromes for direct enumeration of total bacteria in environmental samples: past and present. *Microbiology and Molecular Biology Reviews* 58, 603–615.

Krambeck, C., Krambeck, H.-J. and Overbeck, J. (1981) Microcomputer-assisted biomass determination of plankton bacteria on scanning electron micrographs. *Ambio* 42, 142–149.

Lin, Q. and Brookes, P.C. (1996) Comparison of methods to measure microbial biomass in unamended, ryegrass-amended and fumigated soils. *Soil Biology and Biochemistry* 28, 933–939.

Lindahl, V. and Bakken, L.R. (1995) Evaluation of methods for extraction of bacteria from soil. *FEMS Microbiology Ecology* 16, 135–142.

Liu, J., Dazzo, F.B., Glagoleva, O., Yu, B. and Jain, A.K. (2001) CMEIAS: a computer-aided system for the image analysis of bacterial morphotypes in microbial communities. *Microbial Ecology* 41, 173–194.

Lozan, J.L. and Kausch, H. (1998) *Angewandte Statistik für Naturwissenschaftler*, 2nd edn. Parey, Berlin.

Machulla, G., Barth, N., Heilmann, H. and Pälchen, W. (2001) Bodenmikrobiologische Untersuchungen an landwirtschaftlich genutzten Boden-Dauerbeobachtungsflächen (BDF II) in Sachsen. *Mitteilungen der Deutschen Bodenkundlichen Gesellschaft* 96, 359–360.

Macrae, A., Lucon, C.M.M., Rimmer, D.L. and O'Donnell, A.G. (2001) Sampling DNA from the rhizosphere of *Brassica napus* to investigate rhizobacterial community structure. *Plant and Soil* 233, 223–230.

Mäder, P., Fließbach, A., Dubois, D., Gunst, L., Fried, P. and Niggli, U. (2002) Soil fertility and biodiversity in organic farming systems. *Science* 296, 1694–1697.

Marie, D., Brussard, C.P.D., Thyrhaug, R., Bratbak, G. and Vaulot, D. (1999) Enumeration of marine viruses in culture and natural samples by flow cytometry. *Applied and Environmental Microbiology* 65, 45–52.

Martens, R. (2001) Estimation of ATP in soil: extraction methods and calculation of extraction efficiency. *Soil Biology and Biochemistry* 33, 973–982.

May, K.R. (1965) A new graticule for particle counting and sizing. *Journal of Scientific Instruments* 42, 500–501.

McFeters, G.A., Singh, A., Byun, S., Williams, S. and Callis, P.R. (1991) Acridine orange staining reaction as an index of physiological activity in *Escherichia coli*. *Journal of Microbiological Methods* 13, 87–97.

Moore, S. (1968) Amino acid analysis: aqueous dimethyl sulfoxide as solvent for the ninhydrin reaction. *Journal of Biological Chemistry* 243, 6281–6283.

Moore, S. and Stein, W.H. (1948) Photometric ninhydrin method for use in the chromatography of amino acids. *Journal of Biological Chemistry* 176, 367–388.

Morris, S.J., Zink, T., Conners, K. and Allen, M.F. (1997) Comparison between fluorescein diacetate and differential fluorescent staining procedures for determining fungal biomass in soils. *Applied Soil Ecology* 6, 161–167.

Nagata, T., Someya, T., Konda, T., Yamamoto, M., Morikawa, K., Fukui, M., Kuroda, N., Takahashi, K., Oh, S.-W., Mori, M., Araki, S. and Kato, K. (1989) Intercalibration of the acridine orange direct count method of aquatic bacteria.

Bulletin of the Japanese Society of Microbial Ecology 4, 89–99.

Norland, S. (1993) The relationship between biomass and volume of bacteria. In: Kemp, P.F., Sherr, B.F., Sherr, E.B. and Cole, J.J. (eds) *Handbook of Methods in Aquatic Microbial Ecology*. Lewis Publishers, Boca Raton, Florida, pp. 303–307.

Oberholzer, H.R. and Höper, H. (2001) Reference systems for the microbiological evaluation of soils. *VDLUFA-Schriftenreihe* 55/II, 19–34.

Palmborg, C. and Nordgren, A. (1993) Soil respiration curves, a method to test the abundance, activity and vitality of the microflora in forest soils. In: Torstensson, L. (ed.) *MATS Guidelines: Soil Biological Variables in Environmental Hazard Assessment*. Swedish Environmental Protection Agency, Uppsala, Sweden, pp. 149–153.

Paul, E.A., Harris, D., Klug, M. and Ruess, R. (1999) The determination of microbial biomass. In: Robertson, G.P., Coleman, D.C., Bledsoe, C.S. and Sollins, P. (eds) *Standard Soil Methods for Long-term Ecological Research*. Oxford University Press, New York, pp. 291–317.

Posch, T., Pernthaler, J., Alfreider, A. and Psenner, R. (1997) Cell-specific respiratory activity of aquatic bacteria studied with the tetrazolium reduction method, cytoclear slides and image analysis. *Applied and Environmental Microbiology* 63, 867–873.

Powlson, D.S. and Jenkinson, D.S. (1976) The effects of biocidal treatments on metabolism in soil. II. Gamma irradiation, autoclaving, air-drying and fumigation. *Soil Biology and Biochemistry* 8, 179–188.

Powlson, D.S., Brookes, P.C. and Christensen, B.T. (1987) Measurement of soil microbial biomass provides an early indication of changes in total soil organic matter due to straw incorporation. *Soil Biology and Biochemistry* 19, 159–164.

Pringle, J.R., Preston, R.A., Adams, A.E., Stearns, T., Drubin, D.G., Haarer, B.K. and Jones, E.W. (1989) Fluorescence microscopy methods for yeasts. *Methods in Cell Biology* 31, 357–435.

Rampazzo, N. and Mentler, A. (2001) Influence of different agricultural land use on soil properties along the Austrian–Hungarian border. *Bodenkultur* 52(2), 89–115.

Ross, D.J., Tate, K.R., Cairns, A. and Meyrick, K.F. (1980) Influence of storage on soil microbial biomass estimated by three biochemical procedures. *Soil Biology and Biochemistry* 12, 369–374.

Sachs, L. (1984) *Angewandte Statistik*, 6th edn. Springer-Verlag, Berlin.

Sharma, D.K. and Prasad, D.N. (1992) Rapid identification of viable bacterial spores using a fluorescence method. *Biotechnic and Histochemistry* 67, 27–29.

Stenberg, B., Johansson, M., Pell, M., Sjödahl-Svensson, K., Stenström, J. and Torstensson, L. (1998) Microbial biomass and activities in soils as affected by frozen and cold strorage. *Soil Biology and Biochemistry* 30, 393–402.

Strugger, S. (1949) *Fluoreszenzmikroskopie und Mikrobiologie*. Verlag Schaper, Hanover, Germany.

Suzuki, M.T., Sherr, E.B. and Sherr, B.F. (1993) DAPI direct counting underestimates bacterial abundances and average cell size compared to AO direct counting. *Limnology and Oceanography* 38, 1566–1570.

Taylor, J.P., Wilson, B., Mills, M.S. and Burns, R.G. (2002) Comparison of microbial numbers and enzymatic activities in surface soils and subsoils using various techniques. *Soil Biology and Biochemistry* 34, 387–401.

Torsvik, V., Sørheim, R. and Goksøyr, J. (1996) Total bacterial diversity in soil and sediment communities – a review. *Journal of Industrial Microbiology* 17, 170–178.

Traganos, F., Darzynkiewicz, Z., Sharpless, T. and Melamed, M.R. (1977) Simultaneous staining of ribonucleic and deoxyribonucleic acids in unfixed cells using acridine orange in a flow cytofluorometric system. *Journal of Histochemistry and Cytochemistry* 25, 46–56.

Trolldenier, G. (1973) The use of fluorescence microscopy for counting soil microorganisms. In: Rosswall, T. (ed.) *Modern Methods in the Study of Microbial Ecology*. Bull Ecol Res Comm (Stockholm) 17, 53–59.

Troussellier, M., Bouvy, M., Courties, C. and Dupuy, C. (1997) Variation of carbon content among bacterial species under starvation condition. *Aquatic Microbial Ecology* 13, 113–119.

Turley, C.M. and Hughes, D.J. (1994) The effect of storage temperature on the enumeration of epifluorescence-detectable bacterial cells in preserved sea-water samples. *Journal of the Marine Biological Association of the United Kingdom* 74, 259–262.

Van Veen, J.A. and Paul, E.A. (1979) Conversion of biovolume measurements of soil organisms, grown under various moisture tensions, to biomass and their nutrient content. *Applied and Environmental Microbiology* 37, 686–692.

Vance, E.D., Brookes, P.C. and Jenkinson, D.S. (1987) An extraction method for measuring soil microbial C. *Soil Biology and Biochemistry* 19, 703–707.

West, A.W., Ross, D.J. and Cowling, J.C. (1986) Changes in microbial C, N, P, and ATP contents, numbers and respiration on storage of soil. *Soil Biology and Biochemistry* 18, 141–148.

West, A.W., Sparling, G.P. and Grant, W.D. (1987) Relationships between mycelial and bacterial populations in stored, air-dried and glucose-amended arable and grassland soils. *Soil Biology and Biochemistry* 19, 599–605.

Wilke, B.M. (1988) Long term effects of inorganic pollutants on the microbial activity of a sandy Cambisol. *Zeitschrift für Pflanzenernährung und Bodenkunde* 151, 131–136.

Wilke, B.M. and Winkel, B. (2000) Einsatz mikrobiologischer Methoden in der Bewertung sanierter Böden und in der Ökotoxikologie. *VDLUFA-Schriftenreihe* 55/II, 35–46.

Winkel, B. and Wilke, B.M. (1997) Wirkung von TNT (2,4,6-Trinitrotoluol) auf Bodenatmung und Nitrifikation. *Mitteilungen der Deutschen Bodenkundlichen Gesellschaft* 85, 631–634.

Wu, J., Joergensen, R.G., Pommerening, B., Chaussod, R. and Brookes, P.C. (1990) Measurement of soil microbial biomass C – an automated procedure. *Soil Biology and Biochemistry* 22, 1167–1169.

Wynn-Williams, D.D. (1985) Photofading retardant for epifluorescence microscopy in soil micro-ecological studies. *Soil Biology and Biochemistry* 17, 739–746.

Zelles, L. (1999) Fatty acid patterns of phospholipids and lipopolysaccharides in the characterisation of microbial communities in soil: a review. *Biology and Fertility of Soils* 29, 111–129.

Zimmermann, R. (1975) Entwicklung und Anwendung von fluoreszenz- und rasterelektronen-mikroskopischen Methoden zur Ermittlung der Bakterienmenge in Wasserproben. PhD thesis, Universität Kiel, Kiel, Germany.

Zimmermann, R. (1977) Estimation of bacterial number and biomass by epifluorescence microscopy and scanning electron microscopy. In: Rheinheimer, G. (ed.) *Microbial Ecology of a Brackish Water Ecosystem*. Springer, Heidelberg, Germany, pp. 103–120.

7 Soil Microbial Activity

7.1 Estimating Soil Microbial Activity

OLIVER DILLY

Lehrstuhl für Bodenökologie, Technische Universität München, D-85764 Neuherberg and Ökologie-Zentrum, Universität Kiel, Schauenburgerstraße 112, D-24118 Kiel, Germany; Present address: *Lehrstuhl für Bodenschutz und Rekultivierung, Brandenburgische Technische Universität, Postfach 101344, D-03013 Cottbus, Germany*

Why and How to Estimate Soil Microbial Activity

Microbial communities in soil consist of a great diversity of species exploring their habitats by adjusting population abundance and activity rates to environmental factors. Soil microbial activities lead to the liberation of nutrients available for plants, and are of crucial importance in biogeochemical cycling. Furthermore, microorganisms degrade pollutants and xenobiotics, and are important in stabilizing soil structure and conserving organic matter for sustainable agriculture and environmental quality. Microbial activities are regulated by nutritional conditions, temperature and water availability. Other important factors affecting microbial activities are proton concentrations and oxygen supply.

To estimate soil microbial activity, two groups of microbiological approaches can be distinguished. First, experiments in the field that often require long periods of incubation (e.g. Hatch *et al.*, 1991; Alves *et al.*, 1993) before significant changes of product concentrations are detected, e.g. 4–8 weeks for the estimation of net N mineralization. In this case, variations of soil conditions during the experiment are inevitable, e.g. aeration and site-specific temperature, and may influence the results (Madsen, 1996). Secondly, short-term laboratory procedures, which are usually carried out with sieved samples at standardized temperature, water content and pH value. Short-term designs of 2–5 h minimize changes in community structure during the experiments (Brock and Madigan, 1991). Such microbial activity measurements include enzymatic assays that catalyse substrate-specific transformations and may be helpful to ascertain effects of soil management, land use and specific environmental conditions (Burns, 1978).

Laboratory methods have the advantage of standardizing environmental factors and, thus, allowing the comparison of soils from different geographical locations and environmental conditions, and also data from different laboratories. In contrast, approaches in the field are considered advantageous for integrating site-specific environmental factors, such as temperature, water and oxygen availability and the microbial interactions with plants and animals. Besides net N mineralization measurements, decomposition experiments with litterbags are frequently used. Litterbags of approximately 20 × 20 cm length and width, with 2–5 mm mesh size are filled with site-specific litter (e.g. 10 g dry material) or cellulose (e.g. filter paper) with or without additional N, distributed in the investigated soil and sampled throughout the year to determine remaining mass, physical, chemical and biological characteristics (Dilly and Munch, 1996). Litterbag studies with smaller mesh size are used when estimating the participation of meso- and macrofauna. However, suppressive and stimulating effects of the fauna on the soil microbiota are not considered in such studies (Mamilov *et al.*, 2001). Finally, soil respiration determined in the field suffers from separating the activity of microorganisms and other organisms, such as animals and plants, which vary significantly in different systems and throughout the season (Dilly *et al.*, 2000).

The group of methods on soil microbial activities embraces biochemical procedures revealing information on metabolic processes of microbial communities. They are frequently used to gain information on 'functional groups'. However, laboratory results refer to microbial capabilities, as they are determined under optimal conditions of one or more factors, such as temperature, water availability and/or substrate. These activities have common units: 1/h, 1/day or 1/year.

Here, six methods have been selected for the estimatation of soil microbial activity. Two methods (Section 7.2 'Soil Respiration' and Section 7.3 'Soil Nitrogen Mineralization') refer to C and N cycling, respectively, and no substrate is added. Section 7.4 'Nitrification in Soil' and Section 7.5 'Thymidine and Leucine Incorporation to Assess Bacterial Growth Rate' follow the transformation after addition of substrate and tracer, respectively. 'Nitrification in soil' can also be estimated without substrate (ammonium) addition. Occasionally, assays without substrate addition are identified as 'actual activity' and those with substrate addition 'potential activity'. However, this classification is critical and confusing, since actual activity should refer to activity under the natural environmental conditions and the response to, for example, changing temperature (Q_{10} values) and site-specific water supply. The *in situ* method described in Section 7.6 'N_2O Emissions and Denitrification from Soil' considers specific pathways of the nitrogen cycle and estimates levels of one of the most important radiatively active trace gases in the atmosphere, contributing to at least 5% of observed global warming (Myhre *et al.*, 1998). Enzyme activities in soil are responsible for the flux of carbon, nitrogen and other essential elements in biogeochemical cycles. Measuring 'Enzyme Activity Profiles' (Section 7.7), and understanding the factors that regulate enzyme expression and the rates of substrate turnover, are the first stages in characterizing soil metabolic potential, fertility and quality. The highly abundant and diverse microorganisms in soil have high metabolic potentials. Generally, soil microorganisms are growth limited and, thus, may poorly exploit their capabilities. Combining measurements with reference to both carbon and nitrogen cycling may give information concerning the microbial adjustment to nutritional conditions. Microbial activities related to microbial biomass are used for evaluating environmental conditions, for example the metabolic quotient, which is the ratio between CO_2 production and microbial C content (Anderson and Domsch, 1993). Finally, soil microbial activities of C and N cycles should be related to soil C and N stocks, providing information concerning transformation intensity in labile pools by looking at substrate transformation and product formation.

7.2 Soil Respiration

MIKAEL PELL, JOHN STENSTRÖM AND ULF GRANHALL

Department of Microbiology, Swedish University of Agricultural Sciences, Box 7025, SE-750 07, Uppsala, Sweden

Introduction: Definition of, and Objectives for, Measuring Soil Respiration

Respiration is probably the process most closely associated with life. It is the aerobic or anaerobic energy-yielding process whereby reduced organic or inorganic compounds in the cell serve as primary electron donors and imported oxidized compounds serve as terminal electron acceptors. During the respiration process, the energy-containing compound falls down a redox ladder, commonly consisting of glycolysis, the citric acid cycle (CAC) and, finally, the electron transport chain.

In a less strict sense, respiration can be defined as the uptake of oxygen while, at the same time, carbon dioxide is released. However, in the soil ecosystem CO_2 is also formed by other processes, such as fermentation and abiotic processes, e.g. CO_2 release from carbonate. In addition, several types of anaerobic respiration can take place where, for example, NO_3^- or SO_4^{2-} are used by microorganisms as electron acceptors; hence, O_2 is then not consumed as in aerobic respiration. Thus, when CO_2 or O_2 are used as indices of respiration, they actually represent carbon mineralization or aerobic respiration, respectively.

Basal respiration (BAS) is the steady rate of respiration in soil, which originates from the turnover of organic matter (predominantly native carbon). The rate of BAS reflects both the amount and the quality of the carbon source. BAS may therefore constitute an integrated index of the potential of the soil biota to degrade both indigenous and antropogenically introduced organic substances under given environmental conditions. In the following, respiration, and measurements thereof, refer to BAS unless otherwise stated.

Soil respiration is a key process for carbon flux to the atmosphere. Soil water content, oxygen concentration and the bioavailability of carbon are the main factors that regulate soil respiration. Perhaps the most important regulator is water, since it will dissolve organic carbon as well as oxygen and, by diffusion, control the access rate of these substances to the cell. Hence, water facilitates the availability of organic carbon and energy, while, at the same time, it restricts access to oxygen. Moreover, water will delay the exchange of CO_2 between the soil surface and the atmosphere. The diffusion constraints often obstruct the interpretation of the results in terms of enzyme kinetics. The optimal water content in soil for respiration is thought to be 50–70% of the soil's water-holding capacity (Orchard and Cook, 1983).

Soil respiration is attributed to a wide range of microorganisms, such as fungi, bacteria, protozoa and algae. Moreover, the soil fauna contributes significantly. Generally, the microbial contribution to the total release of CO_2 (excluding root respiration) is thought to be about 90%, compared to 10% released by the fauna (Paustian *et al.*, 1990). Although fungal biomass often dominates microbial biomass (Hansson *et al.*, 1990; Bardgett and McAlister, 1999) the relation fungi:bacteria with respect to respiration may vary considerably, due to, for example, type of ecosystem or soil management (Persson *et al.*, 1980). To complete the picture, plant roots also contribute between 12% and 30% to the total release of CO_2 through respiration in the field (Buyanovsky *et al.*, 1987; Steen, 1990).

When the carbon mineralization capacity is estimated by means of soil respiration, it is important to consider that the soil may act as a sink for CO_2. Both chemolithotrophic bacteria and phototrophic bacteria and plants fix CO_2 into their biomass. Also, different volumes of O_2 are needed for the mineralization of specific amounts of various carbon sources, i.e. the respiration quotient (RQ) is seldom the often-assumed quotient of 1. Dilly (2001) reported that RQ values for BAS in various soil ecosystems are frequently < 1. Moreover, if the oxygen content of the soil is lowered, mineralization can occur through anaerobic respiration or fermentation, meaning that CO_2 is released without O_2 being consumed.

Principle of Measurements

Methods can be divided into those intended for measuring respiration: (i) in the field; and (ii) in the laboratory.

Measurement of soil respiration in the field is usually accomplished by covering a specific soil surface with a gas-tight chamber. During incubation for a specific time, under ambient climate conditions, changes in gas composition (CO_2 or O_2) are monitored. Alternatively, soil probes can be pushed into the soil and gases withdrawn from a desired depth. Field measurements are the only way to assess the general microbial activity under natural conditions. Hence, field methods give the sum of respiration of all organisms (including roots) under conditions that can seldom be controlled by the investigator and therefore often result in large spatial and temporal variations in gas fluxes.

One way to simplify and standardize the work is to sample a large number of intact soil cores and bring them to the laboratory for incubation under constant temperature and/or moisture content. Establishment of a temperature–response curve can partly transform the results into those encountered under field conditions. However, the pore volume of the soil core is probably reduced by the sampling procedure, and the incubation vessel will generate 'wall effects', resulting in altered gas fluxes in a core incubation device.

Laboratory-based techniques, although usually having less resemblance to natural soil conditions, are easier to handle. Besides allowing the measurement of the basal soil respiration under standardized conditions, such techniques also permit well-designed and controlled experiments to be performed, addressing specific questions (Torstensson and Stenström, 1986). Substrates containing inorganic N and P can be added to eliminate limiting factors other than the carbon source, thereby enabling measurement of the enzymatic capacity to mineralize the intrinsic organic material or some added organic test compound.

In all laboratory-based techniques, proper sample treatment and storage are essential for accurate results. Besides the recommendations given in ISO 10381–6 (1993), the soil samples should be transported from the field to the laboratory as quickly as possible (within hours), or be stored in refrigerated containers. At the laboratory, the moist samples should be sieved through a screen with 2–5 mm mesh width. Sometimes the soil must be partially dried, at a constant temperature of +2°C to +4°C, before sieving is possible. Soils from northern countries (Scandinavia), at least, can be stored at −20°C for up to 12 months without affecting the activity (Stenberg *et al.*, 1998a). Before performing an analysis of respiration, the soil should be pre-incubated for at least 1 week, to allow the initial carbon flush to diminish. When handling a soil, it is important to know that all kinds of soil disturbances, such as agitation and cycles of drying–wetting and freezing–thawing, will result in bursts of CO_2.

Whether measurements of soil respiration are based on the analyses of consumption of O_2 or the production of CO_2, two methods could be used (Table 7.1).

1. Static methods, where the gases are collected within a closed incubation system containing the soil, or an incubation chamber placed over the soil surface.
2. Dynamic systems, where CO_2-free air flows continuously through the incubation system and the gas composition is analysed at the outlet.

More complete lists of principles and examples of performances can be found in Stotzky (1965), Anderson (1982), Bringmark and Bringmark (1993), Zibilske (1994), Alef (1995) and Öhlinger *et al.* (1996). International standards for tests of soil quality by respiration are suggested in ISO 16072 (2001) and ISO 17155 (2001).

We have chosen to focus on two static methods, both suitable for routine measurements of large numbers of soil samples. Both methods can be used in agricultural soils (Hadas *et al.*, 1998; Yakovchenko *et al.*, 1998; Goyal *et al.*, 1999; Svensson, 1999; Stenström *et al.*, 2001) as well as the mor layer of forest soils (Palmborg and Nordgren, 1993; Bringmark and Bringmark, 2001a,b).

The first method, 'Basal respiration by titration', is simple and can be performed by most soil laboratories. The method can be modified easily for determination of biomass or degradation capacity of organic ^{14}C compounds. Several variations of this old method exist and the method

Table 7.1. Determination of CO_2 and O_2 in static and dynamic set-ups for the measurement of soil respiration.

Gas	Sample	Principle
CO_2	Absorption in alkali	Gain of weight Titration of remaining OH^- Decrease in conductivity
CO_2	Headspace concentration	IR absorption Gas chromatography
O_2	Headspace concentration	Gas chromatography Electrode Change in partial pressure or gas volume (only static system)

IR, infrared.

described below, based on experience from our lab, is a modification of methods described by Bringmark and Bringmark (1993), Zibilske (1994) and Öhlinger *et al.* (1996).

The second method, 'Basal respiration, substrate-responsive soil microbial biomass, and its active and dormant part', requires special equipment, but is ideal for detailed studies of kinetics and soil microbial subpopulations. This method quantifies:

1. The basal respiration rate (BAS);
2. The substrate-induced respiration rate (SIR); and
3. The distribution between active (r) and dormant (K) microorganisms in the substrate-responsive biomass.

Computerized equipment that allows frequent and automated measurement and storage of data on CO_2 production is needed. The method is based on experience from our laboratory and has been used to study the kinetics of the reversible r ↔ K transition (Stenström *et al.*, 2001) and how the biomass and the r/K distribution are affected by the concentration and contact time of silver in the soil (Johansson *et al.*, 1998), and by the soil water content and the antibiotic cycloheximide (Stenström *et al.*, 1998).

Basal Respiration by Titration

Principles of the method

Carbon dioxide produced from soil is trapped in a sodium hydroxide solution according to the formula:

$$CO_2 + 2NaOH \rightarrow Na_2CO_3 + H_2O$$

As long as the alkaline solution contains a large excess of OH^-, the chemical reaction is forced to the right as CO_2 is dissolved. At the end of the incubation the non-consumed OH^- is titrated with an acid, e.g. hydrochloric acid (HCl).

Materials and apparatus

- Standard laboratory equipment
- The incubation vessel should have a wide opening and hold at least 500 ml. The vessel should be sealable in a rapid and reliable manner. Glass jars for food preserving with a rubber gasket and metal mount are suitable
- Small plastic cups, one (50 ml) with a perforated lid to hold the soil sample and the other (e.g. a scintillation vial) to hold the NaOH solution
- Burette, magnetic stirrer and magnetic stirring bar (4 × 14 mm)

Auto-titration equipment will assist the analysis; for example, in our lab this includes:

- Autoburette ABU 80
- Titrator TTT 60
- Standard pH meter PHM82
- Combined pH electrode E16M306

(supplied by Radiometer, Copenhagen, Denmark).

Chemicals and solutions

- Freshly prepared sodium hydroxide (NaOH) solutions (1 M and 0.1 M)
- Diluted hydrochloric acid (HCl) solution (0.05 M)
- Barium chloride ($BaCl_2$) solution (0.05 M)
- Autoburette or phenolphthalein indicator solution (0.1 g/100 ml 60% (v/v) ethanol)

All chemicals should be prepared using CO_2-free distilled water. Boil the water and, after some cooling, close the flasks with stoppers. Concentrates for preparation of 0.1 M NaOH and 0.05 M HCl can be bought from commercial manufacturers. The concentrations of NaOH should be adjusted according to the rate of BAS and duration of incubation.

Procedure

- 40 g of soil with a water-holding capacity (WHC) of 50–70% is weighed out into a plastic cup. Close the perforated lid and place the sample in a humid place at constant room temperature (see below).
- Pre-incubate for 10 days to let the initial flush of CO_2 cease. During the pre-incubation, loss of water from the soil must be compensated for.
- Alternatively, fill a scintillation vial with 2 ml 1 M NaOH and place the opened sample cup and scintillation vial in the incubation vessel on a filter paper moistened with CO_2-free water. Close the vessel and pre-incubate for 10 days at constant room temperature of 22°C.

- At the start of the basal respiration measurement, open the incubation vessel and replace the scintillation vial with a new vial with 2 ml of fresh 0.1 M NaOH. Make sure that the filter paper is moistened. Close the vessel and incubate for 24 h at 22°C.
- Remove the absorption cup and add 4 ml 0.05 M $BaCl_2$ to precipitate carbonate as $BaCO_3$. Titrate the remaining OH^- to a pH of 8.30 with 0.1 M HCl, using an autoburette; or add three to four drops of phenolphthalein and titrate with 0.05 M HCl to the endpoint of the indicator.
- If respiration measurements are to be continued, repeat the BAS measurement and titration steps above until the study period is finished. At least triplicate samples should be measured. Three incubation vessels without soil should be used as blanks.
- Dispensers for dosing NaOH will increase the start-up speed. A high degree of neutralization of the OH^- solution by CO_2 will result in less reliable results and should therefore be avoided. Continuous incubation for several days can be achieved if additional vials with NaOH are placed in the incubation vial.
- After establishment of the basal respiration, SIR could be determined after the addition of a SIR substrate containing glucose (Alvarez and Alvarez, 2000; Section 6.3, this volume).
- The degradation capacity of a ^{14}C-labelled substance can be determined by amending the soil with this substance (Torstensson and Stenström, 1986). A second scintillation vial with NaOH is then placed into the incubation vessel. After incubation, this vial is analysed for labelled $[^{14}C]CO_2$ using a liquid scintillation counter. Soil without added substance is used as a control.

Calculations

The basal respiration rate (BAS) in units of µg CO_2-C/g DW per hour can be calculated from the formula:

$$BAS = \frac{M_C \times (V_b - V_s) \times 0.05}{S_{dw} \times t \times 2} \times 10^3 \qquad (7.1)$$

where M_C is the molar weight of carbon (Mw = 12.01); V_b and V_s the volume, in ml, of 0.05 M HCl consumed in the titration of the blanks (mean of three replicates) and the sample, respectively; S_{dw} is gram dry weight of soil; and t is the incubation time in hours. Since two OH^- are consumed per CO_2 precipitated, a factor of 2 must be included in the formula. If other concentrations of HCl are used, the formula must be adjusted.

Discussion

Due to the many and diverse groups of respiring microorganisms in soil, even a small change in basal activity must be considered as a serious effect.

Soil respiration is therefore a major component of the soil quality concept (cf. ISO TC 190) and is included in monitoring programmes to assess soil quality (Torstensson *et al.*, 1998), as well as responses to various natural or anthropogenic-induced influences (Brookes, 1995; Franzluebbers *et al.*, 1995; Pankhurst *et al.*, 1995; Yakovchenko *et al.*, 1996; Stenberg *et al.*, 1998b; Benedetti, 2001; Svensson and Pell, 2001).

With regard to soil quality, standard microbial respiration (CO_2 evolved) is included as one of only two indicators suggested for assessing 'soil life and soil biodiversity' (Benedetti, 2001). It is also included in the ISO TC 190 'Soil Quality' parameters necessary to consider regarding soil restoration. Differences between temperate and tropical soils concerning the energy content of their organic matter pool have also been studied by COST members using basal respiration (Grisi *et al.*, 1998). The energetic efficiency of the microbial communities using respiration quotients has been discussed largely by Dilly and co-workers (Dilly, 2001).

Besides being a generally accepted measure of total soil organism activity, basal respiration may, with certain modifications, give additional information regarding environmental impact. For example, small increases in basal respiration that follow small additions of easily metabolizable C compounds can be used to study the resulting priming effects on overall C-mineralization (Anderson and Domsch, 1978; Dalenberg and Jager, 1981, 1989; Falchini *et al.*, 2001). Secondly, microbial-specific respiration (an indicator of stress) has been used particularly for monitoring metal-contaminated soils (Brookes and McGrath, 1984; Chander and Brookes, 1991; Bringmark and Bringmark, 2001a,b; Pasqual *et al.*, 2001).

Basal Respiration, Substrate-responsive Microbial Biomass and its Active and Dormant Part

Principles of the method

Carbon dioxide produced from soil in a closed vessel is trapped in a potassium hydroxide solution. This results in a proportional decrease in its electrical conductivity, which is measured with platinum electrodes in each incubation vessel.

Apparatus and materials

The Respicond respirometer (Nordgren Innovations AB, Umeå, Sweden) is fully computerized and suitable for the measurement of 96 samples at 30-min intervals. Since the conductometric measurements are very temperature sensitive, the respirometer should be installed in a room at constant temperature ($22 \pm 0.02°C$).

Chemicals and solutions

- Potassium hydroxide (KOH) solution (0.2 M), prepared using CO_2-free (boiled) water
- *C:N:P substrate*: a mixture of 7.5 g glucose, 1.13 g $NH_4(SO_4)_2$ and 0.35 g KH_2PO_4 is thoroughly ground and mixed in a mortar and thereafter mixed into 10 g of talcum powder. This substrate can be stored at room temperature

Procedure

- 10 ml of the 0.2 M KOH solution is added to the conductometric cup of the incubation vessel.
- Portions of 20 g (dry weight) of soil, with moisture content of 60% WHC, are weighed into the 250 ml incubation vessels. The vessels are closed and installed in the water bath.
- The programme is started with measurements of accumulated CO_2 twice every hour for each vessel.
- Include at least three vessels without soil in the measurement.
- When the initial flush of CO_2 has ceased, continue the measurement for two more days.
- Replace the KOH solution in the cup of each vessel with 10 ml of a new solution if more than 10% of the KOH has been consumed in any vessel (i.e. if more than 4.4 mg of CO_2 has been produced).
- Thoroughly mix 0.19 g of the substrate into each soil sample. Record the time at which the substrate is added to each vessel. Continue the measurements until the maximum respiration rate has been obtained in all soil samples.

Calculations

Export data on CO_2 accumulation and rate of CO_2 production to a spreadsheet program.

Transform the CO_2 data (in mg) to CO_2-C (in µg)/g dry weight (DW) of soil per hour by multiplying each value by $0.2728 \times 1000/DW = 13.64$.

Due to the disturbance in temperature obtained during addition of the substrate, some initial erratic data have to be removed from the data set (Fig. 7.1). Data belonging to this disturbed period are identified by plotting the rate data for the empty vessels against t.

Calculate the basal respiration rate (BAS), i.e. the slope by linear regression of accumulated CO_2 data against t for the 48 h before substrate addition.

Upon addition of a saturating amount of glucose to a soil sample, active and dormant microorganisms in the soil behave differently. The active

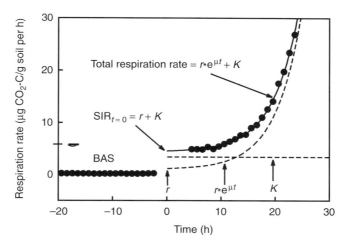

Fig. 7.1. Respiration curve divided into the two phases, basal respiration (BAS) obtained before substrate addition, and the total respiration rate after substrate addition. Substrate-induced respiration (SIR) is the sum of respiration rate of active (r) and dormant (K) soil microorganisms triggered by substrate addition at $t = 0$.

organisms (r) immediately start to grow exponentially, while the dormant ones (K) respond by initiating a constant production rate of CO_2 (Stenström et al., 2001). Thus,

$$\frac{dp}{dt} = re^{\mu t} + K \qquad (7.2)$$

where dp/dt is the total rate of CO_2 formation and μ is the specific growth rate of active organisms. The accumulated amount of CO_2 is obtained by integrating Eq. (7.2):

$$p = p_0 + \frac{r}{\mu}(e^{\mu t} - 1) + Kt \qquad (7.3)$$

where p_0 is a fitting parameter that accounts for the initial conductivity of the KOH solution of each jar at the time of addition of the substrate. The SIR rate (Fig. 7.1) is defined as the respiration rate obtained instantaneously on the addition of the substrate. Thus,

$$SIR = r + K \qquad (7.4)$$

which is obtained by solving Eq. (7.2) for $t = 0$. The value of SIR obtained can be converted to biomass as discussed by Höper (Section 6.3, this volume). Numerical parameter values of Eq. (7.3) are obtained by non-linear regression of accumulated data that belong to the period during which growth is exponential and K is constant. This period is identified as the time from substrate addition until data plotted according to $\ln(dp/dt)$ against t start to fall below the straight line so obtained.

Discussion

This method refers directly to the objectives of the COST Action 831, as follows:

- to improve the effectiveness of microbial and molecular methods so to improve monitoring, conservation and remediation of soil;
- to use new microbial parameters as better indicators of environmental impact;
- to help early detection of any fertility decline of natural ecosystems by the setting up of efficient and rapid methods of soil pollution diagnosis.

The arguments for use of this method are:

- Standardized (basal) respiration measurements are relatively insensitive means for monitoring environmental impact or soil health (Brookes, 1995).
- Microbial and respiratory quotients may be more useful to determine trends with time and to compare soils than measurements of basal or substrate-induced respiration only; but if the quality of the carbon substrates differs greatly, or if soils are of different types, the interpretation of such quotients becomes very difficult (Sparling, 1997; Granhall, 1999).
- The method described here, being more sensitive and accurate, could help in the early detection of soil pollution and its following remediation, as it describes the basic respiration rate, the total glucose-responsive biomass, and the sizes of the active (r) and the dormant (K) microbial populations.

7.3 Soil Nitrogen Mineralization

STEFANO CANALI AND ANNA BENEDETTI

Consiglio per la ricerca e la sperimentazione in Agricoltura, Istituto Sperimentale per la Nutrizione delle Piante, Via della Navicella, 2, 00184 Rome, Italy

Introduction

The process of nitrogen (N) mineralization in the soil can be defined as the conversion of organic N into mineral forms available to plants, which takes place through the biochemical transformation mediated by microorganisms (Stevenson, 1985).

The first step (ammonification) involves the conversion of organic N into NH_4^+ and is performed exclusively by heterotrophic microorganisms, able to operate in both aerobic and anaerobic conditions. The second step (conversion of NH_4^+ into NO_3^-), defined as nitrification, occurs mainly through the activity of two groups of autotrophic aerobic bacteria: *Nitrosomonas* (from NH_4^+ into NO_2^-) and *Nitrobacter* (from NO_2^- into NO_3^-). In cultivated, well-aerated soils, nitrate is the predominant available mineral form of the element. On the other hand, if low O_2 concentrations are present in the soil (i.e. waterlogged conditions), NH_4^+ accumulates (Stevenson, 1985).

Nearly always, the mineralization process is accompanied by the immobilization of N, due to the activities of the soil living biomass and, since the two gross processes take place simultaneously, the increase of mineral N concentration at a defined time indicates the net mineralization (Powlson and Barraclough, 1993).

The possible use of soil N mineralization as an index of soil quality is relevant because of the relation of this process with the capacity of the soil to supply N for crop growth, and also because of the risk of water and atmospheric pollution. Thus, N mineralization is often included in *minimum data sets* set up to evaluate the capacity of a soil to operate within the boundaries of the ecosystem: to promote biological productivity, to maintain environmental quality and safeguard the health of plants and animals (Doran and Parkin, 1994).

The mineralization process is driven by factors able to influence the microbial activity in the soil. The most relevant factors are soil temperature, moisture, pH, O_2 concentration, energy and other nutritive elements available for the accumulation of produced mineral N in the system. If, during the N mineralization measurement, these factors are not controlled, the values may vary within the range and the measurements are performed under 'current conditions'. On the other hand, if one (or more) factor that drives the microbial mineralization process is controlled and set at its optimal value, the measurements are carried out under 'potential conditions'– these are used in laboratory incubation methods.

According to the considerations outlined above, methods for the evaluation of N mineralization in the soil can be divided into those for the evaluation of the gross or the net mineralization activity, and these procedures are classified into two groups: methods for measuring the actual mineralization (field condition) and methods for the evaluation of potential mineralization. The aim of this section is to review the most widespread methods used to measure N mineralization in agricultural soils, describing procedures and discussing the advantages and limitations of their application.

Methods for measuring the actual mineralization activity (field conditions)

Net mineralization measurements

Net nitrogen mineralization can be evaluated by the application of the N balance equation to a defined soil–plant system (Powlson and Barraclough, 1993):

$$M = \Delta NH_4^+ + \Delta NO_3^- + \Delta Plant + loss$$

where ΔNH_4^+ and ΔNO_3^- represent the differences of ammonia and nitrate concentrations at the end and at the beginning of a defined period of time. 'ΔPlant' is the amount of nitrogen uptake by the crop and 'loss' is the N leaving the system via leaching and gaseous emission (ammonia volatilization, denitrification and nitrogen oxide evolution during the nitrification) in the same period.

Although direct measurement of N uptake by annual herbaceous crops is relatively easy, great difficulties arise in the case of perennial tree crops (e.g. fruit crops). In any case, it is probably true that the most important limitation of this procedure is that the measurement of losses under field conditions is an extremely difficult task and, for this reason, the nitrogen balance is often applied without taking into account all the losses in the system in question, which means that the results obtained are not accurate.

An indirect approach to the evaluation of the net rate of N mineralization by applying the N balance consists of the use of some soil contents that prevent the N uptake on the part of the plant and reduce the losses in the system (Hart et al., 1994). These methods employ closed-top solid cylinders or buried bags, carried out in intact soil cores to reduce soil disturbances.

The main advantage of this procedure is that, even when it is performed in a simplified way (no account for crop uptake and losses), it supplies a rough indication of the actual net mineralization, which is useful for the evaluation of the impact of the process on field conditions in the medium or long term.

Gross mineralization measurements: the isotopic dilution technique

Techniques using ^{15}N label and employing the pool dilution principle are based on the measurement of the change in ^{15}N abundance of a labelled ammonium soil pool receiving nitrogen at a natural abundance through mineralization (Powlson and Barraclough, 1993). The measured changes are then described according to a specific set of equations (Kirkham and Bartholomew, 1954; Barraclough *et al.*, 1985; Nishio *et al.*, 1985; Mary *et al.*, 1998) capable of quantifying the size of the soil N pools and the rate of the gross processes (i.e. mineralization, immobilization, nitrification). This method has proven very useful in understanding how soil and crop management affect the N turnover, *in situ* or in intact soil cores (Recous *et al.*, 2001).

In order to obtain useful quantitative estimates, the application of the technique requires four basic assumptions.

- Ammonium-consuming processes (plant uptake and nitrification) do not discriminate between ^{14}N and ^{15}N.
- Added label mixes with native soil ammonium, such that labelled and unlabelled N are used in proportion to the relative amounts present in the system.
- Over the experimental period, all rate processes can be described by a zero-order equation.
- Labelled nitrogen immobilized over the experimental period is not re-mineralized.

These assumptions have been verified exclusively for a short-term period, and the procedure supplies reliable information only for instantaneous rates of nitrogen mineralization.

The limited popularity of this technique is probably due to the (relatively) high costs of the equipment required to perform ^{15}N analyses, and because of the unwieldy formulation of the equations used to describe the processes quantitatively.

Methods for measuring the potential mineralization activity (laboratory incubation methods)

These methods involve the laboratory incubation of the soil for a defined period of time and in conditions that promote the N mineralization from organic sources. The N mineral produced is then measured.

The methods are generally classified on the basis of the length of the incubation period and the incubation conditions. In this section we shall discuss short-term static incubation methods (aerobic and anaerobic) and a long-term aerobic dynamic method.

All laboratory incubation procedures considered measure the net nitrogen mineralization and, since at least the temperature, which is one of the most relevant factors affecting the process, is set to an optimal value, the N mineralization activity is measured under potential conditions. Consequently, the results obtained must be interpreted as a relative indication of the process rate.

Short-term static incubation procedures

The static aerobic incubation procedure was described first by Bremner (1965). The soil sample is mixed with a defined amount of washed sand and the mixture is moistened with water and incubated at 30°C for 15 days. At the end of the incubation, NH_4^+-N, NO_3^--N and NO_2^--N are extracted by means of a concentrated saline solution (KCl) and then evaluated. According to this procedure, the two main factors affecting the mineralization (water contents of the system and the temperature) are set at defined optimal values and the presence of sand should allow adequate O_2 availability.

This is a widespread and well-known method and, in accordance with its general principles, many minor modifications have been proposed by different authors in order to reduce the incubation time and to facilitate laboratory procedures.

Waring and Bremner (1964) and then Keeney (1982) proposed the short-term static anaerobic incubation procedure. The rate of the N mineralization process is determined by measuring the quantity of NH_4^+-N produced during 7 days of waterlogged incubation at a temperature of 40°C.

In the above-mentioned incubation conditions, mineralization of organic N is performed only by microorganisms that are able to operate under anaerobic conditions. This method has been used to study nitrogen transformation in paddy soils and, according to this procedure, Sahrawat and Ponnamperuma (1978) and Sahrawat (1983) measured the net mineralization process in tropical rice soils.

This method can also be applied to agricultural aerated soils. In fact, even in this type of soil, waterlogged conditions can occur for a limited period of time (rainy periods) and anaerobic conditions can be found at any time inside soil aggregates. Indeed, Keeney and Nelson (1982), Stanford (1982) and Meisinger (1984) found satisfactory relationships between the results obtained through the anaerobic and the aerobic procedures. Our hypothesis is that even in the Mediterranean soil of southern Italy, characterized by extreme conditions in which waterlogged situations hardly ever occur, the anaerobic and the aerobic procedures for measuring the net N mineralization process supply substantially the same indicative results.

This method, as compared to the aerobic static procedure, has several advantages that make it an attractive solution when a rapid procedure is needed to supply a relative assessment of N mineralization. These advantages include: (i) simplicity and ease of adaptation in laboratory routines; (ii) a short incubation time (7 days); (iii) a low impact of sample pretreatment on test results; (iv) the elimination of problems related to optimum water contents and water loss during the incubation; and (v) the requirement of only a few machines and reagents (Bremner, 1965; Keeney, 1982; Lober and Reeder, 1993).

The long-term dynamic incubation procedure

This procedure was initially proposed by Stanford and Smith (1972). It is based on aerobic long-term incubation at 35°C, and, within the incubation period, the mineral nitrogen produced (NH_4^+-N, NO_3^--N and NO_2^--N) is leached at predetermined intervals (after 2, 4, 8, 16, 22 and 30 weeks). N mineralization is described by fitting the cumulative experimental data of mineralized N according to a first-order kinetic model.

Soil samples are incubated in a combined filtration–incubation container. Full-potential conditions are imposed by mixing soil with washed quartz sand (1:1 p/p) to allow for an adequate O_2 availability in the system. Water is added to reach the optimum pH value. Mineral nitrogen is removed by leaching to reduce the re-immobilization of mineralized nitrogen and/or to avoid feedback effects on the mineralization process. In order to prevent any limiting effects due to the absence (or reduced concentration) of other elements, after each leaching a nutritive solution, minus N, is applied to the soil.

The long-term incubation should allow problems linked to the influence of pre-treatment of the samples in the N mineralization process to be overcome; for this purpose, Stanford and Smith (1972) suggested that the results obtained during the first 2 weeks of incubation should be ignored. Benedetti *et al.* (1994, 1996) suggested that the results obtained during the first 2–3 weeks of incubation give an indication of the size of the labile mineralizable N pool and/or the biomass nitrogen pool of soils.

Furthermore, after the first weeks of incubation, depending on the specific characteristics of the different soil types, the system generally reaches a steady-state condition. When this new balance is observed, it is possible to obtain indications regarding basal N mineralization (MN_{Bas}). Thus, it is possible to use basal N mineralization to calculate the N mineralization coefficient (MN_{Bas}/N_{tot}) and the N mineralization quotient ($Q_N = MN_{Bas}/N_{bio}$).

When the first-order kinetic model is applied in order to describe the experimental results of potentially mineralizable nitrogen (N_0), and the rate constant (k) is calculated (Stanford and Smith, 1972), the first parameter (N_0) is considered to be a nitrogen availability index and, together with the k constant (the dimension of which is $1/t$), it is capable of describing the long-term N fertility status of the soils.

Benedetti and Sebastiani (1996) determined N_0 and k by the Stanford and Smith method for a number of Italian soils. These authors also confirmed that significantly less reliable estimates of potentially mineralizable N were obtained by using data for up to only 22 weeks of incubation.

Despite the advantages described above, this method is characterized by several disadvantages which are concerned with the substantial time and apparatus requirements (Bundy and Meisinger, 1994). For this reason, this method is usually used only when reliable and long-term N mineralization information (i.e. basal N mineralization) is required.

Conclusions

Actual net and gross N mineralization rate (field conditions) should be determined when the aim of the study is to evaluate the absolute rate of the process (i.e. the capability of soils or organic fertilizers to supply N to the crop).

Methods to assess the potential mineralization (laboratory methods) may be used when performing a comparison of different soil samples or applying several methods for evaluating the same soil.

The anaerobic static incubation procedure is strongly recommended when a quick, work-saving, inexpensive procedure (i.e. field survey approach on a large, regional scale) is needed. We do not yet have confirmation about the potential correlation between the results obtained with the aerobic and the anaerobic static procedure for all climatic and pedological situations.

The aerobic long-term dynamic procedure (Stanford and Smith method) can be applied whenever reliable and exhaustive quantitative information about net N mineralization in potential conditions is necessary.

Anaerobic N Mineralization

Principle of the method

This method is based on incubation of waterlogged soil for 7 days at 40°C. At the end of the incubation, accumulated ammonium is measured.

Materials and apparatus

- Incubator at 40°C
- Shaking apparatus
- Filter paper (N-free)
- Glass or plastic tubes (50 ml) with rubber stoppers and screw caps
- Glass Erlenmeyer flasks (300 ml)
- Plastic bottles (100 ml)

Chemicals and solution

- *4 M KCl solution*: dissolve 298 g of KCl in 750 ml of distilled water in a 1000 ml glass flask; bring up to volume with distilled water

Procedure

- Put 40 ml of distilled water into the tube and add 16 g of soil (three replicates).
- Close the tube by the rubber stopper and the cap, and then shake manually until the soil is completely suspended.
- Incubate for 7 days at 40°C. During the incubation, re-suspend the soil daily by manual shaking.
- After the incubation, transfer the soil–water suspension into the 300 ml Erlenmeyer flask.
- Wash the incubation tube with 10 ml of the 4 N KCl solution and transfer the obtained suspension into the Erlenmeyer flask; repeat the same procedure three times, in order to use 40 ml of the 4 N KCl solution in all. No soil particles should remain in the incubation tube.
- Shake the Erlenmeyer flask containing the soil suspension for 1 h and then filter the supernatant through a filter paper into the plastic 100 ml bottle.
- Use the clear filtrate for ammonium determination, according to the chosen method.
- Samples that are not to be incubated are prepared by putting 16 g of the same soil + 40 ml of distilled water + 40 ml of the 4 N KCl solution into the 300 ml Erlenmeyer flask. Shake and filter according to the procedure reported above.

Calculation

- Mineralized nitrogen during 7 days of incubation is calculated by subtracting the ammonium measured (μg NH_4^+-N/g soil) in the sample that was not incubated from that measured in the incubated sample.
- In order to verify that anaerobic conditions occurred during incubation, the presence of nitrate and nitrite should be assessed – only traces of NO_3^--N and NO_2^--N should be found.

Long-term Aerobic N Mineralization

Principle of the method

The method is based on the aerobic incubation of soil (mixed with quartz sand) in optimal conditions of moisture and temperature for soil microbial activity. Mineralized nitrogen is removed periodically by leaching, and then determined at fixed times for 30 weeks.

Materials and apparatus

- Ceramic Buchner funnel (outer diameter 55 mm)
- Thermostatic system

Chemicals and solutions

- Quartz sand (granulometry: 0.2–0.8 mm)
- Glass-fibre diskettes (diameter chosen according to the Buchner funnel used)
- *Leaching solution*: 0.01 M calcium chloride ($CaCl_2$)
- *Nutritive* N-minus *solution*: 0.002 M $CaSO_4 \cdot 2H_2O$; 0.002 M $MgSO_4$; 0.005 M $Ca(HPO_4)_2 \cdot H_2O$; 0.0025 M K_2SO_4

Procedure

- Mix 10 g of soil (air-dried and sieved at 2 mm) with quartz sand in the ratio 1:1, and put the mixture in a ceramic Buchner funnel (outer diameter 55 mm), on the bottom of which is placed a glass-fibre diskette (three replicates).
- Remove the mineral nitrogen (present in the soil at the beginning of the experiment) by leaching the system 'soil + quartz sand' with 180 ml of 0.01 M $CaSO_4$, followed by the addition of 20 ml of the nutritive *N-minus* solution. Remove excess water with a vacuum system (60 cmHg). Incubate at 30°C. Gaseous exchange through the opened portion of the funnels guarantees the maintenance of aerobic conditions.
- After 2 weeks, remove the mineralized nitrogen by leaching again with 100 ml of 0.01 M $CaSO_4$ and then add 20 ml of the nutritive *'N-minus'* solution, followed by the vacuum application as described above. Incubate again, performing a series of leachings after 2, 4, 8, 12, 16, 22 and 30 weeks. In order to avoid soil dryness during the incubation period, add distilled water to the mixture of soil + quartz sand, until reaching the optimal value of soil moisture.
- Determine NH_4^+-N, NO_3^--N and NO_2^--N in the clear leached solution according to the chosen method.

Calculation

The total mineralized nitrogen is calculated as the sum of NH_4^+-N, NO_3^--N and NO_2^--N content after each leaching.

N_0 (potentially mineralized nitrogen) could be calculated by fitting the cumulative experimental data by the first-order equation:

$$N_t = N_0 (1 - e^{-kt})$$

where: N_t = mineralized nitrogen (cumulated value, in mg N/kg soil); N_0 = potentially mineralizable nitrogen (mg N/kg soil); k = kinetic constant (in 1/week); and t = time (in weeks).

7.4 Nitrification in Soil

ANNETTE BOLLMANN

University of Aarhus, Department of Microbial Ecology, Ny Munkegade Bygning 540, 8000 Aarhus C, Denmark; Present address: *Northeastern University, Department of Biology, 360 Huntington Avenue, 134 Mugar Life Science Building, Boston, MA 02115, USA*

State of the Art

Ammonium (NH_4^+) plays a central role in the terrestrial nitrogen cycle. It is produced by mineralization of organic matter or by nitrogen (N_2) fixation by nitrogen-fixing bacteria. Consumption of NH_4^+ occurs through assimilation by bacteria and plants, and conversion to nitrite and nitrate by nitrification. Nitrification is the oxidation of ammonia (NH_3) to nitrate (NO_3^-) via nitrite (NO_2^-), carried out by two separate groups of specialized bacteria. The ammonia-oxidizing bacteria are obligate aerobic, chemolithoautotrophic bacteria with a pH optimum around 7–8, which convert ammonia into nitrite. They belong to the β-subgroup of the *Proteobacteria*. The nitrite-oxidizing bacteria, catalysing the conversion of nitrite to nitrate, are physiologically and phylogenetically more diverse than the ammonia-oxidizing bacteria. Some are capable of mixotrophic or heterotrophic growth and dissimilatory nitrate reduction, and they belong to different classes of the *Proteobacteria*. These bacteria live in soils, sediments, fresh and marine water, and wastewater treatment systems (Prosser, 1989; Koops and Pommerening-Röser, 2001; Kowalchuk and Stephen, 2001).

There are different ways to determine nitrification: measurement of gross nitrification rates by ^{15}N methods, measurement of net nitrification rates by determining the nitrate accumulation, and measurement of the potential nitrification by measuring rates under optimal conditions. The available methods are reviewed by Schinner *et al.* (1992), Mosier and Schimel (1993), Hart *et al.* (1994) and Schmidt and Belser (1994).

Gross nitrification

^{15}N tracer studies

The substrate pool (NH_4^+) is labelled to follow the fate of the label in the system (Myrold and Tiedje, 1986). This method has been used to determine gross nitrification, but some problems have become evident. The *in situ* ammonium concentration is changed by the addition of the labelled substrate and/or by the *in situ* production of ammonium by other processes. In addition, the product of nitrification (nitrate/nitrite) can be removed by assimilation or dissimilation processes.

^{15}N dilution method

The product pool (NO_3^-) is labelled and, by following the dilution of the ^{15}N NO_3^- pool, the gross nitrification rate can be calculated (Kirkham and Bartholomew, 1954). Good detailed descriptions of this highly recommended method to determine gross nitrification rates are available in Davidson *et al.* (1991), Mosier and Schimel (1993) and Hart *et al.* (1994).

Net nitrification

The measurement of net nitrification in the laboratory and in the field is based on the determination of the accumulation of NO_3^- over time (Beck, 1979; Hart and Binkley, 1985; Davidson *et al.*, 1990). Measuring the accumulation of NO_3^- does not give a good indication about the real size of the nitrification process, because it does not take into account the immobilized NO_3^- fraction, the denitrified NO_3^- fraction and the NO_3^- fraction converted back to NH_4^+. Additionally, the long incubation periods of up to 3 weeks can cause changes in the populations of the ammonia- and nitrite-oxidizing bacteria.

Nitrification potential

The nitrification potential is a method to determine the maximum nitrifying activity under optimal conditions. This method will be described in detail.

Description of the ammonia oxidizer population

Molecular fingerprinting methods, such as denaturing gradient gel electrophoresis (DGGE) and restriction fragment analysis, are available to describe communities of ammonia-oxidizing bacteria, based on the 16S rRNA gene or *amo*A gene (Kowalchuk *et al.*, 1997; Rotthauwe *et al.*, 1997; Stephen *et al.*, 1999; Nicolaisen and Ramsing, 2002).

Principle of the Method

The nitrification potential is a method to measure the maximal nitrifying activity under optimized conditions (Belser, 1979; Hart *et al.*, 1994). By keeping the basic requirements of the ammonia-oxidizing bacteria (NH_4^+, O_2, pH and water) within the optimal range, effects of toxic substances, fertilization or management processes can be determined. Changes in the populations of the ammonia- or nitrite-oxidizing bacteria will not have a significant influence, because potential rates can be determined in short-term experiments (within 24 h). This method, or very similar methods, has been used to determine the nitrification potential in paddy soils (Bodelier *et al.*, 2000), in agricultural soils (Mader, 1994; Stoyan *et al.*, 2000) and in sediments (Bodelier *et al.*, 1996).

Materials and Apparatus

- 250 ml Erlenmeyer flasks, closed with aluminium foil to reduce evaporation
- Shaker
- Eppendorf pipettes and pipette tips with cut-off tips, to prevent blocking of the tip with soil or sediment for sampling
- 1.5 ml Eppendorf cups and Eppendorf cup centrifuge with cooling
- HPLC (high-pressure liquid chromatograph), IC (ion chromatograph) or autoanalyser to determine nitrite and nitrate
- In case no HPLC, IC or autoanalyser is available: photometer and other equipment to determine nitrite and nitrate photometrically (Keeney and Nelson, 1982)

Chemicals and Solutions

- *0.1 M potassium dihydrogen phosphate solution*: dissolve 13.61 g KH_2PO_4 in 1 l of water
- *0.1 M dipotassium hydrogen phosphate solution*: dissolve 17.42 g K_2HPO_4 in 1 l water
- *Ammonium stock solution*: 1 M NH_{4+}: dissolve 66.07 g ammonium sulphate $(NH_4)_2SO_4$ in 1 l of water
- Sodium hydroxide solution and sulphuric acid to adjust the pH value
- *Reagents for photometric NO_2^- determination*:
  – sulphanilamide solution: add 150 ml of ortho-phosphoric acid carefully to 700 ml water, add 10 g sulphanilamide, stir and warm up a little to dissolve the sulphanilamide, add 0.5 g α-naphthyl ethylene diamine dihydrochloride and top up to 1 l (store in a dark bottle in the fridge)
  – calibration stock solution: 10 mM NO_2^-: dissolve 0.69 g sodium nitrite in 1 l of water
- *Reagents for photometric NO_3^- determination*:
  – 0.5% sodium salicylate solution (prepare fresh): dissolve 0.5 g sodium salicylate in 100 ml water
  – concentrated sulphuric acid (95–97%)
  – seignette salt solution: dissolve 300 g sodium hydroxide in 800 ml water, then add 6 g potassium sodium tartrate and fill up to an end volume of 1 l

– calibration stock solution: 10 mM NO_3^-: dissolve 0.849 g sodium nitrate in 1 l of water

Procedure

Nitrification potential

- Prepare a working solution (1 mM PO_4^{3-}, 5 mM NH_4^+, pH 7.5): add per l, 2 ml of the KH_2PO_4, 8 ml of the K_2HPO_4 and 5 ml NH_4^+ stock solution and adjust the pH value to 7.5.
- Put 10 g soil into a 250 ml Erlenmeyer flask (at least three replicates).
- Determine the gravimetric water content of the soil.
- Add 100 ml of the working solutions to the flasks.
- Put on the shaker at 200 rpm.
- Incubate for 26 h and take samples directly after the start and after 1, 3, 6, 22 and 26 h.
- Take a 1.5 ml sample into an Eppendorf cup, by pipetting 2×0.75 ml with a cut-off 1 ml pipette tip on an Eppendorf pipette; shake samples directly before sampling, because the sample should have the same soil:solution ratio as the slurry.
- Put the sample immediately on ice.
- Spin down for 10 min at $12,000 \times g$ at 4°C.
- Transfer the supernatant to a new Eppendorf cup.
- Store the samples overnight at 4°C or for longer storage at –20°C.
- Dilute the samples and determine nitrite and nitrate.

Determination of nitrite

- Put 2 ml of sample (concentration lower than 50 µM NO_2^-) or calibration solution (0–50 µM NO_2^-) into a test tube.
- Add 0.5 ml sulphanilamide solution and mix well.
- Incubate in the dark for 15 min.
- Measure the extinction at 540 nm against water.

Determination of nitrate

- Put 1 ml of sample (concentration lower than 1 mM NO_3^-) or calibration solution (0–1 mM NO_3^-) into a test tube.
- Add 0.5 ml sodium salicylate solution and mix well.
- Dry the mixture in an oven at 110°C overnight.
- The next steps should be performed carefully in a fumehood:

- add 0.5 ml concentrated sulphuric acid to dissolve the dried solution in the test tube;
- add 4 ml water and mix well;
- add 3 ml seignette salt-solution and mix well;
- incubate for 10 min;
- measure the extinction at 430 nm against water.

Calculations

Calibration

- Plot the extinctions against the NO_2^- or NO_3^- concentrations (µM).
- Determine the slope and the intercept with the Y-axis via linear regression.

Determination of the NO_2^- or NO_3^- concentrations in the sample

Determine the NO_2^- and the NO_3^- concentrations (µM) from the calibration curve:

$$NO_2^- \text{ or } NO_3^- \text{ conc}(\mu M) = \frac{(E_{xs} - E_{xb}) - \text{int}}{\text{slo}}$$

where: E_{xs} = extinction of the sample; E_{xb} = extinction of the blank; int = intercept with the Y-axis calculated from the calibration curve; slo = slope of the calibration curve.

Transform the concentrations from (µM) to (µg N/g DW)

$$NO_2^- \text{ or } NO_3^- \text{ conc}(\mu g\,N/g\,DW) = NO_2^- \text{ or } NO_3^- \text{ conc}(\mu M) \times \frac{\left(\frac{100+(10-DW)}{1000}\right) \times 14}{DW}$$

where: DW = weight of the oven-dry soil (g); 14 = conversion factor (1 M nitrogen = 14 g nitrogen); and

$$\left(\frac{100+(10-DW)}{1000}\right) = \text{conversion factor for the soil solution.}$$

Determination of the potential nitrification rate

- Sum up the NO_2^- and NO_3^- concentrations (= NO_x).
- Plot the NO_x concentrations (µg N/g DW) against the time (h) and determine the potential nitrification rate by calculating the slope via linear regression (µg N/g DW/h).

Note: The potential nitrification rate is the maximum nitrification rate of a sample under optimal conditions without growth of the cells during the incubation period. Under certain circumstances ammonia-oxidizing bacteria can have a lag phase until they become fully active, indicating that they reach maximal activity after several hours of incubation. In other cases they are very active – the ammonium can be completely consumed after several hours of incubation and the NO_x concentration no longer increases. Therefore, it is very important to take several samples during the time course of the incubation, to calculate the potential nitrification rate from the maximal increase of NO_x concentration against time and to cross-check the NO_x concentrations with the added ammonium concentrations.

Discussion

Measurement of the potential nitrification rate gives a good insight into the quality of a soil. Nitrification is a process carried out by highly specialized bacteria, which are very sensitive to changes in the environment, environmental stress, and toxic substances. Therefore, nitrification is a useful microbiological parameter for describing soil quality.

Optimal initial ammonium concentration

There is a strong relationship between the ammonia oxidation (potential nitrification) rate and the initial NH_4^+ concentration. This relationship can differ a lot between different soils (Stark and Firestone, 1996). So it would be useful to conduct a pre-experiment to determine the relationship between NH_4^+ concentration and ammonium oxidation rate for the particular soil. Then the nitrification potential at different initial NH_4^+ concentrations (usually between 0.1 mM NH_4^+ and 5 mM NH_4^+) can be determined and the following experiments can be done with an initial NH_4^+ concentration at which the ammonia oxidation rate is maximal.

Addition of chlorate (ClO_3^-) to inhibit nitrite oxidation

ClO_3^- inhibition of nitrite oxidation is a widely used method to determine potential ammonia oxidation rates (Belser and Mays, 1980; Berg and Rosswall, 1985). ClO_3^- acts as a competitive inhibitor of nitrite oxidation. It is most effective at low nitrite concentrations (Belser and Mays, 1980). The use of ClO_3^- is critical because *Nitrobacter winogradskii* (a nitrite oxidizer) is able to convert ClO_3^- to chlorite (ClO_2^-). ClO_2^- inhibits *Nitrosomonas europaea* (Hynes and Knowles, 1983) and other ammonia-oxidizing bacteria. Therefore, the use of the chlorate inhibition method is questionable.

7.5 Thymidine and Leucine Incorporation to Assess Bacterial Growth Rate

JAAP BLOEM AND POPKO R. BOLHUIS

Department of Soil Sciences, Alterra, PO Box 47, NL-6700 AA Wageningen, The Netherlands

Introduction

Besides biomass and respiration, bacterial growth rate is a key parameter involved in microbial functioning in soil food webs and nutrient cycling (Bloem et al., 1997). Moreover, growth or reproduction is very sensitive to contamination, and is thus a useful indicator of stress. Growth rate cannot easily be calculated from increases in biomass or cell number, because often increases are balanced by losses. Losses may be caused by bacterivores such as protozoa and nematodes, or by viruses. Since about 1980, measurement of [^{3}H]thymidine incorporation during short incubations (about 1 h) has become the method of choice to determine bacterial growth rate in water and sediments (Fuhrman and Azam, 1982; Moriarty, 1986). Thymidine is incorporated into bacterial DNA and thus reflects DNA synthesis or cell division. Since about 1990, the method has also been used for bacteria in soil (Bååth, 1990; Michel and Bloem, 1993). Thymidine incorporation has been found to be more sensitive to contamination than biomass and respiration, in both water and soil (Jones et al., 1984; Bååth, 1992). A plausible explanation for a reduced growth rate in contaminated environments is that microorganisms under stress (e.g. heavy metals or pH) divert energy from growth to cell maintenance (Kilham, 1985; Giller et al., 1998). Physiological processes for detoxification require additional energy. Thus, less energy is available for synthesis of new biomass (growth). This may explain why bacterial growth rate appears to be one of the most sensitive indicators of heavy metal stress in contaminated soils (Bloem and Breure, 2003).

Leucine incorporation into proteins has been introduced as an alternative to the commonly used thymidine method, but can also serve as an independent check (Kirchman et al., 1985). Using [^{3}H]thymidine and [^{14}C]leucine in a dual-label approach, both parameters can be measured in a single assay. Here we describe a protocol for simultaneous measurement of thymidine and leucine incorporation into soil bacteria. Differences between these protocols for thymidine incorporation and those published in previous handbooks (Bååth, 1995; Christensen and Christensen, 1995) are discussed. We describe measurement of thymidine incorporation in a soil slurry. Bacteria may also be first extracted from the soil and further treated like a water sample. This relatively simple, but more indirect, method is especially useful for determining soil bacterial community tolerance to heavy metals and pH, for example (Bååth, 1992). A thorough review of thymidine incorporation has been given by Robarts and Zohary (1993).

Principle of the Method

Thymidine is a precursor of thymine, one of the four bases in the DNA molecule. Leucine is an amino acid which is incorporated in proteins. When sufficient thymidine is added, *de novo* synthesis is inhibited, and labelled thymidine is incorporated. Bacterial growth rate is reflected by the incorporation rate of [^{3}H]thymidine and [^{14}C]leucine into bacterial macromolecules during a short incubation of 1 h. If the incubation period is short enough, growth rate is not affected by incubation. Using a dual-label approach (^{3}H and ^{14}C) both parameters are measured in a single assay.

Thymidine versus leucine

Thymidine incorporation is more proportional to growth rate than leucine incorporation because bacterial DNA content (usually 2–5 fg/cell) is more constant than protein content. Fungi do not incorporate thymidine because they lack the key enzyme thymidine kinase. Growth rates of fungi can be estimated by [^{14}C]acetate incorporation into ergosterol (Bååth, 2001). Only bacteria incorporate thymidine, but not all bacteria are able to do so (e.g. pseudomonads, some anaerobes, nitrifiers and sulphate reducers cannot incorporate thymidine). All bacteria incorporate leucine, but leucine can also be incorporated by other organisms. However, the usual concentration of about 2 µM is probably too low, and the incubation time of 1 h is probably too short, to label cells bigger than most bacteria. Bigger cells have a less suitable surface to volume ratio. Incorporation rates of leucine into proteins are an order of magnitude higher than those of thymidine into DNA, thus lower growth rates can be measured with greater precision. Since both methods have advantages and limitations, their use is complementary.

Materials and Apparatus

- Centrifuge
- Polypropylene centrifuge tubes (13 ml) with screw cap (e.g. Falcon) for incubation of samples
- Incubator at 30°C
- Ice
- Cellulose nitrate membrane filters, 0.2 µm pore size: filters are prewashed with unlabelled thymidine and leucine to minimize adsorption of unincorporated labelled thymidine and leucine

- Filtration manifold to handle larger number of samples (e.g. Millipore, for 12 filters)
- Liquid scintillation counter to measure ^{3}H and ^{14}C

Chemicals and Solutions

Radioisotopes

- Methyl-[^{3}H]thymidine (TRK 300, 1 mCi/ml, 25 Ci/mmol = 925 GBq/mmol, 40 µM) and L-[U-^{14}C]leucine (CFB 67, 0.05 mCi/ml, 0.31 Ci/mmol = 11.5 GBq/mmol) (Amersham Ltd, Amersham, UK)
- The isotopes are stored at 4°C before use; we store them no longer than 3 months
- Per sample (tube) we use: 1.5 µl [^{14}C]leucine, 2.0 µl [^{3}H]thymidine and 16.5 µl unlabelled thymidine (2.35 mg/l). This corresponds to 2 µM and 2.78 kBq (= 0.075 µCi) [^{14}C]leucine and 2 µM and 74 kBq (= 2 µCi) [^{3}H]thymidine per tube

PJ mineral solution

Soil is suspended in a mineral solution (Prescott and James, 1955), made up as follows:

- Make up three stock solutions, each with 100 ml of distilled (or Milli Q) water:
  - stock solution A: 0.433 g $CaCl_2.2H_2O$ and 0.162 g KCl
  - stock solution B: 0.512 g K_2HPO_4
  - stock solution C: 0.280 g $MgSO_4.7H_2O$
- The final dilution consists of 1 ml of each stock solution and 997 ml of distilled water

Solution for extraction of macromolecules

5 ml per tube of:

- 0.3 N NaOH (12 g/l; solubilizes DNA, hydrolyses RNA)
- 25 mM EDTA (9.3 g/l; breaks up aggregates)
- 0.1% SDS (1 g/l; sodium dodecyl sulphate, lyses membranes)

Both SDS and EDTA inhibit nucleases and breakdown of DNA

- 1 N HCl (82.6 ml/l; 1.3 ml/tube)
- 29% TCA (trichloroacetic acid; 290 g/l; 1.3 ml/tube)
- 5% TCA (50 g/l; 15 ml/tube)

- 5 mM thymidine (1.21 g/l) and 5 mM leucine (0.65 g/l) (1 ml/tube; unlabelled)
- 0.1 N NaOH (4 g/l; 1 ml/vial)
- Ethylacetate (1 ml/vial)

Procedure

Incubation of soil suspension with labelled thymidine and leucine

- Homogenize 20 g soil in 95 ml mineral medium (PJ) by hand shaking for 30 s.
- Allow settling for 1 min to remove coarse soil particles.
- Add 100 µl soil suspension (20 mg soil) to the centrifuge tubes.
- Add 20 µl labelled [^{3}H]thymidine and [^{14}C]leucine (both 2 µM final concentration).
- Incubate for 1 h at *in situ* temperature; growing cells incorporate thymidine and leucine.
- Stop incorporation by adding 5 ml of extraction mixture (0.3 N NaOH, 25 mM EDTA and 0.1% SDS).
- Blanks are prepared by adding the extraction mixture immediately after the start of the incubation.
- At this stage the samples may be stored. We found no effect of 3 days' storage at 4°C in the dark, on thymidine and leucine incorporation. Dixon and Turley (2000) incubated marine sediment slurries on board ship, and stored them in a freezer until later analysis in the laboratory.

Extraction of labelled macromolecules

- The suspension is left at 30°C overnight (18–20 h) to extract macromolecules in the warm base solution.
- Mix and centrifuge (40 min, 5000 × g, 25°C) to remove soil particles.
- Aspire the supernatant with macromolecules and cool on ice for 5 min.
- Neutralize the base suspension with 1.3 ml ice-cold 1 N HCl.
- Add 1.3 ml ice-cold 29% TCA (w/v) and cool on ice for 15 min to precipitate the macromolecules.
- Collect the precipitated macromolecules on a 0.2 µm cellulose nitrate membrane filter.
- Wash filters three times with 5 ml ice-cold 5% TCA to remove unincorporated label, and transfer to glass scintillation vial.
- Add 1 ml 0.1 N NaOH to dissolve macromolecules and 1 ml ethylacetate to dissolve the filter. Shake, leave for half an hour and shake again.
- Add (15 ml) scintillation fluid. Shake vial and leave for at least 1 h to reduce chemoluminescence.
- Count radioactivity (dpm) of ^{3}H and ^{14}C.

Calculation

The number of blanks equals the number of replicates. Because the blanks have no individual relationship with the replicates, the mean value of the blanks is subtracted from each replicate. After subtraction of blanks, the counted dpm of ^{3}H and ^{14}C, respectively, are multiplied by 0.00284 to calculate pmol thymidine (1 picomol = 10^{-12} mol), and by 0.0759 to calculate pmol leucine incorporated per g soil per hour. These factors are based on the specific activity (Bq/mol, 1 Bq = 1 dps or 60 dpm) of thymidine and leucine as given above.

Using conversion factors, pmol/g per h can be converted to cells and carbon produced. Conversion factors have been established by growing bacteria at well-defined growth rates in continuous cultures (Bloem et al., 1989; Ellenbroek and Cappenberg, 1991; Michel and Bloem, 1993). For thymidine, the mean conversion factor was 0.5×10^{18} cells/mol thymidine. The range was 0.2–1.1×10^{18}. In the literature, much higher factors (range 0.13–7.9×10^{18}) have been published (Bååth and Johansson, 1990; Christensen, 1991). Given the uncertainties, we usually avoid conversion factors and express thymidine incorporation as pmol/g soil/h.

Discussion

Bacterial growth rate appears to be a more sensitive indicator of contamination than biomass or respiration. We compared the effects of copper pollution on bacterial biomass (microscopic measurements), growth rate (thymidine incorporation) and respiration (CO_2 evolution) in an arable sandy soil. The soil was originally unpolluted, with a background concentration of 25 mg Cu/kg. Jars were filled with 180 g gamma-sterilized soil, amended with 360 mg lucerne meal and 40 mg wheat straw meal, and reinoculated with soil organisms (Bouwman et al., 1994). In addition, different amounts of $CuSO_4$ were added, using three microcosms per treatment. Two days after amendment, bacterial growth rate, as determined by thymidine incorporation, was already significantly reduced at a low copper addition of 10 mg/kg (Fig. 7.2). Bacterial biomass and respiration were reduced at higher copper concentrations of 100 mg/kg and 1000 mg/kg.

In this experiment, thymidine incorporation was converted to growth rate using a conversion factor of 0.5×10^{18} cells/mol (Michel and Bloem, 1993), a cell volume of 0.2 µm^3 and a specific carbon content of 3.1×10^{-13} g C/µm^3 (Bloem et al., 1995). This resulted in a calculated growth rate of 50 µg C/g soil/day (Fig. 7.2). However, the respiration rate was only 10 µg C/g soil/day. Assuming a growth efficiency of, at most, 50%, the growth rate should roughly equal the respiration rate. Thus, the growth rate was probably overestimated by at least a factor of 5. Using advanced methods and avoiding conversion factors, Harris and Paul (1994) determined specific rates of DNA synthesis from the specific activities of the DNA precursor deoxythymidine triphosphate (dTTP) and purified bacterial DNA

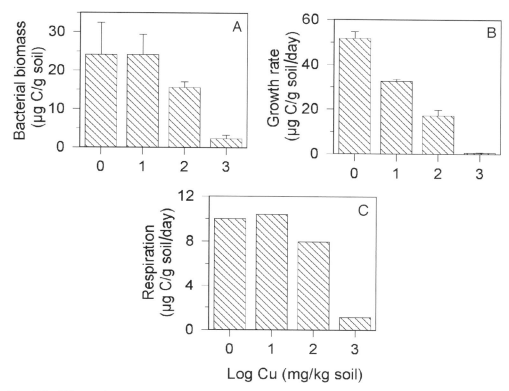

Fig. 7.2. Effects of copper pollution on biomass, growth rate and respiration rate of soil bacteria in microcosms, measured 2 days after amendment. Error bars indicate SD, $n = 3$.

after [^{3}H]thymidine incorporation. Their growth rate estimates were in agreement with respiration rates and much lower than those found in some other studies. This indicates that commonly used conversion factors tend to overestimate bacterial growth rate.

The thymidine incorporation procedure described here is based mainly on Findlay *et al.* (1984) and Thorn and Ventullo (1988), with some modifications (Michel and Bloem, 1993). The procedure can be summarized as follows:

- Incubate soil suspension with [^{3}H]thymidine and [^{14}C]leucine for 1 h.
- Extract macromolecules in warm base overnight.
- Remove soil particles by centrifugation.
- Cool supernatant on ice and precipitate macromolecules with cold acid (TCA).
- Collect macromolecules (including DNA and proteins) on a cellulose nitrate membrane filter.
- Count radioactivity (^{3}H and ^{14}C).

With this procedure, thymidine and leucine incorporation can be determined simultaneously. The procedure differs in some aspects from the procedures for thymidine incorporation described in two previous handbooks. Each procedure probably works well for the conditions it was originally tested for, but it may have shortcomings under other conditions.

Christensen and Christensen (1995):

- Used a concentration of 200 nM thymidine. For some soils this is sufficient, but in other soils at least 1 µM may be needed for maximum incorporation. Ideally, the optimum concentration should be determined by a saturation experiment for each soil, to minimize dilution by unlabelled precursors of thymine which may be present in the environment. Specific techniques to estimate isotope dilution are time consuming and not very accurate (Michel and Bloem, 1993; Robarts and Zohary, 1993). For routine use with large numbers of different samples (in monitoring networks) we use 2 µM thymidine (and leucine).
- Extracted in 0.6 N NaOH for 1 h at 60°C. This may cause degradation of DNA and lower recovery than a milder extraction in 0.3 N NaOH at 30°C for at least 12 h (Findlay *et al.*, 1984; Thorn and Ventullo, 1988).
- Added no EDTA and SDS. EDTA and SDS have been reported to increase recovery by inhibiting nucleases, and promoting cell lysis and breakdown of soil aggregates (Findlay *et al.*, 1984; Thorn and Ventullo, 1988; Bååth and Johansson, 1990).
- Used cellulose acetate filters. Cellulose nitrate is recommended to bind DNA (Robarts and Zohary, 1993).
- Did not pre-wash filters with cold (unlabelled) thymidine: this may cause higher values of blanks (Robarts and Zohary, 1993).

The procedure of Bååth (1995) can be summarized as follows:

- Incubate with [^{3}H]thymidine.
- Centrifuge to collect bacteria and soil.
- Extract macromolecules in warm base overnight at 60°C.
- Centrifuge to remove soil particles.
- Precipitate macromolecules with acid (TCA) on ice.
- Centrifuge twice with 5% TCA to wash macromolecules.
- Hydrolyse DNA in 5% TCA at 90°C for 30 min.
- Centrifuge.
- Collect supernatant ('DNA fraction') and measure radioactivity.

This procedure uses five centrifugation steps and may be more time consuming. Bååth and Johansson (1990) found little effect of extraction temperature on the recovery of added [^{14}C]DNA. In contrast, at 60°C and 90°C, we found 30% and 90% lower thymidine incorporation than at 30°C. This may indicate degradation of DNA at higher temperatures. In Bååth's protocol, the incorporation into total macromolecules is not measured, but the DNA fraction is isolated using hot acid. In principle, it is better to isolate the DNA fraction. However, DNA has been reported to hydrolyse incompletely (about 50%) in hot acid, and this procedure may lead to underestimates of DNA synthesis (Servais *et al.*, 1987; Robarts and Zohary, 1993). Using hot acid hydrolysis, we sometimes found 80% incorporation into the 'DNA fraction'. On other occasions we found as little as 0% incorporation into the 'DNA fraction' of actively growing (and incorporating) bacteria from continuous cultures (Bloem et al., 1989). Robarts and Zohary (1993) criticized (mis)use of acid–base hydrolysis, and alternatively proposed to purify the DNA by rinsing the filters with 50% phenol–chloroform to remove proteins, and with 80% ice-cold ethanol to remove lipids. Using this method, we found between 10% and 100% incorporation into DNA. Since proteins are removed, this method requires separate samples for measuring leucine incorporation.

Thus, a minor fraction of the total thymidine incorporation may be into DNA. Measured incorporation into DNA underestimated actual DNA synthesis by a factor of 6–8 (Jeffrey and Paul, 1988; Ellenbroek and Cappenberg, 1991). This may be caused by a rate-limiting step in the incorporation of labelled thymidine, resulting in intracellular dilution with unlabelled thymine. Part of the (apparently) low incorporation into DNA may be caused by incomplete hydrolysis of DNA in hot acid (Robarts and Zohary, 1993). However, two alternative methods, using enzymatic degradation with DNase or the DNA synthesis inhibitor mitomycin, confirmed a low (about 35%) incorporation into DNA (Ellenbroek and Cappenberg, 1991). Even incorporation into total macromolecules (DNA, proteins, lipids) underestimated real DNA synthesis in continuous culture by at least a factor of two (Bloem et al., 1989; Ellenbroek and Cappenberg, 1991).

Given the difficulties with measuring actual DNA synthesis, and the apparently consistent empirical relationship between (total) thymidine incorporation and bacterial growth rate (Bloem et al., 1989; Ellenbroek and Cappenberg, 1991; Michel and Bloem, 1993), we have chosen to use incorporation of [^{3}H]thymidine and [^{14}C]leucine into total macromolecules. This is generally sufficient to indicate relative effects on bacterial growth rates.

7.6 N$_2$O Emissions and Denitrification from Soil

ULRIKE SEHY,[1] MICHAEL SCHLOTER,[1] HERMANN BOTHE[2] AND
JEAN CHARLES MUNCH[1]

[1]*GSF – National Research Centre for Environment and Health, Institute for Soil Ecology, Ingolstädter Landstraße 1, D-85758 Oberschleissheim, Germany;* [2]*University of Cologne, Institute of Botany, Gyrhofstraße 15, D-50923 Cologne, Germany*

Introduction

The emission of nitrogenous gases from soil plays a crucial role in the global nitrogen cycle, as nitrate or ammonia are turned into gaseous products, which return to the atmosphere. Gases involved include ammonia, nitric oxide, nitrous oxide and dinitrogen. For nitric oxide, nitrous oxide and dinitrogen, two microbial processes, nitrification and denitrification, are considered to be the most important biotic sources (Granli and Bøckman, 1994). Both processes may occur simultaneously in microsites and are controlled by various soil- and weather-dependent parameters.

Denitrification

During denitrification, nitrate is reduced via nitrite, NO and N$_2$O to N$_2$. Denitrification is an energy-yielding process, in which microorganisms utilize nitrate as an alternative terminal respiratory electron acceptor under oxygen-limited conditions. The ability for denitrification is widespread amongst microorganisms. It is known that approximately 50% of all known bacteria can denitrify (or at least carry out some partial denitrification reactions). In addition, some species of fungi and archaebacteria have also been described as denitrifiers. Overall, the occurrence of denitrification in organisms of totally unrelated affiliations suggests that denitrification has been distributed evolutionarily by lateral gene transfer.

The single steps of the denitrification process are catalysed by specific reductases. The first reaction, the conversion of nitrate to nitrite, is catalysed by a molybdenum (Mo)-containing nitrate reductase. All nitrate reductases have a molybdopterin cofactor and contain FeS centres, and, in addition, some possess cytochrome *b* or *c*. Several forms of nitrate reductases exist (membrane-bound, periplasmic and also an assimilatory enzyme with similar characteristics to the dissimilatory enzymes), which make the study of this step interesting but also complex, particularly since some organisms contain more than one type of enzyme.

The next step in the denitrification pathway is the reduction of nitrite to nitric oxide, which is catalysed by nitrite reductase. Bacteria express two different forms of nitrite reductase, containing either a cytochrome (nirS) or copper (nirK) in their prosthetic group. Organisms possess one of the two enzymes. The cytochrome-containing nitrite reductase appears to be more widespread among bacteria, whereas the Cu-enzyme apparently is evolutionarily more conserved. The product of the reaction, NO, is highly reactive and toxic, and is also an important signalling molecule for plant or animal life. Intact denitrifying organisms evolve, at best, minute amounts of this gas, which is effectively utilized by the cytochrome b, c containing nitric oxide reductase. The conversion of NO to N_2O, catalysed by this enzyme, involves the formation of the dinitrogen bond, which is biochemically an extremely interesting, and currently poorly understood, reaction. Finally, nitrous oxide is reduced to the dinitrogen molecule by nitrous oxide reductase, which contains Cu atoms in a novel tetra-nuclear cluster at the active site.

Nitrification

While denitrification generates N, NO and N_2O, nitrification can produce only NO and N_2O. Nitrification is the conversion of inorganic or organic nitrogen from a reduced to a more oxidized state. Chemoautotrophic bacteria are largely, or solely, responsible for the nitrification in soils with a pH above 5.5 (Focht and Verstraete, 1977); at lower pH values there is evidence for acid-tolerant heterotrophic nitrifiers (Schimmel et al., 1984). The heterotrophic nitrifiers (e.g. bacteria, such as strains from *Arthrobacter*, and fungi, such as *Aspergillus*) do not derive energy from the oxidation of NH_4^+. In arable soils, the production of nitrate by heterotrophs appears to be insignificant in relation to that brought about by chemoautotrophs (Paul and Clark, 1989), which is in contrast to potential nitrification levels in acid forest soils (Kilham, 1987).

Consequences of the activity of denitrifying and nitrifying microorganisms

Due to the action of denitrifying and nitrifying microorganisms, the global dinitrogen content in the atmosphere is in balance (due to the formation of the nitric oxide, nitrous oxide and dinitrogen from nitrate and ammonia). On the other hand, nitrogenous oxides, also released from soils and waters, have several impacts on the chemistry of the atmosphere and radiation processes. Nitrous oxide is next to carbon dioxide (CO_2) and methane (CH_4) in its importance as a potent greenhouse gas. Nitric acid and its chemical oxidation product, NO_2, are major constituents of acid rain, and NO and also N_2O interact with ozone in complex reactions and are major causes of the destruction of the protective ozone layer in the stratosphere.

Nitrate is the main source of nitrogen for the growth of plants in agriculture, but it can simultaneously be used by microorganisms in soils.

Denitrification is generally regarded as an anaerobic process, but there are indications that it may also take place in well-aerated soils. The conditions that favour denitrification in soils have been elucidated in detail. It is clear that any use of nitrate by bacteria means a loss of N for the growth of plants. Thus, denitrification also has a severe impact on agriculture.

In addition, products of denitrification (nitrate respiration) have manifold other, mainly adverse, effects on the atmosphere and on waters.

Measurement of NO, N_2O and N_2

While the products N_2O and NO can be easily measured as trace gases, using gas chromatography, the determination of N_2 is not straightforward because comparatively small amounts of N_2 produced during denitrification have to be distinguished from a large background of 78% N_2 in the atmosphere (Aulakh et al., 1992). The methods available for measuring denitrification in the field are based on the use of the stable isotope ^{15}N, or on acetylene for blockage of the enzyme N_2O-reductase.

Principle of the Method

The acetylene inhibition method (AIM) was developed utilizing the fact that acetylene (C_2H_2) blocks the enzymatic reduction of N_2O to N_2 (Balderston et al., 1976; Yoshinari and Knowles, 1977) when present in a range of 1–10% v/v (Granli and Bøckman, 1994). In soils treated with acetylene, the amount of N_2O released thus represents both N_2O and N_2 produced during denitrification.

There are basically two methods available for the application of the AIM in the field:

1. *Chamber methods*: in which enclosures are placed over the soil to separate the soil atmosphere from the ambient atmosphere (Ryden et al., 1979; Burton and Beauchamp, 1984; Mosier et al., 1986; Aulakh et al., 1991). Denitrification measurements can be made either with open chambers, where acetylene is forced through the cover and N_2O is collected at the outlet for analysis, or with closed chambers, where N_2O is allowed to accumulate before withdrawal for analysis.

2. *Soil core methods*: in which undisturbed soil cores are taken and incubated in the laboratory or in the field (Parkin et al., 1985; Ryden et al., 1987; Aulakh et al., 1991; Jarvis et al., 2001). For denitrification measurements, soil cores are collected in the field, by driving a small cylinder into the soil, then incubated together with acetylene. Two methods are commonly used: one is to incubate soil cores statically in containers with acetylene and to collect gas samples through rubber septa (Burton and Beauchamp, 1984). In another method, soil air and acetylene are recirculated through the macropores of the soil and the outgoing air is directly connected to a gas chromatograph for quantification of N_2O (Parkin et al., 1984, 1985).

The following sections will focus on the measurement of N_2O from denitrification. However, with slight modifications of the methods (no acetylene inhibition), it is possible to measure all nitrous gases (see also Fig. 7.3).

Materials and Apparatus

Chamber method

- PVC rings (diameter, 30 cm; area, 0.07 m^2; volume, 15 l)
- Rubber stopper
- Stainless-steel tubes (6 mm diameter)
- 10 ml Vacutainers™ (Becton and Dickinson, Heidelberg, Germany)
- Gas chromatograph equipped with a flame ionization detector (e.g. Shimadzu GC 14, Duisburg, Germany)
- ^{63}Ni-electron capture detector

Soil core method

- Plastic tubes of 9 cm length and 4.5 cm diameter
- Rubber stopper
- Stainless-steel tubes (6 mm diameter)
- 10 ml Vacutainers™ (Becton and Dickinson)
- Gas chromatograph equipped with a flame ionization detector (e.g. Shimadzu GC 14)
- ^{63}Ni-electron capture detector

Chemicals and solutions

- Acetone-free acetylene

Procedure

Chamber method (Mogge *et al.*, 1998, 1999)

For each experimental site, PVC rings are driven into the soil to a depth of 10 cm. During the sampling process the rings are sealed by a lid equipped with a rubber stopper. Gas samples should be taken at the beginning and at the end of the sampling period. Beforehand, a linear increase in N_2O concentration during the sampling period should be proven. *In situ* measurements of N_2O should usually be repeated weekly.

For measurement of denitrification N losses, additional soil chambers are supplied with acetone-free acetylene by diffusion, using four perforated stainless-steel tubes (6 mm diameter) for each cover. The use of flow meters, together with a timer, allows the addition of different amounts of acetylene into different soil types, to ensure equal concentrations of acetylene (0.5–1.5%) during a period of 48 h. Acetylene concentrations are determined by a gas chromatograph equipped with a flame ionization detector (Shimadzu GC 14). *In situ* measurements of denitrification N losses are usually repeated every 14 days.

Evacuated 10 ml Vacutainers™ should be used for sampling and storing gas samples from the chamber atmosphere, after a pretreatment described by Heinemeyer and Kaiser (1996). Gas samples can be analysed for N_2O by a ^{63}Ni-electron capture detector equipped with an automatic sample-injection system (Heinemeyer and Kaiser, 1996).

Soil core methods (Rudaz *et al.*, 1999)

Three to 5 months before the start of the measurements, plastic tubes must be put into the soil and placed in 450 ml jars without removing the soil core. The jars are closed with a lid containing a rubber septum to take gas samples from the headspace. The jars should be placed in the soil and covered with grass in order to avoid a change in temperature. Samples from each site are treated with acetylene (12 kPa) to estimate total N flux from denitrification. To determine N_2O fluxes, controls are not treated with acetylene. Each sample should be incubated for 6 h. Every 2 h, gas samples (7 ml) are withdrawn with a syringe and transferred in a pre-evacuated gas sample flask (9.1 ml). The procedure described above is carried out in the field. In the lab, N_2O is analysed with a gas chromatograph fitted with an electron capture detector. Figure 7.3 shows a possible set-up for the measurement of nitrous oxide emissions with and without acetylene-inhibition techniques (from Mogge *et al.*, 1998).

Calculations

From Bauernfeind (1996).

If a calibration gas mixture (with known mixing ratio) is used for calibration, the µl N_2O must be calculated from the injected volume. To set up the calibration curve, plot the injected µg N_2O under standard conditions (20°C, 293 K and 101,300 Pa).

$$\mu g\ N_2O\ injected = \frac{P \times V_n \times 10^{-9} \times MW \times 10^6}{R \times T}$$

From the calculated µg N_2O of the sample from the calibration curve, the µg N_2O-N/g dm/h can be calculated.

$$\mu g\ N_2O\text{-}N/g\ dm/h = \frac{X \times V \times 0.6363 \times 100}{IV \times t \times SW + \%dm}$$

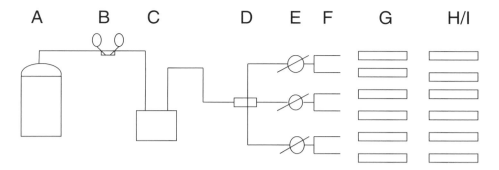

Fig. 7.3. Set-up for the measurement of nitrous oxide emissions with and without the acetylene-inhibition technique. A, gas vessel; B, pressure gauge, magnetic valve, timer; C, acetone trap (sulphuric acid, 95%); D, gas pipe; E, quickfit connector; F, flow meter; G, diffusion tubes; H, soil ring (with acetylene); I, soil ring (without acetylene).

where: P = standard atmospheric pressure (101,300 Pa); V_n = injected volume of N_2O standard (µl); 10^{-9} = conversion factor (1 µl = 10^{-9} m³); MW = molecular weight of N_2O; 10^6 = conversion factor (1 g = 10^6 µg); R = gas constant (8.31 J/mol.K); T = standard temperature (293 K); X = µg N_2O of the injected sample volume; V = total volume of incubation flask minus soil volume (ml); 0.06363 = factor to convert N_2O to N_2O-N; IV = injected sample volume (ml); t = incubation time (h); SW = initial soil weight; and %dm = soil dry matter in percent.

Discussion

General advantages and disadvantages of the acetylene inhibition method

The principles of the described methods, as well as advantages and disadvantages, have been reviewed by Aulakh *et al.* (1992) and Tiedje *et al.* (1989). Major advantages of the AIM are the applicability in undisturbed and in fertilized ecosystems, the low cost of the sampling and the analytical equipment (especially compared to ^{15}N-based methods) and the high sensitivity (a loss of about 1 g N/ha per day can be measured; Duxbury, 1986). The main disadvantage of the acetylene inhibition method is the inhibition of nitrification by acetylene; thus, denitrification can only proceed from nitrate already present in the soil (McCarty and Bremner, 1986). This might lead to underestimation of real denitrification rates, especially in unfertilized, natural ecosystems (Tiedje *et al.*, 1989; Aulakh *et al.*, 1992). Furthermore, the dispersal of acetylene in the soil and diffusion of N_2O out of the soil can be severely hindered in heavy and/or wet soils, or in compacted zones (Klemedtsson *et al.*, 1990; Granli and Bøckman, 1994). Finally, acetylene can be biodegraded after a long time exposure of the microflora. However, this can be avoided by restricting measurements to less than a week (Terry and Duxbury, 1985) and by choosing new points for measurements.

Soil core versus chamber methods

Advantages of chamber methods are the minimal disturbance of natural conditions in the field (Klemedtsson et al., 1990). With chamber methods, actual *fluxes of N gases* from the soil to the atmosphere are measured, while cores give more direct estimates of *N gas production* by biological processes (Tiedje et al., 1989).

Soil core methods have been considered superior to chamber methods in very wet soils, due to better gas diffusion of acetylene into, and of N_2O out of, the smaller soil volume monitored in the former (Ryden et al., 1987).

A major criticism of the soil core method is that the aeration status of the soil is influenced by the coring procedure and the incubation. This has recently been addressed by Jarvis et al. (2001), who designed an incubation box for square-section soil cores with minimum exposure of surfaces to air.

Other methods

The various ^{15}N methods used in denitrification research have been reviewed by Myrold (1990). The principle of the ^{15}N mass balance method is the measurement of the amount of applied ^{15}N-labelled fertilizer in different soil pools and in the plant. The amount of N unaccounted for is assumed to be lost by denitrification (Aulakh et al., 1992). In the ^{15}N gas flux method, highly enriched (20–80%) ^{15}N-labelled fertilizer is applied to soil; the soil is then covered with a chamber, and the flux of N_2 and N_2O following denitrification is measured by quantifying the increase in ^{15}N-labelled gases in the chamber headspace (Stevens and Laughlin, 1998).

It has been stated that ^{15}N methods are preferable to the acetylene inhibition method in heavy-textured soils where acetylene diffusion is hindered (Granli and Bøckman, 1994). However, quantification of denitrification rates using the ^{15}N gas flux method may also suffer from hindered diffusion of N_2O out of the soil, as well as from N_2O being dissolved in soil water (Myrold, 1990). Furthermore, a possible increase in the denitrification rate and change in the N_2O/N_2 ratio might occur after addition of ^{15}N-NO_3^- (Granli and Bøckman, 1994). In addition, dilution of $^{15}NO_3^-$ with soil NO_3^- by various soil processes will cause an underestimate of denitrification if not corrected for. Finally, a uniform distribution of added $^{15}NO_3^-$ is difficult. However, a uniformly labelled soil nitrate pool is the prerequisite for the correct calculation of N_2 flux in the ^{15}N gas flux method (Hauck et al., 1958).

Tiedje et al. (1989) reviewed a number of studies comparing ^{15}N and acetylene inhibition methods. They concluded that both methods gave similar estimates of N loss, but high spatial variability of the denitrification process may be partly responsible for the lack of statistical difference between the methods. Recently, Watts and Seitzinger (2000) showed that denitrification rates measured with a ^{15}N gas flux method and acetylene inhibition methods may differ substantially. Malone et al. (1998) combined the ^{15}N gas flux method with the acetylene inhibition technique in order to confirm the assumption that acetylene had completely blocked N_2O reductase for their particular soil and condition.

7.7 Enzyme Activity Profiles and Soil Quality

LIZ J. SHAW[1] AND RICHARD G. BURNS[2]

[1]*Imperial College London, Department of Agricultural Sciences, Wye Campus, Wye TN25 5AH, UK;* [2]*School of Land and Food Sciences, The University of Queensland, Brisbane, Queensland 4072, Australia*

Introduction

Enzyme activities in soil are primarily the expression of bacteria, fungi and plant roots, and are responsible for the flux of carbon, nitrogen and other essential elements in biogeochemical cycles. Measuring enzymatic catalysis and understanding the factors that regulate enzyme expression and the rates of substrate turnover are the first stages in characterizing soil metabolic potential, fertility and quality, as well as a guide to the resilience of the soil when subjected to various natural and anthropogenic impacts. Furthermore, substrate catalysis and enzyme responses, when combined with the other soil properties described in this handbook, may provide enough information to allow the rational manipulation of soil processes for commercial and environmental benefit. Three obvious advantages of a better understanding of soil enzyme activities are the enhancement of plant nutrient solubilization, the inhibition of phytopathogens, and the stimulation of pollutant degradation. These beneficial influences are likely to be most strongly expressed in the rhizosphere (Shaw and Burns, 2003).

Soil enzyme assays are far from uniform and usually employ incubation conditions that are dissimilar to those encountered in the natural environment. As a consequence, interpretation of the data is difficult, controversial, and sometimes downright misleading. Furthermore, even when the problems associated with the design of meaningful assays appear to have been considered and resolved, there are large variations in activity in both space and time; soil is heterogeneous. For example, enzyme–substrate interactions in macro-aggregates are different from those in micro-aggregates (Ladd *et al.*, 1996), reaction rates in the rhizosphere are likely to be dissimilar to those in the non-rhizosphere (Reddy *et al.*, 1987; Kandeler *et al.*, 2002), and the surface soil will display a different range and level of activities than the sub-surface soil (Taylor *et al.*, 2002). Perhaps introducing even greater complexity is the fact that most biogeochemical processes are mediated, *in toto*, by many enzymes located in one or more locations within the soil matrix.

Most enzymes in soil are directly associated with viable cells and function within the confines of the microbial cell membrane (i.e. are truly intracellular). Others, the extracellular enzymes, are secreted and catalyse reactions at the outer surfaces of the cell wall and in the surrounding environment. In addition, many strictly intracellular enzymes (especially hydrolases) are released from leaking cells and lysing dead cells, but remain functional for a period of time. This is because these, together with the truly extracellular enzymes, become intimately associated with clays and organic (humic) colloids. Clay– and humic–enzyme complexes form a long-term persistent catalytic component of soils, but one that may be poorly correlated with microbial numbers and the biomass (Burns, 1982; Nannipieri et al., 2002).

Thus, estimation of a suite of soil enzyme activities integrates both the intra- and extracellular biogeochemical activities of the soil biological system and should be a key aspect of soil quality assessment.

Numerous methods have been developed to measure the activities of the hundreds of enzymes that catalyse reactions involved in soil nutrient cycling (for an extensive review, see Tabatabai and Dick, 2002). Here we describe the methods for measuring three undeniably important, yet contrasting, enzymatic processes. The first of these is an exclusively intracellular process and depends on the reduction of tetrazolium salts. The second is the result of both intracellular and extracellular microbial activity during which fluorescein diacetate is hydrolysed. In both assays, the products of the reaction are a result of the activities of many different enzymes. For example, fluorescein diacetate can be hydrolysed by the action of lipases, proteases and esterases (Alef and Nannipieri, 1995), and tetrazolium salts are reduced by a number of enzymes integral to the intact cell and reflect the total oxidative potential of the soil microbial community (Dick, 1997). While useful to describe overall microbial activity in soil (in terms of either electron transport activity or hydrolytic capability), these activities do not yield much information regarding the rate of specific catalytic steps, such as nutrient acquisition and biogeochemical cycling, processes which define soil health (Dick, 1997). Therefore, the third type of assay is more appropriate for estimating the activity of hydrolytic enzymes involved in, for example, specific stages in C, N, P and S acquisition. Table 7.2 shows examples of hydrolases commonly assayed, and the soil health function they mediate. As can be seen, a common approach when assaying the activity of soil hydrolases is to use artificial *p*-nitrophenyl-linked substrates. These esters are hydrolysed to *p*-nitrophenol, which is easily determined by colorimetry. Thus, since the principle of the *p*-nitrophenyl-linked substrate hydrolase assay is the same, one enzyme, phosphomonoesterase, is chosen as an example.

Dehydrogenase Activity

Principle of the method

The aerobic microbial oxidation of organic substrates is mediated by the membrane-bound electron transport chain that transfers electrons or

Table 7.2. Examples of commonly assayed soil hydrolases, their function and measurement method (adapted from Alef and Nannipieri (1995); Acosta-Martinez and Tabatabai (2001)).

Enzyme Commission classification	Enzyme	Function	Assay substrate	Assay product
Phosphatases				
3.1.3.1 and 3.1.3.2	Alkaline and acid phosphatase	$RNa_2PO_4 + H_2O \rightarrow R\text{-}OH + Na_2HPO_4$	p-Nitrophenyl phosphate[b]	p-Nitrophenol
3.1.4.1	Phosphodiesterase	$R_2NaHPO_4 + H_2O \rightarrow R\text{-}OH + RNaHPO_4$	Bis-p-Nitrophenyl phosphate[b]	p-Nitrophenol
Sulphatase				
3.1.6.1	Arylsuphatase	$ROSO_3^- + H_2O \rightarrow R\text{-}OH + RNaHPO_4$	p-Nitrophenyl sulphate[b]	p-Nitrophenol
Glycosidases				
3.2.1.20	α-Glucosidase	Glucoside-R + H_2O → glucose + R-OH	p-Nitrophenyl-α-D-glucopyranoside[b]	p-Nitrophenol
3.2.1.21	β-Glucosidase	Glucoside-R + H_2O → glucose + R-OH	p-Nitrophenyl-β-D-glucopyranoside[b]	p-Nitrophenol
3.2.1.22	α-Galactosidase	Galactoside-R + H_2O → galactose + R-OH	p-Nitrophenyl-α-D-galactopyranoside[b]	p-Nitrophenol
3.2.1.23	β-Galactosidase	Galactoside-R + H_2O → galactose + R-OH	p-Nitrophenyl-β-D-galactopyranoside[b]	p-Nitrophenol
Amidohydrolases and arylamidases				
3.5.1.1	L-Aspariginase	L-Asparagine + H_2O → L-Aspartate + NH_3	L-Asparagine[a]	NH_4-N
3.5.1.2	L-Glutaminase	L-Glutamine + H_2O → L-Glutamate + NH_3	L-Glutamine[a]	NH_4-N
3.5.1.4	Amidase	$R\text{-}CONH_2 + H_2O \rightarrow NH_3 + R\text{-}COOH$	Formamide[a]	NH_4-N
3.5.1.5	Urease	Urea + $H_2O \rightarrow CO_2 + 2NH_3$	Urea[a]	NH_4-N

[a] Natural substrate.
[b] Synthetic substrate.

hydrogen from substrates via electron carrier proteins and oxidoreductases to O_2, the final electron acceptor. Thus, dehydrogenases exist as an integral part of intact cells and represent the total oxidative activities of soil microorganisms during the initial stages of organic matter breakdown (Dick, 1997). The concept of estimating microbial activity in soil by quantifying dehydrogenase activity relies on the ability of a tetrazolium salt, iodonitrotetrazolium chloride (INT), to act as an artificial electron acceptor in the place of oxygen. Tetrazolium compounds are characterized by a heterocyclic ring structure which readily accepts hydrogen atoms (and electrons) and becomes reduced. Thus, upon incubation, water-soluble INT becomes biologically reduced to form the purple, water-insoluble, iodonitrotetrazolium formazan (INTF). This can be extracted using an organic solvent and the amount produced determined colorimetrically. The method, detailed below, and recommended as the standard assay, is an INT-based method originally described by Benefield *et al.* (1977), and adapted from Trevors (1984a,b) and vonMersi and Schinner (1991).

Materials and apparatus

- Glass McCartney bottles (28 ml) with rubber-lined screw caps, sterilized by autoclaving
- Semi-micro clear polystyrene spectrophotometer cuvettes (1 ml)
- Microcentrifuge with a rotor to fit 1.5 ml microcentrifuge tubes
- Autoclave
- Spectrophotometer set to 464 nm
- Incubator at 25°C

Chemicals and solutions

- *INT solution*: dissolve INT (Sigma-Aldrich Co. Ltd, Dorset, UK) in distilled water to give a 0.2% (w/v) solution and sterilize by passing through a 0.2 µm filter into a clean sterile container. Note that INT is sparingly soluble in water (solubility limit = 0.3% (w/v), therefore allow for sufficient time (> 2 h) for the INT to become completely dissolved
- *Extractant*: N,N-dimethyl formamide:ethanol (1:1, v/v)
- *INTF master standard solution* (500 µg/ml): dissolve 25.0 mg of INTF (Sigma) in ~40 ml of extractant. Decant to a 50 ml volumetric flask and bring up to the mark with extractant
- *INTF working standard solutions*: prepare INTF working solutions by pipetting the following volumes of the INTF master solution to 10 ml volumetric flasks and make up to the mark with a mixture of extractant:distilled water (5:2, v/v)

Working INTF standard concentration (µg/ml)	Volume of master solution (µl) in 10 ml
0	0
1	20
2	40
4	80
6	120
10	200
15	300
25	500

Procedure

- Soil, freshly collected from the field, should be sieved (< 2.8 mm), field moist and stored at 4°C prior to the assay. (NB. This is essential for all enzyme assays.)
- Weigh replicate soil samples (1 g) into sterile McCartney bottles and add 4 ml of INT solution (0.2%).
- Close the lids and incubate at 25°C for 48 h in the dark.
- Sterile controls (to account for any abiotic INT reduction) should consist of autoclaved soil (121°C, 20 min on three consecutive days). Spectrophotometer blanks for both autoclaved and non-autoclaved treatments should consist of soil with the INT solution replaced with distilled water. Controls and blanks should be treated like the samples.
- After the incubation period, extract soil by addition of 10 ml of the extractant and incubate in the dark with agitation for 1 h.
- After the extraction period, transfer ~1.2 ml of the extractant/soil mixture to 1.5 ml microcentrifuge tubes and remove soil by centrifugation (relative centrifugal force (RCF) = 11,600 × g for 5 min).
- Transfer 1 ml of supernatant fraction to spectrophotometer cuvettes and determine absorbance at 464 nm (OD_{464nm}) against the appropriate blank.
- At the same time, construct a calibration curve by determining OD_{464nm} values for the working standard solutions of INTF (0–25 µg/ml).

Calculation

Using the calibration curve, calculate INTF concentrations from the corresponding OD_{464nm} value.

Dehydrogenase activity (µg INTF/g dry soil/48 h) = $\dfrac{([INTF_s] - [INTF_c]) \times 14}{edw}$

where: $[INTF_s]$ is the INTF concentration (µg/ml) in the sample; $[INTF_c]$ is the INTF concentration in the sterile control; edw is the equivalent dry weight of 1 g of soil (determined by loss of weight of field-moist subsamples after heating at 105°C until constant weight); and 14 is the volume (ml) of solution added in the assay (INT + extractant).

Discussion

Although INT reductase activity is used widely in soil ecotoxicology research (Gong et al., 1999; Welp, 1999; Moreno et al., 2001), no standard method has been adopted, and, consequently, assay parameters, such as pH, incubation temperature and time, substrate concentration and extractant solvent used, vary depending on the soil type and experiment being conducted.

The 48-h incubation period in the method described above is based on the research of Trevors (1984a,b), which allows the accumulation of measurable product and the estimation of dehydrogenase activity at temperatures (i.e. 25°C) more likely to be found in the natural environment. In addition, in the method described above, soil is assayed without use of a buffer and the 'natural' dehydrogenase activity is determined (vonMersi and Schinner, 1991). A pH optimum of 7–7.5 has been determined for INT reduction (vonMersi and Schinner, 1991), with very little activity below a pH of 6.6 or above 9.5 (Trevors, 1984a). Thus, in order to obtain statistically meaningful values for the dehydrogenase actvity of either acidic or alkaline soils, it may be necessary to use a buffered system (vonMersi and Schinner, 1991; Taylor et al., 2002). Other studies (Trevors et al., 1982; Trevors, 1984b) have determined INT reduction in soils incubated with labile carbon sources (glucose and yeast extract) in an attempt to quantify 'potential' as well as 'natural' activity.

The dehydrogenase assay has been a useful parameter in comparable ecotoxicological studies within the same soil type. For example, Welp (1999) used INT reductase activity to examine the toxicity of total and water-soluble concentrations of metals in a loess soil. Gong et al. (1999) also used dehydrogenase activity in the toxicity assessment of TNT-contaminated soil. However, Obbard (2001) has shown that INTF undergoes abiotic interaction with copper, leading to decreased absorbance. Thus, in copper-contaminated soils, biotic complexation between INTF and Cu may lead to an underestimate of dehydrogenase activity.

Fluorescein Diacetate Hydrolysis

Principle of the method

Fluorescein diacetate (FDA) is a relatively non-polar compound. As a result of this it is assumed that it diffuses easily through the cell membrane, where it is hydrolysed by non-specific esterases to the fluorescent compound, fluorescein (Rotman and Papermaster, 1966). In addition, FDA can also be hydrolysed by extracellular enzymes produced by the soil microflora, such as proteases, lipases and esterases. Thus, FDA hydrolysis has been suggested as a measure of the global hydrolytic capacity of soils and a broad-spectrum indicator of soil biological activity (Bandick and Dick, 1999; Perucci et al., 1999).

The method described below is based on that described by Adam and Duncan (2001). Soil is incubated with the substrate, FDA, at 25°C for 30 min. The amount of fluorescein formed is determined colorimetrically following extraction with an organic solvent mixture.

Materials and apparatus

- Glass McCartney bottles (28 ml) with rubber-lined screw caps; sterilized by autoclaving
- Semi-micro clear polystyrene spectrophotometer cuvettes (1 ml)
- Centrifuge(s) with rotor(s) to fit 1.5 ml microcentrifuge tubes and glass McCartney bottles
- Spectrophotometer set to 490 nm
- Shaking incubator at 25°C

Chemicals and solutions

- *Potassium phosphate buffer* (60 mM, pH 7.6): dissolve 8.7 g of K_2HPO_4 and 1.3 g of KH_2PO_4 in 1 l of distilled water; sterilize by autoclaving (121°C, 20 min)
- *FDA solution* (1 mg/ml): dissolve 25 mg fluorescein diacetate (3'6'-diacetyl-fluoresein, Sigma-Aldrich) in 25 ml acetone; store at −20°C
- *Extractant* (2:1 chloroform:methanol): add 666 ml of chloroform (AR grade) to a 1-l volumetric flask, make up to the mark with methanol (AR grade) and mix thoroughly
- *Fluorescein master solution* (2000 μg/ml): dissolve 113.2 mg fluorescein disodium salt in 50 ml of potassium phosphate buffer (60 mM, pH 7.6)
- *Fluorescein working standard solutions*: prepare fluorescein working solutions by pipetting the following volumes of the fluorescein master solution into 100 ml volumetric flasks and make up to the mark with potassium phosphate buffer:

Working fluorescein standard concentration (μg/ml)	Volume of master solution (μl) in 100 ml
0	0
1	50
2	100
3	150
4	200
5	250

Procedure

- Soil, freshly collected from the field, should be sieved (< 2.8 mm), field moist and stored at 4°C prior to the assay.
- Weigh 1 g soil sample into sterile McCartney bottles and add 7.5 ml potassium phosphate buffer (pH 7.6, 60 mM) and allow to equilibrate at 25°C on an end-over-end shaker.
- Start the reaction by the addition of 0.1 ml FDA solution (1000 µg/ml) and return the samples to the shaker. Incubate at 25°C for 30 min.
- Spectrophotometer blanks should consist of the soil and buffer mixture with the FDA solution replaced by 0.1 ml acetone. Incubate as the samples.
- After the 30 min have elapsed, immediately add 7.5 ml of chloroform:methanol (2:1) to samples and blanks to stop the reaction. Replace the lids and mix the contents thoroughly (10 s, vortex mixer).
- Centrifuge the tubes at a low speed (RCF = $300 \times g$ for 2 min) to clarify the phases.
- Transfer 1.2 ml of the upper phase to a 1.5 ml microcentrifuge tube and centrifuge (RCF = $16,500 \times g$ for 5 min) to remove suspended fines.
- Decant 1 ml of supernatant fraction to a 1 ml cuvette and measure absorbance at 490 nm on a spectrophotometer against the soil blank.
- To construct a calibration curve, pipette triplicate 7.5 ml aliquots of each concentration of fluorescein working standards into McCartney bottles, extract with chloroform:methanol and determine the OD_{490nm} of the clarified upper phase as for the samples.

Calculation

Using the calibration curve, calculate the mass of fluorescein produced in each assay from the corresponding OD_{490nm} value and divide by the equivalent dry weight of soil (determined by loss of weight of field-moist sub-samples after heating at 105°C until constant weight). Fluorescein diacetate hydrolysis activity is expressed as µg fluorescein/g dry soil/0.5 h.

Discussion

The method for FDA hydrolysis described above does not use abiotic controls. We tested the FDA hydrolysis assay with four soil types and found no abiotic hydrolysis of FDA using autoclaved soil as the control (Shaw and Burns, unpublished results). Based on this evidence, we suggest that, once established, it may not be necessary to include an abiotic control, despite a single report that FDA can be hydrolysed spontaneously without microbial activity (Guilbault and Kramer, 1964).

Buffer addition can directly or indirectly alter the nutrient conditions of the microbial community and therefore its hydrolytic activity (Battin, 1997).

Consequently, to better estimate the 'natural' FDA hydrolysis capacity of a soil, a non-buffered system should be used. However, sensitivity of the assay may be problematic for acidic soils, due to the decreased fluorescence of fluorescein at low pH. Nevertheless, it may be possible for post-assay adjustment of the pH of the aqueous phase prior to measurement to improve the sensitivity of fluorescein detection. It should also be mentioned that spontaneous hydrolysis of FDA is reported at pH values greater than 8.5 (Guilbault and Kramer, 1964), and this should be considered when making decisions regarding FDA hydrolysis measurements in alkaline soils.

Dumontet et al. (1997) suggested that FDA hydrolysis may be a suitable tool to measure early detrimental effects of pesticides on soil microbial biomass, as it is both sensitive and non-specific. Indeed, the FDA assay has been used in ecotoxicological studies to investigate the influence of pollutants (Vischetti et al., 1997; Perucci et al., 1999) or field management effects (Bandick and Dick, 1999; Haynes and Tregurtha, 1999) on soil microorganisms. However, variable responses have been recorded for pollutant effects on FDA hydrolysis. For example, Vischetti et al. (1997) reported increases in FDA hydrolysis rates in silty clay loam soil treated with rimsulfuron (a sulphonylurea herbicide), whereas Perucci et al. (2000) described a decline in FDA hydrolysis activity after application of the same herbicide to a soil-swelling clay. Since Vekemans et al. (1989) have reported that FDA hydrolysis and microbial biomass content are closely correlated, negative pollutant effects on FDA hydrolysis activity can generally be explained in terms of reduced biomass due to the toxicity of the pollutant. Positive effects of organic pollutants on FDA activity have been interpreted as being either due to the utilization of the molecule as a source of growth by the microbial biomass, or due to the toxic effect of the pollutant, causing cell lysis and release of intracellular hydrolytic enzymes (Perucci et al., 1999). Thus, the challenge is to develop methodology to distinguish between intracellular and extracellular activity. Perucci et al. (2000) introduced the concept of specific hydrolytic activity (qFDA), where the per cent FDA hydrolysed is expressed per unit of microbial biomass carbon. This approach may help distinguish between increases in FDA activity which resulted from the growth of microorganisms on the pollutant or from the lysis of cells, releasing endocellular hydrolases.

Phosphomonoesterase Activity

Principle of the method

Soil humic material contains significant amounts of organic phosphates, in which phosphorus is bound to carbon via ester linkages. Phosphatases (which may be of plant, microbial or animal origin) catalyse the hydrolysis of phosphate esters to inorganic phosphorus. Phosphatases are classified according to the type of substrate upon which they act. For example, phosphomonoesterases (e.g. phytase, nucleotidases, sugar phosphatases) catalyse the hydrolysis of organic phosphomonoesters, whereas phosphodiesterases (e.g. nucleases, phospholipases) catalyse the hydrolysis of organic phosphodiesters (Alef and Nannipieri, 1995) (see Table 7.2). However, it is likely that phosphatases have broad substrate specificity, such that phosphomono- and phosphodiesterases share common substrates (Pant and Warman, 2000). It has been estimated that phosphatase-labile organic P makes up a significant component of the total soil P pool and, therefore, is potentially an important source of P for plants (Hayes et al., 2000). Thus, phosphatase activity is of paramount importance as a soil quality indicator.

The methodology described below involves the use of an artificial substrate, p-nitrophenyl phosphate (p-NPP). The product of phosphomonoesterase activity, p-nitrophenol, is a yellow chromophore under alkaline conditions and can be detected colorimetrically. The much used method of Tabatabai and Bremner (1969) is described.

Materials and apparatus

- Erlenmeyer flasks (50 ml) and stoppers
- Incubator at 37°C
- Spectrophotometer set to 400 nm
- Semi-micro clear polystyrene spectrophotometer cuvettes (1 ml)
- Filter papers (Whatman No. 12) or centrifuge
- Volumetric flasks (1000 ml and 100 ml)

Chemicals and solutions

- Toluene
- *Modified universal buffer (MUB) stock solution*: dissolve 12.1 g Tris, 11.6 g maleic acid, 14 g citric acid and 6.3 g boric acid in 500 ml 1 M NaOH; dilute the solution to 1000 ml with distilled water; store at 4°C
- *MUB (pH 6.5)*: adjust 200 ml of MUB stock solution to pH 6.5 using 0.1 M HCl. Make the volume up to 1000 ml with distilled water
- p-*NPP solution* (115 mM): dissolve 2.13 g of disodium p-nitrophenyl phosphate hexahydrate (Sigma-Aldrich) in 50 ml MUB; store at 4°C
- Sodium hydroxide solution (0.5 M)
- $CaCl_2$ solution (0.5 M)
- p-*Nitrophenol master solution* (1000 µg/ml): dissolve 1 g p-nitrophenol (spectrophotometric grade; Sigma-Aldrich) in 1000 ml of distilled water; store at 4°C
- p-*Nitrophenol working solution* (10 µg/ml): dilute 1 ml of p-nitrophenol master solution to 100 ml with distilled water

Procedure

- Soil, freshly collected from the field, should be sieved (< 2.8 mm), field moist and stored at 4°C prior to the assay.
- Place 1 g soil in a 50 ml Erlenmeyer flask and add 4 ml of MUB, 0.25 ml of toluene and 1 ml of p-NPP solution.
- Swirl contents to mix, stopper and incubate at 37°C for 1 h.
- At the end of the incubation, add 1 ml 0.5 M $CaCl_2$ and 4 ml 0.5 M NaOH and swirl to mix.
- Remove soil by filtration (Whatman No. 12) or centrifugation.
- Transfer the filtrate/supernatant fraction to a spectrophotometer cuvette and determine the absorbance at 400 nm.
- For the controls, add 1 ml of p-NPP solution after the addition of $CaCl_2$ (1 ml) and NaOH (4 ml) but immediately before filtration/centrifugation.
- For the calibration, pipette 0, 1, 2, 3, 4 and 5 ml aliquots of the working p-nitrophenol standard solution into 50 ml Erlenmeyer flasks. Adjust the volume to 5 ml by addition of distilled water and proceed as described for determining p-nitrophenol after the incubation of the soil samples. Plot the mass of p-nitrophenol in each reaction (0–50 µg) against the OD_{400nm} reading.

Calculation

Using the calibration curve, calculate the mass of p-nitrophenol produced in each assay from the corresponding OD_{400nm} value, correct using blank readings, and divide by the equivalent dry weight of soil (determined by loss of weight of field-moist subsamples after heating at 105°C until constant weight).

Phosphomonoesterase activity is expressed as µg p-nitrophenol/g dry soil/h.

Discussion

Phosphomonoesterase is a generic name for a group of enzymes which catalyse the hydrolysis of esters of phosphoric acid. However, both acid (pH optimum of 4–6.5) and alkaline (pH optimum of 9–10) phosphatases have been found in soil (Speir and Ross, 1978). It is suggested that the rates of synthesis of acid and alkaline phosphatases are dependent on the soil pH and that acid phosphatase is predominant in acid soils and alkaline phosphatase is predominant in alkaline soils (Juma and Tabatabai, 1978). The assay described above is buffered to pH 6.5, since this was the maximum activity recorded in the two soils tested by Tabatabai and Bremner (1969). However, the MUB may be adjusted to different pH values by titration with either 0.1 M NaOH or 0.1 HCl (Alef and Nannipieri, 1995). In addition, 'natural' activity may be determined using a non-buffered system, substituting distilled water for the MUB (Tabatabai and Bremner, 1969).

Apart from varying the pH of the assay and other slight modifications, for example using less soil in determinations of phosphatase activity in rhizosphere soil (Tarafdar and Jungk, 1987) or its application to measuring the activity associated with roots (Asmar and Gissel-Nielsen, 1997), the p-NPP method has become standard for most studies. However, in soils with high organic matter content, the method is hampered by large amounts of interfering organic materials (Trasar-Cepeda and Gilsotres, 1987; Freeman et al., 1995). Schneider et al. (2000) were able to get over this problem for high organic matter forest soils by increasing the $CaCl_2$ concentration (to 2 M) and decreasing the NaOH concentration (to 0.2 M). In peat-accumulating wetland soil samples, however, Freeman and co-workers employed the use of methylumbelliferyl (MUF)-linked phosphate coupled to high-performance liquid chromatography to separate the interferences from the compound of interest (Freeman, 1997). This method has the potential to quantify both the MUF-substrate and free-MUF product (Freeman and Nevison, 1999).

Acid and alkaline phosphomonoesterases have been used successfully, in combination with other enzyme activities, to detect heavy metal contamination in Mediterranean soil (Belen Hinojosa et al., 2004).

Interpretation of Assay Data and Conclusions

Definitions of a high-quality soil relate mainly to its ability to produce healthy and abundant crops, but also include the soil's capacity to function as a mature and sustainable ecosystem, capable of degrading organic inputs (Trasar-Cepeda et al., 2000). Although soil physical and chemical properties are important for soil function, it has been suggested that soil biochemical properties are the most useful indicators of soil quality, as they intimately reflect soil nutrient cycles and the ability of soil to break down organic matter (including organic xenobiotics) (Trasar-Cepeda et al., 2000).

However, interpretation of enzyme assay data can be problematic, in particular, dissecting what underlying mechanism(s) is responsible for a measured change in the activity of a particular enzyme. Potential mechanisms relate to microbial growth/death, enzyme de-repression/repression, and enzyme inhibition/activation. For example, an increase in the measured activity of a particular soil enzyme in response to disturbance may be interpreted in terms of five different non-mutually exclusive processes: (i) enzyme induction; (ii) enzyme activation; (iii) a shift in community structure to a species which produces greater quantities of the enzyme; (iv) microbial growth; and (v) cell lysis and release of intracellular enzymes.

Consequently, in order to aid interpretation of enzyme measurements, concurrent measurement of other parameters, such as those describing the relative contribution of intracellular and extracellular enzymes (reviewed in detail by Nannipieri et al., 2002), microbial biomass (Vischetti et al., 1997;

Perucci et al., 1999) and microbial diversity (Waldrop et al., 2000), are extremely valuable.

The existence of methodological artefacts should also be taken into account when interpreting assay data. For example, physical disturbance associated with the collection of the soil sample and sample pretreatment (e.g. air-drying and sieving) will disrupt soil structure at the aggregate and sub-aggregate scale. Consequently, the accessibility of substrates, previously physically inaccessible due to their location within aggregates and small pores, will be increased, thereby altering the environmental parameters that determine the total amount or activity present (Tate, 2002). Thus, before the soil sample is actually used in an enzyme assay, it should be recognized that, compared to the *in situ* activity expressed in the field, sample collection and storage will undoubtedly have resulted in changed activity profiles (Shaw and Burns, 1996).

Artefacts arising from the actual assay methodology itself are numerous and have been reviewed by Speir and Ross (2002). Briefly, the following points should be recognized when interpreting assay data:

- Most enzymes are assayed by using artificial substrates that either may not serve as a substrate for all enzymes catalysing the reaction in nature, or may be more easily transformed than the natural substrate.
- Assays are usually conducted under artificial conditions; the system is buffered to the optimal pH of the enzyme, at enzyme-saturating substrate concentrations, and at temperatures higher than normally encountered in soil.
- Most assay protocols involve the use of water-saturated systems, and often agitation of the samples to negate diffusional constraints. Thus, the interaction between the enzyme and substrate is maximized.

In this section, we have described simple colorimetric methods for the assay of three soil enzyme activities: dehydrogenase, fluorescein diacetate hydrolase and phosphomonoesterase. We highlight studies where the methodology has been employed in soil quality and ecotoxicological assessment. When discussing problems associated with assay interpretation, it should be remembered that there is a methodological revolution under way in which microscopy (e.g. atomic force and confocal microscopy), genomic and post-genomic (stable isotope probing, reporter genes, soil proteomics) techniques are being combined with the more established biochemical methods to solve the long-established problems and anomalies associated with soil enzymology.

Normative References (ISO, DIN) for Chapter 7

Basal respiration by titration

ISO 10381–6 (1993) Soil quality – Sampling – Guidance on the collection, handling and storage of soil for the assessment of aerobic microbial processes in the laboratory.

ISO 11465 (1993) Soil quality – Determination of dry matter and water content on mass basis – Gravimetric method.

ISO 11274 (1998) Soil quality – Determination of the water retention characteristics.

ISO 16072 (2001) Soil quality – Laboratory methods for determination of microbial soil respiration.

ISO 17155 (2001) Soil quality – Determination of abundance and activity of soil microflora using respiration curves.

Basal respiration, substrate-responsive microbial biomass, and its active and dormant part

For handling the soils see above.

ISO 17155 (2001) Soil quality – Determination of abundance and activity of soil microflora using respiration curves.

ISO 16072 (2001) Soil quality – Laboratory methods for determination of microbial soil respiration.

Nitrification

The ISO and DIN norm describes a method very similar to the measurement of net nitrification. As already stated, underestimations of the nitrification rates are possible when using those methods.

Dehydrogenase activity

ISO/DIN 23753–2 (draft standard). Soil quality – Determination of dehydrogenase activity in soil – Part 2 method using iodotetrazolium chloride (INT).

Fluorescein diacetate hydrolysis

None available.

Phosphomonoesterase activity

None available.

References for Chapter 7

Acosta-Martinez, V. and Tabatabai, M.A. (2001) Arylamidase activity in soils: effect of trace elements and relationships to soil properties and activities of amidohydrolases. *Soil Biology and Biochemistry* 33, 17–23.

Adam, G. and Duncan, H. (2001) Development of a sensitive and rapid method for the measurement of total microbial activity using fluorescein diacetate (FDA) in a range of soils. *Soil Biology and Biochemistry* 33, 943–951.

Alef, K. (1995) Soil respiration. In: Alef, K. and Nannipieri, P. (eds) *Methods in Applied Soil Microbiology and Biochemistry*. Academic Press, London, pp. 214–218.

Alef, K. and Nannipieri, P. (eds) (1995) *Methods in Applied Soil Microbiology and Biochemistry*. Academic Press, London.

Alvarez, C.R. and Alvarez, R. (2000) Short-term effects of tillage systems on active microbial biomass. *Biology and Fertility of Soils* 31, 157–161.

Alves, B.L.R., Urquiaga, S., Cadisch, G., Souto, C.M. and Boddy, R.M. (1993) In situ estimation of soil nitrogen mineralization. In: Mulongoy, K. and Merckx, R. (eds) *Soil Organic Matter Dynamics and Sustainability of Tropical Agriculture*. Wiley-Sayce, Leuven, Belgium, pp. 173–180.

Anderson, J.P.E. (1982) Soil respiration. In: Page, A.L. *et al.* (eds) *Methods of Soil Analysis, Part 2. Chemical and Microbiological Properties*, 2nd edn. Agron. Monogr. 9 ASA and SSSA, Madison, Wisconsin, pp. 831–871.

Anderson, J.P.E. and Domsch, K.H (1978) A physiological method for the quantitative measurements of microbial biomass in soils. *Soil Biology and Biochemistry* 10, 215–221.

Anderson, T.H. and Domsch, K.H. (1993) The metabolic quotient for CO_2 (qCO_2) as a specific activity parameter to assess the effects of environmental conditions, such as pH, on the microbial biomass of forest soils. *Soil Biology and Biochemistry* 25, 393–395.

Asmar, F. and Gissel-Nielsen, G. (1997) Extracellular phosphomono- and phosphodiesterase associated with and released by the roots of barley genotypes: a non-destructive method for the measurement of the extracellullar enzymes of roots. *Biology and Fertility of Soils* 25, 117–122.

Aulakh, M.S., Doran, J.W. and Mosier, A.R. (1991) Field evaluation of four methods for measuring denitrification. *Soil Science Society American Journal* 55, 1332–1338.

Aulakh, M.S., Doran, J.W. and Mosier, A.R. (1992) Soil denitrification – significance, measurement and effects of management. *Advances in Soil Sciences* 18, 1–57.

Bååth, E. (1990) Thymidine incorporation into soil bacteria. *Soil Biology and Biochemistry* 22, 803–810.

Bååth, E. (1992) Measurement of heavy metal tolerance of soil bacteria using thymidine incorporation into bacteria after homogenization-centrifugation. *Soil Biology and Biochemistry* 24, 1167–1172.

Bååth, E. (1995) Incorporation of thymidine into DNA of soil bacteria. In: Akkermans, A.D.L., van Elsas, J.D. and DeBruijn, F.J. (eds) *Molecular Microbial Ecology Manual*. Kluwer Academic Publishers, Dordrecht, The Netherlands, pp. 2.8.2, 1–9.

Bååth, E. (2001) Estimation of fungal growth rates in soil using ^{14}C-acetate incorporation into ergosterol. *Soil Biology and Biochemistry* 33, 2011–2018.

Bååth, E. and Johansson, T. (1990) Measurement of bacterial growth rates on the rhizoplane using ^{3}H-thymidine incorporation into DNA. *Plant and Soil* 126, 133–139.

Balderston, W.L., Sherr, B. and Payne, W.J. (1976) Blockage by acetylene of nitrous oxide reduction in *Pseudomonas perfectomarinus*. *Applied and Environmental Microbiology* 31, 504–508.

Bandick, A.K. and Dick, R.P. (1999) Field management effects on soil enzyme activities. *Soil Biology and Biochemistry* 31, 1471–1479.

Bardgett, R.D. and McAlister, E. (1999) The measurement of soil fungal:bacterial biomass ratios as an indicator of ecosystem self-regulation in temperate meadow grasslands. *Biology and Fertility of Soils* 29, 282–290.

Barraclough, D., Geens, E.L., Davies, G.P. and Maggs, J.M. (1985) Fate of fertiliser nitrogen. III. The use of single and double labelled ^{15}N ammonium nitrate to study nitrogen uptake by ryegrass. *Journal of Soil Science* 36, 593–603.

Battin, T.J. (1997) Assessment of fluorescein diacetate hydrolysis as a measure of total esterase activity in natural stream sediment biofilms. *The Science of the Total Environment* 198, 51–60.

Bauernfeind, G. (1996) Actual and potential denitrification rates by acetylene-inhibition technique. In: Schinner, F., Öhlinger, R., Kandeler, E. and Margesin, R. (eds) *Methods in Soil Biology*. Springer Verlag, Berlin, pp. 151–155.

Beck, T.H. (1979) Die Nitrifikation in Böden. *Zeitschrift für Pflanzenernährung und Bodenkunde* 142, 344–364.

Belen Hinojosa, M., Carreira, J.A., Garcia-Ruiz, R. and Dick, R.P. (2004) Soil moisture pre-treatment effects on enzyme activities as indicators of heavy metal-contaminated and reclaimed soils. *Soil Biology and Biochemistry* 36, 1559–1568.

Belser, L.W. (1979) Population ecology of nitrifying bacteria. *Annual Review of Microbiology* 33, 309–333.

Belser, L.W. and Mays, E.L. (1980) Specific inhibition of nitrite oxidation by chlorate and its use in assessing nitrification in soils and sediments. *Applied and Environmental Microbiology* 39, 505–510.

Benedetti, A. (2001) Defining soil quality: introduction to Round Table. In: Benedetti, A. *et al.* (eds) *COST Action 831 – Joint Meeting of Working Groups Biotechnology of Soil Monitoring, Conservation and Remediation*. Round Table: Defining Soil Quality, Rome, 10–11 December 1998. EUR 19225 – COST Action 831 – Report of 1997–1998. European Communities, Luxembourg, pp. 266–269.

Benedetti, A. and Sebastiani, G. (1996) Determination of potentially mineralizable nitrogen in agricultural soil. *Biology and Fertility of Soils* 21, 114–120.

Benedetti, A., Canali, S. and Alianiello, F. (1994) Mineralisation dynamics of the organic nitrogen: soil management effects. *Proceedings of the N-immobilisation workshop*. November, Macaulay Land Use Research Institute, Aberdeen, UK.

Benedetti, A., Vittori Antisari, L., Canali, S., Giacchini, P. and Sequi, P. (1996) Relationship between the fixed ammonium and the mineralization of the organic nitrogen in soil. In: Van Cleemput, O. *et al.* (eds) *Progress in Nitrogen Cycling Studies*. Kluwer Academic Publishers, Dordrecht, The Netherlands, pp. 23–26.

Benefield, C.B., Howard, P.J.A. and Howard, D.M. (1977) The estimation of dehydrogenase activity in soil. *Soil Biology and Biochemistry* 9, 67–70.

Berg, P. and Rosswall, T. (1985) Ammonium oxidizer numbers, potential and actual oxidation rates in two Swedish arable soils. *Biology and Fertility of Soils* 1, 131–140.

Bloem, J. and Breure, A.M. (2003) Microbial indicators. In: Markert, B., Breure, A.M. and Zechmeister, H. (eds) *Bioindicators/Biomonitors – Principles, Assessment, Concepts*. Elsevier, Amsterdam, pp. 259–282.

Bloem, J., Ellenbroek, F., Bär-Gilissen, M.J.B. and Cappenberg, T.E. (1989) Protozoan grazing and bacterial production in stratified Lake Vechten, estimated with fluorescently labelled bacteria and thymidine incorporation. *Applied and Environmental Microbiology* 55, 1787–1795.

Bloem, J., Bolhuis, P.R., Veninga, M.R. and Wieringa, J. (1995) Microscopic methods for counting bacteria and fungi in soil. In: Alef, K. and Nannipieri, P. (eds) *Methods in Applied Soil Microbiology and Biochemistry*. Academic Press, London, pp. 162–173.

Bloem, J., de Ruiter, P.C. and Bouwman, L.A. (1997) Food webs and nutrient cycling in agro-ecosystems. In: van Elsas, J.D., Trevors, J.T. and Wellington, E. (eds) *Modern Soil Microbiology*. Marcel Dekker, New York, pp. 245–278.

Bodelier, P.L.E., Libochant, J.A., Blom, C.W.P.M. and Laanbroek, H.J. (1996) Dynamics of nitrification and denitrification in root-oxygenated sediments and adaptation of ammonia-oxidizing bacteria to low-oxygen or anoxic habitats. *Applied and Environmental Microbiology* 62, 4100–4107.

Bodelier, P.L.E., Hahn, A.P., Arth, I.R. and Frenzel, P. (2000) Effects of ammonium-based fertilisation on microbial processes involved in methane. *Biogeochemistry* 51, 225–257.

Bouwman, L.A., Bloem, J., van den Boogert, P.H.J.F., Bremer, F., Hoenderboom, G.H.J. and de Ruiter, P.C. (1994) Short-term and long-term effects of bacterivorous nematodes and nematophagous fungi on carbon and nitrogen mineralization in microcosms. *Biology and Fertility of Soils* 17, 249–256.

Bremner, J.M. (1965) Nitrogen availability index. In: Black, C.A., Evans, D.D., White, J.L., Ensminger, L.E. and Clark, F.E. (eds) *Methods of Soil Analysis Part 2*. Agronomy 9, American Society of Agronomy, Madison, Wisconsin, pp. 1324–1341.

Bringmark, E. and Bringmark, L. (1993) Standard respiration, a method to test the influence of pollution and environmental factors on a large number of samples. MATS Guideline – Test 03. In: Torstensson, L. (ed.) *Guidelines. Soil Biological Variables in Environmental Hazard Assessment*. Report No. 4262, Swedish EPA, Stockholm, pp. 34–39.

Bringmark, L. and Bringmark, E. (2001a) Soil respiration in relation to small-scale patterns of lead and mercury in mor layers of Southern Swedish forest sites. *Water, Air and Soil Pollution: Focus* 1, 395–408.

Bringmark, L. and Bringmark, E. (2001b) Lowest effect levels of lead and mercury on decomposition of mor layer samples in a long-term experiment. *Water, Air and Soil Pollution: Focus* 1, 425–437.

Brock, T.D. and Madigan, M.T. (1991) *Biology of Microorganisms*. Prentice-Hall, Englewood Cliffs, New Jersey.

Brookes, P.C. (1995) The use of microbial parameters in monitoring soil pollution by heavy metals. *Biology and Fertility of Soils* 19, 269–279.

Brookes, P.C. and McGrath, S.P. (1984) Effects of metal toxicity on size of the soil microbial biomass. *Journal of Soil Science* 35, 341–346.

Bundy, L.G. and Meisinger, J.J. (1994) Nitrogen availability indices. In: Weaver, R.W., Angle, J.S. and Bottomley, P.D. (eds) *Methods of Soil Analysis. Part 2. Microbiological and Biochemical Properties*. Soil Science Society of America, Madison, Wisconsin.

Burns, R.G. (1978) *Soil Enzymes*. Academic Press, London.

Burns, R.G. (1982) Enzyme activity in soil: location and a possible role in microbial ecology. *Soil Biology and Biochemistry* 14, 423–427.

Burton, D.L. and Beauchamp, E.G. (1984) Field techniques using the acetylene blockage of nitrous oxide reduction to measure denitrification. *Canadian Journal of Soil Science* 64, 555–562.

Buyanovsky, G.A., Kucera, C.L. and Wagner, G.H. (1987) Comparative analyses of carbon dynamics in native and cultivated ecosystems. *Ecology* 68, 2023–2031.

Chander, K. and Brookes, P.C. (1991) Effects of heavy metals from past application of sewage sludge on microbial biomass and organic matter accumulation in a sandy loam and silt loam U.K. soil. *Soil Biology and Biochemistry* 23, 927–932.

Christensen, H. (1991) Conversion factors relating thymidine uptake to growth rate with rhizosphere bacteria. In: Kleister, D.L. and Gregan, P.B. (eds) *The Rhizosphere and Plant Growth*. Kluwer Academic, Dordrecht, The Netherlands, pp. 99–102.

Christensen, H. and Christensen, S. (1995) [^{3}H]Thymidine incorporation technique to determine soil bacterial growth rate. In: Alef, K. and Nannipieri, P. (eds) *Methods in Applied Soil Microbiology and Biochemistry*. Academic Press, London, pp. 258–261.

Dalenberg, J.W and Jager, G. (1981) Priming effect of small glucose additions to ^{14}C-labelled soil. *Soil Biology and Biochemistry* 13, 219–223.

Dalenberg, J.W. and Jager, G. (1989) Priming effect of some organic additions to ^{14}C-labelled soil. *Soil Biology and Biochemistry* 21, 443–448.

Davidson, E.A., Stark, J.M. and Firestone, M.K. (1990) Microbial production and consumption of nitrate in an annual grassland. *Ecology* 71, 1968–1975.

Davidson, E.A., Hart, S.C., Shanks, C.A. and Firestone, M.K. (1991) Measuring gross nitrogen mineralisation, immobilisation and nitrification by ^{15}N isotopic pool dilution in intact soil cores. *Journal of Soil Science* 42, 335–349.

Dick, R.P. (1997) Soil enzyme activities as integrative indicators of soil health. In: Pankhurst, C., Doube, B. and Gupta, V. (eds) *Biological Indicators of Soil Health*. CAB International, Wallingford, UK, pp. 121–156.

Dilly, O. (2001) Microbial respiratory quotient during metabolism and after glucose amendment in soils and litter. *Soil Biology and Biochemistry* 33, 117–127.

Dilly, O. and Munch, J.C. (1996) Microbial biomass content, basal respiration and enzyme activities during the course of decomposition of leaf litter in a black alder (*Alnus glutinosa* (L.) Gaertn.) forest. *Soil Biology and Biochemistry* 28, 1073–1081.

Dilly, O., Bach, H.J., Buscot, F., Eschenbach, C., Kutsch, W.L., Middelhoff, U., Pritsch, K. and Munch, J.C. (2000) Characteristics and energetic strategies of the rhizosphere in ecosystems of the Bornhöved Lake district. *Applied Soil Ecology* 15, 201–210.

Dixon, J.L. and Turley, C.M. (2000) The effect of water depth on bacterial numbers, thymidine incorporation rates and C:N ratios in northeast Atlantic surficial sediments. *Hydrobiologia* 440, 217–225.

Doran, J.W. and Parkin, T.B. (1994) Defining and assessing soil quality. In: *Defining Soil Quality for a Sustainable Environment*. Special publication Number 35, Soil Science Society of America, Madison, Wisconsin, pp. 3–21.

Dumontet, S., Perucci, P. and Bufo, S.A. (1997) Les amendments organiques et traitements avec les herbicides: influence sur la biomasse microbienne du sol. In: *Congres du Groupment Francais des Pesticides, 21–22 Mai*. BRGM-Orleans, France.

Duxbury, J.M. (1986) Advantages of the acetylene method of measuring denitrification. In: Hauck, R.D. and Weaver, R.W. (eds) *Field Measurement of Dinitrogen Fixation and Denitrification*. Soil Science Society of America, Madison, Wisconsin, pp. 73–92.

Ellenbroek, F. and Cappenberg, T.E. (1991) DNA synthesis and tritiated thymidine incorporation by heterotrophic freshwater bacteria in continuous culture. *Applied and Environmental Microbiology* 57, 1675–1682.

Falchini, L., Muggianu, M., Landi, L. and Nannipieri, P. (2001) Influence of soluble organic C compounds present in root exudates on soil N dynamics. In: Benedetti, A. *et al.* (eds) *COST Action 831 – Joint Meeting of Working Groups Biotechnology of Soil Monitoring and Remediation*. Round Table: Defining Soil Quality. Rome, 10–11 December 1998. EUR 19225 – COST Action 831 – Report of 1997–1998. European Communities, Luxembourg, pp. 288–289.

Findlay, S., Meyer, J.L. and Edwards, R.T. (1984) Measuring bacterial production via rate of incorporation of [^{3}H]thymidine into DNA. *Journal of Microbiological Methods* 2, 57–72.

Focht, D.D. and Verstraete, W. (1977) Biochemical ecology of nitrification and denitrification. *Advances in Microbial Ecology* 1, 135–214.

Franzluebbers, A.J., Zuberer, D.A. and Hons, F.M. (1995) Comparison of microbiological methods for evaluating quality and fertility of soil. *Biology and Fertility of Soils* 19, 135–140.

Freeman, C. (1997) Using HPLC to eliminate quench interference in fluorogenic substrate assays of microbial enzyme activity. *Soil Biology and Biochemistry* 29, 203–205.

Freeman, C. and Nevison, G.B. (1999) Simultaneous analysis of multiple enzymes in environmental samples using methylumbelliferyl substrates and HPLC. *Journal of Environmental Quality* 28, 1378–1380.

Freeman, C., Liska, G., Ostle, N., Jones, S.E. and Lock, M.A. (1995) The use of fluorogenic substrates for measuring enzyme activity in peatlands. *Plant and Soil* 175, 147–152.

Fuhrman, J.A. and Azam, F. (1982) Thymidine incorporation as a measure of heterotrophic bacterioplankton production in marine surface waters: evaluation and field results. *Marine Biology* 66, 109–120.

Giller, K.E., Witter, E. and McGrath, S.P. (1998) Toxicity of heavy metals to microorganisms and microbial processes in agricultural soils: a review. *Soil Biology and Biochemistry* 30, 1389–1414.

Gong, P., Siciliano, S.D., Greer, C.W., Paquet, L., Hawari, J. and Sunahara, G.I. (1999) Effects and bioavailability of 2,4,6-trinitrotoluene in spiked and field-contaminated soils to indigenous microorganisms. *Environmental Toxicology and Chemistry* 18, 2681–2688.

Goyal, S., Chander, K., Mundra, M.C. and Kapoor, K.K. (1999) Influence of inorganic fertilizers and organic amendments on soil organic matter and soil microbial properties under tropical conditions. *Biology and Fertility of Soils* 29, 196–200.

Granhall, U. (1999) Microbial biomass determinations. Comparisons between some direct and indirect methods. *COST 831 – Working Group 4 Meeting*, Granada, Spain, 15–16 July, 1999.

Granli, T. and Bøckman, O.C. (1994) Nitrous oxide from agriculture. *Norwegian Journal of Agricultural Sciences Supplement* 12, 1–128.

Grisi, B., Grace, C., Brookes, P.C., Benedetti, A. and Dell'Abate, M.T. (1998) Temperature effects on organic matter and microbial dynamics in temperate and tropical soils. *Soil Biology and Biochemistry* 30, 1309–1315.

Guilbault, G.C. and Kramer, D.N. (1964) Fluorometric detection of lipase, acylase, alpha- and gamma-chymotrypsin and inhibitors of these enzymes. *Analytical Chemistry* 36, 409–412.

Hadas, A., Kautsky, L. and Portnoy, R. (1998) Mineralization of composted manure and microbial dynamics in soil as affected by long-term nitrogen management. *Soil Biology and Biochemistry* 28, 733–738.

Hansson, A.J., Andrén, O., Boström, S., Boström, U., Clarholm, M., Lagerlöf, J., Lindberg, T., Paustian, K., Pettersson, R. and Sohlenius, B. (1990) Structure of the agroecosystem. In: Andrén, O., Lindberg, T., Paustian, K. and Rosswall, T. (eds) *Ecology of Arable Land. Organisms, Carbon and Nitrogen Cycling.* (*Ecological Bulletin*) 40, 41–83.

Harris, D. and Paul, E.A. (1994) Measurement of bacterial growth rates in soil. *Applied Soil Ecology* 1, 277–290.

Hart, S.C. and Binkley, D. (1985) Correlations among indices for forest nutrient availability in fertilized and unfertilized loblolly pine plantations. *Plant and Soil* 85, 11–21.

Hart, S.C., Stark, J.M., Davidson, E.A. and Firestone, M.K. (1994) Nitrogen mineralization, immobilisation and nitrification. In: Weaver, R.W., Angle, J.S. and Bottomley, P.D. (eds) *Methods of Soil Analysis. Part 2, Microbiological and Biochemical Properties.* Soil Science Society of America, Madison, Wisconsin, pp. 985–1018.

Hatch, D.J., Jarvis, S.C. and Reynolds, S.E. (1991) An assessment of the contribution of net mineralization to N cycling in grass swards using a field incubation method. *Plant and Soil* 138, 23–32.

Hauck, R.D., Melsted, S.W. and Yankwich, P.E. (1958) Use of N-isotope distribution in nitrogen gas in the study of denitrification. *Soil Science* 86, 287–291.

Hayes, J.E., Richardson, A.E. and Simpson, R.J. (2000) Components of organic phosphorus in soil extracts that are hydrolysed by phytase and acid phosphatase. *Biology and Fertility of Soils* 32, 279–286.

Haynes, R.J. and Tregurtha, R. (1999) Effects of increasing periods under intensive arable vegetable production on biological, chemical and physical indices of soil quality. *Biology and Fertility of Soils* 28, 259–266.

Heinemeyer, O. and Kaiser, E.A. (1996) Automated gas injector system for gas chromatography: atmospheric nitrous oxide analysis. *Soil Science Society America Journal* 60, 808–811.

Hynes, R.K. and Knowles, R. (1983) Inhibition of chemoautotrophic nitrification by sodium chlorate and sodium chlorite: a reexamination. *Appied and Environmental Microbiology* 45, 1178–1182.

Jarvis, S.C., Hatch, D.J. and Lovell, R.D. (2001) An improved soil core incubation method for the field measurement of denitrification and net mineralization using acetylene inhibition. *Nutrient Cycling in Agroecosystems* 59, 219–225.

Jeffrey, W.H. and Paul, J.H. (1988) Underestimation of DNA synthesis by [^{3}H]thymidine incorporation in marine bacteria. *Applied and Environmental Microbiology* 54, 3165–3168.

Johansson, M., Pell, M. and Stenström, J. (1998) Kinetics of substrate-induced respiration (SIR) and denitrification: applications to a soil amended with silver. *Ambio* 27, 40–44.

Jones, R.B., Gilmour, C.C., Stoner, D.L., Weir, M.M. and Tuttle, J.H. (1984) Comparison of methods to measure acute metal and organometal toxicity to natural aquatic microbial communities. *Applied and Environmental Microbiology* 47, 1005–1011.

Juma, N.G. and Tabatabai, M.A. (1978) Distribution of phosphomonoesterases in soils. *Soil Science* 126, 101–108.

Kandeler, E., Marschner, P., Tscherko, D., Gahoonia, T. and Nielsen, N. (2002) Microbial community composition and functional diversity in the rhizosphere of maize. *Plant and Soil* 238, 301–312.

Keeney, D.R. (1982) Nitrogen availability indexes. In: Page, A.L. *et al.* (eds) *Methods of Soil Analysis. Part 2*, 2nd edn. Agronomy Monograph 9, Soil Science Society of America, Madison, Wisconsin, pp. 711–733.

Keeney, D.R. and Nelson, D.W. (1982) Nitrogen – inorganic forms. In: Black, C.A., Evans, D.D., White, J.L., Ensminger, L.E. and Clark, F.E. (eds) *Methods of Soil Analysis, Part 2*. American Society of Agronomy, Madison, Wisconsin, pp. 682–687.

Kilham, K. (1985) A physiological determination of the impact of environmental stress on the activity of microbial biomass. *Environmental Pollution* 38, 283–294.

Kilham, K. (1987) A new perfusion system for the measurement and characterization of potential rates of soil nitrification. *Plant and Soil* 97, 227–232.

Kirchman, D., K'Nees, E. and Hodson, R. (1985) Leucine incorporation and its potential as a measure of protein synthesis by bacteria in natural aquatic ecosystems. *Applied and Environmental Microbiology* 49, 559–607.

Kirkham, D. and Bartholomew, W.V. (1954) Equations for following nutrient transformations in soil utilizing data. *Soil Science Society American Proceedings* 18, 33–34.

Klemedtsson, L., Hansson, G. and Mosier, A.R. (1990) The use of acetylene for the quantification of N_2 and N_2O production from biological processes in soil. In: Revsbech, N.P. and Sorensen, J. (eds) *Denitrification in Soil and Sediment*, Vol. 56. Plenum Press, New York, pp. 167–180.

Koops, H.P. and Pommerening-Röser, A. (2001) Distribution and ecophysiology of the nitrifying bacteria emphasizing cultured species. *FEMS Microbial Ecology* 37, 1–9.

Kowalchuk, G.A. and Stephen, J.R. (2001) Ammonia-oxidizing bacteria: a model for molecular microbial ecology. *Annual Review in Microbiology* 55, 485–529.

Kowalchuk, G.A., Stephen, J.R., de Boer, W., Prosser, J.I., Embley, T.M. and Woldendorp, J.W. (1997) Analysis of ammonia-oxidizing bacteria of the beta subdivision of the class proteobacteria in coastal sand dunes by denaturing gradient gel electrophoresis and sequencing of PCR-amplified 16s ribosomal DNA fragments. *Applied and Environmental Microbiology* 63, 1489–1497.

Ladd, J.N., Forster, R.C., Nannipieri, P. and Oades, J.M. (1996) Soil structure and biological activity. *Soil Biochemistry* 9, 23–78.

Lober, R.W. and Reeder, J.D. (1993) Modified waterlogged incubation method for assessing nitrogen mineralization in soils and soil aggregates. *Soil Science Society American Journal* 57, 400–403.

Mader, T. (1994) Auswirkungen einer praxisüblichen Anwendung von Gardoprim (Terbuthylazin) auf mikrobielle und biochemische Stoffumsetzungen sowie sein abbauverhalten im Feld- und Laborversuch. *Hohenheimer Bodenkundliche Hefte*, 19.

Madsen, E.L. (1996) A critical analysis of methods for determining the composition and biogeochemical activities of soil microbial communities in situ. In: Stotzky, G. and Bollag, J.M. (eds) *Soil Biochemistry*, Vol. 9. Marcel Dekker, New York, pp. 287–370.

Malone, J.P., Stevens, R.J. and Laughlin, R.J. (1998) Combining the ^{15}N and acetylene inhibition techniques to examine the effect of acetylene on denitrification. *Soil Biology and Biochemistry* 30, 31–37.

Mamilov, A.Sh., Byzov, B.A., Zvyagintsev, D.G. and Dilly, O.M. (2001) Predation on fungal and bacterial biomass in a soddy-podzolic soil amended with starch, wheat straw and lucerne meal. *Applied Soil Ecology* 16, 131–139.

Mary, B., Recous, S. and Robin, D. (1998) A model for calculating nitrogen fluxes in soil using ^{15}N tracing. *Soil Biology and Biochemistry* 30, 1963–1979.

McCarty, G.W. and Bremner, J.M. (1986) Inhibition of nitrification in soil by acetylenic compounds. *Soil Science Society American Journal* 50, 1198–1201.

Meisinger, J.J. (1984) Evaluating plant available N in soil crop systems. In: Hauck, R.D. *et al.* (eds) *Nitrogen in Crop Production*. American Society of Agronomy, Madison, Wisconsin, pp 391–416.

Michel, P.H. and Bloem, J. (1993) Conversion factors for estimation of cell production rates of soil bacteria from tritiated thymidine and tritiated leucine incorporation. *Soil Biology and Biochemistry* 25, 943–950.

Mogge, B., Kaiser, E.A. and Munch, J.C. (1998) Nitrous oxide emissions and denitrification N-losses from forest soils in the Bornhöved Lake region (Northern Germany). *Soil Biology and Biochemistry* 30, 703–710.

Mogge, B., Kaiser, E.-A. and Munch, J.C. (1999) Nitrous oxide emissions and denitrification N-losses from agricultural soils in the Bornhöved Lake region: influence of organic fertilizers and land-use. *Soil Biology and Biochemistry* 31, 1245–1252.

Moreno, J., Garcia, C., Landi, L., Falchini, L., Pietramellara, G. and Nannipieri, P. (2001) The ecological dose value (ED50) for assessing Cd toxicity on ATP content and dehydrogenase and urease activities of soil. *Soil Biology and Biochemistry* 33, 483–489.

Moriarty, D.J.W. (1986) Measurement of bacterial growth rates in aquatic systems from rates of nucleic acid synthesis. *Advances in Microbial Ecology* 9, 245–292.

Mosier, A.R. and Schimel, D.S. (1993) Nitrification and denitrification. In: Knowles, R. and Blackburn, T.H. (eds) *Nitrogen Isotope Techniques*. Academic Press, San Diego, California, pp. 181–208.

Mosier, A.R., Guenzi, W.D. and Schweizer, E.E. (1986) Field denitrification estimation by nitrogen-15 and acetylene inhibition techniques. *Soil Science Society American Journal* 50, 831–833.

Myhre, G., Highwood, E.K., Shine, K.P. and Stordal, F. (1998) New estimates of radiative forcing due to well mixed greenhouse gases. *Geophysical Research Letters* 25, 2715–2718.

Myrold, D.D. (1990) Measuring denitrification in soils using ^{15}N techniques. In: Revsbech, N.P. and Sorensen, J. (eds) *Denitrification in Soil and Sediment*. Plenum Press, New York, pp. 181–198.

Myrold, D.D. and Tiedje, J.M. (1986) Simultaneous estimation of several nitrogen cycle rates using ^{15}N: theory and application. *Soil Biology and Biochemistry* 18, 559–568.

Nannipieri, P., Kandeler, E. and Ruggiero, P. (2002) Enzyme activities and microbiological and biochemical processes in soil. In: Burns, R.G. and Dick, R.P. (eds) *Enzymes in the Environment*. Marcel Dekker, New York, pp. 1–33.

Nicolaisen, M.H. and Ramsing, N.B. (2002) Denaturing gradient gel electrophoresis (DGGE) approaches to study the diversity of ammonia-oxidizing bacteria. *Journal of Microbiological Methods* 50, 89–203.

Nishio, T., Kanamori, T. and Fjuimoto, T.

(1985) Nitrogen transformation in an aerobic soil as determined by a $^{15}NH_4^+$ dilution technique. *Soil Biology and Biochemistry* 17, 149–154.

Obbard, J.P. (2001) Measurement of dehydrogenase activity using 2-*p*-iodophenyl-3-*p*-nitrophenyl-5-phenyltetrazolium chloride (INT) in the presence of copper. *Biology and Fertility of Soils* 33, 328–330.

Öhlinger, R., Beck, T., Heilmann, B. and Beese, F. (1996) Soil respiration. In: Schinner, F., Öhlinger, R., Kandeler, E. and Margesin, R. (eds) *Methods in Soil Biology*. Springer-Verlag, Berlin, pp. 93–110.

Orchard, V.A. and Cook, F.J. (1983) Relationships between soil respiration and soil moisture. *Soil Biology and Biochemistry* 15, 447–453.

Palmborg, C. and Nordgren, A. (1993) Soil respiration curves, a method to test the abundance, activity and vitality of the microflora in soils. In: Torstensson, L. (ed.) *Guidelines. Soil Biological Variables in Environmental Hazard Assessment*. Report 4262, Swedish EPA, Stockholm, pp. 149–156.

Pankhurst, C.E., Hawke, B.G., McDonald, H.J., Kirkby, C.A., Buckerfield, J.C., Michelsen, P., O'Brian, K.A., Gubta, V.V.S.R. and Double, B.M. (1995) Evaluation of soil biological properties as potential bioindicators of soil health. *Australian Journal of Experimental Agriculture* 35, 1015–1028.

Pant, H.K. and Warman, P.R. (2000) Enzymatic hydrolysis of soil organic phosphorus by immobilized phosphatases. *Biology and Fertility of Soils* 30, 306–311.

Parkin, T.B., Kaspar, H.F., Sexstone, A.J. and Tiedje, J.M. (1984) A gas-flow soil core method to measure field denitrification rates. *Soil Biology and Biochemistry* 16, 323–330.

Parkin, T.B., Sexstone, A.J. and Tiedje, J.M. (1985) Comparison of field denitrification rates by acetylene-based soil core and nitrogen-15 methods. *Soil Science Society American Journal* 49, 94–99.

Pasqual, J.A., Hernandez, T. and Garcia, C. (2001) In: Benedetti, A. *et al.* (eds) COST Action 831 – Joint Meeting of Working Groups Biotechnology of Soil Monitoring, Conservation and Remediation. Round Table: Defining Soil Quality. Rome, 10–11 December 1998. EUR 19225 – COST Action 831 – Report of 1997–1998. European Communities, Luxembourg, p. 310.

Paul, E.A. and Clark, F.E. (1989) *Soil Microbiology and Biochemistry*. Academic Press, Inc., San Diego, California, 275 pp.

Paustian, K., Bergström, L., Jansson, P.E. and Johnsson, H. (1990) Ecosystem dynamics. In: Andrén, O., Lindberg, T., Paustian, K. and Rosswall, T. (eds) *Ecology of Arable Land. Organisms, Carbon and Nitrogen Cycling*. (*Ecological Bulletin*) 40, 153–180.

Persson, T., Bååth, E., Clarholm, M., Lundkvist, B.E. and Sohlenius, H. (1980) Trophic structure, biomass dynamics and carbon metabolism of soil organisms in a Scots pine forest. In: Ågren, G.I., Andersson, F., Falk, S.O., Lohm, U. and Perttu, K. (eds) *Structure and Function of Northern Coniferous Forests*. (*Ecological Bulletin*) 32, 419–459.

Perucci, P., Vishetti, C. and Battistoni, F. (1999) Rimsulfuron in a silty clay loam soil: effects upon microbiological and biochemical properties under varying microcosm conditions. *Soil Biology and Biochemistry* 31, 195–204.

Perucci, P., Dumontet, S., Bufo, S.A., Mazzatura, A. and Casucci, C. (2000) Effects of organic amendment and herbicide treatment on soil microbial biomass. *Biology and Fertility of Soils* 32, 17–23.

Powlson, D.S. and Barraclough, D. (1993) Mineralization and assimilation in soil–plant system. In: Knowles, R. and Black, T.H. (eds) *Nitrogen Isotope Techniques*. Academic Press, San Diego, California, pp. 209–242.

Prescott, D.M. and James, T.W. (1955) Culturing of *Amoeba proteus* on *Tetrahymena*. *Experimental Cell Research* 8, 256–258.

Prosser, J.I. (1989) Autotrophic nitrification in bacteria. *Advances in Microbial Physiology* 30, 125–181.

Recous, S., Luxhoi, J., Fillery, I.R.P., Jensen,

L.S. and Mary, B. (2001) Soil temperature effects on N mineralisation, immobilisation and nitrification using ^{15}N dilution techniques. *Abstracts of the 11th Nitrogen Workshop, 9–12 September, Reims, France.*

Reddy, G.B., Faza, A. and Bennett, R. (1987) Activity of enzymes in rhizosphere and non-rhizosphere soils amended with sludge. *Soil Biology and Biochemistry* 19, 203–205.

Robarts, R.D. and Zohary, T. (1993) Fact or fiction – bacterial growth rates and production as determined by [methyl-^{3}H]-thymidine? *Advances in Microbial Ecology* 13, 371–425.

Rotman, B. and Papermaster, B.W. (1966) Membrane properties of living mammalian cells as studied by enzymatic hydrolysis of fluorogenic esters. *Proceedings of the National Academy of Science USA* 55, 134–141.

Rotthauwe, J.H., Witzel, K.P. and Liesack, W. (1997) The ammonia monooxygenase structural gene *amoA* as a functional marker: molecular fine-scale analysis of natural ammonia-oxidizing populations. *Applied and Environmental Microbiology* 63, 4704–4712.

Rudaz, A.O., Walti, E., Kyburz, G., Lehmann, P. and Fuhrer, J. (1999) Temporal variation in N_2O and N_2 fluxes from a permanent pasture in Switzerland in relation to management, soil water content and soil temperature. *Agriculture, Ecosystems and Environment* 73, 83–91.

Ryden, J.C., Lund, L.J., Letey, J. and Focht, D.D. (1979) Direct measurement of denitrification loss from soils. II. Development and application of field methods. *Soil Science Society American Journal* 43, 110–118.

Ryden, J.C., Skinner, J.H. and Nixon, D.J. (1987) Soil core incubation system for the field measurement of denitrification using acetylene-inhibition. *Soil Biology and Biochemistry* 19, 753–757.

Sahrawat, K.L. (1983) Mineralization of soil organic nitrogen under waterlogged condition in relation to other properties of tropical rice soils. *Australian Journal of Soil Research* 21, 133–138.

Sahrawat, K.L. and Ponnamperuma, F.N. (1978) Measurement of exchangeable NH_4^+ in tropical land soils. *Soil Science Society of America Journal* 42, 282–283.

Schimmel, E.L., Firestone, M.K. and Killham, K.S. (1984) Identification of heterotrophic nitrification in a sierran forest soil. *Applied and Environmental Microbiology* 48, 802–806.

Schinner, F., Oehlinger, R. and Kandeler, E. (1992) *Bodenbiologische Arbeitsmethoden.* Springer, Berlin.

Schmidt, E.L. and Belser, L.W. (1994) Autotrophic nitrifying bacteria. In: Weaver *et al.* (eds) *Methods in Soil Analysis Part 2, Microbiological and Biochemical Properties*, SSSA Book Series No. 5. Soil Science Society of America, Madison, Wisconsin, pp. 159–177.

Schneider, K., Turrion, M.-B. and Gallardo, J.-F. (2000) Modified method for measuring acid phosphatase activities in forest soils with high organic matter content. *Communications in Soil Science and Plant Analysis* 31, 3077–3088.

Servais, P., Martinez, J., Billen, G. and Vives-Rego, J. (1987) Determining [^{3}H]thymidine incorporation into bacterioplankton DNA: improvement of the method by DNase treatment. *Applied and Environmental Microbiology* 53, 1977–1979.

Shaw, L.J. and Burns, R.G. (1996) Construction and equilibration of a repacked soil column microcosm. *Soil Biology and Biochemistry* 28, 1117–1120.

Shaw, L.J. and Burns, R.G. (2003) Biodegradation of organic pollutants in the rhizosphere. *Advances in Applied Microbiology* 53, 1–60.

Sparling, G.P. (1997) Soil microbial biomass, activity and nutrient cycling. In: Pankhurst, C.E., Doube, B.M. and Gupta, V.V.S.R. (eds) *Biological Indicators of Soil Health.* CAB International, Wallingford, UK, pp. 97–120.

Speir, T.W. and Ross, D.J. (1978) Soil phosphatase and sulphatase. In: Burns, R.G. (ed.) *Soil Enzymes.* Academic Press, London, pp. 197–250.

Speir, T.W. and Ross, D.J. (2002) Hydrolytic enzyme activities to assess soil degradation and recovery. In: Burns, R.G. and Dick, R.P. (eds) *Enzymes in the Environment*. Marcel Dekker, New York, pp. 407–431.

Stanford, G. (1982) Assessment of soil nitrogen availability. In: Stevenson, F.J. (ed.) *Nitrogen in Agricultural Soils*. Agronomy Monograph 22. ASA and SSSA, Madison, Wisconsin.

Stanford, G. and Smith, S.J. (1972) Nitrogen mineralization potentials of soils. *Soil Science Society American Proceedings* 38, 99–102.

Stark, J.M. and Firestone, M.K. (1996) Kinetic characteristics of ammonium-oxidizer communities in a California oak woodland-annual grassland. *Soil Biology and Biochemistry* 28, 1307–1317.

Steen, E. (1990) Agricultural outlook. In: Andrén, O., Lindberg, T., Paustian, K. and Rosswall, T. (eds) *Ecology of Arable Land. Organisms, Carbon and Nitrogen Cycling.* (*Ecological Bulletin*) 40, 181–192.

Stenberg, B., Johansson, M., Pell, M., Sjödal-Svensson, K., Stenström, J. and Torstensson, L. (1998a) Microbial biomass and activities in soil as affected by frozen and cold storage. *Soil Biology and Biochemistry* 30, 393–402.

Stenberg, B., Pell, M. and Torstensson, L. (1998b) Integrated evaluation of variations in biological, chemical and physical soil properties. *Ambio* 27, 9–15.

Stenström, J., Stenberg, B. and Johansson, M. (1998) Kinetics of substrate-induced respiration (SIR): theory. *Ambio* 27, 35–39.

Stenström, J., Svensson, K. and Johansson, M. (2001) Reversible transition between active and dormant microbial states in soil. *FEMS Microbiology Ecology* 36, 93–104.

Stephen, J.R., Chang, Y.J., Macnaughton, S.J., Kowalchuk, G.A., Leung, K.T., Flemming, C.A. and White, D.C. (1999) Effect of toxic metals on indigenous soil p-subgroup proteobacterium ammonia oxidizer community structure and protection against toxicity by inoculated metal-resistant bacteria. *Applied and Environmental Microbiology* 65, 95–101.

Stevens, R.J. and Laughlin, R.J. (1998) Measurement of nitrous oxide and di-nitrogen emissions from agricultural soils. *Nutrient Cycling in Agroecosystems* 52, 131–139.

Stevenson, F.J. (1985) *The Internal Cycle of Nitrogen in Soil. Cycle of Soil C, N, P, S, Micronutrients.* J. Wiley & Sons, New York, pp. 155–215.

Stotzky, G. (1965) Microbial respiration. In: Black C.A. *et al.* (eds) *Methods in Soil Analysis, Part 2*. Agronomy Monograph 9. ASA, Madison, Wisconsin, pp. 1550–1572.

Stoyan, H., de Polli, H., Bohm, S., Robertson, G.P. and Paul, E.A. (2000) Spatial heterogeneity of soil respiration and related properties at the plant scale. *Plant and Soil* 222, 203–214.

Svensson, K. (1999) Effect of different cropping systems on soil microbial activities. *COST 831 – Working Group 4 Meeting*, Granada, Spain, 15–16 July.

Svensson, K. and Pell, M. (2001) Soil microbial tests for discriminating between different cropping systems and fertilizer regimes. *Biology and Fertility of Soils* 33, 91–99.

Tabatabai, M.A. and Bremner, J.M. (1969) Use of p-nitrophenyl phosphate for assay of soil phosphatase activity. *Soil Biology and Biochemistry* 1, 301–307.

Tabatabai, M.A. and Dick, W.A. (2002) Enzymes in soil: research and developments in measuring activities. In: Burns, R.G. and Dick, R.P. (eds) *Enzymes in the Environment*. Marcel Dekker, New York, pp. 567–596.

Tarafdar, J.C. and Jungk, A. (1987) Phosphatase activity in the rhizosphere and its relation to the depletion of soil organic phosphorus. *Biology and Fertility of Soils* 3, 199–204.

Tate, R.L. (2002) Microbiology and enzymology of carbon and nitrogen cycling. In: Burns, R.G. and Dick, R.P. (eds) *Enzymes in the Environment*. Marcel Dekker, New York, pp. 227–248.

Taylor, J., Wilson, B., Mills, M. and Burns, R. (2002) Comparison of microbial numbers and enzymatic activities in surface soils and subsoils using various techniques. *Soil Biology and Biochemistry* 34, 387–401.

Terry, R.E. and Duxbury, J.M. (1985) Acetylene decomposition in soils. *Soil Science Society American Journal* 49, 90–94.

Thorn, P.M. and Ventullo, R.M. (1988) Measurement of bacterial growth rates in subsurface sediments using the incorporation of tritiated thymidine into DNA. *Microbial Ecology* 16, 3–16.

Tiedje, J.M., Simkins, S. and Groffman, P.M. (1989) Perspectives on measurement of denitrification in the field including recommended protocols for acetylene based methods. *Plant and Soil* 115, 261–284.

Torstensson, L. and Stenström, J. (1986) 'Basic' respiration rate as a tool for prediction of pesticide persistence in soil. *Toxicity Assessment: An International Quarterly* 1, 57–72.

Torstensson, M., Pell, M. and Stenberg, B. (1998) Need of strategy for evaluation of soil quality data: arable soil. *Ambio* 27, 4–8.

Trasar-Cepeda, M. and Gilsotres, F. (1987) Phosphatase-activity in acid high organic-matter soils in Galacia (NW Spain). *Soil Biology and Biochemistry* 19, 281–287.

Trasar-Cepeda, C., Leiros, M., Seoane, S. and Gil-Sotres, F. (2000) Limitations of soil enzymes as indicators of soil pollution. *Soil Biology and Biochemistry* 32, 1867–1875.

Trevors, J.T. (1984a) Effect of substrate concentration, inorganic nitrogen, O_2 concentration, temperature and pH on dehydrogenase activity in soil. *Plant and Soil* 77, 285–293.

Trevors, J.T. (1984b) Dehydrogenase activity in soil: a comparison between the INT and TTC assay. *Soil Biology and Biochemistry* 16, 673–674.

Trevors, J.T., Mayfield, C.I. and Inniss, W.E. (1982) Measurement of electron transport system (ETS) activity in soil. *Microbial Ecology* 8, 163–168.

Vekemans, X., Godden, B. and Penninckx, M.J. (1989) Factor-analysis of the relationships between several physiochemical and microbiological characteristics of some Belgian agricultural soils. *Soil Biology and Biochemistry* 21, 53–58.

Vischetti, C., Perucci, P. and Scarponi, L. (1997) Rimsulfuron in soil: effect of persistence on growth and activity of microbial biomass at varying environmental conditions. *Biogeochemistry* 39, 165–176.

vonMersi, W. and Schinner, F. (1991) An improved and accurate method for determining the dehydrogenase activity of soils with iodonitrotetrazolium chloride. *Biology and Fertility of Soils* 11, 216–220.

Waldrop, M.P., Balser, T.C. and Firestone, M.K. (2000) Linking microbial community composition to function in a tropical soil. *Soil Biology and Biochemistry* 32, 1837–1846.

Waring, S.A. and Bremner, J.M. (1964) Ammonium production in soil under waterlogged conditions as an index of nitrogen availability. *Nature* 201, 951–952.

Watts, S.H. and Seitzinger, S.P. (2000) Denitrification rates in organic and mineral soils from riparian sites: a comparison of N_2 flux and acetylene inhibition methods. *Soil Biology and Biochemistry* 32, 1383–1392.

Welp, G. (1999) Inhibitory effects of the total and water-soluble concentrations of nine different metals on the dehydrogenase activity of a loess soil. *Biology and Fertility of Soils* 30, 132–139.

Yakovchenko, V., Sikora, L.J. and Kaufmann, D.D. (1996) A biologically based indicator of soil quality. *Biology and Fertility of Soils* 21, 245–251.

Yakovchenko, V., Sikora, L.J. and Millner, P.D. (1998) Carbon and nitrogen mineralization of added particulate and macroorganic matter. *Soil Biology and Biochemistry* 30, 2139–2146.

Yoshinari, T. and Knowles, R. (1977) Acetylene inhibition of nitrous oxide reduction by denitrifying bacteria. *Biochem Biophys Res Commun* 69, 705–710.

Zibilske, L.M. (1994) Carbon mineralization. In: Weaver, R.W. *et al.* (eds) *Methods of Soil Analysis, Part 2. Microbiological and Biochemical Properties.* SSSA, Madison, Wisconsin, pp. 835–863.

8 Soil Microbial Diversity and Community Composition

8.1 Estimating Soil Microbial Diversity and Community Composition

JAN DIRK VAN ELSAS[1] AND MICHIEL RUTGERS[2]

[1]*Department of Microbial Ecology, Groningen University, Kerklaan 30, NL-9750 RA Haren, The Netherlands;* [2]*National Institute for Public Health and the Environment, Antonie van Leeuwenhoeklaan 9, NL-3721 MA Bilthoven, The Netherlands*

Soils are complex and very heterogeneous environments that may contain as many as 10 billion or more bacterial cells per gram, in addition to large numbers of other microorganisms such as fungi and protozoa, as well as several macroorganisms, collectively called macrofauna. Although microorganisms determine the chemical balance in the soil, interact with plants in positive and negative ways, and even influence soil structure, we have little understanding of the structure and dynamics of these microbial communities, as well as of their enormous diversity. However, over the past decade our perspective with respect to the microbial diversity and community structure of soils has improved enormously. This is in large part due to the rapid development and application of culture-independent methods that allow the characterization of soil microbial communities. For a long period, spanning almost 100 years, plating (counting colony-forming units, CFU) on different media was the technique of choice to investigate the microbial diversity of soil. However, the relative proportion of bacteria that grow readily on agar plates with common bacteriological media to those counted by microscopical approaches varies from 0.1–1% in pristine forest soils to 10% in environments such as arable soil. This implies that assessments of the microbial diversity of these habitats, in terms of species richness and abundance, are grossly underestimated (Amann *et al.*, 1995). Since so few soil microorganisms can be cultivated by standard techniques, the new culture-independent approaches, in particular cloning, sequencing and fingerprinting of ribosomal RNA (rRNA) genes, have already revealed evidence for the existence of an astonishing wealth of novel organisms, of which many are quite different from those known among the cultured isolates (Liesack and Stackebrandt, 1992). Still, as of today, any estimations of the extant numbers of, for example, prokaryotic species are mere guesses, but estimates of 4000 or more species per g of soil have been suggested (Torsvik *et al.*, 1990). In addition to continuously increasing our knowledge of the microbial diversity in soils, culture-independent methods, several of which are described in this chapter, also allow us to better understand the structure and dynamics of soil microbial communities. These methods allow us to monitor changes in either the overall microbial or bacterial community structure, or in more detail, changes in the prevalence of phylogenetic subgroups.

The ultimate goal of most assessments is to understand the overall functioning of soil microbial communities in terms of, for example, the flux of energy, as well as resources, through the system, and how this is influenced by natural environmental changes as well as human impact. Many methods are currently available that aid in the description of the diversity and functioning of soil microbial communities (summarized in Table 8.1). The methods described in this chapter form a subset of this wide range of methods. Together, they allow us to describe soil microbial communities in an indirect way, each with their limitations. However, they may be applied as routine methods in a comparative fashion, in order to provide a picture of soil microbial communities from three angles.

On the other hand, in order to understand the final outcome of the many functions carried out by different microbial populations, or by cooperation between such populations, we need to better understand the structure of the microbial communities of soil. Answers to questions such as: what species are there, what is the relative abundance of the different populations, are the organisms active or dormant and do they interact (via chemical signals, intermediate metabolites or by genetic interactions) are fundamental for a better understanding of the functioning and robustness of the microbial community. To answer these questions, methods are needed that can distinguish between the different populations present, quantify their abundance (population sizes) and, ideally, locate them *in situ*. Thus, there has been a perceived need for sensitive methods that can identify which microbial populations are affected by a certain environmental change (e.g. in time or upon human impact). One example is a recent study that has shown that long-term herbicide applications clearly affect the structure and diversity of specific microbial groups, as shown by 16S rRNA molecular community fingerprinting (Seghers *et al.*, 2003). The fingerprinting method, based on universal bacterial primers, did not show significant differences between the herbicide-treated and the control soil in this study. However, upon zooming in on specific populations or phylogenetic subgroups, such as the methanotrophs, differences in community structure were observed. Whether these differences in community structure also

Table 8.1. Methods to assess the microbial diversity of soils.

Type	Method	Description	Reference
Phenotypic	CLPP	Community-level physiological profiling	Garland and Mills (1991)
Cell components	FAME	Fatty acid methyl ester	Buyer and Drinkwater (1997); Zelles (1999)
	PLFA	Phospholipid fatty acid analysis	Frostegård et al. (1993); Zelles (1999)
Molecular/genetic	Cloning/sequencing of amplified 16S ribosomal RNA genes (rDNA)	Sequence analysis of clone library, resulting in overview of abundant clone types	Akkermans et al. (1995); Kowalchuk et al. (2004)
	16S rDNA-based PCR and fingerprintings: – DGGE/TGGE – SSCP – ARDRA – T-RFLP	PCR may be followed by any one of the fingerprinting methods to determine the diversity of the community on the basis of the 16S rRNA-based phylogenetic marker	Akkermans et al. (1995); Kowalchuk et al. (2004); van Elsas et al. (2000)
	Dot blot hybridization of 16S rRNA genes	Hybridization using short 16S rRNA gene-based fragments as probes; probes generated from V6 region are highly specific per strain	Heuer et al. (1999)
	Base composition profiles	Expressed as mole percent guanine + cytosine (% G + C)	Torsvik et al. (1990); Nüsslein and Tiedje (1998)
	Direct PCR detection	PCR amplification of target gene followed by detection on gel, or after hybridization	van Elsas et al. (1997)
	FISH	Fluorescent in situ hybridization of 16S rRNA	Akkermans et al. (1995); Amann et al. (1995)
	DNA reassociation	'C_0t curves': reassociation time is a measure for the genetic diversity in a sample	Torsvik et al. (1990)

translate into a change in function, or in a community that is less fit to deal with other forms of stress, is not known. It is known that there is a lot of functional redundancy in soils, which may serve as a buffering capacity in cases of severe disturbance. There is, however, no guarantee that the soil microbial community will completely take over the role of populations that have disappeared. How can we estimate or predict whether a slightly changed microbial community will be as resistant to future stressors as before the alteration?

During the past two decades, phenotypic and nucleic acid-based methods have been developed to better characterize the structure and diversity of microbial communities. Table 8.1 lists these methods. Three methods are discussed and described in detail in this chapter. They are:

1. Soil microbial community DNA- or RNA-based profiling methods.
2. Phospholipid-based methods (PLFA, phospholipid fatty acid).
3. Community-level physiological profiling (CLPP) tests based on the use of Biolog™ metabolic response plates.

All three methods are not based on isolated and purified organisms, but rather give a picture of different aspects of the microbial community based on a method that uses total macromolecules or cells extracted from soil. The different methods address different questions and therefore analyse either different sub-groups of the total microbial community or different aspects of the community. Together they form a solid basis for up-to-date soil microbial community analyses.

8.2 Soil Microbial Community Fingerprinting Based on Total Community DNA or RNA

JAN DIRK VAN ELSAS,[1] EVA M. TOP[2] AND KORNELIA SMALLA[3]

[1]*Department of Microbial Ecology, Groningen University, Kerklaan 30, NL-9750 RA Haren, The Netherlands;*
[2]*Department of Biological Sciences, 347 Life Sciences Building, University of Idaho, Moscow, ID 83844–3051, USA;*
[3]*Institute for Microbiology, Virology and Biosafety, Biologische Bundesanstalt, Messeweg 11/12, Braunschweig, Germany*

Introduction

Soil microbial community DNA and/or RNA extraction methods developed in the past decade have paved the way for direct, cultivation-independent, studies of microbial diversity in soils. The direct extraction and analysis of total microbial community DNA or RNA from soil was shown to be fundamental as the basis to describe the *in situ* soil microbiota and its diversity. After the pioneering work of Torsvik and co-workers in the early 1980s (e.g. Torsvik, 1980), the extraction of microbial DNA from soil has found its way into basically every soil microbiology laboratory in the world. During the past decade in particular, a multitude of different extraction protocols has been published. The *Molecular Microbial Ecology Manual*, first and second editions (Akkermans *et al.*, 1995; Kowalchuk *et al.*, 2004), contains examples of all of these protocols. This proliferation of protocols is often explained by the large variations between soils in chemical characteristics, which necessitates the application of different protocols to almost every new soil. However, in the past few years, many laboratories have replaced their often-laborious 'pet' protocols with rapid commercial kit-based protocols, which appear to work comparably well in almost every instance. Two current products, the Bio101 and MoBio soil DNA extraction kits, seem to cover most of the kit-based extraction protocols that are currently in use.

All the different protocols, even the kit-based ones, can be grouped in two basically different types of approaches:

1. Disruption of bonds between microbial cells and soil particles, resulting in the release of largely bacterial or archaebacterial cells, followed by cell lysis and extraction.
2. Direct cell lysis within the soil matrix in a slurry, and DNA extraction from the soil slurry (Ogram *et al.*, 1987).

While the former method provides DNA that is considered to be representative for the prokaryotic (bacterial or archaeal) fractions of the microbial community in soil, the latter has been shown to provide higher DNA yields, but is less specific for prokaryotes, as it also contains eukaryotic (fungal) and extracellular DNA (Steffan et al., 1988; van Elsas et al., 2000). Both methods coextract low levels of extracellular DNA from soil, which should be taken into account in the interpretation of results, i.e. whether positive detection indicates the presence of microbial cells carrying target DNA or merely extracellular target DNA. In spite of these potential problems, the direct lysis and extraction method has become the favourite DNA extraction method in many laboratories working with soil. On the other hand, recent advances have shown definite advantages of the indirect methods, with respect to the specificity of detection (Duarte et al., 1998).

Direct community DNA extraction from soil, as developed by Ogram et al. (1987), has been shown to provide a substantial amount of total soil bacterial DNA, but was also found to be prone to unavoidable yield losses. Moré et al. (1994) showed that many direct protocols are likely to extract DNA mainly from easy-to-lyse cells, whereas, in particular, the minute microbial forms (dwarfs) that are abundant in soil are often excluded. Since the original protocols encompassed tedious caesium chloride (CsCl) gradient and/or hydroxyapatite chromatography purification steps, many more recent protocols have attempted to simplify the original protocol of Ogram et al. (van Elsas et al., 2000). Examples of these simplified protocols can be found in the literature (Pillai et al., 1991; Porteous and Armstrong, 1991; Selenska and Klingmüller, 1991a,b; van Elsas et al., 1991; Tsai and Olson, 1992; Smalla et al., 1993a; Duarte et al., 1998; van Elsas et al., 2000). Some protocols extract RNA in addition to DNA. Others (e.g. Smalla et al., 1993a; Griffith et al., 2000) have been selected as preferred ones in the *Molecular Microbial Ecology Manual* (Akkermans et al., 1995; Kowalchuk et al., 2004).

The nucleic acid extraction methods are thus useful for several purposes (Trevors and van Elsas, 1989). First, they provide insight into the prevalence and/or activity of specific genes in microbial communities in soil ecosystems, resulting in a better understanding of the functioning and selection of such genes under specific soil conditions. Secondly, by using 16S/18S or 23S/25S ribosomal RNA gene sequences as 'signature molecules' (biomarkers), overall community DNA analysis can be the basis for description of microbial community structures. This can be achieved by applying fingerprinting techniques such as temperature or denaturing gradient gel electrophoresis (TGGE or DGGE), terminal restriction fragment length polymorphism (T-RFLP), or single-strand conformational polymorphism (SSCP). These and other methods have been described in the *Molecular Microbial Ecology Manual* (Akkermans et al., 1995; Kowalchuk et al., 2004). In all three methods, PCR products that represent a 'picture' of the soil microbial community are generated with sets of conserved primers, and are subsequently separated by either one of the aforementioned fingerprinting methods. The basis of the fragment separation differs between these methods, as follows. SSCP takes advantage of the conformational changes that occur stochastically in single-stranded DNA (or RNA) molecules, and result in different migration behaviour in polyacrylamide gels. T-RFLP is based on the detection, by a fluorescent label, of a terminal fragment produced by enzymatic restriction of the mixed amplicons, thereby identifying phylotypes with a unique terminal restriction fragment (T-RF). DGGE and TGGE are based on the use of denaturing or temperature gradients, respectively. Both gradient types separate double-stranded DNA molecules by their specific migration distance, which is determined by their melting behaviour resulting from the nucleotide sequence. All of these methods yield a fingerprint pattern that is a representation of the microbial community structure in the soil. Depending on the primers used in the initial PCR amplification, different microbial groups can be targeted. For instance, a range of primer sets has become available that target total bacteria, total fungi or specific bacterial groups, such as the α- or β-subgroups of the proteobacteria, the high-G+C% Gram-positives, the pseudomonads, the bacilli, and *Burkholderia* spp. (Van Overbeek *et al.*, 2005). Primers for specific fungal groups, such as different vesicular arbuscular mycorrhizae, are also under development or have been published (de Souza, personal communication).

Thus, the analysis of the various fingerprints obtained from these microbial community DNA extracts allows us to make inferences about the types of microorganisms present within the extractable cell fraction of the soil. This obviously includes the non- or poorly culturable cells, which represent the largest fraction of cells that can be found in most soils (silent majority). The new angle on soil microbial diversity offered by the direct DNA-based methods has already led to the discovery of a wealth of novel organisms. For instance, several new deep-branching groups of proteobacteria have recently been described, based on the analysis of 16S ribosomal RNA gene clones generated from DNA of total soil communities.

In this section, we discuss the use of two nucleic acid extraction methods, one DNA extraction method developed and used routinely in our laboratories (method I; modified from Smalla *et al.*, 1993a), and one miniaturized DNA/RNA extraction method (method II) recently implemented in our labs (Gomes *et al.*, 2004). We then describe the use of 16S ribosomal RNA gene (rDNA)-based PCR coupled to DGGE analysis to describe the microbial diversity in soil. It has been shown that restrictable and PCR-amplifiable DNA of relatively high molecular weight can be obtained with the majority of nucleic acid extraction protocols. The microbial diversity measurements are, to some extent, dependent on the DNA extraction method used, and hence it is important to use one standard protocol in routine analyses.

Principle of the Methods

In many laboratories, soil nucleic acid extraction is nowadays performed by using commercially available kits. The main driving force of this development has been the relative ease of use and speed offered by these kits.

Several groups have shown that such kits extract microbial community DNA which is roughly, but not completely, similar to that obtained by the more traditional approaches. However, for specific purposes, e.g. when novel results should be compared to previous results obtained by traditional extraction, and in the light of their robustness, the traditional protocols are still in use. Below, we present one robust and highly versatile traditional approach (method I) and one approach modified from a highly accepted commercial kit-based method (method II).

Soil nucleic acid extraction method I

This method will primarily extract DNA from soil, although RNA is sometimes coextracted. The method is based on an efficient lysis of bacterial cells in the soil matrix in a slurry, followed by quick removal of soil particles and humic compounds, using different rapid purification steps. Removal of humic material, proteins, RNA and polysaccharides, as well as other soil compounds (minerals), is required to obtain DNA of sufficient purity for hybridization, restriction or PCR amplification analysis, as well as for cloning purposes (Steffan et al., 1988; Smalla et al., 1993a,b; Tebbe and Vahjen, 1993). The protocol is based on the direct extraction protocol of Ogram et al. (1987) for the extraction of DNA from sediments, as adapted by Smalla et al. (1993a), with omission of the laborious and often inefficient purification via hydroxyapatite chromatography. We adopted bead beating of soil slurries in a Braun's cell homogenizer (B. Braun Diessel Biotech, Melsungen, Germany) as the method of choice for cell lysis, since this strategy was shown to yield higher quantities of DNA than freeze/thaw lysis (Smalla et al., 1993a). However, in laboratories that do not possess a bead beater, freeze/thaw-assisted cell lysis (either using or not using lytic enzymes) may be used as an alternative. Careful control of the bead-beating time and conditions was essential to obtain DNA of large fragment size. This is important, since severely sheared DNA is unsuitable for PCR-based detection of specific genes or analysis of community structure, e.g. using bacterial 16S ribosomal RNA gene sequences as targets. Following cell lysis, extraction with cold phenol in the presence of soil particles separated the DNA from contaminating compounds, while offering protection from nucleases. Subsequent precipitation steps with CsCl and potassium acetate (KAc) were included to further remove impurities (proteins, RNA, humic material) from the DNA. For some soils, e.g. several silt loams, DNA preparations thus obtained are of sufficient purity to serve as targets for restriction, amplification or hybridization analysis. For other soils, e.g. a high organic matter loamy sand, a final clean-up step was required, performed by adsorption/elution over commercially available glassmilk (Geneclean II kit, Bio 101, La Jolla, California, USA) or resin spin columns (Wizard DNA Clean-Up System, Promega, Madison, Wisconsin, USA). In our experience, this flexible protocol allows for the extraction and purification of high-quality DNA from virtually any soil type (van Elsas et al., 1997).

Soil nucleic acid extraction method II

This nucleic acid extraction protocol, which will produce a mixture of DNA and RNA, is based on the use of a commercially available kit (Bio 101, La Jolla, California, USA), which allows an upscaling of the extraction and working in small reaction vessels. The protocol (Gomes *et al.*, 2004) is a modification of the method described by Hurt *et al.* (2001). Because harsh cell lysis is crucial to obtain nucleic acids representing the microbial community in bulk and rhizosphere soils, a bead-beating step was added to the protocol of Hurt *et al.* (2001) to ensure the efficient disruption of the cells. Smalla *et al.* (1993a) had already shown that bead beating yields higher amounts of DNA than freeze/thaw lysis. To process many samples in parallel, a miniaturization of the protocol was required. The whole procedure was scaled down to 0.5 g (wet weight) of soil. Materials, volumes and equipment were adapted to this miniaturized scale, making the protocol simple, fast and suitable for processing large numbers of samples within a short period of time. However, degradation of RNA was still observed with the modifications mentioned above. Therefore, an RNA-protecting substance, ethanol or isopropanol, was added before breaking up the cells (Gomes *et al.*, 2004), as it is known that this reduces the degree of RNA degradation. We recommend the addition of ethanol as the most efficient step to achieve this. The extraction buffer of Hurt *et al.* (2001) was, thus, slightly modified and kept in the incubation step after bead beating.

PCR amplification

Following extraction and purification of the soil nucleic acids, either the DNA is subjected directly to PCR using standard protocols, or RNA is first reverse-transcribed, after which the copy DNA (cDNA) produced is PCR-amplified. PCR has become the method of choice for the enrichment of specific target sequences for subsequent detection, or even cloning (Akkermans *et al.*, 1995; Kowalchuk *et al.*, 2004). The method is based on the cyclic enzymatic 'inward' extension of primers at two opposite ends of a DNA template, resulting in the generation of numerous copies of this template. The amplification cycle, which consists of template denaturing, primer-to-target annealing and primer extension steps, is achieved by concerted changes in reaction temperature, most easily performed in a programmable thermal cycler. Due to the high denaturing temperature (often 94°C), DNA polymerases used in the PCR have to be thermostable, for example the frequently used *Thermus aquaticus Taq* polymerase.

To achieve the desired specificity, primer choice is fundamental, as discussed above. In addition, the primers should be checked as to whether they actually perform well in a soil DNA background, as differences with respect to inhibition of the action of the polymerase have been found between different primer sets. Following PCR, the amplicons obtained should be carefully checked for quality and quantity by electrophoresis in agarose gel, using standard procedures (Sambrook *et al.*, 1989).

Denaturing gradient gel electrophoresis (DGGE)

In order to obtain a fingerprint of the soil microbial community targeted, the PCR products obtained with soil DNA as the target are subjected to electrophoresis over a polyacrylamide gel containing a gradient of denaturing substances via standard procedures. This method is described fully in the *Molecular Microbial Ecology Manual* (Akkermans *et al.*, 1995; Kowalchuk *et al.*, 2004) and is summarized under 'Procedures' (below). The method is able to separate DNA fragments of the same length but with different nucleotide sequences, such as those generated by PCR with 16S or 18S ribosomal RNA gene-based primers. The separation is based on the differentially decreased mobility of the partially melted DNA molecules in a linearly increasing gradient of denaturants (urea and formamide). The melting occurs in discrete melting domains of the molecule. Once the melting condition of a particular region is reached, the helical structure of the double strand turns into a partially melted structure with greatly reduced migration in the gel. Differences in the sequences of the molecular types will cause their migration behaviour to differ. This results in a banding pattern, in which each band, in principle, represents a different molecular type.

In practical terms, a GC-clamp (a 40 bp high-guanine-plus-cytosine-containing stretch) has to be attached to the 5'-end of one primer. The clamp prevents complete melting of the molecules in the denaturing gradient, which results in partially melted molecules of which the migration is almost completely halted.

The DGGE gel has to be run under constant voltage for 4–16 h, after which the gel is stained by ethidium bromide, SYBR green or SYBR gold, or by silver nitrate. The stained gel can be visualized under UV light and gel pictures can be further analysed using relevant computer programs, such as GELCOMPAR (Rademaker, 1995). To calibrate the method and the analysis, a marker containing amplicons of known position in the gel needs to be run in parallel (defining the relative positions of bands). If needed, specific bands from the patterns can be excised from the gel, re-amplified and subjected to sequencing.

Materials and Apparatus

Soil DNA extraction

- Bead beater (Braun's cell homogenizer) or similar
- Glass beads (0.10–0.11 mm)
- Common laboratory ware (glass or plastic tubes, microcentrifuge tubes, vials, pipettes)
- Gel electrophoresis apparatus and electricity source

PCR

- Reaction tubes or arrays
- Thermal cycling apparatus, e.g. Perkin-Elmer (Nieuwerkerk a/d IJssel, NL) or similar
- Gel electrophoresis apparatus and source

Denaturing gradient gel electrophoresis (DGGE)

- DGGE apparatus, e.g. Ingeny (Goes, The Netherlands), PhorU and power
- Gradient maker
- Software program for analysis of banding patterns, e.g. GELCOMPAR (Applied Maths, Sint-Martens-Latem, Belgium)

Chemicals and Solutions

Soil DNA extraction

- Tris-buffered phenol, pH 8.0
- 120 mM sodium phosphate buffer, pH 8.0
- Lysozyme
- 20% sodium dodecyl sulphate (SDS)
- Chloroform:iso-amylalcohol (24:1)
- 5 M NaCl
- Ice-cold 96% ethanol
- TE buffer pH 8.0 (10 mM Tris–HCl pH 8.0, 1 mM EDTA pH 8.0; Sambrook *et al.*, 1989)
- 8 M KAc
- CsCl
- 0.1% diethyl-pyrocarbonate (DEPC) solution

PCR

- Sterile deionized water
- Buffer for *Taq* DNA polymerase Stoffel fragment (Perkin Elmer) 10×
- Each dNTP (1 mM) mix, 5×
- $MgCl_2$ 25 mM
- Primer 1 10 µM
- Primer 2 10 µM
- Formamide
- T4 gene 32 protein (5 mg/ml)
- *Taq* DNA polymerase Stoffel fragment (10 U/µl)

DGGE

- Formamide (Merck, Darmstadt, Germany), deionized using standard procedure (e.g. by 5% AG 501-X8 resin treatment). *Avoid contact with skin and eyes*
- AG 501-XS Resin (BioRad Laboratories, Veenendaal, The Netherlands)
- Tris base (Boehringer, Mannheim, Germany)
- Acetic acid, anhydrous (Merck)
- 30% Acrylamide (4K Mix 37.5:1). *May cause cancer. Toxic if in contact with skin and if swallowed*
- Urea
- Ammonium persulphate (APS) (BioRad) 20% in Milli-Q water
- TEMED (N,N,N,N'-tetramethyl ethylenediamine) (Sigma, Zwijndrecht, The Netherlands)
- SYBR Gold 1, 10,000× concentrated in dimethyl sulphoxide (DMSO) (Molecular Probes, Leiden, The Netherlands), per gel:
  - 250 µl 20× Tris–acetate–EDTA (TAE)
  - 9.75 ml Milli-Q water
  - 2.5 µl SYBR Gold 1
- 20× TAE:
  - 97 g Tris base in 500 ml Milli-Q water; set at pH 7.8 with acetic acid
  - 32.8 g sodium acetate
  - 40 ml 0.5 M EDTA (pH 8)
  - adjust volume to 1 l with Milli-Q water
- Solution A:
  - 100 ml 30% acrylamide 4K mix (37.5:1)
  - 5 ml 50× TAE
  - adjust volume to 500 ml with sterile Milli-Q water
  - store in dark at room temperature
- Solution B:
  - 100 ml 30% acrylamide 4K mix (37.5:1)
  - 5 ml 50× TAE
  - add 50 ml of sterile Milli-Q water
  - 168 g urea (Wm = 60.06 g/mole)
  - 160 ml formamide, deionized
  - adjust volume to 500 ml with sterile Milli-Q water, heat 1 h at 37°C
  - store in dark at room temperature
- Stacking gel solution:
  - 26.67 ml 30% acrylamide (37.5:1)
  - 1 ml 50× TAE
  - adjust volume to 100 ml with sterile Milli-Q water
  - store in dark at room temperature
- Loading buffer (6×):
  - 0.05% (w/v) bromophenol blue
  - 40% (w/v) sucrose
  - 0.1 M EDTA (pH 8.0)
  - 0.5% (w/v) SDS

Procedures

Soil nucleic acid extraction method I

The procedure described will primarily extract microbial DNA from soil. It is based on the extraction of 10 g of soil. It can be scaled down easily to accommodate 1–5 g soil samples. The full procedure is often needed to obtain restrictable and amplifiable DNA from a loamy sand soil with high organic matter content, whereas purification until (and including) purification step I can be sufficient for DNA from soils with low organic matter content (e.g. silt loam types).

1. Resuspend 10 g of soil in 15 ml 120 mM sodium phosphate buffer pH 8.0 (Sambrook *et al.*, 1989) in a 50 ml polypropylene tube.
2. *Lysozyme treatment* (optional):
   (i) Add 75 mg lysozyme to the soil suspension, homogenize and incubate for 15 min at 37°C.
   (ii) Chill on ice.
3. *Bead-beating lysis*:
   (i) Transfer the soil suspension obtained under **1** (above) to a bead-beating vial containing 15 g glass beads (0.09–0.13 mm diameter).
   (ii) Homogenize three times for 90 s in the bead beater (MSK cell homogenizer, B. Braun Diessel Biotech) at 4000 oscillations/min with intervals of 15–30 s.
   (iii) Transfer the lysate to a 45 ml centrifuge or 50 ml polypropylene tube.
   (iv) Add 900 µl of 20% sodium dodecyl sulphate (SDS) and mix well.
   (v) Leave either on ice for 1 h to enhance lysis, or at room temperature for 15 min.
4. *Freeze/thaw lysis*: As an alternative to bead beating, freeze/thaw lysis can be applied. However, we strongly recommend bead beating, as it results in more complete lysis.
   (i) Add 900 µl of 20% SDS to the soil slurry obtained under **2** (above) and mix.
   (ii) Freeze at –20°C (or –80°C) for 1 h, then keep at 37°C for 30–45 min.
   (iii) Repeat freeze/thaw cycle twice.
5. *Extraction and precipitation*:
   (i) Add an equal volume of Tris-buffered phenol pH 8.0 (Sambrook *et al.*, 1989) to the lysed cell slurry obtained under **3** (above).
   (ii) Mix well (manually) and centrifuge for 5 min at $10,000 \times g$ (or 15 min at $3000 \times g$ for polypropylene tubes) at room temperature.
   (iii) Recover the aqueous (upper) phase in a new centrifuge (or polypropylene) tube.
   (iv) Back-extract the phenol/soil mixture with 5 ml 120 mM sodium phosphate buffer (pH 8.0).
   (v) Pool the aqueous phases.
   (vi) Extract the pooled aqueous phases with an equal volume of chloroform/iso-amylalcohol (24:1).

(vii) Recover the upper aqueous phase. If a heavy interphase is present, extract the aqueous phase again with an equal volume of chloroform/iso-amylalcohol.

(viii) Add 0.1 volume of 5 M NaCl and two volumes of ice-cold 96% ethanol. Keep at $-80°C$ for 20 min or at $-20°C$ for at least 1 h (often overnight).

(ix) Centrifuge for 5–10 min at $10,000 \times g$ (or 15–20 min at $3000 \times g$ for polypropylene tubes).

(x) Discard the supernatant and wash the pellet with ice-cold 70% ethanol. Air-dry pellet.

(xi) Resuspend pellet in 1–1.5 ml (sterile) TE buffer pH 8.0 (10 mM Tris–HCl pH 8.0, 1 mM EDTA pH 8.0). Due to the volume of the pellet, the final volume may be larger.

(xii) The solution at this stage is called the crude extract.

6. *Purification step I: CsCl and KAc precipitations.* Perform all steps at room temperature, unless stated otherwise.

(i) Add 0.5 g CsCl to 500 µl crude extract.

(ii) Incubate for 1–3 h.

(iii) Centrifuge for 20 min at maximum speed in an Eppendorf centrifuge.

(iv) Recover the supernatant (≈500 µl) in a 10 ml tube.

(v) Add 2 ml deionized water and 1.5 ml iso-propanol. Mix and incubate for a minimum of 5 min at room temperature.

(vi) Centrifuge for 15 min at $10,000 \times g$. When polypropylene tubes are employed, use maximally at $3000 \times g$ for 20 min. Check degree of pelleting.

(vii) Discard supernatant and resuspend pellet in 500 µl TE buffer (pH 8.0) and transfer the suspension to a new Eppendorf tube.

(viii) Add 100 µl 8 M KAc, mix and incubate for 15 min at room temperature.

(ix) Centrifuge for 15 min at maximum speed in an Eppendorf centrifuge.

(x) Recover supernatant and add 0.6 volume of iso-propanol, mix and incubate for 5 min at room temperature.

(xi) Centrifuge for 15 min at full speed in an Eppendorf centrifuge.

(xii) Wash pellet with ice-cold 70% ethanol, dry and resuspend it in 500 µl TE buffer (pH 8.0).

7. *Purification step II: Wizard DNA Clean-Up System* (Promega, Madison, Wisconsin, USA). The system is based on DNA adsorption to clean-up resin in 6 M guanidine thiocyanate, washing with 80% iso-propanol and elution with TE buffer or deionized water. The reagents can be kept at room temperature protected from exposure to direct sunlight.

(i) Add 1 ml of DNA clean-up resin to 250 µl partially purified DNA extract (after purification step I) in an Eppendorf tube and mix by gently inverting several times.

(ii) Attach the syringe barrel of a 2.5 ml disposable luer-lock syringe to the extension of the Wizard minicolumn.

(iii) Pipette the suspension from the Eppendorf tube into the syringe barrel. Using the plunger, push the slurry slowly into the minicolumn.

(iv) To wash the column, pipette 2 ml of 80% iso-propanol into the syringe barrel. Gently push the solution through the minicolumn.
(v) Remove syringe and place the minicolumn containing the loaded resin on top of an Eppendorf tube. Centrifuge for 20 s at full speed in an Eppendorf centrifuge to dry the resin. Leave the minicolumn at room temperature for 5–15 min to evaporate traces of iso-propanol still present.
(vi) Transfer the minicolumn to a new Eppendorf tube. Apply 125 µl of prewarmed (60–70°C) TE buffer (pH 8.0) or deionized water and leave for 5–10 min. The DNA will remain intact on the minicolumn for up to 30 min. Centrifuge for 20 s at full speed in an Eppendorf centrifuge to elute the bound DNA.
(vii) Repeat step (vi) using the same Eppendorf tube. The total volume of eluate will be about 250 µl.
(viii) Discard minicolumn. The purified DNA may be stored at 4°C or −20°C.
(ix) If needed (as judged by colour and/or suitability for restriction digestion or amplification), repeat steps (i)–(viii).

Soil nucleic acid extraction method II

Before starting the extraction, it is important to prepare RNase-free solutions and materials. Non-disposable materials must be immersed in 0.1% diethyl-pyrocarbonate (DEPC) solution and autoclaved. Alternatively, glassware and metal spatulas can be baked at 400°C for 4 h to inactivate RNases, or cleaned with RNase Away (Molecular Bio-Products, San Diego, California, USA). All solutions must be prepared with deionized water previously treated with 0.1% (vol/vol) DEPC. Add DEPC to the water and incubate for at least 2 h at 37°C to inactivate RNase. The water must be autoclaved after treatment to destroy DEPC.

1. Add soil samples (up to 0.5 g wet weight) into 2 ml microcentrifuge tubes containing 0.4 g glass beads (0.10–0.11 mm). Bacterial pellets (up to 0.5 g wet weight) extracted from environmental samples or microbial cultures can also be processed by following the next steps. Keep samples on ice.
2. Add up to 0.8 ml of ethanol to the samples until they are totally immersed. Homogenize the samples twice by using the FastPrep FP120 bead-beating system (Bio 101, Vista, California, USA) at 5.5 m/s for 30 s. Alternatively, bead beating can be performed by using a cell homogenizer (B. Braun Diessel Biotech) twice for 30 s (4000 oscillations/min). Samples should be kept on ice (30 s) between the bead-beating steps.
3. Centrifuge the tubes for 5 min at $16,000 \times g$. Discard the supernatant and add 1.0 ml of extraction buffer pH 7.0 (1% hexadecyl trimethylammonium bromide (CTAB), 2% SDS, 1.5 M NaCl, 100 mM sodium phosphate buffer pH 7.0, 10 mM Tris–HCl pH 7.0 and 1 mM EDTA pH 8.0). Mix the samples thoroughly and incubate for 30 min at 65°C, mixing carefully every 10 min.

4. Centrifuge the tubes at 16,000 × g for 5 min and transfer the supernatant into 2 ml tubes containing 1 ml aliquots of chloroform:isoamyl alcohol (24:1) previously chilled on ice. Mix the samples thoroughly and centrifuge at 16,000 × g for 5 min.

5. Transfer the upper phase into new microcentrifuge tubes. Precipitate the nucleic acid by addition of 0.6 volume of isopropanol, incubation for at least 30 min at room temperature, and centrifugation at 16,000 × g for 20 min. Pelleted nucleic acids can be either washed with 70% (vol/vol) ethanol for subsequent RNA/DNA separation or directly stocked into 0.5 ml 70% (vol/vol) ethanol at −70°C until use.

6. Before starting the RNA/DNA recovery procedure, centrifuge suspensions at 16,000 × g for 5 min, carefully remove the ethanol and air-dry the pellet containing the nucleic acids for 5 min (do not dry the pellet completely). Dissolve the nucleic acids in 200 µl of DEPC-treated deionized water.

7. Collect 100 µl of the extracted nucleic acids for RNA purification using the RNeasy Mini Kit (QIAGEN GmbH, Germany) according to the manufacturer's instructions. If necessary, this step can be repeated for better RNA purification. Use the other 100 µl aliquot for DNA purification using the GENECLEAN spin kit (Q Biogene, USA) according to the manufacturer's instructions. Alternatively, to obtain higher yields, DNA and RNA can be separated at once from the same sample by using the QIAGEN RNA/DNA mini kit (QIAGEN GmbH, Germany). However, according to our experience, further purification steps might be necessary before using the nucleic acids recovered for PCR applications and gene expression analysis.

PCR

Since DNA amplification via PCR is extremely sensitive, care should be taken not to contaminate the sample material or new PCR reaction mixes with target DNA or PCR products resulting from previous reactions, since this might lead to false positives. Aerosols, which may form when samples containing target DNA or PCR product are handled, are notorious sources of contamination. Therefore, sample preparation and processing, setting up of the PCR reaction mixes, and, in particular, handling of the PCR products, all have to be performed with extreme care. To avoid the occurrence of false-positive results, we find it adequate to have separate sample preparation, PCR and product analysis rooms, with separate equipment, including pipettes. In addition, all glass- and plasticware used in the PCR room should be exclusively handled there. Further, PCR reagents should only be prepared in the PCR room, and divided in aliquots, which are then stored at −20°C.

A notorious problem when using PCR amplification systems based on conserved regions of the 16S ribosomal RNA gene is the presence of these bacterial sequences as contaminants in many commercially available enzymes. To avoid amplification of this instead of target DNA, resulting in false results when amplifying soil DNA with conserved 16S ribosomal RNA gene-based primers, enzyme solutions as well as PCR reaction mixes should be treated so as to remove any 16S ribosomal gene sequences. Several strategies have been developed for this, e.g. treatment with psoralen, UV irradiation or treatment with DNase (Steffan *et al.*, 1988). We commonly treat the PCR reaction mixes and the *Taq* polymerase with DNase I, as outlined below.

The PCR protocol described in the following will address the amplification of soil DNA using eubacterial primers to generate fingerprints of bacterial communities of soil. For amplification of other microbial groups, such as the fungi or specific bacterial groups, the reader is referred to the relevant literature.

Treatment of (Taq) *DNA polymerase with DNase I to remove contaminating bacterial DNA*

1. Add 10 µl DNase I (10 U/µl) to PCR reaction mix (without *Taq* polymerase and primers) for ten reactions.
2. Add 0.1 µl DNase I to 5 µl *Taq* DNA polymerase Stoffel fragment (10U/µl).
3. Incubate tubes for 30 min at 37°C, then inactivate DNase I by heating at 98°C for 10 min. The reaction mix as well as enzyme can be used directly for PCR.

The enzyme used is AmpliTaq DNA Polymerase, Stoffel fragment (Perkin Elmer/Cetus). In our laboratory, this enzyme was shown to be least inhibited by soil impurities. However, several other enzymes may work as well on soil-derived DNA.

Procedure

1. Prepare master mix of reagents for any number of 50 µl reactions. Each mixture contains (final amounts in 50 µl final reaction volume):

Sterile deionized water	23.45 µl
Stoffel buffer 10×	5 µl (final conc. 1×)
Each dNTP (1 mM) mix – 5×	10 µl (final conc. 200 µM)
MgCl$_2$ 25 mM	7.5 µl (final conc. 3.75 mM)
Primer 1 10 µM	1 µl (final conc. 0.2 µM)
Primer 2 10 µM	1 µl (final conc. 0.2 µM)
Formamide*	0.5 µl (final conc. 1%)
T4 gene 32 protein (5 mg/ml)*	0.05 µl (final conc. 0.25 µg/50 µl)
Taq DNA polymerase Stoffel fragment (10 U/µl)	0.5 µl (final conc. 0.1 U/µl)

Total volume 49 µl. Cover with two drops of heavy mineral oil. An aliquot of 1 µl soil DNA extract is added through the oil to the reaction mixture containing all other components in the 'hot start' procedure (see 3 below) prior to starting thermal cycling.

*Formamide (Merck) is added to enhance specific primer annealing. T4 gene 32 protein (Boehringer) is added since it enhances the stability of single-stranded DNA, facilitating primer annealing (Tebbe and Vahjen, 1993). We found a sixfold lower concentration than that used by Tebbe and Vahjen (1993) to be optimal.

2. Prepare positive and negative controls, as indicated above.
3. 'Hot start' procedure: pre-heat the PCR mixture at 95°C for 2 min prior to adding enzyme or target DNA, in order to denature any bonds adventitiously or erroneously formed between primers and target.
4. Run PCR in a programmable thermal cycler, for 25–40 cycles. Often, the machine can be set so as to run the amplification cycles overnight.
5. After thermal cycling, perform a final primer extension at 72°C (e.g. 10 min), and keep reaction mixtures at 4°C until analysis.

Denaturing gradient gel electrophoresis (DGGE) of PCR products

Denaturing gradient gel electrophoresis (DGGE) separates DNA molecules of similar size by virtue of differences in their internal sequence (i.e. their melting behaviour) in a gradient of increasing denaturant (formamide and urea) strength. The gels are run at 60°C in a DGGE apparatus for 16 h.

Denaturing polyacrylamide gels

Polyacrylamide gels are formed by the polymerization of monomeric acrylamide into polymeric acrylamide chains and the cross-linking of these chains by N,N'-methylene bisacrylamide. The polymerization reaction is initiated by the addition of ammonium persulphate (APS), and the reaction is accelerated by TEMED (N,N,N',N'-tetramethylethylenediamine), which catalyses the formation of free radicals from APS. The porosity of the gel is determined by the length of the chains and the degree of cross-linking. The length of the chains is determined by the concentration of acrylamide in the polymerization reaction.

The protocol below is described for use with the Ingeny PhorU system, which is recommended for ease of use.

1. Choose the steepness of the desired gradient, e.g. 50–75%, using the suggested mixtures for the 'upper' solution (e.g. 50%), and for the lower solution (e.g. 75%, see Table 8.2).
2. Prepare 25 ml of each of these solutions in a 50 ml bluecap tube. Also, place 5 ml of solution A in a 12 ml bluecap tube (for stacking gel). Leave at room temperature for at least 30 min.
3. Thaw a tube with 10% APS from freezer (–20°C).

4. Prepare set-up for casting the gel (see 'Step by step' manual Ingeny PhorU-2, pp. 8–9). Clean glass plates and spacers, respectively, with KOH/methanol, soap and water, deminerilized water and ethanol. Dry the plates.
5. Add 150 µl 10% APS to the 'upper' and the 'lower' solution.
6. Add 12 µl TEMED to the 'upper' and the 'lower' solution, mix well, and proceed *immediately* with the following steps (7 and 8).
7. Place 25 ml of 'lower' solution (highest denaturant concentration, e.g. 75%) in the chamber of the gradient mixer connected to the pump. Open tap to the other chamber to let the air in the tap escape, and close. Transfer liquid from the other chamber back to the right chamber using a Pasteur pipette. Add magnetic stirring rod.
8. Place 25 ml of 'upper' solution (lowest denaturant concentration, e.g. 50%) in the other, empty chamber.
9. Switch on magnetic stirrer (position 4) to stir the 'lower' solution. Switch on pump and immediately carefully open the tap connecting the two chambers. Pump speed should be 4–6 ml/min. Gel should be cast in 15 min. Leave for 15–30 min to solidify.
10. Add 50 µl 10% APS and 7 µl TEMED to 5 ml solution A (stacking gel – 8% acrylamide). Using a 5 ml syringe with a long needle, slowly pour the top part of the gel by hand. Avoid air bubbles. Add the comb. Wait for at least 1 h for the gel to polymerize completely.
11. Switch on buffer tank at 60°C (takes 45 min to heat up). The buffer tank contains 15 l of 0.5× TAE.
12. After gel polymerization, all screws should be adjusted to just touch the plexiglass pressure unit. Place the cassette in the buffer tank. Push the U-shaped spacer all the way down now. Little air bubbles under the gel can be removed by holding the cassette at an angle. Remove the comb. Tighten the upper two screws again. Connect buffer flow to connector on cassette. Fill upper buffer tank and close the tap.
13. Connect the electrical plugs. Add the samples. Set power supply to the desired voltage (e.g. 75 V) and time (e.g. 16 h). Wait 10 min, while electrophoresis is in progress, before opening the upper buffer flow again. The reservoir should not overflow.
14. After running, remove gel. Stain with SYBR Gold (1 h) and rinse. Observe on UV transilluminator.

Calculation

The molecular community profiles can be loaded into a software program such as GELCOMPAR – see Rademaker (1995) for details. The program will digitize the profiles and can assign values to each band, in accordance with its intensity. The matrix of band intensity values per lane can then be used in any statistical approach that allows rigid comparison of the profiles. Depending on the purpose of the comparisons, principal components

Table 8.2. Mixtures of solutions A (0% denaturants) and B (80% denaturants) to achieve desired denaturant concentrations.

Denaturant concentration (%)	Solution A (ml)	Solution B (ml)
25	17.2	7.8
30	15.6	9.4
35	14.1	10.9
40	12.5	12.5
45	11	14
50	9.4	15.6
65	4.7	20.3
70	3.1	21.9
75	1.6	23.4

analysis, canonical correspondence/variant analysis or discriminant analysis can be employed on the basis of the collective data. See relevant literature as well as Section 8.3 of this volume.

Discussion

The DNA extraction protocols selected and described here are very suitable for the detection of specific microorganisms and their genes in soil (Smalla *et al.*, 1993a; Duarte *et al.*, 1998; van Elsas *et al.*, 2000). However, as discussed, there may still be some doubt as to the localization of such DNA sequences, i.e. inside cells or extracellularly. As this is a potential problem inherent to all soil DNA extraction protocols, including the cell extraction-based ones, results obtained should be carefully interpreted as to their meaning for the actual cell populations present. Furthermore, even though the bead beater is known to lyse a major part of bacterial species efficiently, including many Gram-positives, there is no absolute certainty that cell lysis is representative for the soil bacterial community. The findings of Moré *et al.* (1994) suggest that lysis may well be confined to the larger cell-size fraction of the microbial community, leaving a major fraction of minute cells unlysed. This feature also has to be taken into account when the protocol is to be used for microbial community structure studies in soil. In fact, it is clear that with this, or any other, DNA extraction protocol, only a limited view of the soil microbiota can be obtained, due to the fact that the extraction/lysis efficiency is commonly below 100% of the total detectable cell populations. The primers used for PCR amplification of soil DNA are obviously determinative for the level at which different microbial groups can be assessed. Thus, diversity indices can be produced for broad microbial groups, such as all bacteria or all fungi. At a finer level of resolution, a picture of the diversity of specific groups (such as the α- or β-proteobacteria, the high-G+C% actinomycetes or the pseudomonads) can be obtained. The advantage of using the group-specific approach is that the enormous complexity of the microflora in soil can be reduced to a level that is better interpretable. In fact, one can focus on those microbial groups that are less numerous in the soil, and would have been overlooked in pictures obtained for the broad groupings, as they would be minority organisms there.

The fingerprints obtained via DGGE separation are often robust and reliable for a given soil or soil treatment, reporting on the microbial diversity status of the sample. The fingerprint obtained quite often provides a representative picture of the numerically dominant types in the microbial group targeted. However, as with any method applied to soil, the DGGE data are not without caveats. These caveats have been described extensively in the literature, and will be discussed only briefly here. First, whereas PCR amplification will often be without bias, that is, each particular sequence will be amplified at about the same rate, sometimes aberrations from this common rate have been observed. This results in a phenomenon called 'preferential amplification', in which a particular sequence is amplified at a (much) higher rate than others, leading to a distorted picture of the relative abundance of the different microbial types present. Secondly, a fraction of the bands that make up the DGGE profile are actually chimeric sequences, which are the result of an aberrant PCR process; the percentage at which chimeras can be formed varies, but low frequencies (a few percent) to frequencies as high as 30% have been reported. To get a handle on this PCR artefact, one can use sequence analysis followed by the feature 'Check_chimera' available in several sequence analysis programs.

Other features of PCR-DGGE analysis of soil DNA are the known facts that some bacterial types can produce multiple bands on DGGE (a result of the presence of several slightly different rRNA operons in the same cell), and that fragments of different sequence can migrate to the same gel positions. Therefore, a note of caution is in place with respect to the interpretation of the profiles generated by DGGE, and the use of these profiles to produce diversity indices.

With all these cautionary remarks in mind, one can safely state that the current protocol opens a wide window for studying the diversity of total microbial populations in soil at various different levels of resolution, an ability that one could only dream of just a decade ago. It is foreseen that developments will continue and that the current methodology will be refined, or even superceded, by other direct methods, such as the use of DNA microarray technology. However, we would like to propose the currently described method as the method of choice for routine assessments of the microbial diversity of soil, possibly at two levels, i.e. that of total bacteria (Kowalchuk *et al.*, 2004) and of total fungi (Vainio and Hantula, 2000).

8.3 Phospholipid Fatty Acid (PLFA) Analyses

ANSA PALOJÄRVI

MTT – Agrifood Research Finland, FIN-31600 Jokioinen, Finland

Introduction

The analysis of ester-linked (EL) phospholipid fatty acids (PLFAs) is an acknowledged biochemical approach to microbial community characterization. PLFAs are constituents of all cell membranes, and have no storage function. Under the conditions expected in naturally occurring communities, phospholipids represent a relatively consistent fraction of cell mass, even though some changes are detected in pure cultures due to changes in growth media composition or temperature (Harwood and Russel, 1984). Lipid analysis offers an alternative that does not rely upon the cultivation of microorganisms. Current extraction and derivatization methods permit effective recovery of PLFAs from living organisms (White, 1988). PLFAs also degrade quickly upon an organism's death (White, 1988), if the degrading enzymatic activity is not inhibited (see Zelles *et al.*, 1997), which allows the detection of rapid changes in microbial populations.

Extraction and subsequent analysis by gas chromatography and mass spectrometry provides precise resolution, sensitive detection and accurate quantification of a broad array of PLFAs. Each analysis yields a profile composed of numerous PLFAs defined on the basis of compound structure and the quantity of each compound present in the sample. The advantage of the method is that the amount of total PLFAs can be used as an indicator for viable microbial biomass, and that further characterization can be done based on specific signature biomarker fatty acids. Taxonomically, the method does not include *Archaea*, since they have ether-linked rather than ester-linked phospholipid fatty acids in their cell membranes. There are no indicated restrictions to the application of the method for any kinds of soil and environmental samples and it is relatively time- and cost-competitive.

The PLFA method has been applied for various environmental questions. Changes in the PLFA patterns have been detected according to the different levels of metal contamination (Pennanen, 2001). Agricultural management practices (Petersen *et al.*, 1997; Schloter *et al.*, 1998) and pesticide use (Widmer *et al.*, 2001) have been shown to cause shifts in the PLFA patterns. Changes in the PLFA patterns have been detected after chloroform fumigation of soil (Zelles *et al.*, 1997). Validation of PLFA in the determination of microbial community structure has been reviewed, e.g. by Zelles (1999) and Pennanen (2001).

Certain PLFAs can serve as unique signatures for specific functional groups of microorganisms. Such biomarkers cannot detect individual species of microorganisms due to overlapping PLFA patterns. Nevertheless, comparison of total community PLFA profiles accurately mirrors shifts in community composition and provides a way to link community composition to specific metabolic and environmental conditions. Signature fatty acids are listed by several authors (Morgan and Winstanley, 1997; Zelles, 1999; Kozdrój and van Elsas, 2001a). For fungal biomarkers see Frostegård and Bååth (1996), Miller *et al.* (1998) and Olsson (1999).

Principle of the Methods

The phospholipid fatty acid (PLFA) analysis is based on the single-phase extraction of lipids described by Bligh and Dyer (1959). The lipids are fractionated into different lipid classes: neutral lipids, glycolipids and phospholipids (Fig. 8.1). Neutral and glycolipids are normally handled as waste fractions and disregarded in microbial community analysis, although they can be used for other purposes, e.g. to describe the nutritional status of microbes (White *et al.*, 1998).

The phospholipid fraction is then methylated to give fatty acid methyl esters (FAME) and analysed by gas chromatograph (GC) with flame-ionization detector (FID) or GC coupled with a mass spectrometer (GC–MS). Due to the very variable contents of different PLFA in cell membranes, the method without further fractionation (Fig. 8.1A) is able to detect the most abundant ester-linked PLFA only. In most cases, 20–40 PLFA are identified. They make up most of the biomass, but not of the number of PLFA in the cells (see Zelles, 1999). This method has been successful in separating microbial communities in various experiments and it is applicable for monitoring.

Extended PLFA analysis discovers a very wide variety of cellular fatty acids, including both ester-linked and non-ester-linked PLFAs (Fig. 8.1B), and offers good potential for the use of signature fatty acids.

Whole-cell fatty acid patterns are based on FAME analysis after direct saponification and methylation of lipids without fractionation to different fatty acid groups (Fig. 8.1C). The method was originally designed for microbial identification of pure cultures (the commercially available Microbial Identification System (MIDI or MIS); Haack *et al.*, 1994; Kozdrój and van Elsas, 2001b). Even though both the whole-cell fatty acid patterns and PLFA patterns are based on FAME analysis, it should be noted that the methods are not comparable. The whole-cell fatty acid patterns (often called the FAME method) comprise lipids derived from non-living organic matter (Petersen *et al.*, 2002), and include storage lipids, which are more sensitive to growth conditions. The PLFA and FAME methods have been reviewed by Zelles (1999).

In this section, a procedure for PLFA analysis is described which is based on White *et al.* (1979), Frostegård *et al.* (1993) and Palojärvi *et al.* (1997), with slight modifications.

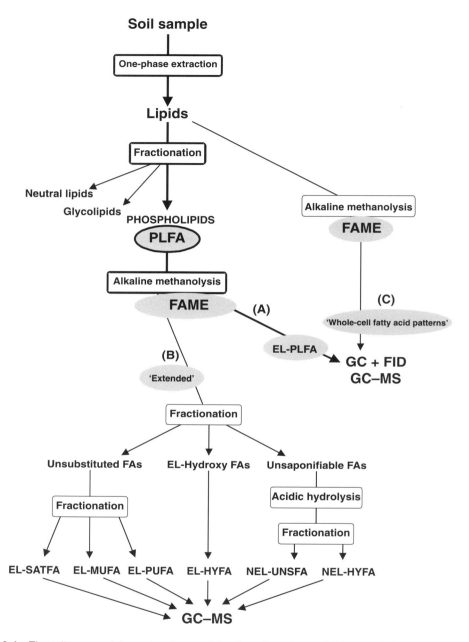

Fig. 8.1. Flow diagram of the extraction and fractionation steps of (A) ester-linked phospholipid fatty acid (EL-PLFA); (B) extended PLFA; and (C) whole-cell fatty acid analyses. Abbreviations: FAME, fatty acid methyl esters; GC, gas chromatograph; FID, flame-ionization detector; MS, mass spectrometer; FA, fatty acid; EL, ester linked; SATFA, saturated FA; MUFA, monounsaturated FA; PUFA, polyunsaturated FA; HYFA, hydroxy FA; NEL, non-ester linked; UNSFA, unsubstituted FA. Figure modified after Zelles and Bai (1994) and Palojärvi and Albers (1998).

Materials and Apparatus

In addition to standard laboratory equipment, the following supplies and apparatus are needed:

Supplies

- Glass test tubes and small glass bottles with Teflon-lined screw caps (ca. 50 ml and 10 ml test tubes for centrifuge, 4–10 ml bottles for various solvent fractions)
- Disposable glass pipettes (Pasteur, Micro pipettes) and compatible pipettors
- Nitrogen gas
- Silicic acid (e.g. Unisil 100–200 mesh) or commercial solid-phase extraction columns (e.g. Varian Bond Elut SILICA Si 500 mg or 2 g)

Apparatus

- Fume hood
- Vortex mixer
- (Orbital) shaker
- Centrifuge
- Nitrogen evaporator
- Gas chromatograph (GC) with flame-ionization detector (FID) or, preferably, GC coupled with a mass spectrometer (GC–MS). The GC should be equipped with a splitless injector and a long non-polar capillary column (e.g. HP-5; 0.2 mm internal diameter, 0.33 μm film thickness, 50 m column length)

Chemicals and solutions

- Citrate buffer (0.15 M $Na_3C_6H_5O_7 \cdot 2H_2O$, pH 4.0), or phosphate buffer (50 mM K_2HPO_4, pH 7.4)
- Solvents (chloroform stabilized with ethanol, methanol, acetone, toluene, hexane, iso-octane); analytical grade or higher
- Bligh-and-Dyer solution (chloroform:methanol:buffer, 1:2:0.8; v/v/v)
- Methanol:toluene (1:1; v/v)
- Methanolic KOH (0.2 M; make fresh daily)
- Hexane:chloroform (4:1; v/v)
- Acetic acid (1.0 M)
- Internal standard 19:0 (methyl nonadeconoate; Sigma)

- Fatty acid methyl ester standards (e.g. Sigma, Supelco, Nu-Chek-Prep)

Note: All chemicals are analytical grade or higher; chloroform should be stabilized with ethanol. Phosphate buffer is widely used for PLFA extractions, but Nielsen and Petersen (2000) showed that citric buffer gives the highest PLFA yields.

Procedure

Lipid extraction

Lipid extractions are carried out on a few grams of fresh, frozen or freeze-dried soil. Fresh soil can be kept in a refrigerator for a few weeks (Petersen and Klug, 1994). The proper amount of soil should always be established beforehand. The amount of PLFA in the sample should be high enough to be above the detection limit. On the other hand, the presence of very high amounts of PLFA may cause problems during the analyses. In experiments with different soils or treatments, it is recommended to standardize the amounts of material to give similar levels of microbial biomass. The single-phase extraction mixture contains chloroform, methanol and buffer in the ratio 1:2:0.8. The water content of the sample is calculated and a sufficient amount of buffer is added. The following procedure is suitable for 1–4 g of soil.

- The soil sample is weighed into a 50 ml test tube.
- Add buffer so that the total water content of the soil is 1.5 ml.
- Add 1.9 ml chloroform, 3.7 ml methanol and 2.0 ml Bligh-and-Dyer solution.
- Mix well using a Vortex mixer.
- Leave the sample in a (orbital) shaker at a low speed (ca. 200 rpm) for at least for 4 h, or overnight, and centrifuge ($c.$ 1000 $\times$ g for 10 min).
- Transfer the supernatant to another 50 ml test tube, wash the soil pellet with 2.5 ml Bligh-and-Dyer solution, repeat the mixing and centrifugation, and combine the supernatants.
- To separate the solvent phase, add 3.1 ml chloroform and 3.1 ml buffer, and mix well with the Vortex ($c.$ 1 min).
- Let the mixture stand overnight to separate the lower organic phase (chloroform) from the upper water–methanol phase.
- Transfer the organic phase into a small glass bottle.
- The samples are placed in a water or sand bath (40°C) and dried under a stream of N_2 gas until near dryness.
- The lipid samples are stored in a freezer.

Lipid fractionation

- The lipids are separated into neutral, glyco- and phospholipids on columns containing silicic acid by eluting with chloroform, acetone and methanol, respectively.
- Prior to separation, commercial columns (500 mg) must be activated with 6 ml chloroform.
- The dried lipid samples are transferred on to the columns in 3×100 µl chloroform.
- Elute neutral lipids with 6 ml chloroform and glycolipids with 12 ml acetone. For PLFA analysis, these are waste fractions.
- Phospholipids are eluted with 6 ml methanol into a 10 ml test tube.
- The methanol fraction is reduced until dryness under N_2 gas in a water or sand bath (maximum temperature 30°C).

Internal standards

- A known amount of methyl nonadecanoate (19:0; Sigma) is added to the phospholipid fraction as an internal standard.
- Methyl tridecanoate (13:0; Sigma) can be used as an additional internal standard to indicate possible losses of short-chain fatty acids.

Mild alkaline methanolysis

- Samples are dissolved in 1 ml methanol:toluene solution.
- Methanolic KOH (1 ml) is added, and the mixture is incubated in a water bath (37°C, 15 min). Add immediately 2 ml hexane:chloroform solution and 0.3 ml acetic acid to neutralize the solution (pH can be checked from the lower phase; pH 5–7).
- Mix well with Vortex (*c.* 1 min) and centrifuge (*c.* $1000 \times g$ for 5 min).
- Transfer the upper, organic phase into a small glass bottle.
- Wash the mixture once with 2 ml hexane:chloroform and combine the upper, organic phase with the former one.
- Reduce the sample with N_2 gas without heating. Place the sample, re-dissolved in 100 µl iso-octane (or hexane), in a GC vial with a glass insert.

GC–MS or GC + FID analysis

The resulting fatty acid methyl esters (FAME) of ester-linked phospholipid fatty acids (EL-PLFA) are separated, quantified and identified by gas chromatography (GC) coupled with a mass spectrometer unit for peak identification (GC–MS). Alternatively, the analysis can be carried out with a GC equipped with a flame-ionization detector (FID); in this case peak identification is based on retention times only. The GC + FID analysis must be confirmed from time to time against GC–MS. Helium is typically used as a carrier gas. The flow rate and temperature programming of the column should be adjusted for the individual instrument. An example of settings: flow rate 0.9 ml/min, initial temperature 70°C for 2 min, increase 30°C/min until 160°C, increase 3°C/min until 280°C, final temperature 280°C for 10 min. The identification and the response factors of different PLFA compounds are based on fatty acid methyl ester standards. Double-bond positions can be determined with their dimethyl disulphide adducts (Nichols *et al.*, 1986).

Fatty acid nomenclature

A short-hand nomenclature used to characterize fatty acids is as follows. In the expression X:YωZ, X indicates the number of carbon atoms in the fatty acid, while Y indicates the degree of unsaturation (= the number of carbon–carbon double bonds). The position of the first double bond from the methyl (or aliphatic; 'ω') end of the molecule is represented by Z. Alternatively, double-bond positions from the carboxyl ('Δ') end are sometimes given. The suffixes 'c' and 't' indicate *cis* and *trans* geometry, respectively. Because *cis* geometry is most common, c is often omitted. The prefixes 'i' and 'a' refer to iso (the second carbon from methyl end) and anteiso (the third carbon from methyl end) branching; 'br' indicates unknown methyl branching position. Other methyl branching is indicated by the position of the additional methyl carbon from the carboxyl end followed by 'Me' (i.e. 10Me18:0). Some authors include the carbon of methyl group in X and use the prefix 'p'. This means, for example, that 10Me16:0 is identical to p10–17:0. The number before the prefix 'OH' indicates the position of a hydroxy group from the carboxyl end (i.e. 3-OH14:0). α and β are sometimes used to indicate a hydroxy substitution at position 2 or 3 from the carboxyl end, respectively. Cyclopropane fatty acids are designated by the prefix 'cy'.

Calculation

The calculation of the concentrations of PLFA is shown in Eq. (8.1):

$$C_x(\text{nmol}/g) = \frac{A_x \times c_i[\mu g] \times f \times 1000}{A_i \times W[g] \times M[\mu g/\mu \text{mol}]} \quad (8.1)$$

where: C_x = concentration of the fatty acid studied; A_x = peak area of the fatty acid studied; A_i = peak area of the internal standard; c_i = absolute amount of internal standard in the vial (μg); f = response factors of different PLFA compounds (peak area to concentration ratio compared to internal standard; if not known, then = 1); W = amount of soil (g); and M = molecule weight of the fatty acid (µg/µmol).

Note: The results can also be expressed as micrograms PLFA per gram dry soil. Often the relative contribution of different PLFA is of major interest ('PLFA fingerprint') and mole fractions (in percentage) are calculated.

Statistical analyses

Multivariate analyses can be applied to PLFA profiles. Principal components analysis (PCA) is most widely used (e.g. Palojärvi *et al.*, 1997; Petersen *et al.*, 1997). Canonical correspondence analysis enables determination of the influence of different environmental factors on the PLFA patterns (Bossio *et al.*, 1998). Non-metric multidimensional scaling (NMDS; Siira-Pietikäinen *et al.*, 2001) and neural computing methods (Noble *et al.*, 2000) have also been applied recently.

Discussion

PLFA analysis is applicable for monitoring and detecting changes in the soil microbial communities. Additionally, viable microbial biomass estimates and information on several biomarker PLFAs can be obtained. Several comparisons have shown that PLFA can detect rapid changes and produce results on microbial characterization comparable to other community-level methods (e.g. Widmer *et al.*, 2001).

Further methodological perspectives: Macnaughton *et al.* (1997) suggested a pressurized hot solvent extraction to enable rapid and improved extraction of lipids from large numbers of environmental samples. Intact phospholipid profiling (IPP), using liquid chromatography/electrospray ionization/mass spectrometry, is an advanced alternative to EL-PLFA analysis by GC–MS (Fang *et al.*, 2000). Further methods have been developed for archaeal ether-linked lipids by Bai and Zelles (1997) and Fritze *et al.* (1999). Different isotope and radiolabelling techniques of PLFA have been applied to study specific biogeochemical processes and microbial activity (Boschker *et al.*, 1998; Roslev *et al.*, 1998).

8.4 Substrate Utilization in Biolog™ Plates for Analysis of CLPP

MICHIEL RUTGERS,[1] ANTON M. BREURE[1] AND HERIBERT INSAM[2]

[1]*National Institute for Public Health and the Environment, Antonie van Leeuwenhoeklaan 9, NL-3721 MA Bilthoven, The Netherlands;* [2]*Universität Innsbruck, Institut für Mikrobiologie, Technikerstraße 25, A-6020 Innsbruck, Austria*

Introduction

For characterization of the microbiology of soil samples, functional aspects related to substrate utilization are as useful as taxonomic or structural investigations based on DNA or RNA analysis (Grayston *et al.*, 1998). An understanding of the functional or metabolic diversity of microbial communities, particularly defined by the substrates used for energy metabolism, is integral to our understanding of biogeochemistry (Hooper *et al.*, 1995). Thus, diversity at the functional level rather than at the taxonomic level may be crucial for the long-term stability of an ecosystem (Pankhurst *et al.*, 1996).

In addition, microbial community analysis on the basis of the best available techniques only addresses a limited number of dominant features. For instance, using PCR-DGGE analysis based on the 16S ribosomal RNA gene sequence, up to about 100 dominant DNA sequences can be recognized. The same holds for other techniques, such as phospholipid fatty acid (PLFA) profiling, morphological characterizations and metabolic diversity analyses (Rutgers and Breure, 1999). Estimates of the numbers of species in microbial communities in soil and sediment range from 10^4–10^5 different species/g dry weight (Torsvik *et al.*, 1990; Dykhuizen, 1998). Consequently, irrespective of the technique, the danger exists that important parts of the community will be overlooked. The use of complementary techniques, such as those offered by the combination of DNA profiling and metabolic diversity analyses, reduces this danger.

Since first proposed by Garland and Mills (1991), *in vitro* community-level physiological profiles (CLPPs) have been used frequently to characterize microbial communities of different habitats, ranging from sediments to seawater, and from oligotrophic groundwater to soils and composts (e.g. Garland and Mills, 1991, 1996; Grayston and Campbell, 1995; Insam *et al.*, 1996; Grayston *et al.*, 1998, 2001).

This section describes the determination of CLPP using 96-well plates for metabolic substrate utilization. Microplate plates allow for easy measurement of many microbial responses, making community analysis practical and open for laboratory automation.

Principle of the method

CLPP involves direct inoculation of environmental samples into Biolog™ microplates, incubation and spectrophotometric detection of heterotrophic microbial activity. Its simplicity and rapidity is attractive to the microbial ecologist, but it requires careful data acquisition, analysis and interpretation. The Biolog™ system was initially developed for characterization of pure isolates of Gram-positive (using GP2 plates, formerly GP plates) and Gram-negative (using GN2 plates, formerly GN plates) heterotrophic isolates (Bochner, 1989). Development of a method for CLPP started with the application of GN2 plates and, to a lesser extent, GP2 plates, both containing 95 different carbon sources. Recently, so-called ECO plates were developed for CLPP of terrestrial communities (Insam, 1997). These plates contain 31 different carbon sources and allow for a triplicate experiment in one plate. GN2 and GP2 plates were recently introduced after GN and GP plates and now include a gellum-forming agent. ECO plates contain this agent too. There are some indications of a different colour-forming regime between GN and GN2 plates (O'Connell et al., 2001).

Several approaches have been used to account for biases related to inoculum density, inoculum activity, incubation time and micro-environment (Garland, 1997; Preston-Mafham et al., 2002). For instance, standardization of initial inoculum density is commonly used (Garland and Mills, 1991), although it is laborious and there is still dispute about an appropriate cell enumeration method. Normalization of optical density (OD) readings by dividing by average well colour development (AWCD) is bound to a number of conditions (Haack et al., 1995; Heuer and Smalla, 1997; Konopka et al., 1998). Single time-point readings and integration of the OD over time are still most widely used, but bear the effect of inoculum density (Garland and Mills, 1991). Following a proposition of continuous plate reading for analysing the kinetics rather than the degree of colour development at a given time, a sigmoidal growth model has been developed (Haack et al., 1995; Garland, 1997; Lindstrom et al., 1998). Again, some kinetic parameters are dependent on inoculum density and need to be normalized prior to statistical analysis. Recently, a normalization procedure that employs integrated OD values derived from a number of dilutions of the same sample, rather than a single dilution level, has been developed as an inoculum-density- and time-independent method of analysis (Gamo and Shoji, 1999; Garland and Lehman, 1999; Breure and Rutgers, 2000; Franklin et al., 2001). The inoculum-density-independent approach is being applied in the Netherlands Soil Monitoring Network. This network consists of about 300 sampling locations (Bloem and Breure, 2003).

In this section, some analytical aspects of sampling and of the analysis of communities by CLPP are described. Considerations for sampling and storing soil, sediment or surface water samples for CLPP analysis are discussed. The overview is neither complete nor exhaustive. For this, the reader is referred to standard textbooks of soil microbial methods. Two different methods for establishing CLPPs are described. The first method is most often used and requires just one plate for a CLPP analysis. This method requires careful inoculum standardization and subsequent data analysis in order to develop a CLPP which is independent of small changes in inoculum size and activity. The second method is based on inoculation of serial sample dilutions, and requires a set of plates for just one CLPP analysis. This method is essentially inoculum-density independent, and can be used for comparing samples of different origin.

Sampling, storage and extraction

As for most biological investigations, fresh samples are often superior to stored samples. However, one has to realize that the microbial communities reflect the climatic conditions just before sampling, especially in shallow soil horizons. This aspect is particularly true for physiology-based methods such as CLPP. Consequently, with remote sampling locations, or for monitoring purposes, it is recommended to include an equilibration period under standardized conditions in the laboratory. For instance, in the Netherlands Soil Monitoring Network, sieved soil samples (2–4 mm mesh) are equilibrated for 4 weeks at 10°C, under 50% water-holding capacity (WHC). This procedure is thought to minimize the effect of gross climatic differences (such as rainfall and temperature) in the weeks before sampling (Bloem and Breure, 2003). If storage is necessary, samples may be stored up to 10 days at 4°C; for longer periods, freezing is recommended.

Extraction procedures range from simple horizontal shaking, head-over-head shaking, or blending of soil or sediment slurries in water, buffer or salt solutions, to elaborate sequential extractions (Hopkins *et al.*, 1991). Usually, a few grams of soil, sediment or compost are sufficient. However, it is strongly recommended that soil sampling, storage and extraction procedures are standardized in order to make comparisons between CLPPs more reliable.

Materials and apparatus

Microplates and their inoculation

Two options are possible, the use of either Biolog™ ECO plates or GN2 plates, containing 31 or 95 different C sources, respectively, plus a water well (Biolog Inc., Hayward, California, USA). ECO plates contain three replicates of the carbon substrate and control. The plates are inoculated with 130 µl suspension per well, diluted (in $\frac{1}{4}$ strength Ringer solution) to obtain a cell density of approximately 1×10^8 cells/ml (determined, for example, by acridine orange direct count (AODC); Bloem *et al.*, 1995). The plates are then incubated at 20°C in the dark and subsequent colour development is measured every 12 h for 5 days (592 nm).

For the inoculum-density-independent approach, serial threefold dilutions of the bacterial suspension are produced (3^0–3^{-11}; with dense bacterial suspensions it is better to use a more diluted series, e.g. 3^{-2}–3^{-13}). The dilution series should essentially give a complete range of colour formation levels from > 95% to < 5% average colour in the plate. Each dilution is inoculated in a section of an ECO plate (Fig. 8.2). The plates are then incubated at 20°C in the dark under 90 ± 5% air humidity to avoid undesired evaporation. Colour development is measured at 590 nm and 750 nm every 8 h or 12 h for 7 days, using an autosampler and a microplate reader or spectrophotometer.

Microplate stackers and autosamplers might be used, since experiments for the analysis of CLPPs usually contain many plates. Commercially available autosamplers and stackers can handle stacks of 20 to more than 100 plates in one run.

Procedures

According to specific needs or availability of equipment, extraction procedures may be modified, inoculation densities may be altered, or single-point data reading may be replaced by continuous readings (alternatives see above). In the following, subheadings preceded by an 'A' relate to the simple method with one Biolog™ plate per sample. Subheadings preceded by a 'B' relate to the inoculum-density-independent method using more than one plate per sample.

Example of an extraction procedure

1. Blend 5 g fresh soil with 20 ml or 50 ml of 0.1% (w/v) sodium cholate solution, 8.5 g cation exchange resin (Dowex 50WX8, 20–50 mesh, Sigma) and 30 glass beads.
2. (Optional) Shake the suspension on a head-over-head shaker (2 h, 4°C).
3. Centrifuge at low speed (500 × g to 800 × g) for 2 min.
4. Decant the supernatant into a sterilized flask.
5. (Optional) Resuspend the pellet in 10 ml Tris buffer (pH 7.4) and shake for 1 h.
6. Centrifuge as above and add the supernatant to the earlier extract.
7. (Optional) If the extract is turbid or dark (due to clay or humic particles) centrifuge the resulting supernatant another time.

A. Dilution, inoculation and incubation (single-plate procedure)

1. Dilute the samples with $\frac{1}{4}$ strength Ringer solution ten (sediment)- to 1000 (composts)-fold and check the cell density by acridine orange direct counting (AODC). As an alternative, substrate-induced respiration

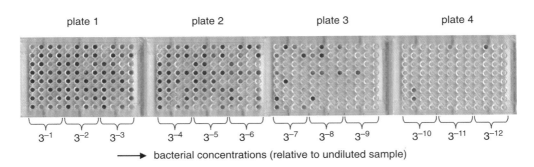

Fig. 8.2. Series of Biolog™ plates (ECO) demonstrating the loss of colour development after stepwise dilution of the inoculum (dilution factor per step is 3 in this case). The loss of colour per well is indicative of the community-level physiological profile (CLPP).

(Anderson and Domsch, 1978) is recommended to determine microbial biomass of the soil samples. Appropriate dilutions of the extracts may then be used to obtain similar inocula.

2. Dilute samples appropriately to obtain a cell density of approximately 1×10^8 cells/ml. In the case of background coloration of the extract, further dilutions are recommended.

3. Inoculate the plate with 100 µl bacterial suspension per well (Biolog™ recommends 150 µl per well, but this is somewhat difficult to handle).

4. Cover the plates with a lid and incubate at 20°C in the dark (other temperatures are possible). Prevent undesired evaporation by putting the plates in polyethylene bags or in a humidified incubation chamber.

5. Measure colour development (592 nm) every 12 h for 5 days using an automated plate reader. Readings may be terminated if the average well colour density reaches an optical density of 2. If curve parameters or the area under the curve are determined, make sure you always have the same reading intervals and the same number of readings (e.g. 10). In case of a long lag-time, or low incubation temperature, incubation may be prolonged (reading interval up to 24 h).

B. Dilution, inoculation and incubation (inoculum-density-independent procedure)

1. Make a dilution series of the bacterial suspension in physiological salts solution or $\frac{1}{4}$ strength Ringer, using threefold dilution steps.

2. Inoculate a minimum of 100 µl of 3^0–3^{-11} diluted bacterial suspension (mineral soil horizons, sandy soils) in 4 ECO plates (1 dilution per section = 32 wells). Depending on the initial concentration of bacteria in the suspension, another set of dilutions might be used (e.g. 3^{-2}–3^{-13} in the case of organic soils and clay).

3. Cover the plates with a lid and incubate at 20°C in the dark (other temperatures are possible). Prevent undesired evaporation by putting the plates in polyethylene bags or in a humidified incubation chamber.

4. Measure colour development (592 nm) every 8 h for 7 days using an automated plate reader. Readings may be terminated if the average well colour density reaches an optical density of 2. If curve parameters or the area under the curve are determined, make sure you always have the same reading intervals and the same number of readings (e.g. 20). In the case of a long lag-time, or low incubation temperature, incubation may be prolonged (reading interval up to 24 h).

Calculation

A. Calculation and data management (single-plate procedure)

Correct raw OD data by blanking response wells against the well showing the minimum absorbance value (R-minimum; Insam et al., 1996). This blanking avoids negative values when compared to subtracting the control well from the response well.

Two alternatives are suggested:

1. Calculate the AWCD from each plate at each reading time. For each plate, those time points of reading are selected that have an AWCD closest to 0.6. Alternatively, other AWCDs (e.g. 0.30, 1.00) may be chosen.
2. Normalize data by dividing each well OD by AWCD (Garland and Mills, 1996). This is particularly important when inoculum densities are not standardized prior to inoculation. Data analysis may be further elaborated by calculating the area under the curve for each well OD for the entire period of incubation (Guckert et al., 1996).

Another procedure that has been used successfully is the estimation of kinetic parameters (K, r, s) by fitting the curve of OD versus time to a density-dependent logistic growth equation (Lindstrom et al., 1998):

$$Y = OD_{592} = \frac{K}{1+e^{-r(t-s)}} \tag{8.2}$$

where: K is the asymptote (or carrying capacity); r determines the exponential rate of OD change; t is the time following inoculation of the microplates; and s is the time when the midpoint of the exponential portion of the curve (i.e. when $Y = K/2$) is reached.

For statistical testing of results, and in particular if the use of MANOVA is planned, inoculate a sufficient number of replicates (one ECO plate contains three replicates) according to $n_i \times q = n \geq 31 + q + 2$, where: q is the number of groups to be compared and n_i is the replicate number required per group (sample sizes are equal in each group). For example, if two groups are compared, an $n_i > 17$ is required (i.e. 6 ECO plates; Insam and Hitzl, 1999).

B. Calculation and data management (inoculum-density-independent procedure)

It is generally recognized that the inoculum density highly affects the outcome of the CLPP, and different methods exist to correct for this (see above). We propose a method to produce a CLPP which is independent of the concentration of microorganisms in the environmental sample or in the inoculum. This is done by inoculating a stack of Biolog™ plates with a series of three- or tenfold dilutions of the inoculum. The pattern of attenuation of colour development along the dilution gradient is regarded as characteristic for the CLPP. The final CLPP is then derived from the relative abundances of all Biolog™ substrate conversions of a sample, which is the difference between the amount of inoculum required for 50% response of a specific substrate conversion in the Biolog™ plate and the amount of inoculum required for 50% response in the Biolog™ plate on average. In this way, the problem of inoculum standardization is solved.

To determine this difference, the following concept has been established (Fig. 8.3A). The amount of inoculum (e.g. in log CFU/ml or in a dilution factor) in the sample that causes 50% of the maximal theoretical average response in a Biolog™ plate is determined from a dilution series. Thereafter, the amount of inoculum that causes 50% of the maximal theoretical response for a specific substrate conversion is compared to that value (both log transformed values), resulting in a value for the relative abundance of that specific substrate conversion. According to this procedure, the final CLPP consists of 95 (in case of GN2 plates) or 31 (in case of ECO plates) relative abundance values, and the average abundance is zero.

The rationale for performing dilution experiments is that the response in the Biolog™ system cannot be related directly to the amount of organisms in the wells, due to non-linearities (Garland, 1997). Consequently, application of a range of cell concentrations in the experimental procedure is advised to escape from estimating the Biolog™ response at extrapolated cell concentrations (Garland and Lehman, 1999; Breure and Rutgers, 2000).

The response (A, for instance normalized well colour development, normalized area under the curve) per substrate (s) is then plotted against the bacterial concentrations in the wells (for instance CFU/ml), which leads to a curve that can be fitted with a log normal distribution (upper line in Fig. 8.3A):

$$A_s = \frac{t}{1+10^{h(\log \text{CFU50}_s - \log \text{CFU})}} \tag{8.3}$$

where: t is the asymptotical maximum of the curve at infinite cell concentration; log CFU is the logarithm of the number of colony-forming units; log CFU50_s is the inflexion point of the curve; and h is the Hill slope (dimensionless). The log CFU50_s gives a measure for the amount of inoculum in the sample necessary for 50% response of the specific activity.

It is recommended to standardize the responses by dividing the observed response by the maximum theoretical response, because t (the maximum) can then be set to 1. The maximum theoretical response can be

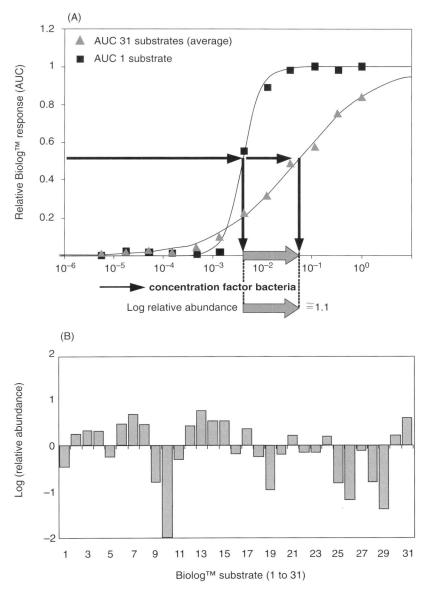

Fig. 8.3. Outline of the procedure to construct a community-level physiological profile (CLPP) which is essentially independent of the inoculum density. (A) A range of inoculum densities is inoculated in a series of Biolog™ plates. Upon dilution, the average response in the Biolog™ plate, or the response of an individual well (e.g. well colour development, or the integrated colour over time), decreases. The dilution of a specific response is fitted with a sigmoidal curve (Eq. 8.3). The log relative abundance (log RA) is given by the difference between the curve for the average response in the plate and the response for a specific substrate; in this example approximately +1.1. (B) In the case of ECO plates, the CLPP consists of 31 relative abundance values; the average abundance is zero. In this example, there was no colour formation in well number 10 at the highest inoculum concentration. Accordingly, this value was artificially set to –2 (see text).

derived from a collation of all previously performed experiments with the same experimental set-up. If this is impossible (e.g. because it is the first experiment) be alert for unrealistic fits with respect to the value of t.

The rationale for using Eq. (8.3) resides in the assumption that maximum response can be reached by infinitely concentrating the sample, while zero response can be reached by an infinite dilution of the sample. The sequential dilution of a discrete number of catalytic units (e.g. microorganisms) is distributed according to Poisson statistics. Although the sigmoidal log-normal distribution curve (Eq. 8.3) does not exactly describe this relationship, it is often used as a robust and adequate approximation (e.g. Haanstra et al., 1985).

The average response in the whole Biolog™ plate ($A_{average}$) was calculated from the responses (A_s). $A_{average}$ was then plotted as a function of the cell concentration and fitted with Eq. (8.3), yielding values for the inflexion point (log $CFU50_{average}$) and the Hill slope (lower line in Fig. 8.3A). The Hill slope gives a measure for the evenness of the Biolog™ substrate conversions in a sample. Discussion of this parameter, although valuable for biodiversity studies, is beyond the scope of this chapter (but see Garland and Lehman, 1999; Breure and Rutgers, 2000; Franklin et al., 2001).

Per well in the Biolog™ plate the log $CFU50_s$ is used to determine the relative abundance (RA) of an activity in the community:

$$\log RA_s = \log CFU50_s - \log CFU50_{average} \tag{8.4}$$

where RA is the number of organisms (in CFU/ml) able to perform 50% activity of the specific substrate conversion relative to the number of organisms converting 50% of all substrates in the Biolog™ plate. Consequently, the average RA is zero, a positive value indicates that the potency to convert this substrate is above average in the particular sample, a negative value vice versa. In cases where there is insufficient colour formation even at the highest inoculum concentrations, this value is artificially set to -2 (substrate number 10 in Fig. 8.3.B). In cases of too much colour formation at the lowest inoculum concentrations, this value is artificially set to $+2$. In this way, unrealistically high and low values for the relative abundance (on a logarithmic scale) will be avoided.

Ultimately, this procedure yields a CLPP of relative abundancies for all 31 (ECO plates) or 95 substrate (GN2 plates) conversions (Fig. 8.3B), which is essentially independent of the number of cells in the inoculum. The only prerequisite is that the highest concentration contains a sufficient amount of cells to colour the plate for at least about 60% of the theoretical maximum. The complete calculation procedure can be automated, for instance by using visual basic programming in Excel. Such visual basic software has been developed in the Netherlands Soil Monitoring Network (Bloem and Breure, 2003).

Data analysis (single-plate and inoculum-density-independent procedure)

The CLPPs of samples can be collated in multivariate analysis using discriminant analysis (DA), principal component analysis (PCA) and/or redundancy analysis (RDA) for exploratory data analysis, and MANOVA for statistical testing (applied on DA, PCA, RDA factors, or single substrates, or on substrate groups (such as carbohydrates, amino acids, carboxylic acids, etc.)).

Discussion

Microbial biomass forms a major part of the total biomass in soil. The number of species in soil is so high that it is not possible to characterize this biomass by determination of the abundance and composition of the species present (e.g. Torsvik *et al.*, 1990; Dykhuizen, 1998). Here, the use of Biolog™ plates to characterize the microbial biomass in soil is described as a specific technique to characterize a sample of the microbial community, i.e. that part which is able to proliferate in the wells of the Biolog™ plate.

By determination of CLPPs in Biolog™ plates, it is possible to distinguish different microbial communities by use of standardized media. It shows the metabolic activity of substrate-responsive microbial cells, after extraction from soil in the medium. The profile, therefore, is dependent on the extraction efficiency, and on how well the organisms are able to metabolize in the medium used. By standardization of the procedure of extraction and incubation, it is possible to distinguish between communities in a reproducible way.

This is especially attractive when stability or succession of communities in time, or due to environmental changes, is investigated. It was seen earlier that CLPP is dependent on the density of the inoculum. Therefore, it was proposed to standardize the amount of cells in the inoculum (Garland and Mills, 1991; Garland, 1997; Preston-Mafham, 2002). An easy method for achieving this is described in this section.

Inoculum standardization does not always provide the solution to differences in cell density, because it is inherently impossible to exactly predict the response in Biolog™ plates from any enumeration technique, and also due to non-linearity between the cell number and the Biolog™ response. The alternative is to use an inoculum-density-independent method, based on a series of inoculum concentrations. Consequently, this method consumes more than one Biolog™ plate per sample. It has been applied in a national survey to determine the quality of soils in The Netherlands and has shown good reproducibility and a strong discriminating power between microbial communities in different soil types and soil management (Breure and Rutgers, 2000; Schouten *et al.*, 2000; Bloem and Breure, 2003).

CLPPs give information on stability of, and changes in, the metabolic capacities of microbial biomass, which may relate to microbial community structure. However, if changes occur, it is not possible to give a causal relationship between (changes in) environmental conditions and changes in community structure. Statistical techniques have to be invoked to couple ecological effects (and stress) and CLPP responses.

In conclusion, the diversity of soil microbial communities is generally so high that it cannot be captured by one single method. The progress made in the use of methods based on ribosomal RNA and their genes has helped to better understand microbial communities, but such data should be carefully evaluated. For example, most methods only look at the 'tip of the iceberg' of the microbial community, or, in other words, only detect or analyse the most dominant populations, in terms of numbers of organisms, biomolecules or physiological characteristics. This is fine as a first attempt to understand the system, as long as the conclusions drawn from these data consider these limitations. Several pitfalls of these methods, especially when based on PCR amplification, have already been recognized and described (Von Wintzingerode *et al.*, 1997). There is clearly a need for more sensitive methods that give an estimate of the entire diversity, including the populations that are present in lower numbers but could play a crucial role in the habitat. In addition, all of these methods do not necessarily tell us which organisms are most strongly involved in the mainstream energy flux of the ecosystem. To achieve this, molecular ecology approaches should be complemented with metabolic mass balance studies and new techniques to correlate soil microbial diversity and structure with soil function. Given the extremely fast development of these newer molecular methods during only the past two decades, current and future research will continue to generate new methods that will allow us to further improve our understanding of the structure and function of soil microbial communities.

References for Chapter 8

Akkermans, A.D.L., van Elsas, J.D. and De Bruijn, F.J. (1995) *Molecular Microbial Ecology Manual.* Kluwer Academic, Dordrecht, The Netherlands.

Amann, R.I., Ludwig, W. and Schleifer, K.H. (1995) Phylogenetic identification and in situ detection of individual microbial cells without cultivation. *Microbiological Reviews* 59, 143–169.

Anderson, J.P.E. and Domsch, K.H. (1978) A physiological method for the quantitative measurement of microbial biomass in soil. *Soil Biology and Biochemistry* 10, 215–221.

Bai, Q.Y. and Zelles, L. (1997) A method for determination of archaeal ether-linked glycerolipids by high performance liquid chromatography with fluorescence detection as their 9-anthroyl derivatives. *Chemosphere* 35, 263–274.

Bligh, E.G. and Dyer, W.J. (1959) A rapid method of total lipid extraction and purification. *Canadian Journal of Biochemistry and Physiology* 37, 911–917.

Bloem, J. and Breure, A.M. (2003) Microbial indicators. In: Markert, B.A., Breure, A.M. and Zechmeister, H.G. (eds) *Bioindicators and Biomonitors.* Elsevier, Amsterdam, pp. 259–282.

Bloem, J., Bolhuis, P.R., Veninga, M.R. and Wieringa, J. (1995) Microscopic methods for counting bacteria and fungi in soil. In: Alef, K. and Nannipieri, P. (eds) *Methods in Applied Soil Microbiology and Biochemistry.* Academic Press, Toronto, pp. 162–191.

Bochner, B. (1989) *Instructions for the use of Biolog GP and GN microplates.* Biolog Inc., Hayward, California.

Boschker, H.T.S., Nold, S.C., Wellsbury, P., Bos, D., de Graaf, W., Pel, R., Parkes, R.J.

and Cappenberg, T.E. (1998) Direct linking of microbial populations to specific biogeochemical processes by ^{13}C-labelling of biomarkers. *Nature* 392, 801–805.

Bossio, D.A., Scow, K.M., Gunapala, N. and Graham, K.J. (1998) Determinants of soil microbial communities: effects of agricultural management, season, and soil type on phospholipid fatty acid profiles. *Microbial Ecology* 36, 1–12.

Breure, A.M. and Rutgers, M. (2000) The application of Biolog plates to characterise microbial communities. In: Benedetti, A., Tittarelli, F., de Bertoldi, S. and Pinzari, F. (eds) *Proceedings of the COST Action 831, Joint Working Group Meeting Biotechnology of Soil, Monitoring, Conservation and Bioremediation.* 10–11 December 1998, Rome, Italy, (EUR 19548), pp. 179–185.

Buyer, J.S. and Drinkwater, L.E. (1997) Comparison of substrate utilization assay and fatty acid analysis of soil microbial communities. *Journal of Microbiological Methods* 30, 3–11.

Duarte, G.F., Rosado, A.S. and van Elsas, J.D. (1998) Extraction of ribosomal RNA and genomic DNA from soil for studying the diversity of the indigenous bacterial community. *Journal of Microbiological Methods* 32, 21–29.

Dykhuizen, D.E. (1998) Santa Rosalia revisited: why are there so many species of bacteria. *Antonie van Leeuwenhoek* 73, 25–33.

Fang, J., Barcelona, M.J. and Alvarez, P.J.J. (2000) A direct comparison between fatty acid analysis and intact phospholipid profiling for microbial identification. *Organic Geochemistry* 31, 881–887.

Franklin, R.B., Garland, J.L., Bolster, C.H. and Mills, A.L. (2001) Impact of dilution on microbial community structure and functional potential: comparisons of numerical simulations and batch culture experiments. *Applied and Environmental Microbiology* 67, 702–712.

Fritze, H., Tikka, P., Pennanen, T., Saano, A., Jurgens, G., Nilsson, M., Bergman, I. and Kitunen, V. (1999) Detection of Archaeal diether lipid by gas chromatography from humus and peat. *Scandinavian Journal of Forest Research* 14, 545–551.

Frostegård, Å. and Bååth, E. (1996) The use of phospholipid fatty acid analysis to estimate bacterial and fungal biomass in soil. *Biology and Fertility of Soils* 22, 59–65.

Frostegård, Å., Bååth, E. and Tunlid, A. (1993) Shifts in the structure of soil microbial communities in limed forests as revealed by phospholipid fatty acid analysis. *Soil Biology and Biochemistry* 25, 723–730.

Gamo, M. and Shoji, T. (1999) A method of profiling microbial communities based on a most-probable-number assay that uses Biolog plates and multiple sole carbon sources. *Applied and Environmental Microbiology* 65, 4419–4424.

Garland, J.L. (1997) Analysis and interpretation of community-level physiological profiles in microbial ecology. *FEMS Microbiology Ecology* 24, 289–300.

Garland, J.L. and Lehman, R.M. (1999) Dilution/extinction of community phenotypic characters to estimate relative structural diversity in mixed communities. *FEMS Microbiology Ecology* 30, 333–343.

Garland, J.L. and Mills, A.L. (1991) Classification and characterization of heterotrophic microbial communities on the basis of patterns of community-level-sole-carbon-source-utilization. *Applied and Environmental Microbiology* 57, 2351–2359.

Garland, J.L. and Mills, A.L. (1996) Patterns of potential C source utilization by rhizosphere communities. *Soil Biology and Biochemistry* 28, 223–230.

Gomes, N.C.M., Costa, R. and Smalla, K. (2004) Rapid simultaneous extraction of DNA and RNA from bulk and rhizosphere soil. In: Kowalchuk, G.A., De Bruijn, F.J., Head, I.M., Akkermans, A.D. and van Elsas, J.D. (eds) *Molecular Microbial Ecology Manual*, 2nd edn. Kluwer Academic, Dordrecht, The Netherlands, pp. 159–170.

Grayston, S.J. and Campbell, C.D. (1995) Functional biodiversity of microbial communities in the rhizospheres of hybrid larch (*Larix eurolepis*) and Sitka spruce (*Picea sitchensis*). *Tree Physiology* 16, 1031–1038.

Grayston, S.J., Wang, S., Campbell, C.D. and Edwards, A.C. (1998) Selective influence of plant species on microbial diversity in the rhizosphere. *Soil Biology and Biochemistry* 30, 369–378.

Grayston, S.J., Griffith, G.S., Mawdsley, J.L., Campbell, C.D. and Bardgett, R.D. (2001) Accounting for variability in soil microbial communities of temperate upland grassland ecosystems. *Soil Biology and Biochemistry* 33, 533–551.

Griffith, R.I., Whiteley, A.S., O'Donnell, A.G. and Bailey, M.J. (2000) Rapid method for coextraction of DNA and RNA from natural environments for analysis of ribosomal DNA- and rRNA-based microbial community composition. *Applied and Environmental Microbiology* 66, 5488–5491.

Guckert, J.B., Carr, G.J., Johnson, T.D., Hamm, B.G., Davidson, D.H. and Kumagai, Y. (1996) Community analysis by Biolog: curve integration for statistical analysis of activated sludge microbial habitats. *Journal of Microbiological Methods* 27, 183–197.

Haack, S.K., Garchow, H., Odelson, D.A., Forney, L.J. and Klug, M.J. (1994) Accuracy, reproducibility, and interpretation of fatty acid methyl ester profiles of model bacterial communities. *Applied and Environmental Microbiology* 60, 2483–2493.

Haack, S.K., Garchow, H., Klug, M.J. and Forney, L.J. (1995) Analysis of factors affecting the accuracy, reproducibility, and interpretation of microbial community carbon source utilizations. *Applied and Environmental Microbiology* 61, 1458–1468.

Haanstra, L., Doelman, P. and Oude Voshaar, J.H. (1985) The use of sigmoidal dose response curves in soil ecotoxicological research. *Plant and Soil* 84, 293–297.

Harwood, J.L. and Russel, N.J. (1984) *Lipids in Plants and Microbes*. George Allen & Unwin Ltd, London.

Heuer, H. and Smalla, K. (1997) Evaluation of community-level catabolic profiling using Biolog GN microplates to study microbial community changes in potato phyllosphere. *Journal of Microbiological Methods* 30, 49–61.

Heuer, H., Hartung, K., Wieland, G., Kramer, I. and Smalla, K. (1999) Polynucleotide probes that target a hypervariable region of 16S rRNA genes to identify bacterial isolates corresponding to bands of community fingerprints. *Applied and Environmental Microbiology* 65, 1045–1049.

Hooper, D., Hawksworth, D. and Dhillion, S. (1995) Microbial diversity and ecosystem processes. In: Heywood, V.H. and Watson, R.T. (eds) *Global Biodiversity Assessment*. Cambridge University Press, Cambridge, pp. 433–443.

Hopkins, D.W., MacNaughton, S.J. and O'Donnell, A.G. (1991) A dispersion and differential centrifugation technique for representatively sampling microorganisms from soil. *Soil Biology and Biochemistry* 23, 217–225.

Hurt, R.A., Qiu, X., Wu, L., Roh, Y., Palumbo, A.V., Tiedje, J.M. and Zhou, J. (2001) Simultaneous recovery of RNA and DNA from soils and sediments. *Applied and Environmental Microbiology* 67, 4495–4503.

Insam, H. (1997) A new set of substrates proposed for community characterization in environmental samples. In: Insam, H. and Rangger, A. (eds) *Microbial Communities*. Springer-Verlag, Heidelberg, pp. 259–260.

Insam, H. and Hitzl, W. (1999) Data evaluation of community level physiological profiles: a reply to the letter of PJA Howard. *Soil Biology and Biochemistry* 31, 1198–1200.

Insam, H., Amor, K., Renner, M. and Crepaz, C. (1996) Changes in functional abilities of the microbial community during composting of manure. *Microbial Ecology* 31, 77–87.

Konopka, A., Oliver, L. and Turco, J.R. (1998) The use of carbon substrate utilization patterns in environmental and ecological microbiology. *Microbiology Ecology* 35, 103–115.

Kowalchuk, G.A., De Bruijn, F.J., Head, I.M., Akkermans, A.D.L. and van Elsas, J.D. (2004) *Molecular Microbial Ecology Manual*, 2nd edn. Kluwer Academic, Dordrecht, The Netherlands.

Kozdrój, J. and van Elsas, J.D. (2001a) Structural diversity of microorganisms in chemically perturbed soil assessed by molecular and cytochemical approaches.

Journal of Microbiological Methods 43, 197–212.

Kozdrój, J. and van Elsas, J.D. (2001b) Structural diversity of microbial communities in arable soils of a heavily industrialised area determined by PCR-DGGE fingerprinting and FAME profiling. *Applied Soil Ecology* 17, 31–42.

Lindstrom, J.E., Barry, R.P. and Braddock, J.F. (1998) Microbial community analysis: a kinetic approach to constructing potential C source utilization patterns. *Soil Biology and Biochemistry* 30, 231–239.

Macnaughton, S.J., Jenkins, T.J., Wimpee, M.H., Cormiér, M.R. and White, D.C. (1997) Rapid extraction of lipid biomarkers from pure culture and environmental samples using pressurized accelerated hot solvent extraction. *Journal of Microbiological Methods* 31, 19–27.

Miller, M., Palojärvi, A., Rangger, A., Reeslev, M. and Kjoller, A. (1998) The use of fluorogenic substrates to measure fungal presence and activity in soil. *Applied and Environmental Microbiology* 64, 613–617.

Moré, M.I., Herrick, J.B., Silva, M.C., Ghiorse, W.C. and Madsen, E.L. (1994) Quantitative cell lysis of indigenous microorganisms and rapid extraction of microbial DNA from sediment. *Applied and Environmental Microbiology* 60, 1572–1580.

Morgan, J.A.W. and Winstanley, C. (1997) Microbial biomarkers. In: van Elsas, J.D., Trevors, J.T. and Wellington, E.M.H. (eds) *Modern Soil Microbiology*. Marcel Dekker, New York, pp. 331–352.

Nichols, P.D., Guckert, J.B. and White, D.C. (1986) Determination of monounsaturated fatty acid double-bond position and geometry for microbial monocultures and complex consortia by capillary GC–MS of their dimethyl disulphide adducts. *Journal of Microbiological Methods* 5, 49–55.

Nielsen, P. and Petersen, S.O. (2000) Ester-linked polar lipid fatty acid profiles of soil microbial communities: a comparison of extraction methods and evaluation of interference from humic acids. *Soil Biology and Biochemistry* 32, 1241–1249.

Noble, P.A., Almeida, J.S. and Lovell, C.R. (2000) Application of neural computing methods for interpreting phospholipid fatty acid profiles of natural microbial communities. *Applied and Environmental Microbiology* 66, 694–699.

Nüsslein, K. and Tiedje, J.M. (1998) Characterization of the dominant and rare members of a young Hawaiian soil bacterial community with small-subunit ribosomal DNA amplified from DNA fractionated on the basis of its guanine and cytosine composition. *Applied and Environmental Microbiology* 64, 1283–1289.

O'Connell, S.P., Lehman, R.M. and Garland, J.L. (2001) Biolog media formulation (GN2 vs. GN) affects microbial community-level physiological profiles. *Proceedings of the 9th International Symposium on Microbial Ecology*. Amsterdam, p. 164.

Ogram, A., Sayler, G.S. and Barkay, T.J. (1987) DNA extraction and purification from sediments. *Journal of Microbiological Methods* 7, 57–66.

Olsson, P.A. (1999) Signature fatty acids provide tools for determination of the distribution and interactions of mycorrhizal fungi in soil. *FEMS Microbiology Ecology* 29, 303–310.

Palojärvi, A. and Albers, B. (1998) Extraktion und Bestimmung membrangebundener Phospholipidfettsäuren. In: Remde, A. and Tippmann, P. (eds). *Mikrobiologische Charakterisierung aquatischer Sedimente–Methodensammlung. Vereinigung für Allgemeine und Angewandte Mikrobiologie (VAAM)*. R. Oldenbourg Verlag, München, pp. 187–195.

Palojärvi, A., Sharma, S., Rangger, A., von Lützow, M. and Insam, H. (1997) Comparison of Biolog and phospholipid fatty acid patterns to detect changes in microbial community. In: Insam, H. and Rangger, A. (eds) *Microbial Communities. Functional versus Structural Approaches*. Springer Verlag, Berlin, pp. 37–48.

Pankhurst, C.E., Ophel-Keller, K., Doube, B.M. and Gupta, V.V.S.R. (1996) Biodiversity of soil microbial communities in agricultural systems. *Biodiversity and Conservation* 5, 197–209.

Pennanen, T. (2001) Microbial communities in boreal coniferous forest humus exposed to heavy metals and changes in soil pH – a summary of the use of phospholipid fatty acids, Biolog® and ^{3}H-thymidine incorporation methods in field studies. *Geoderma* 100, 91–126.

Petersen, S.O. and Klug, M.J. (1994) Effects of sieving, storage, and incubation temperature on the phospholipid fatty acid profile of a soil microbial community. *Applied and Environmental Microbiology* 60, 2421–2430.

Petersen, S.O., Debosz, K., Schjonning, P., Christensen, B.T. and Elmholt, S. (1997) Phospholipid fatty acid profiles and C availability in wet-stable macro-aggregates from conventionally and organically farmed soils. *Geoderma* 78, 181–196.

Petersen, S.O., Frohne, P.S. and Kennedy, A.C. (2002) Dynamics of a soil microbial community under spring wheat. *Soil Science Society of America Journal* 66, 826–833.

Pillai, S.D., Josephson, K.L., Bailey, R.L., Gerba, C.P. and Pepper, I.L. (1991) Rapid method for processing soil samples for polymerase chain reaction amplification of specific gene sequences. *Applied and Environmental Microbiology* 57, 2283–2286.

Porteous, L.A. and Armstrong, J.L. (1991) Recovery of bulk DNA from soil by a rapid, small-scale extraction method. *Current Microbiology* 22, 345–348.

Preston-Mafham, J., Boddy, L. and Randerson, P.F. (2002) Analysis of microbial community functional diversity using sole-carbon source utilisation profiles, a critique. FEMS *Microbiology Ecology* 42, 1–14.

Rademaker, J. (1995) Analysis of molecular fingerprints. In: Akkermans, A.D.L., van Elsas, J.D. and De Bruijn, F.J. (eds) *Molecular Microbial Ecology Manual*. Kluwer Academic, Dordrecht, The Netherlands.

Roslev, P., Iversen, N. and Henriksen, K. (1998) Direct fingerprinting of metabolically active bacteria in environmental samples by substrate specific radiolabelling and lipid analysis. *Journal of Microbiological Methods* 31, 99–111.

Rutgers, M. and Breure, A.M. (1999) Risk assessment, microbial communities, and pollution-induced community tolerance. *Human and Ecological Risk Assessment* 5, 661–670.

Sambrook, J., Fritsch, E.F. and Maniatis, T. (1989) *Molecular Cloning. A Laboratory Manual*, 2nd edn. Cold Spring Harbor Laboratory Press, Cold Spring Harbor, New York.

Schloter, M., Zelles, L., Hartmann, A. and Munch, J.C. (1998) New quality of assessment of microbial diversity in arable soils using molecular and biochemical methods. *Journal of Plant Nutrition and Soil Science* 161, 425–431.

Schouten, T., Bloem, J., Didden, W.A.M., Rutgers, M., Siepel, H., Posthuma, L. and Breure, A.M. (2000) Development of a biological indicator for soil quality. *Setac Globe* 1, 30–33.

Seghers, D., Verthé, K., Reheul, D., Bulcke, R., Siciliano, S.D., Verstraete, W. and Top, E.M. (2003) Effect of long-term herbicide applications on the bacterial community structure and function in an agricultural soil. *FEMS Microbiology Ecology* 46, 139–146

Selenska, S. and Klingmüller, W. (1991a) Direct detection of *nif*-gene sequences of *Enterobacter agglomerans* in soil. *FEMS Microbiology Letters* 80, 243–246.

Selenska, S. and Klingmüller, W. (1991b) DNA recovery and direct detection of *Tn5* sequences from soil. *Letters in Applied Microbiology* 13, 21–24.

Siira-Pietikäinen, A., Haimi, J., Kanninen, A., Pietikäinen, J. and Fritze, H. (2001) Responses of decomposer community to root-isolation and addition of slash. *Soil Biology and Biochemistry* 33, 1993–2004.

Smalla, K., Cresswell, N., Mendonca-Hagler, L.C., Wolters, A.C. and van Elsas, J.D. (1993a) Rapid DNA extraction protocol from soil for polymerase chain reaction-mediated amplification. *Journal of Applied Bacteriology* 74, 78–85.

Smalla, K., Van Overbeek, L.S., Pukall, R. and van Elsas, J.D. (1993b) Prevalence of *npt*II and *Tn5* in kanamycin-resistant bacteria from different environments. *FEMS Microbiology Ecology* 13, 47–58.

Steffan, R.J., Goksoyr, J., Bej, A.K. and Atlas, R.M. (1988) Recovery of DNA from soils and sediments. *Applied and Environmental Microbiology* 54, 2908–2915.

Tebbe, C.C. and Vahjen, W. (1993) Interference of humic acids and DNA extracted directly from soil in detection and transformation of recombinant DNA from bacteria and a yeast. *Applied and Environmental Microbiology* 59, 2657–2665.

Torsvik, V.L. (1980) Isolation of bacterial DNA from soil. *Soil Biology and Biochemistry* 12, 18–21.

Torsvik, V., Goksøyr, J. and Daae, F.L. (1990) High diversity of DNA of soil bacteria. *Applied and Environmental Microbiology* 56, 782–787.

Trevors, J.T. and van Elsas, J.D. (1989) A review of selected methods in environmental microbial genetics. *Canadian Journal of Microbiology* 35, 895–902.

Tsai, Y.-L. and Olson, B.H. (1992) Detection of low numbers of bacterial cells in soils and sediments by polymerase chain reaction. *Applied and Environmental Microbiology* 58, 754–757.

Vainio, E.J. and Hantula, J. (2000) Direct analysis of wood-inhabiting fungi using denaturing gradient gel electrophoresis of amplified ribosomal DNA. *Mycology Research* 104, 927–936.

van Elsas, J.D., Van Overbeek, L.S. and Fouchier, R. (1991) A specific marker, *pat*, for studying the fate of introduced bacteria and their DNA in soil using a combination of detection techniques. *Plant and Soil* 138, 49–60.

van Elsas, J.D., Mantynen, V. and Wolters, A.C. (1997) Soil DNA extraction and assessment of the fate of *Mycobacterium chlorophenolicum* strain PCP-1 in different soils via 16S ribosomal RNA gene sequence based most-probable-number PCR and immunofluorescence. *Biology and Fertility of Soils* 24, 188–195.

van Elsas, J.D., Smalla, K. and Tebbe, C.C. (2000) Extraction and analysis of microbial community nucleic acids from environmental matrices. In: Jansson, J., van Elsas, J.D. and Bailey, M. (eds) *Tracking Genetically Engineered Micro-organisms*. Landes Bioscience, Austin, Texas, pp. 29–52.

Van Overbeek, L.S., Van Vuurde, J.W.L. and van Elsas, J.D. (2005) Application of molecular fingerprinting techniques for studying shifts in bacterial endophytic communities. In: Schultz, B. and Boyle, C. (eds) *Bacterial Endophytes*. Springer Verlag, Heidelberg (in press).

Von Wintzingerode, F., Göbel, U.B. and Stackebrandt, E. (1997) Determination of microbial diversity in environmental samples: pitfalls of PCR-based rRNA analysis. *FEMS Microbiology Reviews* 21, 213–229.

White, D.C. (1988) Validation of quantitative analysis for microbial biomass, community structure, and metabolic activity. *Archiv für Hydrobiologie Beihefte* 31, 1–18.

White, D.C., Davis, W.M., Nickels, J.S., King, J.C. and Bobbie, R.J. (1979) Determination of the sedimentary microbial biomass by extractible lipid phosphate. *Oecologia* 40, 51–62.

White, D.C., Flemming, C.A., Leung, K.T. and MacNaughton, S.J. (1998) In situ microbial ecology for quantitative appraisal, monitoring, and risk assessment of pollution remediation in soils, the subsurface, the rhizosphere and in biofilms. *Journal of Microbiological Methods* 32, 93–105.

Widmer, F., Fließbach, A., Laczko, E., Schulze-Aurich, J. and Zeyer, J. (2001) Assessing soil biological characteristics: a comparison of bulk soil community DNA-, PLFA-, and Biolog™-analyses. *Soil Biology and Biochemistry* 33, 1029–1036.

Zelles, L. (1999) Fatty acid patterns of phospholipids and lipopolysaccharides in the characterisation of microbial communities in soil: a review. *Biology and Fertility of Soils* 29, 111–129.

Zelles, L. and Bai, Q.Y (1994) Fatty acid patterns of phospholipids and lipopolysaccharides in environmental samples. *Chemosphere* 28, 391–411.

Zelles, L., Palojärvi, A., Kandeler, E., von Lützow, M., Winter, K. and Bai, Q.Y. (1997) Changes in soil microbial properties and phospholipid fatty acid fractions after chloroform fumigation. *Soil Biology and Biochemistry* 29, 1325–1336.

9 Plant–Microbe Interactions and Soil Quality

9.1 Microbial Ecology of the Rhizosphere

PHILIPPE LEMANCEAU,[1] PIERRE OFFRE,[1] CHRISTOPHE MOUGEL,[1] ELISA GAMALERO,[2] YVES DESSAUX,[3] YVAN MOËNNE-LOCCOZ[4] AND GRAZIELLA BERTA[2]

[1] UMR 1229 INRA/Université de Bourgogne, 'Microbiologie et Géochimie des Sols', INRA-CMSE, 17 rue Sully, BP 86510 21065, Dijon cedex, France; [2] Università del Piemonte Orientale 'Amedeo Avogadro', Department of Science and Advanced Technology, Corso Borsalino 54, 15100 Alessandria, Italy; [3] Institut des Sciences du Végétal, CNRS, UPR2355, Bâtiment 23, Avenue de la Terrasse, 91198 Gif-sur-Yvette, France; [4] UMR CNRS 5557 Ecologie Microbienne, Université Claude Bernard (Lyon 1), 43 bd du 11 Novembre, 69622 Villeurbanne cedex, France

Soils are known to be oligotrophic environments, whereas the soil microbiota is mostly heterotrophic, and consequently microbial growth in soil is mainly limited by the scarce sources of readily available carbon (Wardle, 1992). Therefore, in soils, the microbiota is mostly in stasis (fungistasis/bacteriostasis) (Lockwood, 1977). In contrast, plants are autotrophic organisms responsible for most of the primary production resulting from photosynthesis. Significant amounts of photosynthates are released from the plant roots to the soil, through a process called rhizodeposition. These products comprise exudates, lysates, mucilage, secretions and compounds released from dead/dying sloughed-off cells, as well as gases including respiratory CO_2 and ethylene. Depending on plant species, age and environmental conditions, rhizodeposits can account for up to 40% of net fixed carbon (Lynch and Whipps, 1990). On average, 17% of net fixed carbon appears to be released by the roots (Nguyen, 2003). This significant release of carbohydrates in the soil by the plant roots stimulates the density and activity of the microbiota located closely to the roots but also affects the physico-chemical properties of the neighbouring soil (Rovira, 1965; Curl and Truelove, 1986; Lynch and Whipps, 1990). Altogether, the modifications of the biological and physico-chemical properties of the soil induced by plant roots are commonly called the rhizosphere effect; the rhizosphere, as proposed one century ago by Hiltner (1904), being the volume of soil surrounding roots in which the microflora is influenced by these roots.

Besides trophic interactions, the relations between plants and microorganisms are also mediated by toxic compounds involved in plant defence reactions against soilborne pathogens (Bais *et al.*, 2004); similarly, various soilborne microbial populations have the ability to produce antibiotics contributing to their ecological fitness, which is especially important in a competitive environment such as the rhizosphere (Mazzola *et al.*, 1992).

Altogether these characteristics of the rhizosphere lead to the fact that not all microbial groups and populations are equally stimulated by the rhizodeposits. Specific microbial groups and populations have clearly been shown to be preferentially associated with plant roots (Mavingui *et al.*, 1992; Lemanceau *et al.*, 1995; Edel *et al.*, 1997). These populations, which are selected by the plant, are better adapted than the others to the rhizosphere environment. As an example, populations of fluorescent pseudomonads associated with the roots differ from soil counterparts in terms of carbon and energetic metabolism: they are more frequently able to use specific organic compounds such as trehalose (electron donors) and they show a higher ability to mobilize ferric iron and to dissimilate nitrogen oxides (electron acceptors) (Lemanceau *et al.*, 1988, 1995; Clays-Josserand *et al.*, 1995; Frey *et al.*, 1997). Populations able to develop strategies to counteract toxic compounds produced by the host plant or by other root-associated microbial populations will have a competitive advantage over the others (Duffy *et al.*, 2003).

In plant-selected microbial populations, the rapid adaptation of the metabolism to the changing conditions of the rhizosphere relies on their ability to perceive the variations of the environment. Among the perception systems described so far, two-component systems allow bacteria to sense and respond to a wide range of environmental changes. The minimal system consists of two proteins: a sensor and a transducer (Stock *et al.*, 1989). Signalling in the rhizosphere also involves microbe–microbe communication. This is the case of the so-called 'quorum sensing system', which allows bacterial populations to sense their densities and to regulate cell physiology, including enzyme and antibiotic synthesis at the population level, making their response to the environmental variation more efficient and concerted (Whitehead *et al.*, 2001). Signalling also occurs between plants and microbes. Various regulatory signals are perceived by receptors and transduced via downstream effectors, and they mediate information that can influence plant–microbe interactions. The corresponding signal molecules are difficult to detect and so far only a few of them have been described (Hirsch *et al.*, 2003). However, well-known examples are the isoflavonoids and flavonoids present in root exudates of various leguminous plants, which activate *Rhizobium* genes responsible for the nodulation process (Peters, 1986; Bais *et al.*, 2004).

Trophic and toxic-mediated interactions, together with communications, lead to the selection of the most adapted populations, to variations of their physiology, and then to shifts in the structure and activity of the microbial community. These variations are known to affect plant health and growth. Depending on its persistence, the rhizosphere effect may not only affect the growth and health of the current crop, but may also affect the soil quality.

The rhizosphere, being a major hot-spot for soil microorganisms, is expected to affect various parameters relevant for soil quality: soil (micro)structure, degradation and mineralization of organic xenobiotic compounds, mobilization and speciation of heavy metals, contribution to the phytoextraction of these toxic elements, enrichment of symbiotic and free-living microbial populations favourable to plant growth and health. The soil quality determined by the indigenous microbial populations will affect the growth and health of the plants. In this way, plant–microbe interactions may be considered as a feedback loop, in which: (i) the plant modifies the environment of soilborne microorganisms through the release of various compounds; (ii) these variations are perceived by the soil microbiota; (iii) this perception by specific populations leads to a variation of their physiology and the selection of the populations most adapted, and in turn to a shift in the microbial community and activity; and (iv) these variations affect the growth and health of the plant and then root exudation patterns, in such a way that the rhizosphere environment is being modified. This feedback loop is a dynamic process submitted to a continuous incrementation, which, in the long term, drives the coevolution of plant and associated microorganisms.

A major challenge in the study of rhizosphere ecology is to make progress in our knowledge of the impact of the plant on the soilborne microbiota. Indeed, the rhizosphere effect appears to be plant- and even cultivar-specific. These studies have to take into account not only specific microbial groups and culturable populations, but also microbial communities, in an untargeted way. The expected output of the corresponding research will be to monitor the indigenous microbiota via the cultivation of specific plant genotypes and the application of particular agricultural practices (crop rotation, intercropping) in order to favour beneficial indigenous populations and disfavour those that are detrimental for the plant and, more generally, for soil quality.

In the present chapter, information will be given on methods to: (i) assess soil quality in relation to the contribution of nodulating symbiotic bacteria promoting growth of legumes via nitrogen fixation; (ii) use arbuscular mycorrhiza for characterization of soil quality in relation to ecotoxicology; (iii) assess the phytosanitary soil quality resulting from both antagonistic and pathogenic microbial populations; and (iv) assess indigenous populations of free-living plant-beneficial microorganisms.

9.2 Nodulating Symbiotic Bacteria and Soil Quality

ALAIN HARTMANN,[1] SYLVIE MAZURIER,[1] DULCE N. RODRÍGUEZ-NAVARRO,[2] FRANCISCO TEMPRANO VERA,[2] JEAN-CLAUDE CLEYET-MAREL,[3] YVES PRIN,[3] ANTOINE GALIANA,[3] MANUEL FERNÁNDEZ-LÓPEZ,[4] NICOLÁS TORO[4] AND YVAN MOËNNE-LOCCOZ[5]

[1]UMR 1229 INRA/Université de Bourgogne, 'Microbiologie et Géochimie des Sols', INRA-CMSE, BP 86510, 21065 Dijon cedex, France; [2]Centro de Investigación y Formación Agraria (CIFA), Las Torres y Tomejil, Aparatdo Oficial, E-41200 Alcalá del Río (Sevilla), Spain; [3]Laboratoire des Symbioses Tropicales et Méditerranéennes, UMR INRA/IRD/CIRAD/ AGRO-M 1063, Campus International de Baillarguet, 34398 Montpellier cedex 5, France; [4]Grupo de Ecología Genética, Estación Experimental del Zaidín, Consejo Superior de Investigaciones Científicas, Profesor Albareda 1, 18008 Granada, Spain; [5]UMR CNRS 5557 Ecologie Microbienne, Université Claude Bernard (Lyon 1), 43 bd du 11 Novembre, 69622 Villeurbanne cedex, France

Introduction

Several bacterial taxa are involved in symbiotic interactions with plant roots, in which the microorganism nodulates the root and fixes nitrogen. These diazosymbiotic bacteria include five genera (*Rhizobium*, *Mesorhizobium*, *Sinorhizobium*, *Azorhizobium* and *Bradyrhizobium*; often collectively referred to as rhizobia) and close to 30 species of the Rhizobiaceae family (Young *et al.*, 2001), which are associated with plants (designated as legumes) from the three botanical families *Papilionoideae*, *Mimosoideae* and *Cesalpinioideae*, as well as an actinobacterial genus (*Frankia*), which nodulates non-legume plants such as *Alnus* or *Casuarina*. The diversity of rhizobial populations in soil has been studied extensively. For a majority of rhizobial species, the ability to engage in a symbiotic relationship with a legume implies plasmid-borne genes. Recently, the documented range of bacterial partners capable of nitrogen fixation within legume nodules has been revised with the inclusion of another α-proteobacterium (*Methylobacterium nodulans*; Sy *et al.*, 2001) and of β-proteobacteria (Moulin *et al.*, 2001). Here, we will focus on diazosymbionts from the α-proteobacteria interacting with legumes (Table 9.1), and adopt the usual term 'rhizobia' to designate these bacteria.

Table 9.1. List of known rhizobia–legume symbiotic interactions. Names of legumes nodulated by more than one rhizobial species are in bold.

Bacterial symbiont (genus, species and biovar)	Principal host legumes
Rhizobium	
R. leguminosarum	
bv. *viciae*	*Pisum* (pea), *Lens*, *Lathyrus*, *Vicia*
bv. *trifolii*	***Trifolium***
bv. *phaseoli*	***Phaseolus*** **(bean)**
R. tropici	***Phaseolus*** **(bean)**, *Leucaena*, etc.
R. etli	
bv. *phaseoli*	***Phaseolus*** **(bean)**
bv. *mimosae*	***Mimosa affinis***
R. gallicum	
bv. *gallicum*	***Phaseolus*** **(bean)**, *Onobrychis*, *Leucaena*, etc.
bv. *phaseoli*	***Phaseolus*** **(bean)**
R. giardinii	
bv. *giardinii*	***Phaseolus*** **(bean)**, *Leucaena*, etc.
bv. *phaseoli*	***Phaseolus*** **(bean)**
R. galegae	*Galega* spp.
R. hainanense	*Desmodium* spp., etc.
R. mongolense	***Medicago ruthenica***
R. huautlense	*Sesbania* spp.
R. undicola	*Neptunia natans*
R. yanglingense	*Amphicarpaea*, *Coronilla*, etc.
Sinorhizobium	
S. meliloti	***Medicago***, *Trigonella*, *Melilotus*
S. fredii	***Glycine max*** **(soybean)**, etc.
S. teranga	
bv. *sesbaniae*	***Sesbania***, etc.
bv. *acaciae*	***Acacia***
S. saheli	
bv. *sesbaniae*	***Sesbania***, etc.
bv. *acaciae*	***Acacia***
S. medicae	***Medicago***
S. morelense	***Leucaena leucocephala***
Mesorhizobium	
M. loti	***Lotus***
M. huakuii	***Astragalus sinicus***
M. ciceri	***Cicer arietinum***
M. mediterraneum	***Cicer arietinum***
M. tianshanense	***Glycine max*** **(soybean)**, *Glycyrrhiza*, etc.
M. plurifarium	***Acacia*, *Leucaena***
M. amorphae	*Amorpha fruticosa*
M. chacoense	***Prosopis alba***
Bradyrhizobium	
B. japonicum	***Glycine max*** **(soybean)**
B. elkanii	***Glycine max*** **(soybean)**
B. liaoningensis	***Glycine max*** **(soybean)**
Azorhizobium	
A. caulinodans	*Sesbania rostrata*

The legume–rhizobia symbiosis is extremely important for the nitrogen cycle in natural terrestrial ecosystems, as well as under agronomic conditions. Indeed, in traditional crop rotation systems, it has been used to enhance soil fertility and the productivity of non-legume crops since ancient times (Tilman, 1998), long before the identification by Hellriegel and Wilfarth (1888) of rhizobia as the source of fixed nitrogen in root nodules. Certain legumes are major crops throughout the world, and grain legumes are grown over almost 1.5 million km^2 of land each year, half of which is cultivated with soybean. Forage legumes cover about 30 million km^2, the three dominant genera being *Trifolium*, *Lotus* and *Medicago*. Legumes are also grown in mixed cropping systems (for example a legume and a cereal), or can be used as green manure. In addition, ligneous legumes are useful for restoring degraded soils or establishing a plant cover on nitrogen-poor mineral substrates. Compared with the use of nitrogen fertilizers, legume-based biological nitrogen fixation (BNF) is more environmentally friendly; thus, optimizing its potential may contribute to improved soil quality and soil preservation.

BNF can fix up to 250 kg N/ha, even more for some legumes, and it is a less expensive source of nitrogen compared with nitrogen fertilizers. In certain situations, however, the symbiosis does not function well, due to the absence, or low numbers, of appropriate rhizobia in soil, or the presence of nodulating rhizobia that are poor nitrogen fixers. Inoculation of the specific and efficient rhizobial symbiont can greatly improve BNF and consequently the yield of the corresponding legume. Soil and climatic conditions (e.g. extreme soil pH, salinity, tillage, temperature stress, drought, availability of mineral nutrients and chemical residues) can affect the numbers and/or the diversity of rhizobial populations, and the functioning of BNF. For instance, the detrimental effects of heavy metals on indigenous rhizobial populations are well established.

In this section, methods available to assess the legume–rhizobia symbiotic interaction are presented. Relevant objectives when addressing the various components of soil quality include the detection (i.e. soil nodulating potential; pp. 233–236) and enumeration (by a MPN approach; pp. 236–240) of nodulating rhizobia indigenous to soil, the assessment of the symbiotic efficacy of indigenous rhizobial populations (pp. 241–243), and the estimation of the amount of plant nitrogen derived from symbiotic fixation (pp. 243–247).

Detection of Indigenous Rhizobia in Soil and Assessment of Soil Nodulating Potential

Introduction

Some legume species are nodulated exclusively by one rhizobial species (e.g. lucerne) or even by a given biovar of one species (e.g. pea, clover). Other legumes are more permissive, as they can be nodulated by strains belonging to different species or even genera of rhizobia (e.g. soybean,

bean). Conversely, rhizobial strains may exhibit a narrow or wide host range (Amarger, 2001). The soil nodulating potential corresponds to the ability of soil to support legume nodulation, which is a prerequisite for the expression of beneficial effects from symbiotic nitrogen fixation. Thus, the soil nodulating potential can be estimated as the ability to nodulate a given legume, or by the occurrence of nodulating strains from a given rhizobial species.

Different methods can be chosen to detect (and/or enumerate) rhizobia from soil. Plant infection tests have been used extensively, since there is no specific medium for direct counting of rhizobia in soil. Experimental designs and protocols for plant infection tests vary with the legume studied, and have been described in the literature (Vincent, 1970; Somasegaran and Hoben, 1994). These methods are presented here, particularly in the case of large-seeded legumes such as soybean or bean, for the detection (and, in the next section, the enumeration) of rhizobial symbionts.

Principle of the methods

When the objective is merely to assess whether or not nodulating rhizobia are present in soil, the occurrence of nodules on roots of the appropriate legume is a sufficient criterion. Since certain rhizobial strains can nodulate, but fix little or no nitrogen (i.e. ineffective strains), it may be useful to use scoring systems that integrate nodule repartition, size and/or internal colour (Dommergues et al., 1999), because these three parameters may be indicative of symbiotic efficacy. For instance, nitrogen fixation is more likely to occur in big, pink nodules than in small, white nodules. Indeed, such nodulation scores often correlate well with indices of plant growth such as foliage dry matter and nitrogen content (Brockwell et al., 1982). When fields are grown with legumes, this approach may be implemented directly by sampling the root system of field-grown legumes and scoring the extent of nodulation. Nodulation scores can be particularly useful to compare nodulation of the same legume in neighbouring fields or when farming practices differ, as well as when comparing legume cultivars or characterizing the impact of plant protection products (Fettell et al., 1997; Moënne-Loccoz et al., 1998). If the legume is not grown in the soil of interest, the assessment can be done after sampling the soil and performing plant infection tests, based on the capacity of specific rhizobia to nodulate a given species of legume: the legume is grown in the soil sampled and the extent of plant nodulation is recorded.

Materials and apparatus

When assessing field-grown legumes:
- Digging equipment such as shovels, etc.
- Scalpels (to split the nodules open)

When assessing soil with plant infection tests:

- Calcinated clay (Chemsorb; CONEX) or perlite
- 16 mm diameter clay beads (Argi-16TBF, TBF, France)
- 1-l plastic containers
- Legume seeds (soybean or bean)
- Controlled-environment cabinet (or growth chamber)
- Sieve

Chemicals and solutions

When assessing soil with plant infection tests:

- Sterile deionized water
- Saturated calcium hypochlorite solution

Procedure and calculation

When assessing field-grown legumes: plants (at least 15 individual plants) need to be sampled when the legumes are expected to be fully nodulated, e.g. shortly before flowering. The root systems and surrounding soil are carefully dug up (to a depth of at least 20 cm, preferably more). The soil is removed by shaking, followed, if need be, by dipping in water, and nodulation is scored (Table 9.2). The average nodulation score is computed.

When assessing soil by use of plant infection tests: in order to avoid inconsistent results due to the spatial heterogeneity of rhizobial soil populations, assessment needs to be carried out on composite soil samples to be agronomically meaningful. Composite soil samples are obtained by pooling and mixing six to ten soil cores taken at random over the field plot. All equipment used for sampling must be free of rhizobia, and thus the use of disinfected instruments and disposable plastic bags is recommended. Soil samples are sieved (2–4 mm). They may be used immediately or stored at 4°C for up to 1 month, if necessary. The method described hereafter is optimized for soybean and bean, but may also be used for other legumes. Plants are grown in soil, or in a mixture of soil and sterile, substrate-like calcinated clay or perlite (i.e. siliceous sand expanded at 1200°C). One-litre plastic containers are filled with the following layers (from bottom to top): 200 ml of 16 mm diameter Argi-16TBF clay beads, 400 ml of rhizobia-free substrate (calcinated clay or perlite), 100 g of soil, 200 ml of rhizobia-free substrate. Seeds (soybean or bean) are disinfected for 5 min in a saturated calcium hypochlorite solution and rinsed six times with sterile deionized water. Four seeds are sown per container. Containers are watered with sterile deionized water. Plants are grown in a controlled-environment cabinet (16 h light at 240 µE/s, 8 h dark, at 22°C). Nodulation of the plants, which may be achieved even when soil contains low population levels of

Table 9.2. Scoring system to assess the extent of nodulation (derived from Brockwell et al., 1982).

Nodule score	Number of presumably effective[a] nodules	
	Crown[b]	Elsewhere
7	> 7	> 9
6	> 7	5–9
5	> 7	0–4
4	1–7	> 9
3	1–7	5–9
2	1–7	1–4
1	1–7	0
0	0	0

[a]Presumably effective nodules are identified on the basis of size and internal pigmentation.
[b]The crown is defined as the top 5 cm of the root system.

nodulating rhizobia, is recorded at 1 month, by counting nodules or using a scoring table.

Discussion

The occurrence of indigenous rhizobia in most soils has been extensively documented (Amarger, 1980). For instance, rhizobia nodulating peas and clovers are ubiquitous in most French and Spanish soils. Conversely, rhizobia nodulating lucerne are absent in soils with a pH below 6, whereas rhizobia nodulating lupin are not found in soil with a pH above 6 (Amarger, 1980). Different analyses of the *Sinorhizobium meliloti* population in agricultural soils of Spain indicate that highly competitive strains, genetically similar to the well-studied strain GR4, are often present, noticeably in soils with a high lucerne yield (Villadas et al., 1995; Velázquez et al., 1999). In similar soils from the same geographical region, but with poor lucerne yield, the GR4-like subpopulation is typically replaced by other, less-effective strains (our unpublished results), which suggests that the presence of competitive, effective strains can be used as a soil quality indicator. The growth of non-native legumes often requires the inoculation of appropriate rhizobial symbionts (e.g. for soybean outside Asia), since they are naturally absent from soils. These introduced rhizobia are likely to survive and adapt in soil (Revellin et al., 1996).

Enumeration of Indigenous Rhizobia by MPN Plant Infection Counts

Introduction

The number of indigenous rhizobia can vary to a large extent when comparing different soils or farming practices. Unfortunately, selective growth media are not available to enumerate rhizobia directly from soil. Therefore, alternative procedures, based on most probable number (MPN) determination, are needed. The MPN is a statistical estimate derived from the number of nodulated plants obtained following inoculation with aliquots from a dilution series of the soil under study.

Principle of the methods

The plant infection count is based on the most probable number (MPN) method. Axenic plantlets are inoculated with diluted soil suspensions, tubes containing nodulated plants are recorded as positive, and rhizobial numbers are deduced from a statistical table (Fisher and Yates, 1963).

Materials and apparatus

- Erlenmeyer flasks
- Rotary shaker
- Seeds
- Sterile perlite
- Gibson glass tubes (220 × 22 mm)
- Filter paper strip
- Aluminium foil
- Growth chamber or a greenhouse

Chemicals and solutions

- Sterile deionized water
- Saturated calcium hypochlorite solution
- Nitrogen-free plant nutrient solution

Procedure

1. A soil suspension is prepared by transferring 10 g of soil to a sterile flask containing 90 ml of sterile water, and flasks are shaken for 30 min at 150 rpm.
2. A tenfold dilution series is prepared (ten dilution steps, including the initial soil suspension, are necessary to provide dilutions with and without rhizobia). One ml of each dilution is used to inoculate axenic seedlings grown in test tubes (four tubes per dilution), prepared as follows:

- In the case of soybean or bean, seeds are disinfected for 5 min in a saturated calcium hypochlorite solution and rinsed six times with sterile deionized water.
- Seeds are germinated in sterile perlite for 3 days at 28°C.
- Axenic seedlings are planted in Gibson tubes (Gibson, 1963) containing a strip of filter paper as a wick and root support. The tubes are filled with quarter-strength Jensen nitrogen-free solution (Jensen, 1942) and capped with aluminium foil, as described by Vincent (1970) (Fig. 9.1).

3. Plants are grown for 4 weeks in a growth chamber (16 h light at 240 µE/s, 8 h dark, at 22°C) or a greenhouse. Plants are scored as positive (at least one nodule present) or negative (no nodule). Narrowing the dilution series (four- or twofold dilutions) decreases the confidence limit of the MPN.

Calculation

The MPN of rhizobia can be deduced from the McCrady table (Vincent, 1970) (Table 9.3).

Discussion

Depending on the legume considered, numbers of nodulating rhizobia below 10^2–10^4 (10^4 for soybean) per g of dry soil can be a limiting factor for the symbiotic growth of legumes, and inoculation is recommended. However, it needs to be kept in mind that MPN counts are not sufficient to assess whether nodulating rhizobia are effective at fixing nitrogen (see pp. 241–243). One limitation to the use of plant infection counts is that it takes several weeks to obtain the results, and a growth chamber or greenhouse is required. In addition, it must be kept in mind that experimental conditions do not well reflect those under field conditions, and soil, climatic and biotic factors prevailing in the soil environment also influence nodulation (as well as the efficiency of symbiotic nitrogen fixation). Molecular tools may also be used to enumerate rhizobia, e.g. by plating soil suspensions on semi-selective media and identifying rhizobia by colony hybridization, but the procedure is tedious (Laguerre *et al.*, 1993; Bromfield *et al.*, 1995; Hartmann *et al.*, 1998a). Species-specific molecular markers (detected by real-time quantitative PCR) seem promising for direct estimation of *Bradyrhizobium japonicum*, but they require further calibration.

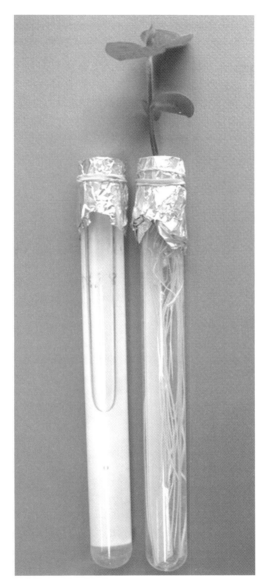

Fig. 9.1. Most probable number (MPN) plant infection counts of rhizobia nodulating soybean in Gibson tubes (220 × 22 mm). The tubes are filled with quarter-strength Jensen nitrogen-free solution and capped with aluminium foil. The filter-paper wick can be seen clearly in the left tube, which does not contain a plant. The right tube contains a soybean seedling.

Table 9.3. Number of rhizobia estimated by the plant infection count (from Vincent, 1970). Tenfold dilutions, four tubes per dilution step.

Positive tubes	Number of dilution steps (s)			
	$s = 10$	$s = 8$	$s = 6$	$s = 4$
40	$> 7 \times 10^8$			
39	$> 7 \times 10^8$			
38	6.9×10^8			
37	3.4×10^8			
36	1.8×10^8			
35	1.0×10^8			
34	5.9×10^7			
33	3.1×10^7			
32	1.7×10^7	$> 7 \times 10^6$		
31	1.0×10^7	$> 7 \times 10^6$		
30	5.8×10^6	6.9×10^6		
29	3.1×10^6	3.4×10^6		
28	1.7×10^6	1.8×10^6		
27	1.0×10^6	1.0×10^6		
26	5.8×10^5	5.9×10^5		
25	3.1×10^5	3.1×10^5		
24	1.7×10^5	1.7×10^5	$> 7 \times 10^4$	
23	1.0×10^5	1.0×10^5	$> 7 \times 10^4$	
22	5.8×10^4	5.8×10^4	6.9×10^4	
21	3.1×10^4	3.1×10^4	3.4×10^4	
20	1.7×10^4	1.7×10^4	1.8×10^4	
19	1.0×10^4	1.0×10^4	1.0×10^4	
18	5.8×10^3	5.8×10^3	5.9×10^3	
17	3.1×10^3	3.1×10^3	3.1×10^3	
16	1.7×10^3	1.7×10^3	1.7×10^3	$> 7 \times 10^2$
15	1.0×10^3	1.0×10^3	1.0×10^3	$> 7 \times 10^2$
14	5.8×10^2	5.8×10^2	5.8×10^2	6.9×10^2
13	3.1×10^2	3.1×10^2	3.1×10^2	3.4×10^2
12	1.7×10^2	1.7×10^2	1.7×10^2	1.8×10^2
11	1.0×10^2	1.0×10^2	1.0×10^2	1.0×10^2
10	5.8×10^1	5.8×10^1	5.8×10^1	5.9×10^1
9	3.1×10^1	3.1×10^1	3.1×10^1	3.1×10^1
8	1.7×10^1	1.7×10^1	1.7×10^1	1.7×10^1
7	1.0×10^1	1.0×10^1	1.0×10^1	1.0×10^1
6	5.8	5.8	5.8	5.8
5	3.1	3.1	3.1	3.1
4	1.7	1.7	1.7	1.7
3	1.0	1.0	1.0	1.0
2	0.6	0.6	0.6	0.6
1	< 0.6	< 0.6	< 0.6	< 0.6

The confidence interval at a 0.95 probability is calculated by multiplying and dividing the number by 3.8.

Assessment of the Symbiotic Efficacy of Indigenous Soil Rhizobial Populations

Introduction

Successful nitrogen fixation is expected to translate into better plant development, and this may be assessed by analysing plant development and biomass. In the method presented on pp. 233–236, the plant may obtain significant amounts of combined nitrogen from the soil and effective nitrogen fixation does not always lead to significantly higher plant biomass, which prevents satisfactory assessment of the symbiotic efficacy of the indigenous rhizobia unless dealing with chemically poor soils. To circumvent this limitation, assessment can be carried out using a rhizobia-free soilless system, in which legumes are inoculated with diluted suspensions of the soil under study (whole soil inocula technique; Somasegaran and Hoben, 1994). Thus, both nodulation and the symbiotic effectiveness of the nodulating population can be characterized, thereby providing a more complete estimate of the symbiotic potential of rhizobial populations present in the soil.

Principle of the method

The method involves extracting indigenous microorganisms from soil and using the resulting soil suspension directly as an inoculum for legumes grown in a soil-less system from which the plant cannot take up significant amounts of nitrate and/or ammonium (Fig. 9.2). Plant biomass is assessed and compared with that in the absence of inoculation or in the non-inoculated positive control (inoculation with an efficient strain).

Material and apparatus

- 5-l plastic containers
- 16-mm diameter clay beads (Argi-16TBF)
- Perlite
- Surface-sterilized seeds
- Greenhouse
- Oven (105°C)
- Balance
- Basic microbiology equipment

Fig. 9.2. Efficiency tests for *Bradyrhizobium japonicum* strains in 5-l soil-less containers. The two containers at the front are non-inoculated controls (pale plants), whereas the containers at the back are inoculated with efficient strains.

Chemicals and solutions

- N-free mineral nutritive solution (per litre: K_2HPO_4, 0.14 g; $MgSO_4.7H_2O$, 0.25 g; $CaCl_2$, 0.28 g; K_2SO_4, 0.12 g; Sequestrene 138 Fe Novartis, 10 mg; H_3BO_3, 2 mg; $MnSO_4.H_2O$, 1.80 mg; $ZnSO_4.7H_2O$, 0.2 mg; $CuSO_4.5H_2O$, 0.08 mg; $Na_2MoO_4.2H_2O$, 0.25 mg)
- Fresh culture of an efficient rhizobial strain

Procedure

- The N-free soil-less system is prepared as follows (Fig. 9.2). 5-l plastic containers are filled with the following layers (from bottom to top): 1 l of 16-mm diameter clay beads (Argi-16TBF), 4 l of rhizobia-free substrate (perlite).
- Surface-sterilized seeds (eight in the case of soybean, thinned to four plants after germination) are sown in each container.
- 10^{-1} to 10^{-3} diluted soil suspensions are used as inocula, and, for the positive control, a fresh culture of an efficient rhizobial strain is prepared (Somasegaran and Hoben, 1994) and diluted in sterile water to give an inoculum of 10^6 cells per seed.

- Four containers (i.e. four replicates) are used per treatment (including the two controls).
- Plants are watered (e.g. through a hole at the bottom of the containers) with an N-free mineral nutritive solution, taking care to avoid cross-contamination.
- The experiment follows a randomized block design (four Fisher blocks).
- Plants are grown for 10 weeks in the greenhouse.
- Shoots are harvested, dried at 105°C for 24 h and weighed. Foliage dry matter production is considered a good indicator of total nitrogen uptake (Brockwell *et al.*, 1982), but results may be complemented by analysing total foliage nitrogen (Kjeldahl method).

Calculation

Variance analysis is performed to determine strain efficiency by comparison with the efficient reference strain and the non-inoculated control.

Discussion

A critical point is to include adequate controls, since diluted soil suspensions contain nutrients (including mineral nitrogen) that will not be present in non-inoculated and reference strain treatments. Variance analysis or other statistical treatment of the data should take this bias into account. This method may also be implemented in soil, but plant biomass determinations may be of limited interest under soil conditions where legumes can also acquire significant amounts of combined nitrogen by direct uptake of soil nitrate and/or ammonium. In this case, the amount of plant nitrogen specifically derived from biological fixation can be estimated using the approach presented below.

Assessment of Plant Nitrogen Derived from Symbiotic Fixation

Introduction

The isotopic methods based on ^{15}N variation are the most reliable for evaluating nitrogen fixation by plants under field conditions. Other methods can be used in parallel as useful indicators of nitrogen fixation activity *in situ*, such as acetylene reduction activity by nitrogenase or analysis of the nitrogen content of xylem sap (Dommergues *et al.*, 1999). The latter indirect methods will not be discussed hereafter. Isotopic methods are based on the assumptions that the root systems of the N_2-fixing legume and the non-fixing reference species have similar architectures and explore the same soil horizons, and that they assimilate from the same nitrogen pools (Danso *et al.*, 1993). The occurrence and type (endo- or ectomycorrhizae) of mycorrhizae should also be similar, as they can affect plant $\delta^{15}N$ (Högberg, 1990), and plants should have a similar phenology. These assumptions are rarely fully valid in most field studies, which highlights the importance of the choice of the non-fixing reference plant species. The natural ^{15}N abundance and the ^{15}N enrichment methods are presented here.

Principle of the methods

The natural ^{15}N abundance method is based on the isotopic discrimination process occurring in most chemical reactions involved in gaseous loss, such as ammonia volatilization or denitrification. Indeed, the latter processes favour a progressive disappearance of the lightest, main nitrogen isotope (^{14}N), and thus the heaviest one (^{15}N) tends comparatively to accumulate in soil (Dommergues et al., 1999). Consequently, since the $^{15}N/^{14}N$ ratio of atmospheric N_2 is very low and stable, it is expected that nitrogen immobilized in N_2-fixing legumes will contain a lower proportion of ^{15}N compared with that of plants that do not fix nitrogen and take this element up from soil nitrate and/or ammonium. This means that the comparison of the relative proportion of ^{15}N (i.e. $\delta^{15}N$) in N_2-fixing legumes and non-fixing reference plants can be used to estimate the amount of nitrogen derived from the atmosphere by nitrogen fixation (i.e. % Ndfa) in the former.

The ^{15}N enrichment method, also called isotopic dilution method, is based on the comparison of ^{15}N excess measured in N_2-fixing legumes and non-fixing reference plants grown in soils enriched in ^{15}N through the incorporation of ^{15}N-labelled fertilizers (containing urea, nitrate or ammonium) to the soil. Since the N_2-fixing legumes will both take up combined nitrogen from soil and fix atmospheric nitrogen, their ^{15}N excess will be lower than that of the non-fixing reference plant, which has access to soil nitrogen (enriched in ^{15}N) only.

Materials and apparatus

A highly sensitive mass ratio spectrometer is required for the determination of natural abundance; ^{15}N measurement can be subcontracted to specialized service laboratories equipped with the relevant apparatus.

Procedure

The natural ^{15}N abundance method is only applicable when the soil $\delta^{15}N$ is superior or equal to +2 (Unkovich and Pate, 2001), and when spatial variability of soil $\delta^{15}N$ (as assessed through standard deviation) is not too high. Indeed, soil $\delta^{15}N$ may vary from site to site, according to the successive biological processes that occurred during a given soil history. Therefore, a preliminary sampling of leaves on different local species (including legumes and non-legumes) is required, followed by $\delta^{15}N$ analysis, to determine whether the natural ^{15}N abundance method can be used. If so, a more extensive sampling is then carried out. For herbaceous species, many experimental designs can be chosen, since sampling of plant material is easy to perform on both the N_2-fixing legume and the non-fixing species. The entire plant can be harvested and plant parts (leaves, fruits, stem, roots, nodules) analysed to determine dry weight (after drying for 3 days at 60°C), total nitrogen content and $\delta^{15}N$. $\delta^{15}N_a$ (the $\delta^{15}N$ of the N_2-fixing legume when deriving all its nitrogen from nitrogen fixation) is obtained by growing the legume (after inoculation with the most efficient rhizobial strain available) in an N-free system, e.g. in greenhouse pots containing an artificial substrate such as perlite or vermiculite and watered with an N-free nutrient solution. Plants (usually up to ten plants per treatment) are collected at 3–12 months and oven-dried at 60°C for 3 days. Plant parts are analysed for dry weight, total nitrogen and $\delta^{15}N$ assessments, as above. The best strategy is to compare the $\delta^{15}N$ of different non-fixing reference plant species and to calculate the percentage of N derived from biological fixation (%Ndfa) according to each of them. The different %Ndfa obtained are then presented as different assumptions. The total amount of nitrogen fixed by the legume (Ndfa) is obtained by multiplying %Ndfa with the total N content of the plant (see below for the formula).

In the ^{15}N enrichment method, %Ndfa is calculated from the percentages of ^{15}N excess in the N_2-fixing plant and the non-fixing reference plant (see below for the formula). Several protocols have been described for the ^{15}N enrichment method (Dommergues *et al.*, 1999). The sampling procedure proposed for the natural ^{15}N abundance method can be extrapolated to the ^{15}N enrichment method (except for the determination of $\delta^{15}N_a$, which is not required).

Calculation

In the natural ^{15}N abundance method, $\delta^{15}N$ is computed as follows:

$$\delta^{15}N(‰) = 1000 \times (\%^{15}N_{sample} - 0.3663)/0.3663$$

where 0.3663 is the percentage of ^{15}N in the atmosphere; the value is constant worldwide (Junk and Svec, 1958; Mariotti, 1983).

Then, %Ndfa in the N_2-fixing plant is computed as follows (Amarger *et al.*, 1977; Bardin *et al.*, 1977; Shearer and Kohl, 1986):

$$\%Ndfa = 100 \times (\delta^{15}N_{nf} - \delta^{15}N_f)/(\delta^{15}N_{nf} - \delta^{15}N_a)$$

where: $\delta^{15}N_{nf}$ corresponds to the $\delta^{15}N$ of the non-fixing reference plant; $\delta^{15}N_f$ to the $\delta^{15}N$ of the N_2-fixing legume studied; and $\delta^{15}N_a$ (also known as the β factor of isotopic discrimination) to the $\delta^{15}N$ of the N_2-fixing legume when deriving all its nitrogen from nitrogen fixation.

In the ^{15}N enrichment method, %Ndfa is calculated from the percentages of ^{15}N excess in the N_2-fixing plant and the non-fixing reference plant, as follows:

$$\%Ndfa = 100 \times (1 - E_i/E_o)$$

where: E_i corresponds to the ^{15}N excess measured in the N_2-fixing plant and E_o to the ^{15}N excess of the non-fixing reference plant. These values are generally determined using an emission spectrometer.

Discussion

The ^{15}N enrichment method displays several drawbacks. First, a correct estimation is only reached when uptake of soil nitrogen by both the N_2-fixing legume and the non-fixing reference plant follows the same kinetics, which is not often the case (Danso et al., 1993). Secondly, gaseous loss (as, for example, N_2O or NH_3) or leaching of the labelled fertilizer can be other sources of error for the final calculation of %Ndfa. Thirdly, it is sometimes difficult to obtain a homogeneous labelling of the soil (especially with woody species), whereas the natural abundance method does not require any handling of soil and is thus less disturbing for the ecosystem. Fourthly, the use of ^{15}N-labelled fertilizer remains costly, making the natural abundance method much cheaper. Fifthly, the measurement is performed over a short period of time, which limits the significance of the results when the method is performed on perennial legumes. Consequently, the enrichment method is generally used in the framework of short-term field experiments, and thus is more adapted to quantification studies on annual crops. Finally, this remains the only applicable method when the soil δ^{15}N of a given site is too low.

Conclusion

As nitrogen is a key plant nutrient, its availability is a crucial parameter of soil quality. Development and optimization of symbiotic BNF can limit the expensive, energy-consuming and polluting industrial transformation required for production of chemical nitrogen fertilizers. Symbiotic nitrogen fixation performed by rhizobia–legume associations exhibits the highest rate of N fixation, and it is important to detect situations where rhizobial BNF is not optimal and propose solutions to improve it. In particular, inoculation has now been used successfully for years, mainly when the bacterial symbiont is absent from the soil or poorly efficient. Beside BNF, rhizobia are soil and rhizosphere bacteria that could have other roles in improving soil quality. For example, many rhizobia produce plant hormones or modify plant hormonal balance via 1-aminocyclopropane-1-carboxylate deaminase activity, and they might have a potential use as plant-growth-promoting rhizobacteria (PGPR) (Sessitsch et al., 2002; Ma et al., 2003). Certain rhizobia are involved in phosphate solubilization and thus could influence phosphorus nutrition of plants (Chabot et al., 1996). The size and diversity of indigenous rhizobial populations may be affected by soil pollutants, farming practices, etc. (McGrath et al., 1995; Morrissey et al., 2002; Chen et al., 2003; Walsh et al., 2003), which highlights the usefulness of these bacteria as bioindicators of soil quality.

9.3 Contribution of Arbuscular Mycorrhiza to Soil Quality and Terrestrial Ecotoxicology

SILVIO GIANINAZZI,[1] EMMANUELLE PLUMEY-JACQUOT,[1,2] VIVIENNE GIANINAZZI-PEARSON[1] AND CORINNE LEYVAL[2]

[1]*UMR INRA1088/CNRS 5184/UB PME, CMSE-INRA, 17 rue Sully, BP 86510, 21065 Dijon cedex, France;* [2]*LIMOS, Laboratoire des Interactions Microorganismes-Minéraux-Matière Organique dans les Sols, CNRS FRE 2440, 17 rue N.D. des Pauvres, B.P. 5, 54501 Vandoeuvre-les-Nancy cedex, France*

Introduction

Although evolution has produced a general state of resistance to 'non-self' in plants, more than 95% of plant taxa in fact form compatible root associations, mycorrhizas, with certain soil fungi. These symbiotic associations are no doubt the most frequent examples of susceptibility of plants to fungi. Consequently, the root systems of a very large number of plants, and in particular those of many cultivated plants, whether they are agricultural, horticultural or fruit crops, do not exist simply as roots, but as complex mycorrhizal associations (Harley and Harley, 1987).

Mycorrhizas are usually divided into three morphologically distinct groups, depending on whether or not there is fungal penetration of the root cells: endomycorrhizas, ectomycorrhizas and ectendomycorrhizas. However, the most widespread plant root symbiosis is represented by (arbuscular) endomycorrhiza (AM) and is formed by more than 80% of plant families. The fungi involved all belong to the Glomeromycota (*Glomus, Acaulospora, Gigaspora, Entrophospora, Sclerocystis, Scutellospora*).

An AM association is generally mutualistic, in that the fungi obtain a carbon source from the host, while the latter benefits from enhanced nutrient uptake through transfer of mineral elements from the soil via the fungal hyphae (Smith and Read, 1997). In fact, there is an external network of ramifying mycelium of the symbiotic fungus into the soil from mycorrhizal root systems. This external mycelium supplies the plant with an extensive supplementary pathway for absorbing mineral nutrients and water from the soil, and it can facilitate mobilization by the plant of poorly mobile elements such as phosphate, ammonium, zinc and copper. The distance over which the fungal hyphae can translocate these nutrients exceeds the radius of any depletion zone likely to develop around an actively absorbing root, enabling plants to better exploit soil resources. It has been estimated that mycorrhizal hyphae can explore volumes of soil that are a hundred to a thousand times greater than the volume exploited by roots alone (Gianinazzi-Pearson and Smith, 1993). Strong evidence supports the existence of a direct link between the level of AM fungal biodiversity in soil and that of the plant species above ground (van der Heijden *et al.*, 1998).

Precolonization of roots by AM fungi can also alleviate stress due to metal and organic acid toxicity in soils of low pH (Leyval *et al.*, 1997), and reduce damage by soil-borne pathogens such as nematodes, *Fusarium*, *Pythium* or *Phytophthora*. The mechanism involved in this protection against a pathogen is complex, but there is strong evidence that AM fungi activate plant defence mechanisms against microbial pathogens (Gianinazzi, 1991; Cordier *et al.*, 1998; Dumas-Gaudot *et al.*, 2000). Furthermore, AM formation stimulates the development in the mycorrhizosphere of microorganisms with antagonistic activity towards soil-borne pathogens (Linderman, 2000). Therefore, the contribution of AM to soil quality and terrestrial ecotoxicology is of primary importance.

Mycorrhizal Inoculation to Improve Plant Health and Growth

The AM association is the most common type of mycorrhiza in agricultural systems. The importance of AM for plant growth and health is now widely demonstrated. It has become clear that they form an integral part of many cultivated plants and positively influence several aspects of plant physiology: mineral nutrition, water uptake, hormone production and resistance to root diseases (Smith and Read, 1997). Plants often grow badly in soils where AM fungi have been eliminated and their presence appears to be a factor determining soil fertility. Because of these characteristics, it is considered that AM fungi can be used as biological tools for increasing plant productivity in the field without creating problems of environment degradation, and by reducing chemical fertilizer or pesticide input (Gianinazzi and Gianinazzi-Pearson, 1988).

Given the effects of AM fungal inoculation on plants, it is generally accepted that appropriate management of this symbiosis should permit reduction of agrochemical inputs, and thus provide for sustainable and low-input plant productivity. Maximum benefits will only be obtained from inoculation with efficient AM fungi and a careful selection of compatible host/fungus/soil combinations (Gianinazzi *et al.*, 2002).

AM fungi cannot be grown in pure culture and must therefore be multiplied on living roots. This requires the use of techniques for inoculum production that are different from those employed for other mycorrhizal fungi. This, usually considered as a major disadvantage, appears in our experience to be an advantage, since the risk of multiplying AM fungi that have lost their symbiotic properties is very low, because culture collections have to be maintained on living host plants. On the other hand, producing symbiotic fungi on living plants raises difficulties because of precautions that have to be taken in order to obtain 'clean' inoculum. It is possible to produce AM inoculum on either excised roots or whole plants under axenic conditions using disinfected spores as inoculum (Gianinazzi et al., 1989). However, for the moment, only one company is proposing commercial inoculum produced *in vitro*.

In order to use any of these types of inoculum rationally and successfully, it is necessary to define a strategy of inoculation, that is, to determine whether for a given situation it is necessary to inoculate and which fungi to use. We have developed biological tests for this. Without going into details about these, the aim is first to estimate the soil mycorrhizal potential, which indicates the number of fungal propagules, including spores, roots or hyphae, present in a given soil or substrate, and secondly, to evaluate the potential effectiveness of indigenous fungi for plant production and determine soil receptivity to fungi that are to be introduced into the system.

Information obtained using these biotests for mycorrhizal potential and fungal effectiveness, together with current knowledge about the mycorrhizal dependency of different plant genera and species, will provide the essential tools for a strategy of inoculation (Gianinazzi et al. 2002). The creation of 'La Banque Européenne des Glomales (BEG)' under the impulsion of the EU-COST programme (Dodd et al., 1994), and its recent development into the international bank of Glomeromycota (http://www.kent.ac.uk/bio/beg/), will greatly help in this task. AM biotechnology is feasible for many crop production systems and the recent development of AM fungal-specific molecular probes provides tools for monitoring these microsymbionts in soil and roots (Van Tuinen et al., 1998; Jacquot et al., 2000).

A perspective for the near future should be the development of integrated biotechnologies (plant biotization) in which not only AM fungi, but also other microorganisms capable of promoting plant growth and/or protection – such as symbiotic or associative bacteria, plant-growth-promoting rhizobacteria (PGPR), pathogen antagonists (*Trichoderma*, *Gliocladium*, *Bacillus*, etc.), or hypovirulent strains of pathogens – will be used in synergy.

Application of Arbuscular Mycorrhiza as Bioindicators of Soil Quality

Introduction

Arbuscular mycorrhizal fungi are abundant soil fungi and are associated with many plant species. They are an integral part of the plant (Gianinazzi et al., 1982), and many plants are highly dependent on mycorrhizal colonization for their growth. Not only can they be considered as a powerful extension of the plant-root system, providing access to a larger soil volume for nutrient uptake and playing a role in root protection against disease or drought, but they also affect the transfer of pollutants, such as heavy metals, to plants and can be affected by elevated metal concentrations (Gildon and Tinker, 1983; van der Heijden et al., 1998) before toxicity symptoms can be seen on plants. Because of this, and because they are a direct link between the soil and plant roots, AM fungi have been proposed as bioindicators of heavy metal toxicity (Leyval et al., 1995), as a supplement to chemical extraction procedures commonly used to assess availability of metals. Toxicity tests have been performed using techniques based on spore germination (Leyval et al., 1995; Weissenhorn and Leyval, 1996; Jacquot-Plumey et al., 2003), and on AM root colonization using MPN (Leyval et al., 1995; Jacquot-Plumey et al., 2002), or using Ri-T-DNA transformed roots (Wan et al., 1998) or nested PCR, which allows monitoring of the effect of sewage sludges on the diversity of AM fungi *in planta* and in soil (Jacquot et al., 2000; Jacquot-Plumey et al., 2002). Very recently, the potential use of proteomics to study changes in protein expression of AM plants under stress conditions has also been investigated (Bestel-Corre, 2002). Two bioassays have been developed in collaboration with ADEME (Agence de l'Environnement et de la Maîtrise de l'Energie – Contract number 02750021) at LIMOS-CNRS (Nancy, France) and UMR BBCE-IPM, INRA/University of Burgundy (Dijon, France), based on spore germination and root colonization, respectively, and have been submitted to the Technical Committee T95^E on Terrestrial Ecotoxicology of the French Association of Normalisation (AFNOR) as a bioassay for ecotoxicological effects of chemicals or wastes and for soil quality.

Principle of the method

These bioassays are based on the germination of spores of an AM fungus (*Glomus mosseae*) sensitive to metals (Weissenhorn et al., 1993), and on colonization of roots of *Medicago truncatula* inoculated with the same fungus. They are complementary since the spore germination assay concerns the AM fungus alone and the initial stage of the symbiotic cycle, while the root colonization assay reflects a further stage of the symbiosis when the fungus interacts with the host plant. Both bioassays are direct contact tests, for acute and chronic toxicity, respectively. Contaminated soils, sewage sludge, wastes or solutions can be tested for their potential toxicity to beneficial microbial activity. AM spore germination and root colonization within a test substrate containing a substance, or a contaminated soil, are compared to a control test substrate alone or a non-contaminated soil comparable to the soil sample to be tested.

Materials and apparatus

- Nitrocellulose filter membranes (0.45 µm, 47 mm in diameter)
- Petri dishes
- Filter paper
- Seeds of *Medicago truncatula* Gaertn (line J5)
- *G. mosseae* inoculum (Biorize, Dijon, France)

- Sand (0.5–1 mm particle size)
- Growth chamber
- Binocular magnifier
- Photonic microscope

Chemical and solutions

- Agar water
- Mycofert nutrient solution (Biorize, Dijon, France)

Procedure

AM spore germination:

1. Thirty *G. mosseae* spores (Nicol. & Gerd.) Gerdemann & Trappe (BEG12) (obtained from Biorize, Dijon, France) are placed between two nitrocellulose filter membranes (0.45 µm, 47 mm in diameter) with gridlines, held together by a slide frame as described by Weissenhorn *et al.* (1993).
2. This so-called sandwich with spores is placed in a Petri dish containing 50 g of test substrate (sand, particle size 0.5–1 mm) mixed with, for example, a chemical compound solution or a contaminated soil to be tested (Fig. 9.3).
3. The sandwich is then covered with another 50 g layer of sand/soil.
4. Soil and sand are allowed to moisten up to water-holding capacity. A filter paper, lining the bottom of the dish and extending outside the dish into a water-filled larger dish, moistens the substrate by capillarity when soil or sludges are used. When sand is used, it is moistened by watering.
5. The Petri dish is sealed and kept at 24°C in the dark for 2 weeks.
6. The spore sandwich is then removed from the soil/sand, stained using tryptan blue for 1 h (Bestel-Corre, 2002) and carefully rinsed with tap water before opening.
7. The number of spores recovered and of germinated spores are recorded at ×32 magnification.
8. Six replicates are made for each sample.
9. The dose effect can be studied using sample dilution in sand and estimation of the mean inhibitory concentration at 50% (MIC50).

Root colonization:

1. Pre-germinated seeds of *M. truncatula* Gaertn. (line J5) on water-agar (0.7%) are transferred individually into pots, and each plant is inoculated with 15 g of a *G. mosseae* inoculum placed in the planting hole of 200 g of a mixture of sand (0.5–1 mm particle size) with the solution, soil or sewage sludge to be tested.
2. Plants are cultivated under controlled conditions (photoperiod 16 h, 19–22°C, relative humidity 60–70%, light intensity 320 $\mu E/m^2/s^1$), and receive weekly 10 ml Mycofert nutrient solution.

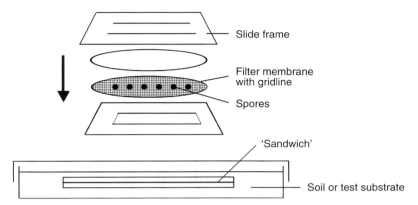

Fig. 9.3. Arbuscular mycorrhiza (AM) spore germination test for pollutants in soil, sewage sludge or wastes.

3. Frequency of root colonization by *G. mosseae* is estimated after 4 weeks according to Trouvelot *et al.* (1985) and a simplified determination is presently made based on a + (fungal detection)/− notation.
4. Six replicate plants are used for each sample.
5. The dose effect can be studied using sample dilution in sand to measure the mean inhibitory concentration at 50% (MIC50).

Discussion

The spore germination bioassay has been used to compare the toxicity of different soils (Leyval *et al.*, 1995) and sewage sludges, and to compare the tolerance to cadmium of different AM fungi (Weissenhorn *et al.*, 1993). The comparison of composted sewage sludges and other wastes from different origins showed a decrease in spore germination with a composted sludge amended with polycyclic aromatic hydrocarbons (PAH) (B31), with combustion ashes (C1) or combusted household refuse (C2), but not when the sludge was unamended or amended with metallic pollutants (B32) (Fig. 9.4A). Dilution of these sludges and ashes showed that C1 and C2 were more toxic to AM spore germination than B31 sludge (Fig. 9.4B).

The root colonization bioassay has been used to compare, *ex situ*, the toxicity of different non-composted sewage sludges and to study, *in situ*, the effects of three successive sewage sludge amendments (Jacquot *et al.*, 2000, Jacquot-Plumey *et al.*, 2002). Comparison of the sewage sludges showed a slight decrease in root colonization with the addition of B3 sewage sludge to sand and even more when the sludge was amended with polycyclic aromatic hydrocarbons (B31) and heavy metals (B32) (Fig. 9.5).

Neither of the bioassays based on AM fungi is specific: a reduction in germination or in root colonization may be due to different pollutants, and possibly also to other soil characteristics, such as very low pH, or high P

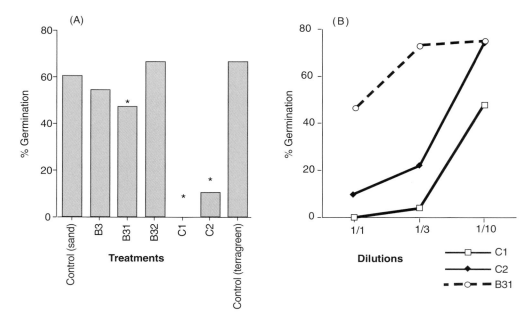

Fig. 9.4. Germination (%) of *Glomus mosseae* spores (A) in different composted sewage sludges and wastes and (B) after dilution of the sludge and wastes in sand using the sandwich bioassay (Fig. 9.3). * indicates significant differences from control (sand or terragreen, chi-squared test, $P \leq 0.05$). B3, composted sludge; B31, B3 composted sludge spiked with polycyclic aromatic hydrocarbons (PAH); B32, B3 composted sludge spiked with heavy metals; C1, combustion ash; C2, combusted household refuse.

content. Therefore, when there is such a reduction, further investigations should be performed to identify its origin.

Early in the 1980s, Gildon and Tinker reported the presence of heavy-metal-tolerant AM fungi in polluted soils and their effect on metal uptake by plants (Gildon and Tinker, 1983). Since then, the presence of AM fungi in heavy-metal-polluted sites, and the contribution of AM fungi to the transfer of heavy metals and radionuclides from soil to plants and to the translocation from root to shoots, has been addressed in many studies. They have concerned different metallic trace elements, such as Cd, Zn and Ni, and radionuclides such as Cs (Leyval and Joner, 2001). Results showed that heavy-metal-tolerant AM fungi may reduce metal transfer to plants and protect them against heavy metal toxicity (Leyval *et al.*, 1997). However, results are not always consistent, probably due to different metal concentrations, fungi and availability in soils (Leyval and Joner, 2001; Leyval *et al.*, 2002; Rivera-Becerril *et al.*, 2002). They also cannot be generalized and may differ with plants, with heavy metals/radionuclides and their availability, and possibly, although this has been less investigated, with other microbial components of the soils.

Remediation of heavy metal and radionuclide contaminated soils includes immobilization and extraction techniques. Different plants can be

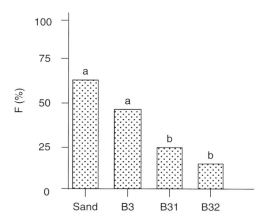

Fig. 9.5. Frequency of root colonization (F%) by *Glomus mosseae* grown in sand and in sand mixed with an unamended sewage sludge (B3), or amended with polycyclic aromatic hydrocarbons (PAH) (B31) and metallic pollutants (B32). Values in columns headed by the same letter are not significantly different ($P \leq 0.05$).

used for phytoextraction (extraction of metals by plants accumulating high metal concentrations) and phytostabilization (use of plants to reduce heavy metal availability, erosion and leaching). For phytoextraction, hyperaccumulative plants from the *Brassicaceae* family are often used, but most plants of this family are reported as non-mycotrophic. However, other plants accumulating lower metal concentrations, but producing higher biomass, such as sunflower and willow, are also receiving attention and these are mycorrhizal. There has been much work done, and much progress made, in the field of AM fungi in soil remediation and restoration. AM inoculum production on a commercial scale is available and is no longer a limit to applied studies (Von Alten *et al.*, 2002). Field studies and field applications of AM fungi have been reported and are currently running, but they concern mainly revegetation studies, horticultural and agricultural applications. There is still a lack of demonstration and of *in situ* projects using AM fungi in polluted sites. However, preliminary field trials have indicated that dual inoculation with a selected AM fungus and *Rhizobium* can improve the yield and nodulation of pea genotypes in a metal contaminated site (Borisov *et al.*, unpublished results; EU project INCO-Copernicus IC15-CT98–0116). The persistence of the inoculated fungi, and the relative diversity of AM fungi in roots *in situ* are poorly known and need to be investigated further.

Arbuscular mycorrhizal fungi provide a direct link between soil and roots and are key components of soil–plant ecosystems. Their presence and beneficial activity can be affected by soil perturbation and contamination, and can be used as an early sign of soil disfunctioning. Besides improving soil fertility and plant productivity, they affect the fate of pollutants such as heavy metals and radionuclides, and therefore bring many advantages to mycorrhizal plants in polluted soils. However, there is still much research needed, especially to predict the conditions in which a beneficial effect of AM fungi can be expected. This requires a better understanding of the mechanisms involved in heavy metal tolerance, taking into account not only AM fungi but also the complexity of the soil–plant–microbe systems.

9.4 Concepts and Methods to Assess the Phytosanitary Quality of Soils

CLAUDE ALABOUVETTE,[1] JOS RAAIJMAKERS,[2] WIETSE DE BOER,[3] RÉGINA NOTZ,[4] GENEVIÈVE DÉFAGO,[4] CHRISTIAN STEINBERG[1] AND PHILIPPE LEMANCEAU[1]

> [1] *UMR 1229 INRA/Université de Bourgogne 'Microbiologie et Géochimie des Sols', INRA-CMSE, BP 86510, 21065 Dijon cedex, France;* [2] *Wageningen University, Laboratory of Phytopathology, Binnenhaven 5, 6709 PG Wageningen, The Netherlands;* [3] *NIOO-CTE, Department of Plant–Microorganisms Interactions, PO Box 40, 6666 ZG Heteren, The Netherlands;* [4] *Institut für Pflanzenwissenschaften/Phytopathologie, LFW B25, Universitätsstraße 2, ETH Zentrum, CH-8092 Zurich, Switzerland*

Concepts

A plant disease results from the intimate interaction between a plant and a pathogen, and today there is a great effort of research devoted to the study of plant–pathogen interactions at the cellular and molecular levels. The importance of these direct interactions should not hide the role of environmental factors that influence disease severity. These indirect interactions are particularly important in the case of diseases induced by soilborne pathogens. The existence of soils that suppress diseases provides an example of biotic and abiotic factors affecting the pathogen, the plant and/or the plant–pathogen interaction. Indeed, in suppressive soils, disease incidence or severity remains low in spite of the presence of a virulent pathogen, a susceptible host plant and climatic conditions favourable for disease development.

Soils suppressive to diseases caused by a range of important soilborne pathogens have been described; they include fungal and bacterial pathogens, but also nematodes (Schneider, 1982; Cook and Baker, 1983; Schippers, 1992). These soils control root rot and wilt diseases induced by: *Aphanomyces euteiches*, *Cylindrocladium* sp., several formae speciales of *Fusarium oxysporum*, *Gaeumannomyces graminis*, *Pythium* spp., *Phytophthora* spp., *Rhizoctonia solani*, *Ralstonia solanacearum*, *Streptomyces scabies*, *Thielaviopsis basicola* (*Chalara elegans*) and *Verticillium dahliae*. This large diversity of pathogens controlled by suppressive soils shows that soil suppressiveness is not a rare phenomenon. On the contrary, every soil has some potential of disease suppression, leading to the concept of *soil receptivity* to diseases.

The receptivity of a soil to soilborne diseases is its capacity to suppress more or less the saprophytic growth and infectious activity of the pathogenic populations present in the soil. Indeed, the soil is not a neutral milieu where pathogenic microorganisms interact freely with the roots of the host plant. On the contrary, the soil interferes in several ways with the relationships between and among microorganisms, pathogens and plants, and it can modify the interactions among microorganisms themselves. Soil receptivity (or soil suppressiveness) is a continuum, going from highly conducive soils to strongly suppressive soils (Alabouvette *et al.*, 1982; Linderman *et al.*, 1983).

Soil suppressiveness to some diseases can be related to another fundamental phenomenon affecting the soil microorganisms: microbiostasis. Well studied by Lockwood (1977) in the case of fungal spores, *fungistasis* is defined as the global effect of the soil that restricts the germination and growth of fungi. In general, the germination and saprophytic growth of fungi are more restricted in soil than would be expected from their behaviour *in vitro* under similar environmental conditions of temperature, moisture and pH. Based on this definition, fungistasis only concerns the saprophytic growth of the fungi, without taking into account their interaction with the plant, but in some cases fungistasis has been associated with soil suppressiveness to diseases.

The concept of soil receptivity to diseases was already evoked in the definition of *inoculum potential* proposed by Garrett (1970) as 'the energy of growth of a parasite available for infection of a host at the surface of the host organ to be infected'. One of the most important words in this definition is 'energy' of growth. It clearly states that the presence of the inoculum, although necessary, is not sufficient to explain the disease. Among the factors that affect the 'energy of growth' from the inoculum, Garrett (1970) pointed to 'the collective effect of environmental conditions', and indicated that 'the endogenous nutrients of the inoculum might be augmented by exogenous nutrients from the environment'.

Applied to soilborne pathogens, this concept of inoculum potential led to that of 'soil inoculum potential', which was at the origin of both theoretical and practical studies. Baker (1968) gave a definition of inoculum potential as the product of inoculum density per capacity. Louvet (1973) proposed to define inoculum capacity as the product of innate inoculum energy and the effects of the environment on this inoculum. Thus, in this definition, the effects of the environment on the inoculum corresponds to what we have defined above as the soil receptivity to diseases.

At the same time, the soil inoculum potential was defined by Bouhot (1979) as the pathogenic energy present in a soil. This inoculum potential depends on three main factors: (i) the inoculum density; (ii) the pathogenic capacity of this inoculum; and (iii) the soil factors which influence both the inoculum density and capacity. This last factor, again, corresponds to the soil receptivity as defined above.

Thus, whatever the definition, all these authors acknowledge that the soil plays a major role in influencing the interactions between a susceptible host plant and its specific pathogens present in soil. It is therefore very important to take into consideration both the inoculum potential of a naturally infested soil and its level of suppressiveness, when elaborating control strategies.

Indeed, the traditional approach to control soilborne disease consisted of trying to eradicate the pathogens from the soil. This led to the use of very dangerous biocides, such as methyl bromide, use of which will be totally banned in the near future. Opposite to that, a new approach consists of either enhancing the natural suppressive potential that exists in every soil, or introducing specific biological control agents. In both cases it is necessary to characterize and, if possible, quantify the soil inoculum potential and the soil receptivity of the soil to the disease, and also the soil receptivity to the biological control agents.

To characterize the phytosanitary quality of a soil, it is important to detect the presence or absence of the main pathogens of the crop to be cultivated. But this knowledge might not be sufficient to predict the risk for the crop to be severely diseased. Therefore, it is preferable to assess the soil inoculum potential, which will give a better view of the capacity of the soil to provoke the disease. A low inoculum potential may result from a very low inoculum density in a conducive environment, or a high inoculum density in a suppressive environment. Therefore, it is very interesting to assess the level of both soil inoculum potential and receptivity, to better estimate the probability of a healthy crop.

Assessment of soil inoculum potential and receptivity requires bioassays, and a few of them will be described in this section. Fungistasis, which is assessed *in vitro*, may be considered as a bioindicator of the phytosanitary quality of the soil, a few examples will be described below. Another strategy to assess the phytosanitary soil quality, which will not be described in detail in this section, is to quantify specific populations and/or specific genes shown to be involved in suppressiveness (Weller *et al.*, 2002), for example, phloroglucinol-producing pseudomonads, shown to be involved in the take-all decline phenomenon (Raaijmakers and Weller, 1998).

Bioassays for Assessing Soil Suppressiveness to *Fusarium* Wilts

Principle of the method

A standardized method has been proposed by Alabouvette *et al.* (1982). It involves infesting soil samples with increasing concentrations of a pathogenic strain of *Fusarium oxysporum*, to grow a susceptible plant under well-controlled conditions, and to establish the disease progress curve in relation to inoculum concentrations. This method is usually applied using flax (*Linum usitatissimum*) and its specific pathogen *F. oxysporum* f. sp. *lini*.

Materials and apparatus

- Seeds of 'Opaline', a highly susceptible cultivar of flax (*Linum usitatissinum*)
- Stock culture of a virulent strain of *F. oxysporum* f. sp. *lini* (Foln3)
- Vessels to produce fungal inoculum in shake culture
- Rotary shaker
- Sintered glass funnel (pore size 40–100 µm)
- Growth chamber to perform the bioassay
- Polystyrene trays (60 × 40 × 5 cm)
- Calcinated clay (Chemsorb®, CONEX Damolin GmbH, Peckhauser Str. 11, D. 40822 Mettmann, Germany)

Chemicals and solution

- Liquid malt extract (10 g/l, pH 5.5)
- Hydrokani® nutrient solution

Procedure

1. The pathogenic strain of *F. oxysporum* f. sp. *lini* (Foln3) is cultivated in liquid malt extract (10 g/l, pH 5.5) for 5 days on a rotary shaker (150 rpm) at 25°C.
2. The whole culture is filtered through a sterile sintered glass funnel (pore size 40–100 µm) and the conidial suspension in the filtrate is adjusted to different concentrations in order to infest the soil samples at concentrations of 1×10^3, 10^4, 10^5 propagules/ml soil.
3. The bioassay is set up in polystyrene trays (60 × 40 × 5 cm). Each tray has 12 rows consisting of eight holes of 7 cm diameter. The holes are closed at the bottom with a plastic grid, topped with a layer of small beads, in order to maintain the soil in the hole and permit excess water to drain.
4. 800 g soil is distributed in two rows of eight holes (i.e. 50 g/hole) and is infested with 4 ml of inoculum suspension at the concentrations needed to reach 1×10^3, 10^4, 10^5 propagules/ml soil.
5. Several seeds of flax cultivar (Opaline) are placed in-between two layers of calcinated clay (Chemsorb®) on the top of the soil, to prevent seeds from damping off. Uninoculated rows (4 ml sterile water) serve as a control.
6. After emergence, the number of seedlings is reduced to one per hole, making a total of 16 plants per treatment.
7. Each treatment is replicated three times.
8. The experiment is carried out in a growth chamber, with 17°C night and 22°C day temperature for 2 weeks, then with 20°C night and 25°C day temperature with a daily 15-h photoperiod of 16,000 lux for the rest of the experiment.

9. The plants are irrigated regularly with nutrient solution (Hydrokani®) and water alternatively, and the number of healthy plants is recorded twice a week from 21 days to 58 days after sowing.

Calculation

Depending on the number of plants used in the bioassay, two procedures of calculation might be used.

When a small number of plants per replicate is used, each plant is identified (as it is the case with the flax, for example) and survival data analysis is used (Hill *et al.*, 1990). This permits estimation of the duration of life of the population and the comparison of the soil receptivity to *Fusarium* wilts (Fig. 9.6). This procedure uses a Kaplan Meier estimate to calculate the survival function $S(t)$, which provides the probability that the failure time (typical wilt symptom appearance) is at least t for each plant, t being the time elapsed since the day of inoculation. A mean survival time (MST) is then evaluated for each subpopulation of plants (i.e. the set of 16 plants making a replicate) inoculated with the respective doses of the pathogen, provided that at least one plant among the 16 of a subpopulation of flax exhibits symptoms (Tu, 1995). In the example given, the MST of the control with no pathogen could not be evaluated because no mortality occurred during the bioassay.

When large numbers of plant are required for the experimental set-up (as might be the case with the assessment of soil suppressiveness to other diseases), or when a disease index is used to evaluate the disease incidence, the area under the disease progress curve (AUDPC) is calculated for each replicate by plotting the cumulative value of the disease index (symptom occurrence, symptom intensity or dead plants, etc.) against time.

Both procedures are applicable to evaluate natural soil suppressiveness to diseases and to evaluate the impact of any treatment on the level of soil suppressiveness or the pathogenic infectious activity (Höper *et al.*, 1995; Migheli *et al.*, 2000). MST values and AUDPC values are then analysed by analysis of variance (ANOVA) and tests of multiple comparison of means (as Fisher's LSD test and Student *t*-test, for instance, with $P = 0.05$). All statistical analyses are performed using StatView® software issued by SAS Institute Inc. (Cary, North Carolina, USA).

Discussion

Soil suppressiveness to *Fusarium* wilts, whatever the plant and the corresponding forma specialis, can be assessed by this procedure using flax and its specific pathogen, *F. oxysporum* f. sp. *lini* (Abadie *et al.*, 1998). But it is also possible to adapt this procedure to other plant–formae speciales models, such as tomato–*F. oxysporum* f. sp. *lycopersici* or melon–*F. oxysporum* f.

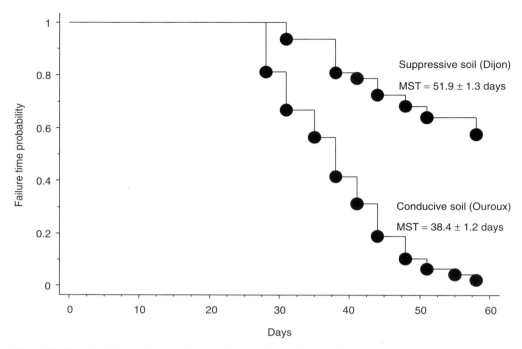

Fig. 9.6. Survival time of populations of flax cultivated in a soil suppressive to *Fusarium* wilt (Dijon) and in a soil conducive to the same disease (Ouroux). Both soils were inoculated with *F. oxysporum* f. sp. *lini* Foln3.

sp. *melonis*. Procedures based on the same principle have been proposed to measure soil suppressiveness to *Pythium* or *Rhizoctonia solani* damping-off (Camporota, 1980), *Aphanomyces* root rot (Persson *et al.*, 1999), take-all (Raaijmakers *et al.*, 1997), *Thielaviopsis basicola* (*Chalara elegans*) root rot (Stutz and Défago, 1985), etc.

Assessment of Soil Inoculum Potential

Principle

Inoculum potential being the pathogenic energy present in a soil, assessment of inoculum potential consists of growing susceptible host plants in the soil under environmental conditions chosen to be very favourable to disease expression.

To quantify this potential, it is necessary to dilute the naturally infested soil in a disinfested soil in different proportions, in order to obtain a dose–response relationship. This principle will be illustrated by the procedure described by Bouhot (1975a,b,c) to estimate the inoculum potential of soils infested by *Pythium* spp.

Materials and apparatus

- Seeds of a variety of cucumber (*Cucumis sativus*), for example 'Le généreux', highly susceptible to *Pythium* damping-off but not to other fungi, such as *Phytophthora* spp., also responsible for damping-off
- A mortar
- A sieve with mesh of 1000 µm
- Pots containing 300 ml of soil
- Sterile soil
- A blender to mix soil and oatmeal
- A growth chamber in which to perform the bioassay

Chemicals and solutions

- Oatmeal

Procedure

1. The soil to be analysed is first air-dried, then ground in a mortar and sieved through a mesh of 1000 µm.
2. This soil is amended with oatmeal at a rate of 20 g soil/litre.
3. The amended soil is diluted into sterile soil to obtain the following concentrations: 30%, 10%, 3%, 1%, 0.3% and 0.1% of soil in sterile soil.
4. Cucumber plants, cultivar 'Le généreux', are produced by sowing ten seeds in steamed, disinfested soil, in 10-cm diameter pots.
5. Plants are cultivated at 25°C under only 4000 lux for 15 h/day.
6. After 5 days, when the hypocotyls are 3–4 cm long, water is added in order to reach the water-holding capacity of the soil.
7. 60 ml of the infested soil amended with oatmeal is spread on the surface of the disinfested soil.
8. This soil layer is adjusted to 80% of its water-holding capacity.
9. The pots with the cucumber plants are then placed for 24 h at 15°C in the dark, then transferred again to 22°C during the day, 19°C at night, with a photoperiod of 15 h under 9000 lux.
10. Damping-off symptoms appear from the second to the seventh day.

Calculations

A correlation can be calculated between the number of dead plants and the concentration of the amended soil. From the regression lines it is possible to determine an inoculum potential unit, defined as the weight of soil needed to provoke the death of 50% of the plants. But, in many cases, the distribution is not normal; it is therefore necessary to utilize different transformations before calculating the inoculum potential units.

Discussion

Based on the same principle, other procedures have been proposed to assess the inoculum potential of soil infested by *Rhizoctonia solani* and *Aphanomyces euteiches* (Camporota, 1982; Williams-Woodward et al., 1998; Reverchon, 2001). It is important to choose the experimental conditions and the test plant carefully, in order to avoid symptoms produced by other fungi than that of interest. Depending on the pathogen studied, the dose–response relationship is not always evident; thus, it might be difficult to quantify the soil inoculum potential.

Soil Fungistasis

Introduction

Following the definition given by Lockwood (1977), the terms 'fungistasis' and 'mycostasis' are used to describe the phenomenon that germination and growth of fungi in most natural soils are more restricted than would be expected from their behaviour under similar environmental conditions of temperature, moisture and pH *in vitro*. Fungistasis affects both plant-pathogenic and saprophytic fungi, but the former are generally more sensitive (Garrett, 1970; De Boer et al., 1998). The high sensitivity of plant-pathogenic fungi to fungistasis can have both positive and negative consequences for disease development (Lockwood, 1977). A clearly negative aspect of fungistasis is that it protects propagules of plant pathogens from germination under unfavourable conditions, e.g. the absence of a host plant. On the other hand, continuing exposure to fungistasis results in loss of vitality of plant-pathogenic propagules. Furthermore, fungistasis limits the distance over which plant pathogens can reach host roots. In general, the intensity of fungistasis for a pathogenic fungal species is found to be positively correlated with the receptivity or disease suppressiveness of soils, but lack of correlation has also been reported (Lockwood, 1977; Hornby, 1983; Larkin et al., 1993; Knudsen et al., 1999; Peng et al., 1999).

The intensity of fungistasis is dependent on several factors, such as physical and chemical soil properties, fungal life-history characteristics, soil microbial activity and soil microbial community composition (Schüepp and Green, 1968; Lockwood, 1977; Hornby, 1983; Toyota et al., 1996; De Boer et al., 1998). The importance of a biological factor is indicated by the relief of fungistasis by (partial) sterilization treatments and addition of antibiotics (Lockwood, 1977; Toyota et al., 1996; De Boer et al., 1998, 2003).

The explanation given most often for the microbial cause of fungistasis is that soil microorganisms limit nutrient availability to the germinating spores or invading hyphae (Ho and Ko, 1986; Lockwood, 1988). Fungistasis has also been attributed to the presence of antifungal compounds of microbiological origin (Romine and Baker, 1973; Liebman and Epstein, 1992; De Boer *et al.*, 1998, 2003). The distinction between nutrient limitation and antibiosis as main mechanisms for fungistasis is not so easily made, as both can be the result of competition for nutrients (Fravel, 1988; Paulitz, 1990). Yet, although the actual mechanism of fungistasis is still open to debate, it is clear that it is a function of activities of soil microorganisms and their interactions with fungi. Therefore, fungistasis can be considered as a soil ecosystem function.

Principle of the methods

The methods used to quantify fungistasis focus on inhibition of either germination of fungal spores or extension of fungal hyphae. To determine differences in fungistasis between soils, it is important to use fixed amounts of soil (dry weight basis) and identical moisture conditions. In general, soil samples are wetted to pF 1 (–1 kPa) or 2 (–10 kPa) to establish optimal conditions for diffusion of nutrients or toxic compounds. For germination tests, spores are mixed directly into the soil or inoculated on a membrane or gel that is in contact with the soil (Schüepp and Green, 1968; Wacker and Lockwood, 1991; Knudsen *et al.*, 1999). After incubation, the amount of germinated spores and/or the length of the germination tubes is determined microscopically. The basic methods have been modified when the aim of the study has been to elucidate the mechanism of fungistasis, e.g. to establish the contribution of toxic solutes or volatiles (Romine and Baker, 1973; Liebman and Epstein, 1992).

Quantification of the effect of soil on hyphal growth has been given much less attention than that on spore germination. The methods used, so far, report on production of fungal biomass, growth rate or colonizing ability (Hsu and Lockwood, 1971; De Boer *et al.*, 1998).

Almost all studies on soil fungistasis have been done using one method, i.e. either spore germination or hyphal extension. For a dune soil isolate of *Fusarium oxysporum*, both methods were compared during incubation of a dune soil that was subjected to a sterilization + soil inoculum treatment (Fig. 9.7; W. de Boer and P. Verheggen, unpublished results). Both methods indicated a rather constant level of fungistasis in the untreated soil and a quick (within a few weeks) return to this level in the re-inoculated sterile soil.

Differences in intensity of fungistasis between soils can remain obscured when undiluted soils are used. To overcome this, dilution series of soil in sterilized quartz sand have been used (Wacker and Lockwood, 1991; Knudsen *et al.*, 1999).

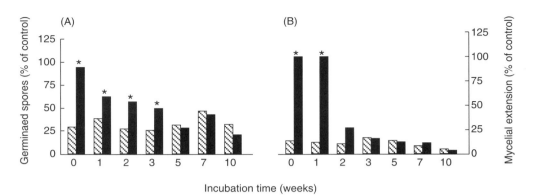

Fig. 9.7. (A) Spore germination and (B) mycelial extension of a dune isolate of the fungus *Fusarium oxysporum* in a dune soil that was incubated for 10 weeks after sterilization + soil inoculation (black bars). Results of similarly incubated untreated dune soil (hatched bars) are also given.* indicates significant differences (two sample *t*-test) between treated and untreated soil. Spore germination and mycelial extension were determined as described in the text.

Materials and apparatus

- Petri dishes (13.5 cm diameter)
- Acid-washed sand (or another nutrient-free, soil-like matrix)
- Glass wool
- Microcentrifuge
- Large membrane filters (polycarbonate; pore 0.2 µm; diameter 9 cm)
- Small membrane filters (polycarbonate or cellulose ester; pore 0.2 µm; diameter 2.5 cm)
- Glass slides
- Binocular microscope
- Epifluorescence microscope

Chemicals and solutions

- Calcofluor White M2R (Sigma Aldricht, St Louis, Missouri, USA) staining solution: 2.3 mg/ml in sterile water
- Potato dextrose agar (Oxoid: UNIPATH S.A., 6 rue de Paisy, 69570 Dardilly cedex, France)

Procedure for the spore germination test

For the sake of comparison, the procedure for isolation and storage of the spores should be standarized, as it has been shown that their germination ability is affected by age, storage conditions, medium composition and strain characteristics (Garrett, 1970; Mondal and Hyakumachi, 1998). Since growth requirements and sporulation conditions differ among fungal species, it is not possible to give one general applicable protocol. Most often, fungi are grown for 2–3 weeks on a solid medium such as potato dextrose agar or oatmeal agar. The spores are isolated by applying a small amount of sterile water or salt solution and gently rubbing the surface with a bent glass rod. The spore suspension is then filtered through sterile glass wool to remove clumps of spores and mycelial fragments. Further cleaning of spores is done by centrifugation. The density of the spores is determined microscopically and adjusted, if necessary. Spore suspensions are used immediately or stored at –80°C in an aqueous solution of glycerol (Liebman and Epstein, 1992).

As indicated above, several methods can be used for the germination test. Here we propose to inoculate the spores on top of a filter, as this is easiest for subsequent microscopic inspection. The following protocol is given as an example:

1. Petri dishes are filled with 90 g moist (–10 kPa) soil, which is spread evenly to obtain a smooth surface (bulk density mineral soil about 1 g/cm^3).
2. A sterile, water-saturated polycarbonate membrane filter (9 cm diameter) is placed on top of the soil. Two sterile glass slides are put on top of the membrane filter (4 cm apart) to keep it in close contact with the soil. The Petri dish is sealed with Parafilm and incubated for 48 h at 20°C to allow diffusion of soil solutes into the membrane filter.
3. Small membrane filters containing the spores (added by vacuum filtration) are placed in the area between the slide glasses. The Petri dish is sealed again and incubated for 16–48 h (depending on fungal species).
4. Spores are stained by floating the membranes for 2 min in a solution of Calcofluor White M2R (Sigma) in demineralized water (2.3 mg/ml). After destaining with demineralized water, at least 100 spores/filter are checked for germination using an epifluorescence microscope.
5. As control for germination ability, soil is replaced by nutrient-free, acid-washed sand.

Procedure for the hyphal extension test

The method described here tests the ability of fungal hyphae to invade soils from nutrient-rich agar (De Boer *et al.*, 1998, 2003).

1. Petri dishes (8.5 cm diameter) are filled with 50 g moist (–10 kPa) soil, which is spread evenly to obtain a smooth surface (bulk density mineral soil about 1 g/cm^3).
2. Potato dextrose agar discs (1 cm diameter; 0.4 cm thick) from the growing margin of the fungal colony are inverted and placed centrally on top of the soil in a Petri dish.

3. After 3 weeks of incubation at 20°C, the extension of the mycelium is determined using a binocular microscope, and the area of hyphal extension is estimated (Fig. 9.8). Image analysis can be used to obtain a more accurate estimate of the mycelial area.

4. The hyphal extension on the soil samples is compared with that on nutrient-free, acid-washed beach sand (−10 kPa).

As a modification, the agar disc containing fungal inoculum may be separated from soil by inert material (e.g. stainless-steel discs) to prevent stimulation of growth of (antagonistic) soil microorganisms by agar.

Calculations

For both tests (germinated spores or invaded area) fungistasis is expressed as the percentage of the control (acid-washed sand).

Discussion

A simple estimate of intensity of soil fungistasis is obtained by measuring the difference between hyphal extension or spore germination on a nutrient-free, soil-like control and on the soil under study. In fact, the nutrient availability in the control should be the same as in the soil, but this is very hard to realize. Therefore, given the fact that most soils are carbon-limited for microbial growth, a nutrient-free control seems a reasonable option. In addition, effect of soil sterilization on growth and germination may be determined to differentiate between biotic and abiotic causes of fungistatis (Dobbs and Gash, 1965; De Boer *et al.*, 1998, 2003). Yet, care should be taken that the sterilization procedure does not introduce undesired (inhibiting) side-effects.

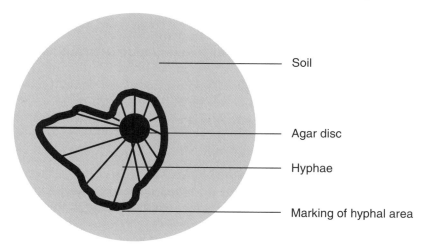

Fig. 9.8. Determination of the area of hyphal extension.

For comparison of fungistastic properties of a wide range of soils, it would be best to use a fixed set of type fungal species/strains. However, from an ecological or phytopathological point of view this has little meaning, as subspecies or strains can be adapted to specific climatic and environmental conditions. Hence, the specific questions to be answered determine what will be the most appropriate fungal strains to use for fungistasis measurement.

It is important to realize that soil fungistasis is dynamic. It can be decreased strongly by enhanced carbon availability, e.g. caused by roots. This explains why even highly fungistatic soils do not necessarily prevent pathogenic fungi from causing disease. Hence, measurements of fungistasis must be completed by bioassays allowing the assessment of receptivity and inoculum potential.

9.5 Free-living Plant-beneficial Microorganisms and Soil Quality

YVAN MOËNNE-LOCCOZ,[1] SHERIDAN L. WOO,[2] YAACOV OKON,[3] RENÉ BALLY,[1] MATTEO LORITO,[2] PHILIPPE LEMANCEAU[4] AND ANTON HARTMANN[5]

[1] UMR CNRS 5557 Ecologie Microbienne, Université Claude Bernard (Lyon 1), 43 bd du 11 Novembre, 69622 Villeurbanne cedex, France; [2] Universita degli Studi di Napoli 'Federico II', Dip. di Arboricoltura, Botanica e Patologia Vegetale, Sez. Patologia Vegetale, Lab. di Lotta Biologica, Portici (NA), Italy; [3] Hebrew University of Jerusalem (HUJI), Department of Plant Pathology and Microbiology, Faculty of Agricultural, Food and Environmental Quality Sciences, Rehovot, Israel; [4] UMR 1229 INRA/Université de Bourgogne 'Microbiologie et Géochimie des Sols', INRA-CMSE, BP 86510, 21065 Dijon cedex, France; [5] GSF-National Research Center for Environment and Health, Department of Rhizosphere Biology, Ingolstädter Landstraße 1, 85764 Neuherberg/München, Germany

Introduction

Many microorganisms living in the rhizosphere and benefiting from root exudates can have positive effects on plant growth and health. The relationship between these plant-beneficial microorganisms and the plant host corresponds to a symbiosis or an associative symbiosis (cooperation). The first case often involves differentiation of one partner or both, which facilitates identification of such symbiotic interactions; in this handbook, symbiotic interactions with the plant are covered in the sections dealing with nodulating, nitrogen-fixing bacteria (Section 9.2) and mycorrhizal fungi (Section 9.3). The second case corresponds to microorganisms designated plant-growth-promoting rhizobacteria (PGPR) and plant-growth-promoting fungi (PGPF), and these are the focus of this section.

PGPR and PGPF can exert positive effects on plants by various mechanisms, some of them implying a directly positive effect on seed germination, root development, mineral nutrition and/or water utilization (i.e. phytostimulation) (Jacoud *et al.*, 1998, 1999; Dobbelaere *et al.*, 2001). Indirect effects can also take place, and typically involve suppression of phytopathogenic bacteria or fungi, and/or phytoparasitic nematodes (i.e. biological control) (Cronin *et al.*, 1997; Walsh *et al.*, 2001; Burdman *et al.*, 2002). In certain cases, the biocontrol effect mediated by indigenous free-living plant-beneficial microorganisms results in the suppression of plant disease (Moënne-Loccoz and Défago, 2004), which is an emerging ecosystemic property (see Section 9.4). In most cases, however, the plant-beneficial effects of indigenous free-living microorganisms remain unnoticed because of the absence of visible differentiation (in contrast to root nodules in the nitrogen-fixing symbiosis, for example), and the fact that these effects add up with a multitude of other effects related to variability in space of plant genotypes, soil composition, farming practices, microclimatic conditions and the composition of the soil biota. This is particularly true for phytostimulatory effects due to indigenous free-living plant-beneficial microorganisms, because they do not necessarily lead to differences in plant health. However, the largely unnoticed contribution of indigenous free-living plant-beneficial microorganisms to plant growth and health is important. Furthermore, since chemical inputs into farming can have deleterious effects on environmental health and food product quality, effective PGPR and PGPF inoculants have the potential to be used as a replacement for, or in combination with reduced rates of, chemical fertilizers (phytostimulation) and pesticides (biocontrol).

Many taxa of free-living plant-beneficial microorganisms are known, for both bacteria and fungi. Some of them, e.g. fluorescent pseudomonads and non-pathogenic *Fusarium oxysporum*, have been discussed in Section 9.4, due to their contribution to disease suppressiveness of certain soils. In this section, the emphasis will be placed on free-living plant-beneficial microorganisms belonging to the bacterial genus *Azospirillum* and the fungal genera *Trichoderma* and *Gliocladium*.

Azospirillum (group 1 of the α-proteobacteria) are PGPR found in close association with the roots of plants, particularly *Gramineae* (Tarrand *et al.*, 1978; Bally *et al.*, 1983), and they mainly colonize the root elongation zone. They exert beneficial effects on plant growth and yield of many crops of agronomic importance (Okon and Labandera-Gonzalez, 1994), and represent one of the best-characterized PGPR (Okon, 1994). During the early days of the investigation on *Azospirillum*–plant associations, plant growth promotion was thought to derive from the contribution of biological N_2 fixation by the bacterial partner. However, further studies demonstrated that the positive effects of *Azospirillum* are due mainly to morphological and physiological changes of the roots of inoculated plants, which lead to an enhancement of water and mineral uptake, especially when plants grow in suboptimal conditions (Dobbelaere *et al.*, 2001). Indeed, inoculation with *Azospirillum* increases the density and length of root hairs, as well as the appearance and elongation rates of lateral roots, thus increasing the root surface area. These effects are linked to the secretion of plant growth hormones such as auxins, gibberellins and cytokinins by the bacterium (Dobbelaere *et al.*, 2001), a property shared with a variety of other plant-beneficial root-colonizing bacteria. In addition to phytostimulation, certain *Azospirillum* strains display biocontrol properties towards phytopathogenic bacteria (Bashan and de-Bashan, 2002) or parasitic plants (Bouillant *et al.*, 1997; Miché *et al.*, 2000).

Trichoderma and *Gliocladium* are fast-growing, spore-producing fungi, commonly found in soil throughout the world (Klein and Eveleigh, 1998). They are resistant to many xenobiotic compounds, and can catabolize a wide range of natural and synthetic organic compounds (including complex polymers), thus having an effect on the nitrogen and carbon cycles (Danielson and Davey, 1973b; Kubicek-Pranz, 1998). Most importantly, biocontrol interactions exist with several phytopathogens, such as *Rhizoctonia*, *Pythium*, *Sclerotinia*, *Fusarium*, *Verticillium*, *Phytophthora*, *Phomopsis*, *Gaeumannomyces* and *Sclerotium* (Weindling, 1932; Jeffries and Young, 1994). The biocontrol mechanisms implicated are diverse. First, *Trichoderma* and *Gliocladium* species are relatively unspecialized, disruptive or necrotrophic mycoparasites, and may also be parasitic towards nematodes. Constitutive secretion of cell-wall-degrading enzymes (CWDEs), such as chitinases and cellulase, plays an important role (Lorito *et al.*, 1996; Lorito, 1998; Zeilinger *et al.*, 1999). The subsequent release of cell-wall degradation products enables chemotactical location of the host (Zeilinger *et al.*, 1999). Physical contact with the phytopathogen triggers coiling, attachment and host penetration by *Trichoderma* (Inbar and Chet, 1992, 1995). Secondly, *Trichoderma* and *Gliocladium* species produce antibiotics, which can affect bacteria and/or pathogenic fungi (Howell, 1998; Kubicek *et al.*, 2001). Thirdly, *Trichoderma* and *Gliocladium* can compete with phytopathogenic species. Although competition is less important than other biocontrol mechanisms, it is a prerequisite for efficient plant colonization (Lo *et al.*, 1996; Harman and Bjorkman, 1998). Fourthly, *Trichoderma* and *Gliocladium* may have direct effects on the plant, e.g. via solubilization of inorganic nutrients or induced resistance in the plant (Windham *et al.*, 1986; Harman, 2000), resulting in better seed germination, enhanced plant growth and development, and increased yield (Lindsey and Baker, 1967; Windham *et al.*, 1986; Harman 2000). The population levels at which one can expect beneficial effects from these indigenous biocontrol fungi are not known, and they probably depend on the phytopathogen, since biocontrol effects are host-dependent. However, there is a link to soil quality, in that environmental conditions that stimulate growth and subsequent colonization and sporulation of biocontrol fungi (e.g. high nutrient availability) will also reduce pathogen populations in the soil. *Trichoderma* and/or *Gliocladium* are used in many commercial inoculants worldwide. Websites that can be consulted for information include http://www.agrobiologicals.com, http://www.oardc.ohio-state.edu/apsbcc/productlist.htm, http://attra.ncat.org/attra-pub/orgfert.html and http://www.epa.gov/pesticides/biopesticides.

In this section, methods available to assess indigenous populations of free-living plant-beneficial microorganisms (*Azospirillum*, *Trichoderma* and *Gliocladium*) are described. Since these microorganisms may be present in different microbial habitats in the soil ecosystem (from bulk soil to root tissues), a strategy to separate the relevant microhabitats/compartments and validated in the case of bacteria is presented (pp. 273–274). The resulting samples can be processed via cultivation-based methods (pp. 274–278 and 281–284), which can only recover a minority of individuals (the culturable ones), but are needed to obtain strains to be used as inoculants. Alternatively, methods for cultivation-independent analysis of free-living plant-beneficial bacteria are described (pp. 291–281).

Extraction of Indigenous Free-living Plant-beneficial Bacteria from Soil and Roots

Introduction

Desorption of bacteria from soil and roots is important for qualitative and quantitative monitoring, regardless of whether indigenous bacteria or inoculants are considered. A standardized protocol for the separation and differentiation of different rhizosphere compartments (e.g. rhizosphere soil/rhizoplane versus root tissues) and the extraction of bacterial cells adsorbed to the root surface has been described in the case of *Medicago sativa* cv. Europae (Mogge *et al.*, 2000; Hartmann *et al.*, 2004). The procedure followed most recommendations made by Macdonald (1986) and Herron and Wellington (1990) and is presented hereafter. Fluorescent *in situ* hybridization (FISH) in combination with confocal scanning laser microscopy (CSLM) is useful to confirm successful desorption of bacteria in root surface studies (see pp. 279–281).

Principle of the method

The method consists of separating roots from soil by physical means, followed by the maceration of root tissues to free endophytic microorganisms. Microbial cells are then extracted from the samples.

Materials and apparatus

- Sterile tweezers
- Stomacher 80 (Seward Medical, Thetford, UK)
- Laboratory glassware (Erlenmeyer flasks, etc.)
- Gauze (40 µm mesh size)
- 5-µm syringe filter (Sartorius No. 17549, Göttingen, Germany)

Chemicals and solutions

- 0.01 M phosphate buffer (Na_2HPO_4/KH_2PO_4, pH 7.4)
- 0.1% sodium cholate buffer
- Polyethylene glycol 6000 (PEG 6000; Sigma, Deisenhofen, Germany)
- Cation exchange polystyrene beads (Chelex 100; Sigma)

Procedure

1. Roots are carefully separated from the soil using sterile tweezers. All steps are conducted with sterile solutions on ice.
2. Non-rhizosphere soil (bulk soil compartment) and root-attached soil particles collected by shaking the roots (rhizosphere compartment) are each suspended in 0.01 M phosphate buffer in a 1:9 (w/v) ratio and dispersed for 1 min at highest speed in a Stomacher 80.
3. To extract rhizoplane and endophytic bacteria (root compartment), 1 g of fresh roots previously cleaned from adhering soil particles (see above) and washed in phosphate buffer are suspended in 20 ml of 0.1% sodium cholate buffer (Macdonald, 1986). The suspension is treated in a Stomacher 80 at highest speed for 4 min to disrupt polymers.
4. After transfer into an Erlenmeyer flask, 0.5 g of PEG 6000 and 0.4 g of cation exchange polystyrene beads (Chelex 100) are added and the suspension is stirred for 1 h at 50 rpm and 4°C.
5. The Stomacher/stirring procedure is repeated three times, the roots being transferred to fresh 0.1% sodium cholate buffer with PEG 6000 and Chelex 100 after each extraction step. The suspensions obtained after each step are pooled.
6. Root and soil particles are removed by filtration through a gauze (40 μm mesh size) and subsequently a 5-μm syringe filter. The resulting sample is used to study bacteria from the root compartment.

Discussion

The suspensions thus obtained from bulk soil, the rhizosphere and the root compartment (i.e. rhizoplane and root tissues) can be used for analysis of indigenous free-living plant-beneficial bacteria (see below) as well as monitoring of bacterial inoculants.

Cultivation Approach to Enumerate Indigenous *Azospirillum* spp.

Introduction

The presence of *Azospirillum* spp. in the rhizosphere can be shown by enrichment and cultivation on semi-solid nitrogen-free media, and this approach will also target other root-associated nitrogen-fixing bacteria (Döbereiner, 1995). Table 9.4 summarizes the different media suitable for the enrichment of *Azospirillum* spp. on the basis of different pH requirements and carbon substrate preferences. *Azospirillum* spp. differ in their physiology, as summarized in Table 9.5, which can be used to identify the isolates using physiological criteria. In addition, routine API® and Biolog™ test systems are used for convenient physiological identification purposes. Here, protocols based on 16S rDNA hybridization to identify *Azospirillum* species are presented.

Table 9.4. Four media used for the isolation and cultivation of *Azospirillum* spp.

Ingredients (per litre)	NFb[a,b]	LGI[a,b]	Modified NFb	Potato agar[c]
DL-Malic acid	5 g	–	5 g	2.5 g
Sucrose	–	5 g	–	2.5 g
K_2HPO_4	0.5 g	0.2 g	0.13 g	–
KH_2PO_4	–	0.6 g	–	–
$MgSO_4.7H_2O$	0.2 g	0.2 g	0.25 g	–
NaCl	0.1 g	–	1.2 g	–
$CaCl_2.2H_2O$	0.02 g	0.02 g	0.25 g	–
$Na_2MoO_4.2H_2O$	–	0.002 g	–	–
Na_2SO_4	–	–	2.4 g	–
$NaHCO_3$	–	–	0.22 g	–
Na_2CO_3	–	–	0.09 g	–
K_2SO_4	–	–	0.17 g	–
Minor element solution[d]	2 ml	–	2 ml	2 ml
Bromothymol blue solution[e]	2 ml	2 ml	–	–
Fe-EDTA, 1.64%	4 ml	4 ml	4 ml	–
pH-value (adjusted with KOH)	6.8	6.0	8.5	6.8
Vitamin solution[f]	1 ml	1 ml	1 ml	1 ml
Agar	1.75 g	1.75 g	1.75 g	15 g

[a] Ingredients should be added to the medium in the stated order; [b] Semi-solid medium; [c] 200 g fresh potatoes are peeled, cooked for 30 min and filtered through cotton before other ingredients are added; [d] $CuSO_4.5H_2O$, 0.4 g; $ZnSO_4.7H_2O$, 0.12 g; H_3BO_3, 1.4 g; $Na_2MoO_4.2H_2O$, 1 g; $MnSO_4.H_2O$, 1.5 g; H_2O, 1000 ml; [e] 0.5% bromothymol blue in 0.2 N KOH; [f] Biotin, 10 mg; Pyridoxol/HCl, 20 mg; H_2O, 100 ml.

Principle of the methods

Colony hybridization with 16S rDNA-targeted oligonucleotides (Kabir et al., 1994, 1995; Chotte et al., 2002) can be used to identify *Azospirillum* species, and the procedure is usually carried out on nitrogen-fixing isolates able to fix the dye Congo Red, because this is a typical *Azospirillum* attribute (Rodriguez Caceres, 1982). Species-specific probes include Al (for *A. lipoferum*, 5'-CGTCGGATTAGGTAGT-3'), used at a hybridization/washing temperature of 43°C, and Aba (for both *A. brasilense* and *A. amazonense*, 5'-CGTCCGATTAGGTAGT-3'), used at a hybridization/washing temperature of 51°C (Chotte et al., 2002).

Materials and apparatus

- Sterile membranes (GeneScreen Plus; Nen Life Science Products, Boston, Massachusetts, USA)
- 3MM paper
- UV table
- Hybridization tubes and oven (65°C)
- Hyper film TM-MP (Amersham Labs, Amersham, UK)
- Autoradiography set-up

Table 9.5. Main physiological characteristics of *Azospirillum* spp.

	A. doebereinerae	*A. lipoferum*	*A. largimobile*	*A. brasilense*	*A. amazonense*	*A. irakense*	*A. halopraeferens*
Carbon utilization test (API)							
N-Acetylglucosamine	−[a]	+	+	−	d	+	ND
D-Glucose	d	+	+	d	+	+	−
Glycerol	+[b]	+	+	+	−	−	+
D-Mannitol	+	+	+	−	−	−	+
D-Ribose	−	+	+	−	+	d	+
D-Sorbitol	+	+	+	−	−	−	−
Sucrose	−	−	ND[d]	−	+	+	−
Acid formation (API 50 anaerobe)							
From glucose	d[c]	+	+	−	−	−	−
From fructose	+	+	+	−	−	−	+
Miscellaneous							
Biotin requirement	−	+	−	−	−	−	+
Optimum growth temperature	30°C	37°C	28°C	37°C	35°C	33°C	41°C
Optimum pH for growth	6.0–7.0	5.7–6.8	ND	6.0–7.8	5.7–6.5	5.5–8.5	6.8–8.0
Occurrence of pleomorphic cells	+	+	+	−	+	+	+

[a] sign (−) means less than 10% of the investigated strains showed a positive response;
[b] sign (+) means more than 90% of the investigated strains showed a positive response;
[c] d (depends) means between 11% and 89% of the investigated strains showed a positive response; [d] Not determined.

For the alternative hybridization protocol:

- Glass slides
- Epifluorescence microscope

Chemicals and solutions

- Tryptone-yeast extract (TY; per litre: tryptone, 5 g; yeast extract, 3 g; $CaCl_2.H_2O$, 0.5 g)
- 10% (w/v) sodium dodecyl sulphate (SDS)
- Denaturing solution (NaOH 0.5 M and NaCl 1.5 M)
- Neutralizing solution (NaCl 1.5 M, Tris 1 M, pH 7.4)
- 2× standard saline citrate (SSC; NaCl 0.1 M, sodium citrate 15 mM, pH 7.0)
- Pre-hybridization solution (i.e. 16.5 ml sterile distilled water, 3 ml dextran 50%, 1.5 ml SDS 10%, 0.58 g NaCl)

- ^{32}P
- Denatured herring sperm DNA

For the alternative hybridization protocol:

- Phosphate-buffered saline (PBS; 0.13 M NaCl, 7 mM Na_2HPO_4 and 3 mM NaH_2PO_4 (pH 7.2))
- Paraformaldehyde
- Agarose
- Ethanol
- FISH hybridization buffer (20 mM Tris-hydrochloride, pH 7.2, 0.01% SDS and 5 mM EDTA)
- NaCl
- Formamide
- DAPI
- Citifluor AF1 (Citifluor Ltd, London, UK)

Procedure

1. The bacteria are grown for 48 h at 28°C on sterile membranes (GeneScreen Plus) previously placed on TY plates. Two membranes are used for each probe.
2. The membranes are then treated successively for 6 min in 10% (w/v) SDS, 10 min in denaturing solution, 9 min in neutralizing solution and 9 min in 2× standard saline citrate to promote cell lysis.
3. The membranes are left drying for 1 h on 3MM paper, followed by a 4-min UV treatment for DNA binding.
4. The hybridization procedure consists of moistening the membranes in 2× SSC, rolling and transferring them into hybridization tubes containing pre-hybridization solution.
5. Each probe is labelled at the 5' end with ^{32}P (as described by Sambrook et al., 1989).
6. The hybridization tubes are then put for 2 h in an oven (65°C) before adding the labelled probe and 600 µl of denatured herring sperm DNA.
7. After hybridization for 12 h, the membranes are rinsed at room temperature for 5 min (twice) using 2× SSC, 30 min (twice) using 2× SSC and 1% SDS, and 30 min (three times) using 0.1× SSC.
8. The membranes are left to dry for 1 h at room temperature on 3MM paper sheets and are then packed with Hyper film TM-MP for at least 12 h at −80°C for autoradiography.

An alternative hybridization protocol involves:

1. Fixing bacteria overnight at 4°C in PBS containing 3% paraformaldehyde.
2. Then they are washed in PBS, mixed with 0.3% agarose, dropped on to glass slides and dried at room temperature.

3. These glass slides are immersed successively in 50%, 80% and 96% ethanol for 3 min each and stored at room temperature.
4. Oligonucleotide probes (Table 9.6) labelled with Cy3, Cy5 or 5(6)-carboxyfluorescein-N-hydroxysuccinimide ester (FLUOS) at the 5' end are used.
5. The oligonucleotides are stored in distilled water at a concentration of 50 ng/µl (Amann *et al.*, 1990).
6. FISH is performed (Wagner *et al.*, 1993) at 46°C for 90 min in FISH hybridization buffer containing 0.9 M NaCl and formamide at the percentages shown in Table 9.6.
7. Hybridization is followed by a stringent washing step at 48°C for 15 min.
8. The washing buffer is removed by rinsing the slides with distilled water.
9. Counterstaining with DAPI and mounting in Citifluor AF1 is performed as described previously (Aßmus *et al.*, 1995).
10. Observations are made by epifluorescence microscopy.

Table 9.6. 16S rRNA-targeted oligonucleotide probes for FISH analysis of the *Azospirillum* cluster (Stoffels *et al.*, 2001).

Probes and competitors	Sequence (5'–3')	Stringency[a]	Specificity
AZO440a +	GTCATCATCGTCGCGTGC	50	*Azospirillum* spp.
AZO440b	GTCATCATCGTCGTGTGC	50	*Conglomeromonas* spp., *Rhodocista* spp.
AZOI665	CACCATCCTCTCCGGAAC	50	*Azospirillum* species cluster[b]
Abras1420	CCACCTTCGGGTAAAGCCA	40	*A. brasilense*
Alila1113	ATGGCAACTGACGGTAGG	35	*A. lipoferum*, *A. largimobile*
Adoeb587	ACTTCCGACTAAACAGGC	30	*A. doebereinerae*
Ahalo1115	ATGGTGGCAACTGGCAGCA	45	*A. halopraeferens*
Aama1250	CACGAGGTCGCTGCCCAC	50	*A. amazonense*
Airak985	TCAAGGCATGCAAGGGTT	35	*A. irakense*
Rhodo654	ACCCACCTCTCCGGACCT	65	*Rhodocista centenaria*
Sparo84	CGTGCGCCACTAGGGGCG	20	*Skermanella parooensis*
Abras1420C	CACCTTCGGGTAAAACCA	40	Competitor[c]
Alila113C	ATGGCAACTGGCGGTAGG	35	Competitor
Ahalo1115C	ATGATGGCAACTGGCAGTA	45	Competitor

[a] Amount of formamide (%, v/v) in hybridization buffer; [b] *lipoferum*, *brasilense*, *halopraeferens*, *doebereinerae*, and *largimobile*; [c] the competitor oligonucleotide (without fluorescent label) is used in the FISH analysis to prevent false-positive hybridizations, which could be possible in rare cases of indigenous bacteria harbouring very close oligonucleotide similarity according to the sequence analysis.

Cultivation-independent Approach to Monitor Indigenous *Azospirillum* spp. in Extracted Root Compartments or in the Rhizosphere

Introduction

Culture-dependent techniques may not always enable recovery of all targeted bacteria, even when considering a particular strain belonging to an easily culturable taxon (Défago *et al.*, 1997; Mascher *et al.*, 2003). Even the efficacy of the PCR-based method may depend on the physiological status of the cells (Rezzonico *et al.*, 2003). Here, a FISH method enabling detailed localization of cells and an *in situ* approach is presented.

Principle of the methods

Concomitant staining with the general DNA stain DAPI and FISH enables counting total and hybridizing bacteria in the three compartments outlined in Fig. 9.9, after collection on polycarbonate filters. The main advantage of FISH analysis is that the bacteria can also be identified and localized directly in the rhizosphere, provided that they are present in a physiologically active state (i.e. harbouring a high ribosome content).

Materials and apparatus

- 0.2 µm polycarbonate filters
- Zeiss Axiophot 2 epifluorescence microscope (Zeiss, Jena, Germany)
- Filter sets F31–000, F41–001 and F41–007 (Chroma Tech. Corp., Battleboro, Vermont, USA)

Chemicals and solutions

For analysis of cell suspensions:

- Formaldehyde
- Ethanol
- Citifluor AF1 (Citifluor Ltd, London, UK)
- DAPI
- Species-specific DNA probes (see Table 9.6)

For analysis *in situ*:

- PBS (0.13 M NaCl, 7 mM Na_2HPO_4 and 3 mM NaH_2PO_4 (pH 7.2))
- Formaldehyde
- A confocal laser scanning microscope, e.g. microscopes LSM 410 or LSM 510 (Zeiss, Jena, Germany).

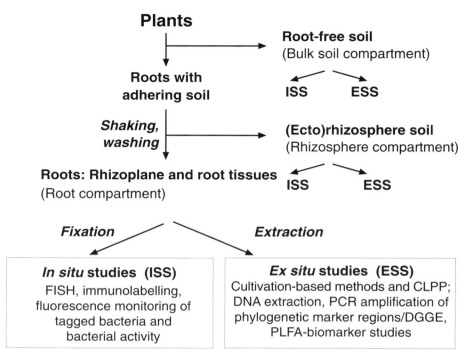

Fig. 9.9. Strategy to separate different microhabitats/compartments in soil–plant systems for detailed analysis of indigenous free-living plant-beneficial microorganisms and monitoring of microbial inoculants by *in situ* studies (ISS) and *ex situ* studies (ESS).

Procedure

Concomitant staining with DAPI and FISH is done as follows.

1. The cell suspensions (for extraction details, see pp. 273–274) are fixed overnight at 4°C with 3% formaldehyde and concentrated on to 0.2 µm polycarbonate filters (100 µl aliquots).
2. Dehydration of cells is performed successively with 50%, 80% and 96% ethanol for 3 min each.
3. The slides are mounted with Citifluor AF1 to reduce photobleaching.
4. A Zeiss Axiophot 2 epifluorescence microscope equipped with filter sets F31–000, F41–001 and F41–007 can be used for the enumeration of bacteria on filters.
5. Total cell counts (DAPI) and hybridizing bacteria using a set of domain-specific to species-specific probes (see Table 9.6) are determined by evaluating at least ten microscopic fields with 20–100 cells per field.

When FISH is used for *in situ* assessment, root samples are fixed overnight at 4°C in PBS containing 3% paraformaldehyde, washed in PBS and treated as described above. For *in situ* identification of *Azospirillum* on the root surface, the autofluorescence problem (Hartmann *et al.*, 1998b) makes it necessary to use a confocal laser scanning microscope, such as the inverted Zeiss microscopes LSM 410 or LSM 510, which are equipped with lasers (Ar-ion UV; Ar-ion visible; HeNe) supplying excitation wavelengths at 365 nm, 488 nm, 543 nm and 633 nm. Usually, a general cell DNA staining with DAPI is combined with FISH using probes specific for the domain bacteria, group-specific probes (Amann *et al.*, 1995), and genus- or species-specific probes.

Calculation

For *in situ* assessment, sequentially recorded images are assigned to the respective fluorescence colour and then merged to obtain a true colour display. All image combining and processing is performed with the standard software provided by Zeiss.

Discussion

These analyses may be completed as follows. When cell suspensions are used, PCR amplification of the 16S rDNA from the samples and subsequent electrophoretic fingerprinting of the amplification products or clone bank analysis can be performed (Weidner *et al.*, 1996; Muyzer and Smalla, 1998). Structural and functional microbial diversity aspects can also be assessed using community-level fatty acid analysis (Zelles, 1997; White and Ringelberg, 1998) and physiological profiling (Garland *et al.*, 1997).

When the assessment is performed *in situ*, other specific oligonucleotide probes available for a number of root-associated and symbiotic bacteria (Kirchhof *et al.*, 1997; Ludwig *et al.*, 1998; Hartmann *et al.*, 2000) can also be used, which enables extension of the characterization of rhizosphere bacteria to other root-associated bacteria of interest.

Cultivation Approach to Monitoring Indigenous *Trichoderma* and *Gliocladium* Species

Introduction

A range of different methods have been assessed for monitoring of *Trichoderma* and *Gliocladium* species, including direct PCR (Lieckfeldt *et al.*, 1998) and methods based on the use of fluorogenic substrates (Miller *et al.*, 1998) and specific monoclonal antibodies (Thornton *et al.*, 2002). Hereafter, only cultivation methods will be presented. *Trichoderma* and *Gliocladium* species are ubiquitous in soil, and they can be readily isolated due to their rapid growth and profuse sporulation. Samples may be obtained from agricultural soils growing the crop of interest for a biocontrol application, or from soils naturally suppressive to the target pathogen. Particular soil microhabitats can be targeted (see pp. 273–274).

Principle of the methods

The fungi are isolated by homogenizing soil in water, diluting the soil suspension and plating on to solid media containing various compounds to selectively suppress growth of bacteria, oomycetes, mucorales and other fungi. Many different media have been formulated for selective isolation of *Trichoderma* and *Gliocladium* (Davet, 1979; Elad *et al.*, 1981; Johnson *et al.*, 1987; Park *et al.*, 1992; Askew and Laing, 1993), and examples of commonly used media are provided below. Identification of fungal colonies is performed by microscopic analysis of key morphological properties.

Materials and apparatus

- Analytical balance
- Magnetic stirrer and stir bars
- pH meter
- Blender
- Autoclave
- 50°C water bath
- Vortexer
- Laminar flow hood
- Bunsen burner
- Incubator or growth chamber with a constant temperature of 25°C and light
- 180°C oven
- Light microscope
- Glassware: Erlenmeyer flasks (1 l) with autoclavable lids or aluminium foil, beakers (250 ml), 10–15 ml tubes in rack, graduated cylinder (1 l)
- Parafilm
- Pipettes (P10, P200, P1000) and sterile disposable tips
- Petri plates (90 mm) and plate spreader
- Spatulas
- Microscope slides and coverslips

Chemicals and solutions

Petri plates are prepared in advance:

- *Trichoderma* Selective Medium (TSM; Elad and Chet, 1983) is prepared from a 1 l solution containing 200 mg $MgSO_4.7H_2O$, 900 mg KH_2PO_4, 150 mg KCl, 1 g NH_4NO_3, 3 g glucose, 20 g agar. Autoclave and cool to 50°C, then add 250 mg chloramphenicol, 300 mg fenaminosulf, 200 mg pentachloronitrobenzene, 200 mg Rose Bengal, 20 mg Captan (50% wettable powder).

- Modified *Trichoderma* Selective Medium (Smith *et al.*, 1990) is prepared from a 1 l solution containing 260 mg KNO_3, 260 mg $MgSO_4.7H_2O$, 120 mg KH_2PO_4, 50 mg citric acid, 1 g $Ca(NO_3)_2$, 1 g $CaCl_2.2H_2O$, 2 g sucrose, 20 g agar, Igepal CA-630 (pH adjusted to 4.5). Autoclave and cool to 50°C, then add 50 mg chlortetracycline, 40 mg Captan (50% wettable powder), 2.5 mg Vinclozolin.
- Selective Basal Medium (Papavizas and Lumsden, 1982) is prepared by mixing 200 ml V-8, 800 ml water, 1 g glucose, 20 g agar. Autoclave and cool to 50°C, then add 500 µg sodium propionate, 100 µg neomycin sulphate, 100 µg Bacitran, 100 µg penicillin G, 100 µg chloroneb, 25 µg chlortetracycline, 20 µg nystatin, 2 ml alkylaryl polyether alcohol.

Procedure

1. Soil (10 g) is added to 100 ml distilled water and homogenized for 1 min.
2. A dilution series (0, 10^{-1}, 10^{-2}, 10^{-3}) is prepared by transferring 1 ml of the soil suspension into 9 ml distilled water and vortexing well before each pipetting. If the medium contains Igepal CA-630, the dilution series can be reduced by one.
3. Add 100 µl of each dilution per plate, spread the sample evenly over the surface with a sterile spreader and cover the plates (three plates per dilution).
4. The plates are incubated 5–7 days with light, at 25°C. In addition, particular temperature or pH conditions or the presence of chemical pesticides may be used as additional selective criteria for obtaining potential inoculant for, for example, post-harvest applications (Johnson *et al.*, 1987; Widden and Hsu, 1987; Faull, 1988).
5. Sporulating fungal colonies are isolated from the plates and identified. Identification of fungal isolates is confirmed microscopically by morphology examination using taxonomic keys such as those of Gams and Bissett (1998).
6. The *Trichoderma* or *Gliocladium* isolates can be purified. They are maintained on potato dextrose, oatmeal or cornmeal agar at 25°C or as a spore suspension prepared in 20% glycerol for storage at –20°C.

Calculation

Colony-forming units (CFU) are computed and expressed per g of soil.

Discussion

The CFUs obtained from direct soil plating are more likely indicative of the number of dormant spores than the amount of active mycelial biomass in the soil. In general, *Trichoderma* and *Gliocladium* species are relatively adaptable to diverse ecological habitats. However, the distribution of diverse species and isolates depends upon varying abiotic environmental factors, such as temperature, moisture availability and nutrients, as well as different biotic factors, such as crop type and the presence of other microorganisms (Hjeljord and Tronsmo, 1998). *Trichoderma* spp. were found to comprise up to 3% and 1.5% of the total quantity of fungal propagules obtained from a range of forest and pasture soils, respectively (Brewer *et al.*, 1971; Danielson and Davey, 1973a).

Conclusion

Populations of free-living plant-beneficial microorganisms are naturally present in soil ecosystems, colonizing plant roots and benefiting the plant. Therefore, they make a very significant contribution to soil quality, even if this contribution is often difficult to quantify. The only case where the effects of indigenous free-living plant-beneficial microorganisms can be 'visualized' corresponds to disease-suppressive soils, but even then these soils are not so easy to identify and therefore most of them remain unnoticed. Free-living plant-beneficial microorganisms can be found in a wide range of bacterial and fungal taxa, and their effect on the plant can involve many different modes of action. Indigenous free-living plant-beneficial microorganisms can be monitored in the soil ecosystem, and this has been illustrated in the case of the PGPR *Azospirillum*, and the PGPF *Trichoderma* and *Gliocladium*. Although culture-independent methods such as FISH are available to detect certain free-living plant-beneficial microorganisms, much remains to be done to develop these approaches (e.g. PCR methods) for detection and quantification of a wider range of free-living plant-beneficial bacteria and fungi.

Normative references for Chapter 9

Results from a nationally organized evaluation of the bioassays in several laboratories has led to an experimental national standardization of the germination bioassay in France (XP X 31–205–1).

References for Chapter 9

Abadie, C., Edel, V. and Alabouvette, C. (1998) Soil suppressiveness to fusarium wilt: influence of a cover-plant on density and diversity of *Fusarium* populations. *Soil Biology and Biochemistry* 30, 643–649.

Alabouvette, C., Couteaudier, Y. and Louvet, J. (1982) Comparaison de la réceptivité de différents sols et substrats de culture aux fusarioses vasculaires. *Agronomie* 2, 1–6.

Amann, R.I., Binder, B.J., Olson, R.J., Chisholm, S.W., Devereux, R. and Stahl, D.A. (1990) Combination of 16S rRNA-targeted oligonucleotide probes with flow cytometry for analyzing mixed microbial populations. *Applied and Environmental Microbiology* 56, 1919–1925.

Amann, R.I., Ludwig, W. and Schleifer, K.-H. (1995) Phylogenetic identification and *in* situ detection of individual microbial cells without cultivation. *FEMS Microbiology Reviews* 59, 43–169.

Amarger, N. (1980) Aspect microbiologique de la culture des légumineuses. *Le Sélectionneur Français* 28, 61–66.

Amarger, N. (2001) Rhizobia in the field. *Advances in Agronomy* 73, 109–168.

Amarger, N., Mariotti, A. and Mariotti, F. (1977) Essai d'estimation du taux d'azote fixé symbiotiquement chez le lupin par le traçage isotopique naturel (^{15}N). *Comptes-Rendus de l'Académie des Sciences série D* 284, 2179–2182.

Askew, D.J. and Laing, M.D. (1993) An adapted selective medium for the quantitative isolation of *Trichoderma* species. *Plant Pathology* 42, 686–690.

Aßmus, B., Hutzler, P., Kirchhof, G., Amann, R.I., Lawrence, J.R. and Hartmann, A. (1995) *In situ* localization of *Azospirillum brasilense* in the rhizosphere of wheat with fluorescently labeled, rRNA-targeted oligonucleotide probes and scanning confocal laser microscopy. *Applied and Environmental Microbiology* 61, 1013–1019.

Bais, H.P., Park, S.-W., Weir, T.L., Callaway, R.M. and Vivanco, J.M. (2004) How plants communicate using the underground information superhighway. *Trends in Plant Science* 9, 26–32.

Baker, R. (1968) Mechanisms of biological control of soilborne pathogens. *Annual Review of Phytopathology* 6, 263–294.

Bally, R., Thomas-Bauzon, D., Heulin, T., Balandreau, J., Richard, C. and de Ley, J. (1983) Determination of the most frequent N_2-fixing bacteria in the rice rhizosphere. *Canadian Journal of Microbiology* 29, 881–887.

Bardin, R., Domenach, A.M. and Chalamet, A. (1977) Rapports isotopiques naturels de l'azote. II. Application à la mesure de la fixation symbiotique de l'azote *in situ*. *Revue d'Ecologie et de Biologie des Sols* 14, 395–402.

Bashan, Y. and de-Bashan, L.E. (2002) Protection of tomato seedlings against infection by *Pseudomonas syringae* pv. tomato by using the plant growth-promoting bacterium *Azospirillum brasilense*. *Applied and Environmental Microbiology* 68, 2637–2643.

Bestel-Corre, G. (2002) La protéomique: un outil d'étude des interactions dans la rhizosphère-Identification de protéines reliées à la mycorhization et à la nodulation et évaluation de l'impact de boues d'épuration sur les symbioses. Doctoral thesis, University of Burgundy, France.

Bouhot, D. (1975a) Recherches sur l'écologie des champignons parasites dans le sol. V. Une technique sélective d'estimation du potentiel infectieux des sols, terreaux et substrats infestés par *Pythium* sp. Etudes qualitatives. *Annales de Phytopathologie* 7, 9–18.

Bouhot, D. (1975b) Recherches sur l'écologie des champignons parasites dans le sol. VII. Quantification de la technique d'estimation du potentiel infectieux des sols, terreaux et substrats infestés par *Pythium* sp. *Annales de Phytopathologie* 7, 147–154.

Bouhot, D. (1975c) Technique sélective et quantitative d'estimation du potentiel infectieux des sols, terreaux et substrats infestés par *Pythium* sp. Mode d'emploi. *Annales de Phytopathologie* 7, 155–158.

Bouhot, D. (1979) Estimation of inoculum density and inoculum potential: techniques and their values for disease prediction. In: Schippers, B. and Gams, W. (eds) *Soil-borne Plant Pathogens*. Academic Press, London, pp. 21–34.

Bouillant, M.L., Miché, L., Ouedraogo, O., Alexandre, G., Jacoud, C., Sallé, G. and Bally, R. (1997) Inhibition of *Striga* seed germination associated with sorghum growth promotion by soil bacteria. *Comptes-Rendus de l'Académie des Sciences* 320, 159–162.

Brewer, D., Calder, F.W., MacIntyre, T.M. and Taylor, A. (1971) Ovine ill-thrift in Nova Scotia: I. The possible regulation of the rumen flora in sheep by the fungal flora of permanent pasture. *Journal of Agriculture Science* 76, 465–477.

Brockwell, J., Diatloff, A., Roughley, R.J. and Date, R.A. (1982) Selection of rhizobia for inoculants. In: Vincent, J.M. (ed.) *Nitrogen Fixation in Legumes*. Academic Press, Sydney, pp. 173–191.

Bromfield, E.S.P., Barran, L.R. and Wheatcroft, R. (1995) Relative genetic structure of a population of *Rhizobium meliloti* isolated directly from soil and from nodules of lucerne (*Medicago sativa*) and sweet clover (*Melilotus alba*). *Molecular Ecology* 4, 183–188.

Burdman, S., Kadouri, D., Jurkevitch, E. and Okon, Y. (2002) Bacterial phytostimulators in the rhizosphere: from research to application. In: Bitton, G. (ed.) *Encyclopedia of Environmental Microbiology: Soil Microbiology* (Volume 1). John Wiley & Sons, New York, pp. 343–354.

Camporota, P. (1980) Recherches sur l'écologie des champignons parasites dans le sols. XV.-Choix d'une plante piège et caractérisation de souches de *Rhizoctonia solani* pour la mesure du potentiel infectieux des sols et substrats. *Annales de Phytopathologie* 12, 31–44.

Camporota, P. (1982) Recherches sur l'écologie des champignons parasites dans le sol. XVII: Mesure du potentiel infectieux de sols et substrats infestés par *Rhizoctonia solani* Kühn, agent de fontes de semis. *Agronomie* 2, 437–442.

Chabot, R., Antoun, H. and Cescas, M.P. (1996) Growth promotion of maize and lettuce by phosphate-solubilizing *Rhizobium leguminosarum* biovar *phaseoli*. *Plant and Soil* 184, 311–321.

Chen, Y.X., He, Y.F., Yang, Y., Yu, Y.L., Zheng, S.J., Tian, G.M., Luo, Y.M. and Wong, M.H. (2003) Effect of cadmium on nodulation and N_2-fixation of soybean in contaminated soils. *Chemosphere* 50, 781–787.

Chotte, J.-L., Schwartzmann, A., Bally, R. and Jocteur Monrozier, L. (2002) Changes in bacterial communities and *Azospirillum* diversity in soil fractions of a tropical soil under 3 or 19 years of natural fallow. *Soil Biology and Biochemistry* 34, 1083–1092.

Clays-Josserand, A., Lemanceau, P., Philippot, L. and Lensi, R. (1995) Influence of two plant species (flax and tomato) on the distribution of nitrogen dissimilative abilities within fluorescent *Pseudomonas* spp. *Applied and Environmental Microbiology* 61, 1745–1749.

Cook, R. and Baker, K.F. (1983) *The Nature and Practice of Biological Control of Plant Pathogens*. APS, St Paul, Minnesota.

Cordier, C., Pozo, M.J., Barea, J.M., Gianinazzi, S. and Gianinazzi-Pearson, V. (1998) Cell defense responses associated with localized and systemic resistance to *Phytophthora parasitica* induced in tomato by an arbuscular mycorrhizal fungus. *Molecular Plant Microbe Interactions* 11, 1017–1028.

Cronin, D., Moënne-Loccoz, Y., Dunne, C. and O'Gara, F. (1997) Inhibition of egg hatch of the potato cyst nematode *Globodera rostochiensis* by chitinase-producing bacteria. *European Journal of Plant Pathology* 103, 433–440.

Curl, E.A. and Truelove, B. (1986) *The Rhizosphere*. Advanced series in agricultural sciences, vol. 15. Springer-Verlag, Berlin.

Danielson, R.M. and Davey, C.B. (1973a) The abundance of *Trichoderma* propagules and distribution of species in forest soils. *Soil Biology and Biochemistry* 5, 485–494.

Danielson, R.M. and Davey, C.B. (1973b) Carbon and nitrogen nutrition of *Trichoderma*. *Soil Biology and Biochemistry* 5, 505–515.

Danso, S.K.A., Hardarson, G. and Zapata, F. (1993) Misconceptions and practical problems in the use of ^{15}N soil enrichment techniques for estimating N fixation. *Plant and Soil* 152, 25–52.

Davet, P. (1979) Technique pour l'analyse de populations de *Trichoderma* et de *Gliocladium virens* dans le sol. *Annales de Phytopathologie* 11, 529–534.

De Boer, W., Klein Gunnewiek, P.J.A. and Woldendorp, J.W. (1998) Suppression of hyphal growth of soil-borne fungi by dune soils from vigorous and declining stands of *Ammophila arenaria*. *New Phytologist* 138, 107–116.

De Boer, W., Verheggen, P., Klein Gunnewiek, P.J.A., Kowalchuk, G.A. and van Veen, J.A. (2003) Microbial community composition affects soil fungistasis. *Applied and Environmental Microbiology* 69, 835–844.

Défago, G., Keel, C. and Moënne-Loccoz, Y. (1997) Fate of released *Pseudomonas* bacteria in the soil profile: implications for the use of genetically-modified microbial inoculants. In: Zelikoff, J.T., Lynch, J.M. and Shepers, J. (eds) *EcoToxicology: Responses, Biomarkers and Risk Assessment*. SOS Publications, Fair Haven, New Jersey, pp. 403–418.

Dobbelaere, S., Croonenborghs, A., Thys, A., Ptacek, D., Vanderleyden, J., Dutto, P., Labandera-Gonzalez, C., Caballero-Mellado, J., Aguirre, J.F., Kapulnik, Y., Brener, S., Burdman, S., Kadouri, D., Sarig, S. and Okon, Y. (2001) Responses of agronomically important crops to inoculation with *Azospirillum*. *Australian Journal of Plant Physiology* 28, 871–879.

Dobbs, C.G. and Gash, M.J. (1965) Microbial and residual mycostasis in soils. *Nature* (London) 207, 1354–1356.

Döbereiner, J. (1995) Isolation and identification of aerobic nitrogen-fixing bacteria from soil and plants. In: Alef, K. and Nannipieri, P. (eds) *Methods in Applied Soil Microbiology and Biochemistry*. Academic Press, London, pp. 134–141.

Dodd, J.C., Gianinazzi-Pearson, V., Rosendahl, S. and Walker, C. (1994) European Bank of *Glomales* – an essential tool for efficient international and interdisciplinary collaboration. In: Gianinazzi, S. and Schüepp, H. (eds) *Impact of Arbuscular Mycorrhizas on Sustainable Agriculture and Natural Exosystems*. Birkhäuser Verlag, Basel, Switzerland, pp. 41–45.

Dommergues, Y.R., Duhoux, E. and Diem, H.G. (1999) In: Ganry, F. (ed.) *Les Arbres Fixateurs d'Azote. Caractéristiques Fondamentales et Rôle dans l'Aménagement des Ecosystèmes Méditerranéens et Tropicaux*. CIRAD/Editions Espaces 34/FAO/IRD, Montpellier, France, pp. 163–165.

Duffy, B., Schouten, A. and Raaijmakers, J.M. (2003) Pathogen self-defense: mechanisms to counteract microbial antagonism. *Annual Review of Phytopathology* 41, 501–538.

Dumas-Gaudot, E., Gollotte, A., Cordier, C., Gianinazzi, S. and Gianinazzi-Pearson V. (2000) Modulation of host defense systems. In: Kapulnik, Y. and Douds, D.D. Jr (eds) *Arbuscular Mycorrhizas: Physiology and Function*. Kluwer Academic, Dordrecht, The Netherlands, pp. 173–200.

Edel, V., Steinberg, C., Gautheron, N. and Alabouvette, C. (1997) Populations of nonpathogenic *Fusarium oxysporum* associated with roots of four plant species compared to soilborne populations. *Phytopathology* 87, 693–697.

Elad, Y. and Chet, I. (1983) Improved selective media for isolation of *Trichoderma* spp. or *Fusarium* spp. *Phytoparasitica* 11, 55–58.

Elad, Y., Chet, I. and Henis, Y. (1981) A selective medium for improving quantitative isolation of *Trichoderma* spp. from soil. *Phytoparasitica* 9, 59–68.

Faull, J.L. (1988) Competitive antagonism of soil-borne plant pathogens. In: Burge, M.N. (ed.) *Fungi in Biological Control Systems*. Manchester University Press, Manchester, UK, pp. 125–140.

Fettell, N.A., O'Connor, G.E., Carpenter, D.J., Evans, J., Bamforth, I., Oti-Boateng, C., Hebb, D.M. and Brockwell, J. (1997) Nodulation studies on legumes exotic to Australia: the influence of soil populations and inocula of *Rhizobium leguminosarum* bv. viciae on nodulation and nitrogen fixation by field peas. *Applied Soil Ecology* 5, 197–210.

Fisher, R.A. and Yates, F. (1963) *Statistical Tables for Biological, Agricultural and Medical Research*, 6th edn. Oliver and Boyd, Edinburgh, UK.

Fravel, D.R. (1988) Role of antibiosis in the biocontrol of plant diseases. *Annual Review of Phytopathology* 26, 75–91.

Frey, P., Frey-Klett, P., Garbaye, J., Berge, O. and Heulin, T. (1997) Metabolic and genotypic fingerprinting of fluorescent pseudomonads associated with the Douglas Fir–*Laccaria bicolor* mycorrhizosphere. *Applied and Environmental Microbiology* 63, 1852–1860.

Gams, W. and Bissett, J. (1998) Morphology and identification of *Trichoderma*. In: Kubicek, C.P. and Harman, G.E. (eds) *Trichoderma and Gliocladium: Basic Biology, Taxonomy and Genetics* (Volume 1). Taylor & Francis, London, pp. 3–34.

Garland, J.L., Cook, K.L., Loader, C.A. and Hungate, B.A. (1997) The influence of microbial community structure and function on community-level physiological profiles. In: Insam, H. and Rangger, A. (eds) *Microbial Communities: Functional versus Structural Approaches*. Springer, New York, pp. 171–183.

Garrett, S.D. (1970) *Pathogenic Root-Infecting Fungi*. Cambridge University Press, Cambridge, UK.

Gianinazzi, S. (1991) Vesicular-arbuscular endomycorrhizas: cellular, biochemical and genetic aspects. *Agriculture, Ecosystems and Environment* 35, 105–119.

Gianinazzi, S. and Gianinazzi-Pearson, V. (1988) Mycorrhizae: a plant's health insurance. *Chimica Oggi*, pp. 56–58.

Gianinazzi, S., Gianinazzi-Pearson, V. and Trouvelot, A. (1982) Mycorrhizae, an integral part of plants: biology and perspectives for their use. *Les colloques de l'INRA*, no. 13.

Gianinazzi, S., Gianinazzi-Pearson, V. and Trouvelot, A. (1989) Potentialities and procedures for the use of endomycorrhizas with special emphasis on high value crops. In: Whipps, J.M. and Lumsden, R.D. (eds) *Biotechnology of Fungi for Improving Plant Growth*, vol. 3. Cambridge University Press, Cambridge, UK, pp. 41–54.

Gianinazzi, S., Schüepp, H., Barea, J.M. and Haselwandter, K. (2002) *Mycorrhizal Technology in Agriculture: From Genes to Bioproducts*. Birkhäuser Verlag, Basel, Switzerland.

Gianinazzi-Pearson, V. and Smith, S.E. (1993) Physiology of mycorrhizal mycelia. In: Ingram, D.S., Williams, P.H. and Tommerup, I.C. (eds) *Advances in Plant Pathology*. Academic Press, London, pp. 55–82.

Gibson, A.H. (1963) Physical environment and symbiotic nitrogen fixation. I. The effect of root temperature on recently nodulated *Trifolium subterraneum* L. plants. *Australian Journal of Biological Sciences* 16, 28–42.

Gildon, A. and Tinker, P.B. (1983) Interactions of vesicular-arbuscular mycorrhizal infection and heavy metal in plants: I: The effect of heavy metals on the development of vesicular arbuscular mycorrhizas. *New Phytolologist* 95, 247–261.

Harley, J.L. and Harley, E.L. (1987) A check list of mycorrhiza in the British Flora. *New Phytologist* 105, 1–102.

Harman, G.E. (2000) Myths and dogmas of biocontrol – changes in perceptions derived from research on *Trichoderma harzianum* T-22. *Plant Disease* 84, 377–393.

Harman, G.E. and Bjorkman, T. (1998) Potential and existing uses of *Trichoderma* and *Gliocladium* for plant disease control and plant growth enhancement. In: Harman, G.E. and Kubicek, C.P. (eds) Trichoderma *and* Gliocladium*: Enzymes, Biological Control and Commercial Applications* (Volume 2). Taylor & Francis, London, pp. 229–265.

Hartmann, A., Giraud, J.J. and Catroux, G. (1998a) Genotypic diversity of *Sinorhizobium* (formerly *Rhizobium*) *meliloti* strains isolated directly from a soil and from nodules of lucerne (*Medicago sativa*) grown in the same soil. *FEMS Microbiology Ecology* 25, 107–116.

Hartmann, A., Lawrence, J.R., Aßmus, B. and Schloter, M. (1998b) Detection of microbes by laser confocal microscopy. In: Akkermans, A.D.L., van Elsas, J.D. and de Bruijn, F.J. (eds) *Molecular Microbial Ecology Manual*. Kluwer Academic Publishers, Dordrecht, The Netherlands, pp. 1–34.

Hartmann, A., Stoffels, M., Eckert, B., Kirchhof, G. and Schloter, M. (2000) Analysis of the presence and diversity of diazotrophic endophytes. In: Triplett, E.W. (ed.) *Prokaryotic Nitrogen Fixation: A Model System for Analysis of a Biological Process*. Horizon Scientific Press, Wymondham, UK, pp. 727–736.

Hartmann, A., Pukall, R., Rothballer, M., Gantner, S., Metz, S., Schloter, M. and Mogge, B. (2004) Microbial community analysis in the rhizosphere by *in situ* and *ex situ* application of molecular probing, biomarker and cultivation techniques. In: Varma, A., Abbott, L., Werner, D. and Hampp, R. (eds) *Plant Surface*

Microbiology. Springer-Verlag, Berlin, pp. 449–469.

Hellriegel, H. and Wilfarth, H. (1888) *Untersuchungen über die Stickstoff-nahrung der Gramineen und Leguminosen. Beilageheft zu der Zeitschrift des Vereins für die Rübenzucker-Industrie des Deutschen Reiches, Buchdruckerei der "Post"*. Kayssler & Co., Berlin.

Herron, P.R. and Wellington, E.M.H. (1990) New method for the extraction of streptomycete spores from soil and application to the study of lysogeny in sterile amended and nonsterile soil. *Applied and Environmental Microbiology* 56, 1406–1412.

Hill, C., Com-Nougué, C., Kramar, A., Moreau, T., O'Quigley, J., Senoussi, R. and Chastang, C. (1990) *Analyse Statistique des Données de Survie*. Edited by M.-S. INSERM. Flammarion, Paris.

Hiltner, L. (1904) Über neuere erfahrungen und problem auf dem gebeit der bodenbakteriologie und unter besonderer berucksichtigung der grundungung und brache. *Arbeit und Deutsche Landwirschaft Gesellschaft* 98, 59–78.

Hirsch, A.M., Bauer, W.D., Bird, D.M., Cullimore, J., Tyler, B. and Yodder, J.I. (2003) Molecular signals and receptors: controlling rhizosphere interactions between plants and other organisms. *Ecology* 84, 858–868.

Hjeljord, L. and Tronsmo, A. (1998) *Trichoderma* and *Gliocladium* in biological control: an overview. In: Harman, G.E. and Kubicek, C.P. (eds) Trichoderma *and* Gliocladium: *Enzymes, Biological Control and Commercial Applications* (Volume 2). Taylor & Francis, London, pp. 131–151.

Ho, W.C. and Ko, W.H. (1986) Microbiostasis by nutrient deficiency shown in natural and synthetic soils. *Journal of General Microbiology* 132, 2807–2815.

Högberg, P. (1990) ^{15}N natural abundance as a possible marker of the ectomycorrhizal habit of trees in mixed African woodlands. *New Phytologist* 115, 483–486.

Höper, H., Steinberg, C. and Alabouvette, C. (1995) Involvement of clay type and pH in the mechanisms of soil suppressiveness to fusarium wilt of flax. *Soil Biology and Biochemistry* 27, 955–967.

Hornby, D. (1983) Suppressive soils. *Annual Review of Phytopathology* 21, 65–85.

Howell, C.R. (1998) The role of antibiosis in biocontrol. In: Harman, G.E. and Kubicek, C.P. (eds) Trichoderma *and* Gliocladium: *Enzymes, Biological Control and Commercial Applications* (Volume 2). Taylor & Francis, London, pp. 173–183.

Hsu, S.C. and Lockwood, J.L. (1971) Responses of fungal hyphae to soil fungistasis. *Phytopathology* 61, 1355–1362.

Inbar, J. and Chet, I. (1992) Biomimics of fungal cell–cell recognition by use of lectin-coated nylon fibers. *Journal of Bacteriology* 174, 1055–1059.

Inbar, J. and Chet, I. (1995) The role of recognition in the induction of specific chitinases during mycoparasitism by *Trichoderma harzianum*. *Microbiology* 141, 2823–2829.

Jacoud, C., Faure, D., Wadoux, P. and Bally, R. (1998) Development of a strain-specific probe to follow inoculated *Azospirillum lipoferum* CRT1 under field conditions and enhancement of maize root development by inoculation. *FEMS Microbiology Ecology* 27, 43–51.

Jacoud, C., Job, D., Wadoux, P. and Bally, R. (1999) Initiation of root growth stimulation by *Azospirillum lipoferum* CRT1 during maize seed germination. *Canadian Journal of Microbiology* 45, 339–342.

Jacquot, E., van Tuinen, D., Gianinazzi, S. and Gianinazzi-Pearson, V. (2000) Monitoring species of arbuscular mycorrhizal fungi *in planta* and in soil by nested PCR: application to the study of the impact of sewage sludge. *Plant and Soil* 226, 179–188.

Jacquot-Plumey, E., van Tuinen, D., Chatagnier, O., Gianinazzi, S. and Gianinazzi-Pearson, V. (2002) Molecular monitoring of arbuscular mycorrhizal fungi in sewage sludge treated field plots. *Environmental Microbiology* 3, 525–531.

Jacquot-Plumey, E., Caussanel, J.P., Gianinazzi, S., van Tuinen, D. and Gianinazzi-Pearson, V. (2003) Heavy metals in sewage sludges contribute to their adverse effects on the arbuscular mycorrhizal fungus *Glomus mosseae*. *Folia Geobotanica* 38, 167–176.

Jeffries, P. and Young, T.W.K. (1994) *Interfungal Parasitic Relationships*. CAB International, Wallingford, UK.

Jensen, H.L. (1942) Nitrogen fixation in leguminous plants. I. General characters of root nodule bacteria isolated from species of *Medicago* and *Trifolium* in Australia. *Proceedings of the Linnean Society of New South Wales* 66, 98–108.

Johnson, L.F., Bernard, E.C. and Peiyuan, Q. (1987) Isolation of *Trichoderma* spp. at low temperatures from Tennessee and Alaska (USA) soils. *Plant Disease* 71, 137–140.

Junk, G. and Svec, H.J. (1958) The absolute abundance of the nitrogen isotopes in the atmosphere and compressed gas from various sources. *Geochimica et Cosmochimica Acta* 14, 234–243.

Kabir, M.M., Chotte, J.L., Rahman, M., Bally, R. and Jocteur Monrozier, L. (1994) Distribution of soil fractions and location of soil bacteria in a vertisol under cultivation and perennial grass. *Plant and Soil* 163, 243–255.

Kabir, M.M., Faure, D., Haurat, J., Jacoud, C., Normand, P., Wadoux, P. and Bally, R. (1995) Oligonucleotide probes based on 16S rRNA sequences for the identification of four *Azospirillum* species. *Canadian Journal of Microbiology* 41, 1081–1087.

Kirchhof, G., Schloter, M., Aßmus, B. and Hartmann, A. (1997) Molecular microbial ecology approaches applied to diazotrophs associated with non-legumes. *Soil Biology and Biochemistry* 29, 853–862.

Klein, D. and Eveleigh, D.E. (1998) Ecology of *Trichoderma*. In: Kubicek, C.P. and Harman, G.E. (eds) Trichoderma *and* Gliocladium: *Basic Biology, Taxonomy and Genetics* (Volume 1). Taylor & Francis, London, pp. 57–74.

Knudsen, I.M.B., Debisz, K., Hockenhull, J., Jensen, D.F. and Elmholt, S. (1999) Suppressiveness of organically and conventionally managed soils towards brown foot rot of barley. *Applied Soil Ecology* 12, 61–72.

Kubicek, C.P., Mach, R.L., Peterbauer, C.K. and Lorito, M. (2001) *Trichoderma*: from genes to biocontrol. *Journal of Plant Pathology* 83, 11–23.

Kubicek-Pranz, E.M. (1998) Nutrition, cellular structure and basic metabolic pathways in *Trichoderma* and *Gliocladium*. In: Kubicek, C.P. and Harman, G.E. (eds) Trichoderma *and* Gliocladium: *Basic Biology, Taxonomy and Genetics* (Volume 1). Taylor & Francis, London, pp. 95–119.

Laguerre, G., Bardin, M. and Amarger, N. (1993) Isolation from soil of symbiotic and nonsymbiotic *Rhizobium leguminosarum* by DNA hybridization. *Canadian Journal of Microbiology* 39, 1142–1149.

Larkin, R.P., Hopkins, D.L. and Martin, F.N. (1993) Ecology of *Fusarium oxysporum* f. sp. *niveum* in soils suppressive and conducive to *Fusarium* wilt of watermelon. *Phytopathology* 83, 1105–1116.

Lemanceau, P., Samson, R. and Alabouvette, C. (1988) Recherches sur la résistance des sols aux maladies. XV. Comparaison des populations de *Pseudomonas* fluorescents dans un sol résistant et un sol sensible aux fusarioses vasculaires. *Agronomie* 8, 243–249.

Lemanceau, P., Corberand, T., Gardan, L., Latour, X., Laguerre, G., Boeufgras, J.-M. and Alabouvette, C. (1995) Effect of two plant species, flax (*Linum usitatissimum* L.) and tomato (*Lycopersicon esculentum* Mill.), on the diversity of soilborne populations of fluorescent pseudomonads. *Applied and Environmental Microbiology* 61, 1004–1012.

Leyval, C. and Joner, E. (2001) Bioavailability of heavy metals in the mycorrhizosphere. In: Gobran, G.R., Wenzel, W.W. and Lombi, E. (eds) *Trace Elements in the Rhizosphere*. CRC Press, Boca Raton, Florida, pp. 165–185.

Leyval, C., Singh, B.R. and Joner, E. (1995) Occurrence and infectivity of arbuscular-mycorrhizal fungi in some Norwegian soils influenced by heavy metals and soil properties. *Water Air Soil Pollution* 84, 203–216.

Leyval, C., Turnau, H. and Haselwandter, K. (1997) Effect of heavy metal pollution on mycorrhizal colonization and function: physiological, ecological and applied aspects. *Mycorrhiza* 7, 139–153.

Leyval, C., Joner, E.J., del Val, C. and Haselwandter, K. (2002) Potential of arbuscular mycorrhizal fungi for bioremediation. In: Gianinazzi, S., Barea, J.M., Schuepp, H. and Haselwandter, K. (eds) *Mycorrhizal Technology in Agriculture: From Genes to Bioproducts*. Birkhäuser Verlag, Basel, Switzerland, pp. 175–186.

Liebman, J.A. and Epstein, L. (1992) Activity of fungistatic compounds from soil. *Phytopathology* 82, 147–153.

Lieckfeldt, E., Kuhls, K. and Muthumeenakshi, S. (1998) Molecular taxonomy of *Trichoderma* and *Gliocladium* and their teleomorphs. In: Kubicek C.P. and Harman, G.E. (eds) *Trichoderma and Gliocladium: Basic Biology, Taxonomy and Genetics* (Volume 1). Taylor & Francis, London, pp. 35–56.

Linderman, R.G. (2000) Effects of mycorrhizas on plant tolerance to diseases – mycorrhiza–disease interactions. In: Kapulnik, Y. and Douds, D.D. Jr (eds) *Arbuscular Mycorrhizas: Physiology and Function*. Kluwer Academic, Dordrecht, The Netherlands, pp. 345–365.

Linderman, R.G., Moore, L.W., Baker, K.F. and Cooksey, D.A. (1983) Strategies for detecting and characterizing systems. *Plant Disease* 67, 1058–1064.

Lindsey, D.L. and Baker, R. (1967) Effect of certain fungi on dwarf tomatoes grown under gnotobiotic conditions. *Phytopathology* 57, 1262–1263.

Lo, C.T., Nelson, E.B. and Harman, G.E. (1996) Biological control of turfgrass diseases with a rhizosphere competent strain of *Trichoderma harzianum*. *Plant Disease* 80, 736–741.

Lockwood, J.L. (1977) Fungistasis in soils. *Biological Reviews* 52, 1–43.

Lockwood, J.L. (1988) Evolution of concepts associated with soilborne plant pathogens. *Annual Review of Phytopathology* 26, 93–121.

Lorito, M. (1998) Chitinolytic enzymes and their genes. In: Harman, G.E. and Kubicek, C.P. (eds) *Trichoderma and Gliocladium: Enzymes, Biological Control and Commercial Applications* (Volume 2). Taylor & Francis, London, pp. 87–115.

Lorito, M., Woo, S.L., D'Ambrosio, M., Harman, G.E., Hayes, C.K., Kubicek, C.P. and Scala, F. (1996) Synergistic interaction between cell wall degrading enzymes and membrane affecting compounds. *Molecular Plant–Microbe Interactions* 9, 206–213.

Louvet, J. (1973) Les perspectives de lutte biologique contre les champignons parasites des organes souterrains des plantes. *Perspectives de lutte biologique contre les champignons parasites des plantes cultivées et des tissus ligneux. Symposium International, Lausanne*, pp. 48–56.

Ludwig, W., Amann, R., Martinez-Romero, E., Schönhuber, W., Bauer, S., Neef, A. and Schleifer, K.-H. (1998) rRNA based identification and detection systems for rhizobia and other bacteria. *Plant and Soil* 204, 1–19.

Lynch, J.M. and Whipps, J.M. (1990) Substrate flow in the rhizosphere. *Plant and Soil* 129, 1–10.

Ma, W., Sebestianova, S.B., Sebestian, J., Burd, G.I., Guinel, F.C. and Glick, B.R. (2003) Prevalence of 1-aminocyclopropane-1-carboxylate deaminase in *Rhizobium* spp. *Antonie van Leeuwenhoek* 83, 285–291.

Macdonald, R.M. (1986) Sampling soil microfloras: dispersion of soil by ion exchange and extraction of specific microorganisms from suspension by elutriation. *Soil Biology and Biochemistry* 18, 399–406.

Mariotti, A. (1983) Atmospheric nitrogen is a reliable standard for natural ^{15}N abundance measurements. *Nature* 303, 685–687.

Mascher, F., Schnider-Keel, U., Haas, D., Défago, G. and Moënne-Loccoz, Y. (2003) Persistence and cell culturability of biocontrol *Pseudomonas fluorescens* CHA0 under plough pan conditions in soil and influence of the anaerobic regulator gene *anr*. *Environmental Microbiology* 5, 103–115.

Mavingui, P., Laguerre, G., Berge, O. and Heulin, T. (1992) Genetic and phenotypic diversity of *Bacillus polymyxa* in soil and in the wheat rhizosphere. *Applied and Environmental Microbiology* 58, 1894–1903.

Mazzola, M., Cook, R.J., Thomashow, L.S., Weller, D.M. and Pierson, L.S.D. (1992)

Contribution of phenazine antibiotic biosynthesis to the ecological competence of fluorescent pseudomonads in soil habitats. *Applied and Environmental Microbiology* 58, 2616–2624.

McGrath, S.P., Chaudri, A.M. and Giller, K.E. (1995) Long-term effects of metals in sewage sludge on soils, microorganisms and plants. *Journal of Industrial Microbiology* 14, 94–104.

Miché, L., Bouillant, M.-L., Rohr, R. and Bally, R. (2000) Physiological and cytological studies of the inhibitory effect of soil bacteria of the genus *Azospirillum* on striga seeds germination. *European Journal of Plant Pathology* 106, 347–351.

Migheli, Q., Steinberg, C., Davière, J.M., Olivain, C., Gerlinger, C., Gautheron, N., Alabouvette, C. and Daboussi, M.J. (2000) Recovery of mutants impaired in pathogenicity after transposition of *impala* in *Fusarium oxysporum* f. sp. *melonis*. *Phytopathology* 90, 1279–1284.

Miller, M., Palojärvi, A., Rangger, A., Reeslev, M. and Kjoller, A. (1998) The use of fluorogenic substrates to measure fungal presence and activity in soil. *Applied and Environmental Microbiology* 64, 613–617.

Moënne-Loccoz, Y. and Défago, G. (2004) Life as a biocontrol pseudomonad. In: Ramos, J.L. (ed.) *Pseudomonas: Genomics, Life Style and Molecular Architecture* (Volume 1). Kluwer Academic/Plenum Publishers, New York, pp. 457–476.

Moënne-Loccoz, Y., Powell, J., Higgins, P., McCarthy, J. and O'Gara, F. (1998) An investigation of the impact of biocontrol *Pseudomonas fluorescens* F113 on the growth of sugarbeet and the performance of subsequent clover–*Rhizobium* symbiosis. *Applied Soil Ecology* 7, 225–237.

Mogge, B., Lebhuhn, M., Schloter, M., Stoffels, M., Pukall, R., Stackebrandt, E., Wieland, G., Backhaus, H. and Hartmann, A. (2000) Erfassung des mikrobiellen Populationsgradienten vom Boden zur Rhizoplane von Luzerne (*Medicago sativa*). In: Hartmann, A. (ed.) *Biologische Sicherheit: Biomonitor und Molekulare Mikrobenökologie*. Projektträger BEO, Jülich, Germany, pp. 217–224.

Mondal, S.N. and Hyakumachi, M. (1998) Carbon loss and germinability, viability, and virulence of chlamydospores of *Fusarium solani* f. sp. *phaseoli* after exposure to soil at different pH levels, temperatures, and matric potentials. *Phytopathology* 88, 148–155.

Morrissey, J.P., Walsh, U.F., O'Donnell, A., Moënne-Loccoz, Y. and O'Gara, F. (2002) Exploitation of genetically modified inoculants for industrial ecology applications. *Antonie van Leeuwenhoek* 81, 599–606.

Moulin, L., Munive, A., Dreyfus, B. and Boivin-Masson, C. (2001) Nodulation of legumes by members of the β-subclass of Proteobacteria. *Nature* 411, 948–950.

Muyzer, G. and Smalla, K. (1998) Application of denaturing gradient gel electrophoresis (DGGE) and temperature gradient gel electrophoresis (TGGE) in microbial ecology. *Antonie van Leeuwenhook* 73, 127–141.

Nguyen, C. (2003) Rhizodeposition of organic C by plants: mechanisms and controls. *Agronomie* 23, 375–396.

Okon, Y. (ed.) (1994) *Azospirillum/Plant Associations*. CRC Press, Boca Raton, Florida.

Okon, Y. and Labandera-Gonzalez, C.A. (1994) Agronomic applications of *Azospirillum*: an evaluation of 20 years world-wide field inoculation. *Soil Biology and Biochemistry* 26, 1591–1601.

Papavizas, G.C. and Lumsden, R.D. (1982) Improved medium for isolation of *Trichoderma* spp. from soil. *Plant Disease* 66, 1019–1020.

Park, Y.-H., Stack, J.P. and Kenerly, C.M. (1992) Selective isolation and enumeration of *Gliocladium virens* and *G. roseum* from soil. *Plant Disease* 76, 230–235.

Paulitz, T.C. (1990) Biochemical and ecological aspects of competition in biological control. In: Baker, R. and Dunn, P.E. (eds) *New Directions in Biological Control: Alternatives for Suppressing Agricultural Pests and Diseases*. Alan R. Liss, New York, pp. 713–724.

Peng, H.X., Sivasithamparam, K. and Turner, D.W. (1999) Chlamydospore germination and *Fusarium* wilt of banana plantlets in suppressive and conducive soils are affected by physical and chemical factors. *Soil Biology and Biochemistry* 31, 1363–1374.

Persson, L., Larsson Wikström, M. and Gerhardson, B. (1999) Assessment of soil suppressiveness to *Aphanomyces* root rot of pea. *Plant Disease* 83, 1108–1112.

Peters, N.K. (1986) A plant flavone, luteolin, induces expression of *Rhizobium meliloti* genes. *Science* 233, 977–980.

Raaijmakers, J.M. and Weller, D.M. (1998) Natural protection by 2,4-diacetylphloroglucinol-producing *Pseudomonas* spp. in take-all decline soils. *Molecular Plant–Microbe Interactions* 11, 144–152.

Raaijmakers, J.M., Weller, D.M. and Thomashow, L.S. (1997) Frequency of antibiotic producing *Pseudomonas* spp. in natural environments. *Applied and Environmental Microbiology* 63, 881–887.

Revellin, C., Pinochet, X., Beauclair, P. and Catroux, G. (1996) Influence of soil properties and soya bean cropping history on the *Bradyrhizobium japonicum* population in some French soils. *European Journal of Soil Science* 47, 505–510.

Reverchon, S. (2001) Un diagnostic adapté aux champignons du sol. *Phytoma – La défense des Végétaux* 542, 20–23.

Rezzonico, F., Moënne-Loccoz, Y. and Défago, G. (2003) Effect of stress on the ability of a *phlA*-based quantitative competitive PCR assay to monitor biocontrol strain *Pseudomonas fluorescens* CHA0. *Applied and Environmental Microbiology* 69, 686–690.

Rivera-Becerril, F., Calantzis, C., Turnau, K., Caussanel, J.P., Belimov, A.A., Gianinazzi, S., Strasser, R.J. and Gianinazzi-Pearson, V. (2002) Cadmium accumulation and buffering of cadmium-induced stress by arbuscular mycorrhiza in three *Pisum sativum* L. genotypes. *Journal of Experimental Botany* 53, 1–9.

Rodriguez Caceres, E.A. (1982) Improved medium for isolation of *Azospirillum* spp. *Applied and Environmental Microbiology* 44, 990–991.

Romine, M. and Baker, R. (1973) Soil fungistasis: evidence for an inhibitory factor. *Phytopathology* 63, 756–759.

Rovira, A.D. (1965) Interactions between plant roots and soil microorganisms. *Annual Review of Microbiology* 19, 241–266.

Sambrook, J., Fritsch, E.F. and Maniatis, T. (1989) *Molecular Cloning: A Laboratory Manual.* Cold Spring Harbor Laboratory Press, Cold Spring Harbor, New York.

Schippers, B. (1992) Prospects for management of natural suppressiveness to control soilborne pathogens. In: Tjamos, E.C., Papavizas, G.C. and Cook, R.J. (eds) *Biological Control of Plant Diseases.* Plenum Press, New York, pp. 21–34.

Schneider, R.W. (1982) *Suppressive Soils and Plant Disease.* American Phytopathological Society, St Paul, Minnesota, p. 96.

Schüepp, H. and Green, R.J. Jr (1968) Indirect assay methods to investigate soil fungistasis with special consideration of soil pH. *Phytopathologische Zeitschrift* 61, 1–28.

Sessitsch, A., Howieson, J.G., Perret, X., Antoun, H. and Martínez-Romero, E. (2002) Advances in *Rhizobium* research. *Critical Reviews in Plant Sciences* 21, 323–378.

Shearer, G. and Kohl, D.H. (1986) N_2-fixation in field settings: estimations based on natural ^{15}N abundance. *Australian Journal of Plant Physiology* 13, 669–756.

Smith, S.E. and Read, D.J. (1997) *Mycorrhizal Symbiosis.* Academic Press, London, 605 pp.

Smith, V.L., Wilcox, W.F. and Harman, G.E. (1990) Potential for biological control of *Phytophthora* root and crown rots of apple by *Trichoderma* and *Gliocladium* spp. *Phytopathology* 80, 880–885.

Somasegaran, P.H. and Hoben, J. (1994) *Handbook for Rhizobia: Methods in Legume–Rhizobium Technology.* Springer, New York.

Stock, J.B., Ninfa, A.J. and Stock, A.M. (1989) Protein phosphorylation and regulation of adaptive responses in bacteria. *Microbiological Reviews* 53, 450–490.

Stoffels, M., Castellanos, T. and Hartmann, A. (2001) Design and application of new 16S rRNA-targeted oligonucleotide probes for the *Azospirillum–Skermanella–Rhodocista* cluster. *Systematic and Applied Microbiology* 24, 83–97.

Stutz, E.W. and Défago, G. (1985) Effect of parent materials derived from different geological strata on suppressiveness of soils to black root rot of tobacco. In: Parker, C.A., Rovira, A.D., More, K.J. and Wong, P.T.W. (eds) *Ecology and Management of Soilborne Plant Pathogens.* APS, St Paul, Minnesota, pp. 215–217.

Sy, A., Giraud, E., Jourand, P., Garcia, N., Willems, A., de Lajudie, P., Prin, Y., Neyra, M., Gillis, M., Boivin-Masson, C. and Dreyfus, B. (2001) Methylotrophic *Methylobacterium* bacteria nodulate and fix nitrogen in symbiosis with legumes. *Journal of Bacteriology* 183, 214–220.

Tarrand, J.J., Krieg, N.R. and Dobereiner, J. (1978) A taxonomic study of the *Spirillum lipoferum* group, with descriptions of a new genus, *Azospirillum* gen. nov. and two species, *Azospirillum lipoferum* (Beijerinck) comb. nov. and *Azospirillum brasilense* sp. nov. *Canadian Journal of Microbiology* 24, 967–980.

Thornton, C.R., Pitt, D., Wakley, G.E. and Talbot, N.J. (2002) Production of a monoclonal antibody specific to the genus *Trichoderma* and closely related fungi, and its use to detect *Trichoderma* spp. in naturally infested composts. *Microbiology* 148, 1263–1279.

Tilman, D. (1998) The greening of the green revolution. *Nature* 396, 211–212.

Toyota, K., Ritz, K. and Young, I.M. (1996) Microbiological factors affecting the colonisation of soil aggregates by *Fusarium oxysporum* f. sp. *raphani*. *Soil Biology and Biochemistry* 28, 1513–1521.

Trouvelot, A., Kough, J.L. and Gianinazzi-Pearson, V. (1985) Mesure du taux de mycorhization VA d'un système radiculaire. Recherche de méthodes d'estimation ayant une signification fonctionnelle. In: *Physiological and Genetical Aspects of Mycorrhizae.* Proceedings of the 1st European Symposium on Mycorrhizae, INRA, Paris, pp. 217–221.

Tu, X.M. (1995) Nonparametric estimation of survival distribution with censored initiating time, and censored and truncated terminating time: application to transfusion data for acquired immune deficiency syndrome. *Applied Statistics* 44, 3–16.

Unkovich, M. and Pate, J.S. (2001) Assessing N_2 fixation in annual legumes using ^{15}N natural abundance. In: Unkovich, M., Pate, J., NcNeill, A. and Gibbs, D.J. (eds) *Stable Isotope Techniques in the Study of Biological Processes and Functioning of Ecosystems.* Kluwer Academic, Dordrecht, The Netherlands, pp. 103–118.

van der Heijden, M.G.A., Klironomos, J.N., Ursic, M., Moutoglis, P., Streitwolfengel, R., Boller, T., Wiemken, A. and Sanders, I.R. (1998) Mycorrhizal fungal diversity determines plant biodiversity, ecosystem variability and productivity. *Nature* 396, 69–72.

van Tuinen, D., Jacquot, E., Zhao, B., Gollotte, A. and Gianinazzi-Pearson, V. (1998) Characterization of root colonization profiles by a microcosm community of arbuscular mycorrhizal fungi using 25S rDNA-targeted nested PCR. *Molecular Ecology* 7, 879–887.

Velázquez, E., Mateos, P.F., Velasco, N., Santos, F., Burgos, P.A., Villadas, P.J., Toro, N. and Martínez-Molina, E. (1999) Symbiotic characteristics and selection of autochthonous strains of *Sinorhizobium meliloti* populations in different soils. *Soil Biology and Biochemistry* 31, 1039–1047.

Villadas, P.J., Velázquez, E., Martínez-Molina, E. and Toro, N. (1995) Identification of nodule-dominant *Rhizobium meliloti* strains carrying pRmeGR4b-type plasmid within indigenous soil populations by PCR using primers derived from specific DNA sequences. *FEMS Microbiology Ecology* 17, 161–168.

Vincent, J.M. (1970) *A Manual for the Practical Study of Root Nodule Bacteria.* I.B.P. Handbook No. 15. Blackwell Scientific Publications, Oxford, UK.

Von Alten, H., Blal, B., Dodd, J.C., Feldmann, F. and Vosatka, M. (2002) Quality control of arbuscular mycorrhizal fungi inoculum in Europe. In: Gianinazzi, S., Barea, J.M., Schuepp, H. and Haselwandter, K. (eds) *Mycorrhizal Technology in Agriculture: From Genes to Bioproducts.* Birkhäuser Verlag, Basel, Switzerland, pp. 281–296.

Wacker, T.L. and Lockwood, J.L. (1991) A comparison of two assay methods for

assessing fungistasis in soils. *Soil Biology and Biochemistry* 23, 411–414.
Wagner, M., Amann, R., Lemmer, H. and Schleifer, K.-H. (1993) Probing activated sludge with oligonucleotides specific for proteobacteria: inadequacy of culture-dependent methods for describing microbial community structure. *Applied and Environmental Microbiology* 59, 1520–1525.
Walsh, U.F., Morrissey, J.P. and O'Gara, F. (2001) *Pseudomonas* for biocontrol of phytopathogens: from functional genomics to commercial exploitation. *Current Opinion in Biotechnology* 12, 289–295.
Walsh, U.F., Moënne-Loccoz, Y., Tichy, H.-V., Gardner, A., Corkery, D.M., Lorkhe, S. and O'Gara, F. (2003) Residual impact of the biocontrol inoculant *Pseudomonas fluorescens* F113 on the resident population of rhizobia nodulating a red clover rotation crop. *Microbial Ecology* 45, 145–155.
Wan, M.T., Rahe, J.E. and Watts, R.G. (1998) A new technique for determining the sublethal toxicity of pesticides to the vesicular-arbuscular mycorrhizal fungus *Glomus intraradices*. *Environmental Toxicological Chemistry* 17, 1421–1428.
Wardle, D.A. (1992) A comparative assessment of factors which influence microbial biomass carbon and nitrogen levels in soil. *Biological Reviews* 67, 321–342.
Weidner, S., Arnold, W. and Pühler, A. (1996) Diversity of uncultured microorganisms associated with the seagrass *Halophila stipulacea* estimated from restriction fragment length polymorphism analysis of PCR-amplified 16S rRNA genes. *Applied and Environmental Microbiology* 62, 766–771.
Weindling, R. (1932) *Trichoderma lignorum* as a parasite of other fungi. *Phytopathology* 22, 837–845.
Weissenhorn, I. and Leyval, C. (1996) Spore germination of arbuscular-mycorrhizal (AM) fungi in soils differing in heavy metal content and other physicochemical properties. *European Journal of Soil Biology* 32, 165–172.
Weissenhorn, I., Leyval, C. and Berthelin, J. (1993) Cd-tolerant arbuscular mycorrhizal (AM) fungi from heavy-metal polluted soils. *Plant and Soil* 157, 247–256.

Weller, D.M., Raaijmakers, J.M., Mc-Spadden-Gardener, B.B. and Thomashow, L.S. (2002) Microbial populations responsible for specific soil suppressiveness to plant pathogens. *Annual Review of Phytopathology* 40, 309–348.
White, D.C. and Ringelberg, D.B. (1998) Signature lipid biomarker analysis. In: Burlage, R.S., Atlas, R., Stahl, D., Geesey, G. and Sayler, G. (eds) *Techniques in Microbial Ecology*. Oxford University Press, New York, pp. 255–272.
Whitehead, N.A., Barnard, A.M.L., Slater, H., Simpson, N.J.L. and Salmond, G.P.C. (2001) Quorum-sensing in Gram-negative bacteria. *FEMS Microbiology Review* 25, 365–404.
Widden, P. and Hsu, D. (1987) Competition between *Trichoderma* species: effects of temperature and litter type. *Soil Biology and Biochemistry* 19, 89–94.
Williams-Woodward, J.L., Pfleger, F.L., Allmaras, R.R. and Fritz, V.A. (1998) *Aphanomyces euteiches* inoculum potential: a rolled-towel bioassay suitable for fine-textured soils. *Plant Disease* 82, 386–390.
Windham, M.T., Elad, Y. and Baker, R. (1986) A mechanism for increased plant growth induced by *Trichoderma* spp. *Phytopathology* 76, 518–521.
Young, J.M., Kuykendall, L.D., Martinez-Romero, E., Kerr, A. and Sawada, H. (2001) A revision of *Rhizobium* Frank 1889, with an amended description of the genus, and the inclusion of all species of *Agrobacterium* Conn 1942 and *Allorhizobium undicola* de Lajudie *et al.* 1998 as new combinations: *Rhizobium radiobacter, R. rhizogenes, R. rubi, R. undicola* and *R. vitis. International Journal of Systematic and Evolutionary Microbiology* 51, 89–103.
Zeilinger, S., Galhaup, C., Payer, K., Woo, S.L., Mach, R.L., Fekete, C., Lorito, M. and Kubicek, C.P. (1999) Chitinase gene expression during mycoparasitic interaction of *Trichoderma harzianum* with its host. *Fungal Genetics and Biology* 26, 131–140.
Zelles, L. (1997) Phospholipid fatty acid profiles in selected members of soil microbial communities. *Chemosphere* 35, 275–294.

10 Census of Microbiological Methods for Soil Quality

OLIVER DILLY

Lehrstuhl für Bodenökologie, Technische Universität München, D-85764 Neuherberg and Ökologie-Zentrum, Universität Kiel, Schauenburgerstraße 112, D-24118 Kiel, Germany; Present address: *Lehrstuhl für Bodenschutz und Rekultivierung, Brandenburgische Technische Universität, Postfach 101344, D-03013 Cottbus, Germany*

Introduction

An extensive spectrum of soil microbiological methods has been developed to investigate and evaluate soil quality. Classical soil microbiological methods referring to activity rates and biomass content can be separated from modern methods related to molecular techniques and isotope determinations. Several textbooks containing methods of the two groups have been published. These textbooks frequently contain an enormous set of methods selected by authors or editors without clearly indicating the field of application, or they apply more specifically to modern techniques that are still under development for routine analyses. This handbook aims to give a selection of methods for soil quality determination. Some information should also be given to rank the use of the methods. Therefore, a questionnaire was developed in order to evaluate the use of these methods. In addition, the database should provide information to enable location of research partners and laboratory expertise; thus, the questionnaire was not anonymous. Scientists involved in COST Action 831, and also other laboratories working in this area, were encouraged to complete the questionnaire.

Method

The questionnaire was developed to be completed via the Internet and was connected to a database. The database expertise was provided by http://www.kaufraum.de. The Internet address of the questionnaire was located at http://www.soiltechnology.de (accessed 26 October 2004) and was available for more than 1 year; evaluation was carried out at the end of 2003.

The questionnaire asked for information on country, city, laboratory, scientist, postal address, phone, fax and e-mail address. Using a radio button, it was possible to select 1 of 18 methods dealing with biomass, community structure and activity, carried out either in the laboratory or in the field. For each method, one form was to be filled in. Whether the methods were used as routine analysis, in monitoring programmes or for research only was to be noted. In addition, information was collected about application of the method with reference to: (i) soil use; (ii) soil type according ISSS/ISRIC/FAO (1998); (iii) soil texture; (iv) data use; (v) specific comments; and (vi) references. In some categories, 'unspecified' could be selected. The field 'unspecified' referred either to broad variability or to vague knowledge.

Results

This homepage was initiated by, and announced at, meetings or via e-mail to those on the distribution lists of COST Action 831, so it was visited mainly by scientists actively involved in this Action. Only a few people not actively involved in COST Action 831 found the homepage and filled in the questionnaire.

Overall, 159 questionnaires were completed by 32 laboratories in 14 countries. The highest number of questionnaires were returned from Germany, followed by Austria and Italy (Table 10.1). Most methods refer to bacterial community structure, unspecified activity measures and microbial biomass estimated by fumigation–extraction (Table 10.2). The methods were used mainly for research; approximately 50% were also used for monitoring and one-third for routine analysis (Table 10.2).

The soils studied were mainly under agricultural land use. One-third were industrial or urban soils. Eighteen different soil types were studied with the techniques. However, most of the scientists did not specify soil type when completing the questionnaire. In contrast, soil texture was specified in more detail, showing that the contributors were dealing mainly with loamy or sandy soil samples and, to a lesser extent, clay soils.

Conclusions

Although the database represents only an initial picture of European laboratories dealing with microbiological methods for evaluating soil quality, it is clear that several methods for bacterial community structure, a variety of activity methods, and microbial biomass estimated by fumigation–extraction are mainly used. Furthermore, agricultural soils and soils with loamy texture were investigated most frequently and information on soil units was generally not given.

The questionnaire was not completed by all scientists involved in COST Action 831 over the 12 months that it was available, most likely due to the fact that these scientists did not visit the website or found the procedure

Table 10.1. Number of questionnaires returned from the countries involved in COST Action 831.

Country	Amount
Germany	36
Austria	27
Italy	18
United Kingdom	17
Switzerland	12
The Netherlands	9
Spain	9
Sweden	9
Slovenia	8
Denmark	6
Belgium	2
Norway	2
France	2
Hungary	2

difficult and time consuming, e.g. each method needed to be reported in a separate questionnaire. In contrast, some laboratories offered a broad spectrum of methods, because they found the information useful.

Perspectives

To obtain more completed questionnaires for a more representative census, we hope that readers of this book will communicate the Internet address further to their scientific communities, or to societies dealing in this field, such as soil science societies, societies in the field of microbiology and private laboratories.

To enhance the acceptance of, and interest in, completing the questionnaire, some incentives may be provided, e.g. raffle of books or laboratory tools. Interest may also be encouraged when the webpage is officially supported by national authorities and the European Commission. The ability to refer to personal homepages, or to obtain information about specific contributing scientists, may further stimulate interest.

On the technical side, the actual display of graphics with statistics of methods used for the respective purposes of monitoring, research or routine analysis may be of interest. The simultaneous selection of several methods with the respective information would make the repeated filing of the data related to one laboratory unnecessary. Finally, a button for the submission of comments for improving the questionnaire may be included.

Regarding the content of the questionnaire, some questions related to soil–plant interactions, such as microorganisms and plants (in *Latin*), should be considered in the questionnaire, since these aspects are included in this book. Furthermore, the following categories may be included.

Census of Microbiological Methods for Soil Quality

Table 10.2. Results of the questionnaire on methods used to study soil quality involved in COST Action 831.

Method		Soil Use		Soil type[a]		Soil texture		Use for	
Microbial community structure: Bacteria	21	Agricultural	133	Not specified	77	Loam	119	Research	140
Microbial activity: Others	18	Forest	79	Cambisol	35	Sand	96	Monitoring	74
Microbial biomass: Fumigation–extraction	18	Grassland	73	Luvisol	25	Clay	77	Routine analysis	43
Microbial activity: Basal respiration	16	Urban/industrial	53	Histosol	24	Silt	59		
Microbial biomass: Substrate-induced respiration	15	Not specified	20	Podzol	22	Others	53		
Microbial activity: N mineralization	13			Gleysol	21				
Molecular tools: Bacteria	12			Fluvisol	19				
Microbial biomass: ATP	10			Anthrosol	16				
Microbial biomass: Fungi	6			Regosol	11				
Microbial activity: C mineralization	5			Acrisol	9				
Molecular tools: Others	5			Arenosol	9				
Field experiments: Others	4			Leptosol	9				
Microbial biomass: Bacteria	4			Andosol	8				
Microbial community structure: Fungi	3			Umbrisol	8				
Microbial biomass: Others	3			Chernozem	7				
Field experiments: Litter bag	2			Ferralsol	7				
Field experiments: *in situ* C mineralization	2			Phaeozem	2				
Field experiments: *in situ* N mineralization	2			Vertisol	1				
				Durisol	1				

[a] According to ISSS/ISRIC/FAO (1998).

Soils under extreme environmental factors:

- soils under cold or high temperature;
- soils under limited or excess of water;
- soil polluted with heavy metals;
- soil polluted with organic compounds;
- salty soils;
- soil with extreme soil pH value;

and information may be included on soil pH value, for example:

- highly acidic soil;
- acid soil;
- neutral soil;
- alkaline soil.

To get more insight into the use of the respective methods, it would be interesting to know how many samples per year are analysed with each technique.

References

ISSS/ISRIC/FAO (1998) *World Reference Base for Soil Resources*. World Soil Resources, Report 84. FAO, Rome.

Index

accessibility: of indicators 17
acetylene inhibition method (AIM)
    advantages and disadvantages 155–156
    calculation 154–155
    equipment and reagents 153
    principle 152–153
    procedure 153–154
acridine orange 94, 98–99
activity, microbial 7–8
    decomposition experiments 115
    effects of fertilization 35
    grasslands *vs.* horticultural farms 38
    measured by means of soil respiration *see* respiration
    N mineralization *see* mineralization, nitrogen
    nitrification and denitrification *see* denitrification; nitrification
    overview of methods used 30–31
        field *vs.* laboratory methods 114–115
        see also enzymes, activity of; growth, bacterial
agriculture
    effects of agricultural practices on soil quality indicators 40–44, 66–68
        grasslands *vs.* horticultural farms 35–37(tab), 38
        microbial community structure and function 55–59
agrochemicals
    effects on microbial community structure 55–56, 184, 186
    see also fertilization

AMOEBA method: for data presentation 38–40
arbuscular mycorrhiza *see* mycorrhiza, arbuscular
ATP (adenosine triphosphate)
    as index of microbial biomass 9, 75
Azospirillum spp.
    cultivation-based monitoring
        equipment and reagents 275–277
        media 275(tab)
        oligonucleotide probes for FISH 278(tab)
        principles 274–275
        procedure 277–278
    cultivation-independent monitoring 279–281
    extraction from soil and roots 273–274
    positive effects on soil 271

bacteria
    nodulating symbiotic *see* rhizobia
    plant-growth promoting *see* rhizobacteria, plant-growth promoting
basal respiration 30
    definition 117
    importance of measurement 122–123
    substrate-responsive biomass method
        calculations 124–125
        justification for method 126
        materials and procedure 123–124
        normative references 171

basal respiration *continued*
   titration method
      calculations 122
      equipment and reagents 121
      normative references 170–171
      principles 120
      procedure 121–122
biological nitrogen fixation *see* rhizobia
Biolog™ *see* CLPP method
biomass and number, microbial 7
   as 'early warning' system 54
   ATP as index 9, 75
   cell counts *see* counting: microbial cells
   chloroform-fumigation extraction *see*
      chloroform-fumigation
      extraction (CFE)
   combined indicator with carbon
      mineralization 54
   effects of contamination 34,
      106–107(fig), 146–147(fig)
   effects of crops 55, 67–68
   effects of fertilization 35
   effects of sample storage 28–29(fig)
   effects of tillage systems 67
   fatty acid analyses 75
   methods standardization 76
   methods: overview 30
   nucleic acid analyses 75–76
   relation with substrate-induced
      respiration 85(fig), 87(fig)
   substrate-induced respiration *see*
      substrate-induced respiration
      (SIR) method
biomass, overall
   effects of agricultural practices 24–25,
      35, 38, 41
   effects of soil types 41
biovolume, bacterial 101–102

Canada 20
carbon
   biomass: organic ratio 4
   mineralization *see* mineralization,
      carbon
   rhizodeposition of fixed carbon 228–229
   soil organic carbon pools 53–54
categories: of soil quality evaluation
   methods 6–7, 12(tab)
census: of methods
   methodology and results 296–297

   methods used and soils tested 297,
      299(tab)
   number of responses 298(tab)
   suggested enhancements to
      questionnaire 298, 300
chloroform-fumigation extraction (CFE) 30,
   74–75
   biomass C by dichromate oxidation
      78–80
   biomass C by UV-persulphate oxidation
      80–81
   biomass ninhydrin *vs*. biomass C 83
   fumigation–extraction procedure 78
   materials and reagents 78
   ninhydrin-reactive nitrogen
      determination 81–83
   principle 77
   relation with substrate-induced
      respiration 85(fig)
chromium 34
CLPP method 11, 31
   Biolog™ system 213, 216(fig)
   calculation and data management
      217–220
   data analysis 221
   different substrates in Biolog™ plates
      33(fig)
   dilution, inoculation and incubation
      215–217
   equipment and reagents 214–215
   outline of profile construction 219(fig)
   principle 213–214
   sampling, storage and extraction 214
   standardization 221
colonization: of roots by mycorrhiza
   251–253, 255(fig)
communities, microbial *see* diversity and
   community structure, microbial
community fingerprinting *see* fingerprinting,
   community
community-level physiological profile *see*
   CLPP method
copper 34, 146–147(fig)
COST Action 831 (EU) 3–4, 64
counting: microbial cells
   biomass 102–103
   biovolume 101–102
   chemicals and solutions 97–98
   counting and calculations 100–105
   direct cell counts 74, 96
   drawbacks of plate counts 73–74

errors
    maximum error estimation 105–106
    sources and propagation 103–104(tab)
examination of slides 100
filters 96
microscopy 96–97
preparation of soil suspension 98–99
results: example 106–107(fig)
sample collection, preparation and storage 95–96
staining procedures 99–100
stains used 93–95
statistical considerations 103–106
counting: soil organisms 30–31
crops: effects on microbial community structure 55, 67–68
Czech Republic 32

DAPI (4′,6-diamidino-2-phenylindole-dihydrochloride) 94, 96–99, 280–281
data: presentation methods 38–40
decomposition: as measure of microbial activity 115
definitions 17–18, 24, 51, 117–118
dehydrogenases 159, 161–163
    normative references 171
denaturing gradient gel electrophoresis (DGGE) 32(fig), 192–194, 200–201, 202(tab), 203
denitrification
    acetylene inhibition method 152–156
    consequences of nitrifying microorganisms 151–152
    description 150–151
    isotope methods 156–157
    vs. nitrification 151
Denmark 28
DGGE *see* denaturing gradient gel electrophoresis (DGGE)
diazosymbionts *see* rhizobia
dichromate: oxidation by 78–80
differential fluorescent stain (DFS) 95, 98, 100
diseases, plant: effects of soil *see* fungistasis, soil; soil inoculum potential; suppressiveness, soil
diversity and community structure, microbial 8
    effects of agricultural practices 35–36, 42(fig), 43

    agricultural chemicals 55–56, 184, 186
    crops 55
    precision farming *vs.* conventional management 56–57
    transition to organic farming 58–59
effects of contamination 34
methods
    advantages of non-culture techniques 55, 183–184
    CLPP method *see* CLPP method
    community fingerprinting *see* fingerprinting, community
    obstacles to measurement and research needs 44–45
    overview 31–34, 185(tab)
    phospholipid fatty analyses *see* phospholipid fatty acid analyses
prokaryotic species: estimated number 184, 212
DNA
    as biomass indicator 76
    bacterial profiles 31
        effects of agricultural practices 35–36, 40–41
        effects of contamination 32(fig), 34
        effects of storage 29
    fingerprinting *see* fingerprinting, community
DTAF (5-(4.6-dichlorotriazin-2-yl) aminofluorescein 94, 98, 100
Dutch Soil Quality Network
    impediments and research needs 44–45
    parameters measured 33–34
    results
        contaminated and experimental reference sites 34–36
        data presentation 38–40
        first year pilot study 36–38
        five-year cycle 40–44
    sampling methodology 26–30
    Soil Quality Index 39–40
    structure and activity 26

EcoBiolog® system 58–59
ecosystems, health of 24
ecotoxicology: risk assessment 92
electrophoresis, denaturing gradient gel *see* denaturing gradient gel electrophoresis (DGGE)

enumeration *see* counting
Environmental Protection Agency (EPA, USA) 18
environmental quality 18
enzymes, activity of 31, 158–159
    commonly assayed hydrolases 160(tab)
    dehydrogenase activity 159, 161–163
    effect of precision farming 57
    fluorescein diacetate hydrolysis 163–166
    interpretation of assay data 169–170
    normative references 171
    phosphomonoesterase activity 166–169
European Commission
    COST actions 3–4, 64
    environmental measures 50–51

farming, organic 35, 58–59
farming, precision 56–57
fatty acids, phospholipid *see* phospholipid fatty acid analyses
fertilization
    effects of different practices 35–36
    see also agrochemicals
fingerprinting, community
    calculation 201–202
    denaturing gradient gel electrophoresis (DGGE) 32(fig), 192–194, 200–201, 202(tab), 203
    fingerprinting techniques: overview 188–189
    limitations 212
    nucleic acid extraction
        DNA method 190, 195–197
        DNA/RNA method 191, 197–198
        equipment and reagents 192–193
        limitations of methodology 202
        types of methods 187–188
    PCR amplification 191–193, 198–200, 202–203
Finland 28
fixation, biological nitrogen *see* rhizobia
flax: in soil suppressiveness assay 259–262
fluorescein diacetate, hydrolysis of 163–166
fluorescence *in situ* hybridization (FISH): for *Azospirillum* spp. 277–278, 280–281
food webs, soil 25, 36, 38
fungi *see also* mycorrhiza, arbuscular
    assessment of soil inoculum potential 262–264
    assessment of soil suppressiveness to *Fusarium* wilts 259–262

    effect of precision farming on diversity 57
    fungal to bacterial ratio 44
    plant-growth promoting species
        cultivation-based monitoring 282–284
        effects 270–272
        separation and extraction 280(fig), 282
fungistasis, soil 264–268
Fusarium oxysporum
    soil suppressiveness assay 259–262
    spore germination and mycelial extension 266(fig)

GELCOMPAR program 201–202
genetically modified plants 68–69
Germany 28–29, 31
germination:
    of mycorrhizal spores as measure of soil toxicity 251–254(fig)
    spore germination test of soil fungistasis 266–267
Gliocladium spp.
    cultivation-based monitoring 281–284
    positive effects on soil 272
Glomus mossease: use in bioassays 251–253
grasslands 35–37(tab), 38
    AMOEBA presentation of indicator values 39(fig)
greenhouse gases 58
growth, bacterial 30–31
    assessment by thymidine and leucine incorporation
        calculation and conversion factors 146–147
        comparison with other procedures 147–149
        equipment and procedure 143–145
        overview and principle 142–143
    effects of contamination 106–107(fig), 146–147(fig)
    effects of sample storage 28–29(fig)
    grasslands *vs.* horticultural farms 38

health: as criterion for soil quality 18
heavy metals 106–107(fig)
    mycorrhizal assay 250–254
Heinemeyer soil biomass analyser 86–90

hybridization: methodology: for *Azospirillum* spp. 277–278, 280–281
hydrolases 160(tab)
hyphal extension test 267–268

indexes, soil quality 17–18, 39–40, 64–65
indicators and parameters, soil quality
    criteria 4–5, 15–18, 53, 69
    definitions 17–18
    importance of biological parameters 20–21
    minimum data set 66(fig)
    regulatory agency proposals 18–21
    relationships between parameters 9–12
    suitability depends on land type and use 53
infection counts, plant (for enumeration of rhizobia) 236–240
inoculation, mycorrhizal 249–250
inoculum potential, soil *see* soil inoculum potential
ISO Technical Committee 19(tab), 123
isotopes
    for assessment of plant nitrogen from symbiotic fixation 243–246
    in denitrification research 156–157
    isotopic dilution technique 129
Italy 28–29, 66–68

leucine, incorporation of *see under* growth, bacterial
liming 35
Linum usitatissinum: in soil suppressiveness assay 259–262
litterbags: in decomposition experiments 115

markers, molecular 75–76
McCrady table 240(tab)
measurability 16
Medicago truncatula: use in bioassays 251–253
metals, heavy 106–107(fig)
mineralization, carbon 36
    combined with biomass 54
    effects of storage 28–29(fig)
    limitations of estimation by means of soil respiration 118
    see also respiration
mineralization, nitrogen 31–33, 36
    description 127–128

    disadvantages of field method 114
    effects of fertilization 35, 41–42(fig)
    effects of sample storage 28–29
    effects of soil types 41–42(fig)
    gross mineralization: isotopic dilution technique 129
    net mineralization measurements 128
    potential mineralization activity
        anaerobic N mineralization method 132–133
        long-term aerobic N mineralization method 133–135
        short-term *vs.* long-term procedures 130–132
MPN plant infection counts 236–240
mycorrhiza, arbuscular
    characteristics 248–249
    germination in sewage sludge and waste 254(fig)
    in soil remediation and plant protection 249, 254–255
    in tests of heavy metal contamination 250–254
    mycorrhizal inoculation 249–250
    tolerance to heavy metals 254

Netherlands, The *see* Dutch Soil Quality Network
New Zealand 20–21(tab), 32
nickel: effects on bacterial DNA profiles 34
nitrification 31
    consequences of nitrifying microorganisms 151
    gross nitrification 136–137
    net nitrification 137
    nitrification potential
        calculations 140–141
        chlorate inhibition method 141
        equipment and reagents 138–139
        optimum initial ammonia concentration 141
        principle 137–138
        procedure 139–140
    normative references 171
    vs. denitrification 151
nitrogen
    biological nitrogen fixation *see* rhizobia
    mineralization *see* mineralization, nitrogen
    ninhydrin-reactive: determination 81–83

nitrous oxide: determination *see* acetylene
 inhibition method (AIM)
nodulation scores 234, 236(tab)
nucleic acids *see* fingerprinting, community

OECD 16–18, 64
operativeness 17
organic farming 35, 58–59
organic matter, soil 53–54
oxidation
    dichromate 78–80
    UV-persulphate 80–81

parameters, soil quality *see* indicators and
 parameters, soil quality
pathogens, plant: effects of soil *see*
 fungistasis, soil; soil inoculum
 potential; suppressiveness, soil
PCR amplification 191–193, 198–200,
 202–203
persulphate: in UV oxidation reaction 80–81
PGPF *see* fungi, plant-growth promoting
 species
PGPR *see* rhizobacteria, plant-growth
 promoting
phospholipid fatty acid analyses 31, 75
    advantages and applications 204–205
    calculation and statistical analysis
 210–211
    equipment and reagents 207–208
    fatty acid nomenclature 210
    gas chromatography 209–210
    internal standards 209
    limitations 212
    lipid extraction and fractionation
        flow diagram 206(fig)
        methodology 208–209
    methanolysis 209
    other methodologies 211
    principle of the methods 205
phosphomonoesterases 166–169
plant infection counts (for enumeration of
 rhizobia) 236–240
plants: interactions with microbes *see*
 mycorrhiza, arbuscular;
 rhizobacteria, plant-growth
 promoting; rhizobia; rhizosphere
PLFA analyses *see* phospholipid fatty acid
 analyses

pollution 106–107(fig)
precision farming 56–57
productivity, soil 18

quality, soil
    definitions 18, 24, 51
    phytosanitary quality *see* fungistasis,
 soil; soil inoculum potential;
 suppressiveness, soil

relevance, political 16
reliability 17
remediation: of soil 254–255
representativeness 16–17
respiration
    basal *see* basal respiration
    definition 117–118
    effects of contamination 146–147(fig)
    field *vs.* laboratory methods 118–119
    static *vs.* dynamic methods 119–120(tab)
    substrate-induced *see* substrate-induced
 respiration (SIR) method
    see also mineralization, carbon
rhizobacteria, plant-growth promoting
    Azospirillum
        cultivation-based monitoring
 274–278
        cultivation-independent monitoring
 279–281
    effects 270–271
    separation and extraction 273–274,
 280(fig)
rhizobia (nodulating bacteria)
    detection of rhizobia and assessment of
 nodulating potential
        equipment and materials 234–235
        principles 233–234
        procedure and calculation 235–236
    enumeration by MPN plant infection
 counts 236–240
    occurrence in different soils 236
    other roles in improving soil quality
 246–247
    symbiosis with legumes (biological
 nitrogen fixation)
        assessment of plant nitrogen from
 symbiotic fixation 243–246
        assessment of symbiotic efficacy
 241–243
        importance 233
        known interactions 232(tab)

rhizosphere 9
    effects on soil quality 230
    microbial adaptation to environmental
        changes 229–230
    rhizodeposition of fixed carbon 228–229
    see also mycorrhiza, arbuscular;
        rhizobacteria, plant-growth
        promoting; rhizobia
risk assessment, ecotoxicological 92
RNA *see* fingerprinting, community
Rodale Institute 18
root colonization bioassay 251–253, 255(fig)

sampling
    depth 27
    for analysis of respiration 88, 119
    for CLPP method 214
    for enzyme assays 162
    for phospholipid fatty analyses 208
    need for replicates 26–27
    sample storage 28–30
    timing 27
    use of sieving 27
signalling: microbial and plant 229–230
SINDI (Soil Indicators scheme New Zealand)
    20–21(tab)
Sixth Environmental Action Programme (6th
    EAP, EU) 50–51
soil inoculum potential 258–259, 262–264
soil organic matter (SOM): living *vs.* non-
    living 53–54
Soil Quality Index: Dutch Soil Quality
    Network 39–40
Soil Science Society of America (SSSA)
    20(tab)
soil: effect of type on quality indicators 40–44
spore germination bioassays 251–253,
    266–267
stains: for microbial counting 93–95, 99–100
    see under counting: microbial cells
stress: time dependence 4
substrate-induced respiration (SIR) method
    9, 11, 30
    disadvantages and limits 92

materials and equipment 86–88
overview 84–85
principle 85–86
procedure and calculation 88–90
relation with microbial biomass 85(fig),
    87(fig)
sources of variability 91(tab)
technical advantages 90
use in ecotoxicological risk assessment
    92
use in soil quality monitoring 91
suppressiveness, soil
    concept 257–259
    Fusarium bioassay 259–262
sustainability, agricultural 51–52, 59–60
Sweden 27–30
Switzerland 28–29, 32, 35
symbiosis *see* mycorrhiza, arbuscular;
    rhizobia

temperature: for sample storage 28–29
terminology 17–18
The Netherlands *see* Dutch Soil Quality
    Network
thymidine, incorporation of *see under*
    growth, bacterial
tillage: effects on soil quality 67
toxicity: mycorrhizal assay 251–254
Trichoderma spp.
    cultivation-based monitoring 281–284
    positive effects on soil 272

ultraviolet light: oxidation by 80–81
United Kingdom 28–29, 32
United States 18, 20(tab)

validity 16
variation 38

zinc: effects on bacterial DNA profiles 34

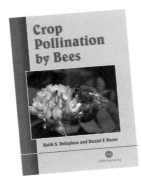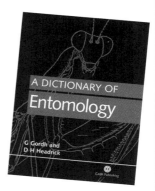

NO FEAR

BUSINESS LEADERSHIP IN THE AGE OF DIGITAL COWBOYS

Pekka A. Viljakainen has been an ambassador between business executives and technology teams, handling complex governance and political topics inside large international organizations and delivering results for over twenty years.

Mark Mueller-Eberstein (Müller-Eberstein) is a business accelerator, coach, speaker and the author of *Agility: Competing and Winning in a Tech-Savvy Marketplace* (Wiley, 2010), and has over fifteen years of global leadership experience in the IT industry, helping companies, governments and their leaders to succeed.

Marshall Cavendish publishes an exciting range of books on business, management and self-development.

If you would like to:
- Find out more about our titles
- Take advantage of our special offers
- Sign up to our e-newsletter

Please visit our special website at: www.business-bookshop.co.uk

NO FEAR

BUSINESS LEADERSHIP IN THE AGE OF DIGITAL COWBOYS

PEKKA A. VILJAKAINEN
with
MARK MUELLER-EBERSTEIN

AND EIGHT VISITING FRIENDS QUESTIONING OUR THOUGHTS ...

Copyright © 2011 WSOYpro Ltd. (WSOY)

First published in Finnish by WSOYpro in 2011, Helsinki, Finland

Translated from the original Finnish by Arttu Tolonen

First published in English 2011 by Marshall Cavendish Business
An imprint of Marshall Cavendish International

PO Box 65829
London EC1P 1NY
United Kingdom
info@marshallcavendish.co.uk

and

1 New Industrial Road
Singapore 536196
genrefsales@sg.marshallcavendish.com
www.marshallcavendish.com/genref

Marshall Cavendish is a trademark of Times Publishing Limited

Other Marshall Cavendish offices:
Marshall Cavendish International (Asia) Private Limited, 1 New Industrial Road, Singapore 536196 •
Marshall Cavendish Corporation, 99 White Plains Road, Tarrytown NY 10591-9001, USA •
Marshall Cavendish International (Thailand) Co Ltd. 253 Asoke, 12th Floor, Sukhumvit 21 Road,
Klongtoey Nua, Wattana, Bangkok 10110, Thailand •
Marshall Cavendish (Malaysia) Sdn Bhd, Times Subang, Lot 46, Subang Hi-Tech Industrial Park,
Batu Tiga, 40000 Shah Alam, Selangor Darul Ehsan, Malaysia

The right of Pekka A. Viljakainen and Mark Mueller-Eberstein to be identified as the authors of this
work has been asserted by them in accordance with the Copyright, Designs and Patents Act 1988.

All rights reserved

No part of this publication may be reproduced, stored in a retrieval system or transmitted, in any form
or by any means, electronic, mechanical, photocopying, recording or otherwise, without the prior
permission of the copyright owner. Requests for permission should be addressed to the publisher.

The author and publisher have used their best efforts in preparing this book and disclaim liability
arising directly and indirectly from the use and application of this book.

All reasonable efforts have been made to obtain necessary copyright permissions. Any omissions or
errors are unintentional and will, if brought to the attention of the publisher, be corrected in future
printings.

A CIP record for this book is available from the British Library

ISBN 978-981-434-666-5

Printed and bound in Great Britain by CPI William Clowes

Contents

Foreword and acknowledgements		vii
About the contributing authors		xi
A note about trademarks		xiii
Introduction		1

1 Fear equals failure — 8
Today's status: changing business fundamentals and orbiting leaders — 12
Left side of the brain versus the right side of the brain — 25
The Digital Cowboys of the PlayStation generation — 32
Your contribution to **www.nofear-community.com** — 34

2 Why should I follow you? The challenge from the leader's perspective — 37
Leaders from the Industrial Revolution enter the digital age — 37
What is the fundamental change when leading an enterprise? — 40
What makes you authentic? — 41
What is your value creation? I mean, really? — 48
Welcome to the Professional Service Firm culture — 58
Clarity, risk management and simplification — 72
Bill Fischer: Leaders and cowboys: unleashing talent in the digital age — 75

3 Hey, old man. What do you know about me? The challenge from the digital cowboy's perspective — 87
Hey, Boss — this is not just work — 88
What do you expect from a leader? — 89
Digital Cowboys as part of an organization — 95
What do we mean by a truly international experience? — 99
Philipp Rosenthal: Avoiding negative organizational gravity while becoming a Digital Cowboy — 103

4	How does an industrial-age relic turn into an authentic leader?	111
	How many mistakes did I share today?	114
	Anatomy, physiology, psychology or psychology, physiology, anatomy?	117
	A leader is a producer, not an invisible delegator	120
	Is Facebook my value creation network?	123
	Choosing people: old world versus new world	127
	Me and my chairman	131
	Kari Hakola: The CEO as a number-one change agent	133
5	What I as an executive should change in my company, in practice	142
	Theoretical versus practical models	142
	Organizational change in the age of the Digital Cowboy	150
	Victor Orlovsky: What will the future be? Is the world vertical?	172
6	Places of magnificent growth and magnificent failure – emerging markets	191
	The amazing leap from second to fourth generation	192
	How to add value for Digital Cowboys in emerging markets	196
	Arkady Dvorkovich: New leadership for new leaders	206
	Birger Steen: Leading a team of Russian super-professionals	213
	Alex Lin: The Internet is speeding up the integration between China and the world	223
7	Technology – your saviour or your nemesis	232
	Information versus intelligence	234
	The leader's role in all this change	237
	The consumerization of IT	243
	Simplify, amplify and the IT infrastructure	250
	Simplification versus customization	254
	Learning from previous technology cycles	257
	The next technology cycle and new leadership	263
	Mårten Mickos: Building the next-generation enterprise	279
8	Fearless means stupidity. NO FEAR can mean success	288
	The burden of doing only the right things	291
	Complete transparency and trust are key	294
	The grand finale: conclusions	296
	Bibliography	297
	Notes	300

Foreword

IN JUNE 2009, I started to experience some health-related problems. After thousands of flights and meetings around the world, it was time to stop. It sounds dramatic, but at the ICU I realized we each have a finite amount of time on Earth. More importantly, I decided I would spend the rest of mine concentrating on those things I cared for most.

I'm a 38-year old father of two children, a nerd past his sell-by date, a helicopter pilot and a corporate leader. I could list my critical battles and my areas of focus for each of these roles. I could also list all the mistakes I've made in each role, all the times I totally screwed up.

In this book, I will concentrate on role number four – corporate leader. Throughout my career, I've had the honour of meeting very talented people and working with them. I've seen and experienced good leadership, both in what I do and what others have done. I want to understand leadership and, most of all, grow as a leader. That's why I started working on this book. I've learned that you grow the most when in the company of your superiors, when you have the courage to state your opinion and are prepared to discuss it. That's how I worked when writing this book. I didn't do it alone. I did it with friends. I certainly don't consider this the final truth about leadership, but I do hope we can present opinions that are good enough to warrant further discussion and development. To be totally honest, I decided to write this book for very selfish reasons. Before I retire in 2040, my most important task as a leader is to recruit and train dozens, maybe hundreds, of new leaders. In order to do this, I need to learn and I need to do it fast.

This is not a textbook. Reading this book from cover to cover will not turn you into an Alexander the Great for the PlayStation generation. This book does not contain academically proven truths or carefully documented interviews or any other statistical data. This book is based on informal interviews, several dozen leadership training and coaching sessions and, most importantly, on all the mistakes I've made in my career thus far. I'll share some quotes and thoughts from colleagues and friends that I found interesting. Some are anonymous, others not.

The reason why I wanted to document my findings in one book is that transparency and sharing are at the heart of the change the leadership culture is undergoing. Knowledge is no longer power, even though knowledge workers are at the top of the value creation hierarchy. Power is no longer based on organizational hierarchy or the mandate you were given. Your power and your worth (horrible words both, by the way) are determined by the concrete value you can create for your team and organization. The constantly changing nature of your organization makes this harder. The digital revolution accelerates the change. In this book, I talk a lot about the PlayStation generation (as you know, PlayStation® is a registered trademark owned by Sony) and the people who best symbolize and embrace change – the Digital Cowboys. They are change agents who determine how we need to lead in the future. The change is most apparent in the so-called "emerging markets". In this book we pay special attention to international networks and markets, as well as what it takes to find success in them.

The biggest step I took in my career was when I got into a 100 per cent international situation five years ago. When I left bucolic Joroinen in Finland in the early 1990s, I'd visited six countries and five US states. I thought I was a real man of the world. I thought I knew something about what it means to be international. Two life-changing moments made me realize how totally wrong I was and that I had so much more to learn. First, I fell in love with a Russian pianist whose family lives in Columbia, Peru, Russia and Kazakhstan. Soon after, I moved from my nice glasshouse in Sipoo to Munich, where I was offered the opportunity to lead a team of 8,000 people with members in 20 different countries.

After a few months I realized I knew about 0.1 per cent of what I needed to know to lead a truly international team. I'd gone through several dozen training sessions aimed at leaders. I'd read a lot of business books and could order a Campari and orange juice in 30 different languages. Regardless of all this, I felt naked and exposed. Even though I was a technically skilled leader, my obstinacy and my clearly limited worldview were unpleasant surprises. I had a choice to make: was I going to pretend I was capable of generating added value everywhere I went, or was I ready to admit to my ignorance and start learning, fast?

As I prepared to write this book, I decided I'd take a page from the Digital Cowboy playbook – work as a network and be open and ready for critical discussions. It was my ambition to get some of my talented friends to take part in the writing of the book and in openly questioning some of my simple thoughts. I succeeded. Each chapter in the book includes a guest columnist from a different business background and country. I'm genuinely happy about their contributions.

I'm not a writer. I'm a corporate leader. I'd like to apologize for my at times clumsy way of expressing myself. My aim is to draw as clear a picture as possible of the challenges we face, as well as my partial solutions for them. I hope they'll be of use to you some day.

I'm always a little disappointed if in my interactions I am unable to spark off some fiery debate and exchange of ideas. The same goes for this book. To make this technically as easy as possible, we've built a discussion forum where you can share your questions and even your most controversial thoughts. If a brilliant idea or relevant criticism pops into your head while reading this, go to **www.nofear-community.com** and share them with the other readers. For me the only key performance indicator of this book is whether it moves you, me or the other leaders a step in some direction – preferably forward.

Dear fellow leaders, have no fear and enjoy.

Pekka
Joroinen, 1 March 2011

Acknowledgements

We believe that NO FEAR is more than a book; it is a community. This book is only the foundation. The community will continue to grow and increase in value and insights through the ongoing discussions and contributions beyond this printed text.

Creating the content and the book itself was only made possible by the generous sharing of insights and experiences from people around the world. Trying to have a detailed acknowledgement in this book would be either far too long or inexcusably incomplete. To support the spirit of the growing community, we turned our humble acknowledgements and sincere thanks into a living document. You can see the individuals (and their social media connections) who have been and are instrumental in developing "Business Leadership for the Digital Age" at **www.nofear-community.com/acknowledgments**.

In addition to all the leaders, experts, Digital Cowboys, publishing professionals and personal support system who created this book and the community with us, more thinkers and leaders are joining the community every day. Even before the book manuscript was finished, more than 100 people contributed their comments, support and insights to our project. The release of the NO FEAR book is just a beginning of our journey and debate. The story continues at **www.nofear-community.com**.

Thanks to all.

The NO FEAR team

About the contributing authors

Arkady Dvorkovich (*Аркадий Владимирович Дворкович*) is a recognized Russian economist who was appointed Aide to the President of the Russian Federation in May 2008. He was also selected by President Medvedev to be the "Russian Sherpa" to the G8 and G20 summits. Fluent in English and German, Dvorkovich has played a prominent role as an international voice in Russian economic policy. Being responsible for internal and international economic affairs, he is pushing President Medvedev's plans of Russia's economic modernization forward.

Bill Fischer is Professor of Technology Management at IMD in Lausanne, Switzerland. Professor Fischer has been involved in technology-related activities his entire professional career. He was a development engineer in the American steel industry and an officer in the US Army Corps of Engineers, and has consulted on R&D and technology issues in industries such as pharmaceuticals, telecommunications, textiles and apparel, and packaging. Additionally, he has served as a consultant to a number of government and international-aid agencies on issues relating to the management of science and technology.

Kari Hakola, former senior vice president of Tieto, has been a business leader in IT services and management consultancy for four decades. His clients ranged from retail and telecom to banking and insurance, with a focus on enabling business changes through IT and business services. During his forty years of experience, Kari has witnessed the extensive influence technology has had on business models, practices and competencies.

Alex Lin (Lin Yong Qing), founder and CEO of ChinaValue.Net, is a widely regarded international business and information technology authority. Alex is the founding partner of SuperValue International Inc., the premier investment-consulting firm in China. He was a senior executive at Intel Corporation and the Deputy General Manager and VP of Marketing for TsingHua TongFang Computer Ltd. He is a top business expert on new media in China.

Mårten Mickos is the CEO of Eucalyptus Systems, the leader in open source cloud computing platforms for on-premise use. In his previous position as CEO of MySQL AB, Marten guided the company from a garage start-up to the second largest open source company in the world. After Sun Microsystems acquired MySQL AB for $1bn, he served as Senior Vice President of Sun's Database Group.

Victor Orlovsky (*Орловский Виктор Михайлович*). As Senior Vice President and CIO of Sberbank, Victor Orlovsky has an extensive record of expertise in both the Russian and international finance and banking sectors. Recognized for his knowledge, Orlovsky is also a member of several supervisory boards and boards of directors. He has degrees from Tashkent Electro-technical Institute of Communication, Moscow State University of Economy Statistic and Informatics and an MBA from the University of Warwick.

Philipp Rosenthal is a marketing and industry sales professional and an enthusiast about the future of a social media inspired workplace. In his current role as "Future Office Evangelist", he is leading Tieto's international solution area for the digital workplace of information and knowledge workers.

Birger Steen is President of Parallels, the world's leading enabler of cloud computing services. Birger started his professional career in 1992 as an oil trader at Norwegian Oil Trading, where he opened the company's first office in the former Soviet Union. He joined McKinsey & Company in Norway and then moved to Schibsted ASA as VP of Business Development. In March 2000, he was named CEO of Scandinavia Online AB. In 2002, Birger joined Microsoft, working in Norway, Russia and Redmond, Washington.

A note about registered trademarks

THE FOLLOWING TRADE NAMES are mentioned in this book, some only occasionally, others frequently. All are registered trademarks and the legal property of the owners named below.

Amazon® is a registered trademark of Amazon.com, Inc.

American Idol® is a registered trademark of 19 TV Ltd and FremantleMedia North America, Inc.

Apple® and iPad® are registered trademarks of Apple Inc.

Facebook® is a registered trademark of Facebook, Inc.

Google Android™ and Google Chrome™ are trademarks of Google, Inc.

GroupOn® is a registered trademark of Groupon, Inc.

LinkedIn® is a registered trademark of LinkedIn Corporation.

McDonald's® is a registered trademark of the McDonald's Corporation.

Mercedes® is a registered trademark of the Mercedes-Benz Division of Daimler AG.

Microsoft® Windows and Microsoft Windows® Phone 7 are registered trademarks of Microsoft Corporation.

MTV® is a registered trademark of Viacom International Inc.

Netflix® is a registered trademark of Netflix, Inc.

No Fear® is a registered trademark of No Fear Corporation.

Oracle® is a registered trademark of Oracle Corporation.

PlayStation® is a trademark or registered trademark of Sony Computer Entertainment Inc.

Qualcomm® is a registered trademark of Qualcomm Incorporated.

Samsung® is a registered trademark of Samsung Electronics Co., Ltd.

SAP® is a registered trademark of SAP AG in Germany.

Siemens® is a registered trademark of Siemens AG.

Starbucks Coffee® is a registered trademark of the Starbucks Corporation.

Tata Nano® is a registered trademark of Tata Sons Limited.

Tieto® is a registered trademark of Tieto Oyj.

Toy Story® and Walt Disney® are registered trademarks of the Walt Disney Company.

Twitter® is a registered trademark of Twitter, Inc.

Volkswagen® is a registered trademark of Volkswagen Aktiengesellschaft.

Volvo® is a registered trademark of Volvo Car Corporation.

Wikipedia® is a registered trademark of the Wikipedia Foundation, Inc.

Introduction

DURING THE FIRST 30 YEARS of my life, I never gave a thought to leading and improving people. I was primarily interested in computers, programming and doing something no one had done before. At the Lappeenranta University of Technology, I concentrated on courses dealing with corporate management, accounting and financial reporting. I completed the courses with great gusto and dedication, but – as far as I can recall – no one ever said anything about the challenges of leadership.

I can't say I was really frustrated at school but I had other things going on which preoccupied me. I had started my own business five years earlier, thanks to a sizeable loan guaranteed by my father. It was the early nineties and Finland was in the grip of a deep recession. The cash flow was very much in the red. I ended up having to use my and my girlfriend's student loans to pay my employees. Clearly, I had to make some hard choices. So I ended my academic career, hired some key personnel from school and got serious about being an entrepreneur.

My great-grandfather, A. J. Viljakainen, was a scrap dealer in the early years of the twentieth century. He sold everything from sewing machines to cars. He was impulsive by nature and, as a young man, left home with nothing but his confirmation suit. His life's work was dealing and he was good at it. A. J. was a talker and never shied away from a good debate or argument. My grandfather,

Kyösti Viljakainen, was an engineer and primarily interested in technology. At the end of his career, he retired from a vice-president's position at TelevaOy. Televa later became TeleNokia, which in turn turned into Nokia. Kyösti was an excellent grandpa, but a little withdrawn. He hated crowds and I'm pretty sure he'd think Facebook was the stupidest thing he'd ever seen. Who the hell would be interested in how someone else is doing?

I grew up in a very different environment. My father was a psychologist and my mother a psychiatric nurse. When I was a small child, we moved to the family estate out in the country. My four siblings and I literally grew up in the middle of the forest. Things like corporate management, entrepreneurship or improving people were never discussed at home. My father always dreamed of running a small company, but – as a humanist – chose a different road to travel.

As a small child, I wanted to be a performing artist – a ringmaster in a circus or a rock musician. I had a strong narcissistic bent very early on. When I was twelve or thirteen, a teacher told me that young people are supposed to have hobbies. Since we lived in the middle of the forest, my choices were limited. I decided to go to the library. I asked the librarian about books related to hobbies and was directed to a shelf with maybe a dozen books on it. The tomes about gardening and hunting didn't appeal to me at all. So I ended up holding the manual to a Tandy 80 computer. I read it, got hooked and became a nerd.

Like all good nerds, I had zits, wore loose, ill-fitting pants and never combed my hair. What made me different was the fact that I had received an open-minded and social upbringing at home. From my great-grandfather, I inherited a burning desire to sell. My grandfather gave me an avid interest in technology and business. I was seven years old when he started reading Finnish business magazines with me and telling me about how to close a deal. I still vividly remember a particular lesson my grandfather stressed back then: "Try to avoid entering into contracts with Siemens in

Germany or IBM. They have so many lawyers and their contracts are so thick, they'll cheat you every time." I was always an obedient boy, but this is one commandment I have had to break later in my career.

Since being a performing artist was not the most likely career trajectory for an overweight, zit-faced boy living near the Soviet border, I decided to find another route to the limelight. I started training people in the use of the technology which I so dearly loved at various events, courses and seminars. To get around to my business appointments, I bought a '69 Volvo Amazon and hired a chauffeur. At the age of fourteen (four years before I could get a driver's licence in my home country), I became an entrepreneur.

I've since often thought about my early years in business. I was good at teaching and training because I loved getting people excited about technology. At the heart of leading an expert organization and managing a business is the ability to get people excited about what they're doing, about each other as experts and about the team as a whole. You can't order your team to be excited about the work they do. You have to cultivate the environment. A fancy title might buy you two minutes of concentrated listening, but little else. I believe that the leader of an expert organization can only succeed if he or she is willing to personally choose, train and challenge his or her team. When I say train, I'm not referring to developing the individual in their specific area of expertise. I'm talking about supporting them in working really hard – together. I think a leader is primarily a teacher and coach. Making decisions and taking responsibility follow.

Several of my junior and high-school teachers were models for what a good teacher could achieve. Working with them left an indelible mark. We worked on our first software projects nights and weekends. I was a bohemian nerd and liked to fly by the seat of my pants. My instructors, Jäppinen and Suur-Askola, were merciless about even the smallest detail. We might work on a couple of screens of animations for weeks on end, striving for perfection in details that to me seemed completely ridiculous. But they kept me involved and excited about the work and taught me to appreciate not only the result but the path.

Their leadership moved me. These teachers were 40 years older than me. Instead of just articulating the problem and retreating into the teachers' lounge to see what I could come up with, they were present from beginning to end. Despite the vast difference in age and life-experience, I always felt like an equal partner in the project and the team. I was respected for my skills and I, in turn, respected the other people in the team. Those were good times.

Fast-forward to the mid-1990s. I have my own company (Oy Visual Systems Ltd), a good 50-person team, positive cash flow and a phenomenon called the Internet is about to happen. A huge stroke of luck came my way when my company got the opportunity to design a first-generation Internet bank. We were competing against some heavy hitters. Our team worked non-stop and really believed in what we were doing. We won the design competition, got the job from a major Finnish bank and designed the next four generations of their online bank. The rest is history. I had learned many of the things that contributed to this victory in those computer classes at my small rural school.

Building an Internet bank had technical challenges but the true challenge of the project was making the system usable for both the end-user and the bank's staff. This was the first time that consumers were given real-time access to information systems the banks had been developing for decades. The bank teller was no longer there to serve as a translator. If consumers didn't "get it" right away, they simply weren't going to use online banking.

Strong leadership and management were instrumental in the process. A bank – a very traditional and hierarchical organization – had to learn to knuckle down and work hard as a team. A new battle tactic was clearly needed. Strict orders percolating down the hierarchical chain would've worked over time, but there wasn't enough time for the traditional approach. We had to create a collaboration model that was fundamentally different to their previous way of working. A bank manager friend of mine stated his experience clearly: "I'd worked in a bank for thirty years. The netbank development project was the first project where all the different departments had to work together." The project itself not only became the foundation for our business, the insights the team and I gained on the technical, organizational and people challenges started my journey to becoming the leader I am today.

To really understand the operational environment, let's look at some facts:

- The client was a strong, traditional bank over 100 years old with 11,000 employees.
- The development team consisted of 90 people with an average age of 26 – not your traditional bank employee.
- The daily work was on a secret project doing something completely new – no best practices on which to rely.
- The bank's employees feared losing their jobs to automation. A deep banking crisis was still fresh in their memory.
- Many practices (security, consumer privacy, legal questions) had to be submitted for official approval during the process without bringing the project's progress to a crawl.
- There was a true *fear* of failure at all levels of the organization and within our team.

Leadership – getting everyone at the bank excited about what was happening – was a challenge. During the entire development process, we invested innumerable resources in explaining the coming change to various target groups. We spent countless days as travelling preachers, telling people openly about the impact of the changes that were coming. This was no walk down the red carpet, hands held aloft in victory. It was only a few years since thousands of bank employees had lost their jobs in the banking crisis of the early nineties, and we were coming along to tell people that many of their jobs would be transferred online.

Management challenges came in the form of coordinating small teams of 10 to 50 people that were working hard on various facets of the development project with various bank departments. These teams were given a lot of freedom in their work, but also the responsibility to produce solutions that fulfilled the stringent security and quality requirements of the banks.

Despite all this, we succeeded. We easily surpassed the goals set by the customer and created something we could be proud of. For me as a manager, this process produced a model for creating new, business-critical and customer-centric solutions. Without going into detail about the solution itself, the process changed my notions of leadership and its challenges. First of all, it was plain that leadership had to extend beyond our own "gang". During a fire, you concentrate on putting out the fire in your neighbour's house with as much energy as you would a fire in your own home. Otherwise, you will stand by and watch both houses burn to the ground. Leadership requires excursions into the neighbour's – your customer's – yard. You're in this together. Naturally, some people are annoyed by this sort of thing. I am not. For me, it is about getting the right results with the right people. Hence, I have picked up the nickname "the Bulldozer".

Based on this success, the company grew to 150 employees in the mid-nineties and we were on top of the world. We had challenging projects. We had our pick of the best experts on the market. Our financial success made it possible to really invest in people in a variety of ways. The one thing I didn't invest in was bringing up leaders and expert managers in my own organization. I should have.

In the twelve years that followed that initial success, many of these questions critical to growth started to dawn on me in phases. I also got to learn from some great leaders. For example, CEO Matti Lehti, the man who created the Tieto group (Oy Visual Systems Ltd merged with Tieto in 2000) and led it to rapid growth, was exceptionally interested in developing people. His own background in finance precluded a deep technological dialogue between us, but as team builder and leader he was in a class of his own.

Under Matti's leadership the company invested in developing leaders. For the first time, I led teams with non-Finnish members. I had teams of truly diverse and experienced employees and I took advantage of this. I made it a point to discuss matters unrelated to technology and service development with leaders of our customer organizations for a very simple reason. In addition to technical innovation, many of our projects were more and more tied to the deep changes our customers were going through – both in their own organizations and their customers'. All this forced me and my team to think about how management and leadership would have to learn and adjust, particularly in international organizations.

During our strategy-setting in 2004, our management team wrote for the first time about the PlayStation generation entering the job market. We predicted that digitalization and customer-driven technology trends – the currency of the PlayStation generation – would permanently alter the way companies, including ours, worked going forward. And this would have a major impact on how management and leadership developed.

We were wrong on one count only. We underestimated the speed and depth of the coming change. For this book, I built on these experiences and pulled together thought-leaders across the globe to help you understand and leverage this change instead of being swept away by the tidal wave of transformation.

1 Fear equals failure

WHEN WE WERE YOUNG, my brother was quite into motocross and NO FEAR was one of his favourite companies. He and I shared a room and I'd stare at his NO FEAR poster on the wall every night as I fell asleep. That poster has stuck with me throughout my life. When I was thinking about this book, NO FEAR really illustrated the beliefs and experiences I wanted to share. The idea has a profound connection to leadership and improving people. Thus, I borrowed the name of the company for this book.

It is depressing that – at least in the European corporate world – fear kills new innovations. Executives seem to have a general fear of failure and it holds them and their organizations back. This is especially depressing knowing our continent's great history of innovation. We have the capital to finance innovation. Globalization provides Europe with human resources at as low a rate as everyone else. There is a highly developed educational system. Companies have access to support networks and globally competitive amounts of general knowledge, language skills and benevolence. The societies are, on the whole, very democratic and transparent. There's not much corruption to slow us down. So why are we constantly in danger of being left behind?

Some consider the idea of executive fear naive. But consider the conclusions that the team behind this book discovered when analysing their own companies and clients:

- *All companies must become expert organizations to succeed.* If this conclusion is true, not even something as basic as a mine can generate billions in revenue for its owners just by digging and processing more matter. Instead, the company needs to gather and distribute the expertise and thoughts of all their employees so fast that no new claim or mineral that may be relevant a year from now is missed.

- *You can't lead an expert organization as you would an industrial one.* An executive can no longer take refuge in the safety of highly developed processes. Communist societies had a rule and a regulation for every eventuality under the sun. That model failed to bear much worthwhile fruit. Exceptional individuals received their due when it suited the leadership, but they failed to transform individual achievement into global success. This point is particularly important when we look at our next conclusion.

- *The PlayStation generation – workers born after 1985 – will produce the billion-dollar innovations in the coming years.* No executive can dodge the fact that his or her most important task is integrating the PlayStation generation into the organization. The language, values, freedom and social context for this generation are very different from what executives learned in the 1970s, 1980s or 1990s. It's easy to misjudge this generation. On the surface, they look like a bunch of individualists and online gamers who march to their own drum. From a management perspective, that conclusion is a fatal flaw because networking and collaborating on shared projects drive these people to top results.

- *Leading the PlayStation generation will require community, authenticity and leadership at the forefront of the troops.* A leader can no longer stay on the hilltop, shouting down encouragement to the toiling troops: "Hey, guys, it's me, your fearless leader! I really appreciate what you're doing down there. Do it even harder ... whatever it is ..." The leader has to be visible on the front line – as a translator, a go-between and someone who takes responsibility.

If any of these four conclusions is true, executives have been unceremoniously tossed out of their comfort zones. The idea behind this book is to make you think about how to get used to that uncomfortable zone, how to stay there in good times and bad, and how to generate value for new talent and the organizations we lead.

Years ago on an aeroplane over the Pacific Ocean, I was cranky and bored. I couldn't sleep so I turned on the video monitor. A movie about Alexander the Great was on. It was one of those moments when ideas come together and crystallize into a bigger picture. Over the next few months, I looked at the legend of this conqueror from all angles and from all possible sources. Without getting too deep into the history of it all, I found many interesting parallels between Alexander's life and the theories we had been working on.

Alexander was an arrogant young aristocrat born with a silver spoon in his mouth. According to legend, he was not one to fear a superior enemy. His contemporaries probably thought he was overly arrogant and foolhardy. Had he been a different person, he could easily have ascended to the throne that was his inheritance and spent his life engaging in border skirmishes with his neighbours and playing the political games that are a part of every court. But no, he wanted to conquer the known world.

The history of the world is full of conquering megalomaniacs. What set Alexander apart – making him a legend during his lifetime – was his desire to ride in the vanguard with his troops. But even that wasn't enough. He wanted to personally lead the critical troops that were most vulnerable to being killed but who also had the most decisive strategic significance. He could make unpredictable strategic decisions in the middle of the battle because he was right there and had to live with his decisions. He could see the expressions on the faces of his men and use his highly developed situational awareness to make corrections when necessary.

Which parts of the Alexander the Great legend are true and which parts were written by his corporate communications department is irrelevant. The important part is that in an expert organization, a leader's most important task is to be where the battle rages – in the middle of the throng consisting of the troops, the customers and the competition. And like Alexander, a leader has to be present as an ordinary mortal, ready to take the same blows the troops take. As soon as you delegate that role, your troops get nervous, your customers start looking at other service providers and your competition realizes they're holding an ace in the hole.

Many good organizations come to a point where protecting the status quo for the leading individuals becomes the main motivation and reward. Taking risks and failing are more frightening for those who are supposed to lead the troops than foreseeable mid-term consequences for the core business and long-term health of the organization. Organizations that can take calculated risks, that chart new territories and take advantage of new opportunities, are the ones that stay ahead. And they need the leaders who are not afraid to fail.

Summary: Fear of failure and an unhealthy distancing of leadership from the front line are a cancer that eats away our growth potential and competitiveness. The West – Europe in particular – can't compete against the rest of the world with demography or cost structures. We face an overwhelming army of experts as Alexander did in Persia. We have to be able to do an hour's work in fifteen minutes and do it better than the others. We have to advance with purpose and without fear.

At home, I was taught never to belittle other people's feelings – especially fear. The purpose of NO FEAR is not to belittle these feelings, but to process them and work to find concrete solutions to dealing with them. If you are not willing to deal with fear and find a solution to these challenges, you have no business being a leader. There are many other challenging and interesting jobs in the world.

▶ Today's status: changing business fundamentals and orbiting leaders

In my own cosmology, today's business firmament has six planets. All affect people's development and organizational success.

- Productivity is stalled.
- Globalization impacts competitiveness and leadership.
- Capital is flowing in different directions.
- Consumer behaviour changes.
- The network economy is now the mainstream.
- The innovation race has changed.

In addition to these six planets, there are many stationary stars in the sky that define the nature of leadership. However, they do not fall within the scope of this book.

Stalled productivity

The point of technology was to make life better and increase productivity. Corporate investments in IT combined with the depression of the early nineties increased the productivity of the knowledge worker. The millennium focused people's attention on whether companies could keep their computers running. The millennium was also used to sell a considerable amount of new software and additional functionalities. I did this, too. After the millennium economic growth accelerated, but in many industries the growth of productivity stalled. Ask yourself why.

For one, measuring the productivity of knowledge workers is no easy task. It is not like measuring the productivity of a telephone service centre. At a service centre, you can easily measure how long it takes to resolve the customer issues, as well as customer satisfaction. You can measure the time it takes to process invoices and the amount of human resources tied to this process. These are all simple transactions. But how do you measure the efficiency with which a doctor, hospital or insurance company handles information related to a patient? How do you optimize the use of the expert resources tied to that process? How do you even identify the processes? The more creative the job, the more multifaceted the communication and sharing of information between people becomes. However, just because measuring productivity is hard, it is not a sufficient excuse for abandoning real productivity development and measurements. And measurement is possible. Together with John Henderson and his team from Boston University, Mark Mueller-Eberstein looked into the effectiveness of teams in managing and growing knowledge and generating productive results. Their research paper and Mark's book *Agility* clearly outlines what processes and tools and technologies those highly effective teams utilize and emphasizes that the impact is actually and clearly measurable.[1, 2]

Globalization and the networked economy have also had a huge impact on productivity. Ten years ago, a construction project could be managed within a company, often by a team of people within shouting distance of each other. Today, construction projects involve a diverse and dispersed network. Yet projects still require the management of processes and precise agreements that cover the smallest details. Add to this the fact that as soon as the network is given a new product, a new material, a new continent on which to work or maybe a completely new architecture, productivity plummets. While construction is a traditional business, the industry has faced these challenges and become a productivity pioneer of the network economy. I'm convinced that the way global construction companies work today will – within the next twenty years – become the model, more or less, of every industry and all companies.

The global nature of today's networks contributes to making productivity challenges just a little more of a hard slog. As long as you can stroll through a construction site and see all the members of the network face to face, there's the potential for understanding. Global networks make leadership more complex. You have to optimize a production process in which the members will never smell each other or talk to and challenge each other face to face.

Maintaining productivity development is the most important measurement of a leader's effectiveness, and since we're all engaged in a network-like expert business, the productivity of our *experts* determines who wins. As a leader, your challenge is to create a culture that supports these experts – a culture based on total transparency and two-way value creation. The key to restarting stalled productivity lies in the real-time and transparent interaction and leadership of expert *networks*. There are many leaders who have spent their entire careers blowing up walls *between divisions, countries and profit centres*. The challenge for the next decade of leadership is to do away with the walls *between individual experts and micro-teams*. This simplification tends to drive organizational structures and their leaders headlong into an impossible situation: while a leader is supposed to be in the front line with the troops, the corporation tends to make it impossible by creating units much too large for the front line. We will discuss this topic more in Chapter 2.

The impact of globalization on competitiveness and leadership

Most of the time when people talk about globalization, the only thing they talk about is the price of work – salary levels. To me this is a crass simplification. Adding 15–30 per cent in extra administrative, freight and travel expenses on top of the hourly rate of a business might work if we're talking about industrial production – such as making can openers. For the leader of an expert organization, globalization must have a more profound meaning than just cheap labour.

Countries like Russia, India and China historically offered really cheap, top-flight know-how. In the past, this was a primary reason for establishing a base in these markets. But in the coming years, the list of reasons for establishing a base in these countries is going to look like this:

1. Knowledge of growth markets
2. Availability

3. Level of expertise
4. Demographics
5. Network economy prowess

Globalization and internationalization of your existing and potential workforce puts your expert networks on steroids and you need to rethink how to navigate this broader world. The PlayStation generation – experts in global networking – has infiltrated the ranks of most international organizations. Years ago, they learned to ideate, share and argue across international networks playing online games. While language and cultural barriers haven't disappeared, these experts do not have a problem cooperating in an international network – unlike your average executive in a Western company.

From a leader's perspective the challenge culminates in two ways. First, you have to create your authenticity as a leader and show real added value without necessarily ever seeing the other person. I like to talk about creating a trusted relationship. It's the challenge of using a small, smudgy picture conveyed via webcam to build a relationship that lasts years and is, in some ways, comparable to marriage.

The second challenge is establishing equality. The first decade of globalization created the A team and the B team – expensive teams in the West and cheap teams in developing countries. This is no longer the case, but some hopelessly incompetent leaders, to this day, like to speak of "low-cost units". Implementing equality has been hard in the past, but without it an organization can neither be transparent nor tap into their insanely broad knowledge bank. Chatting around the proverbial coffee machine in your mother tongue is no longer the easiest way to get things done.

In my own work, I've made it a point to treat colleagues working 20 kilometres away as if they were thousands of kilometres away. Instead of getting in my car and driving to the office, I stay at home and take part in meetings via video. Showing off the technology was not the point. The point was to level the playing field for the members of the team who were in Russia, Latvia, Germany and my own office. The value of each member's contribution is thus dependent on their individual expertise. Behaviour of this sort – no matter how soft a value you consider equality – should be foremost in a manager's mind when working.

And I am not alone in this work style. Companies like Shell have fully embraced a culture of virtual work wherever possible. And they understand that technology alone is not sufficient to unlock the potential. Guidance, best practices and processes need to be clearly thought through, adapted and communicated to make the new technologies a success. At Shell, for example, the rule is that even if only one meeting participant is not in the same physical location as the others, rather than meeting in a conference room all should "dial in" from their offices. This way the "remote" participant is treated in exactly the same way as the "local" colleagues and can much more easily contribute his or her expertise to the task at hand.[3]

In the coming years, the competition for talent will be fierce. It will be especially gruesome in international know-how centres. If I'm right, no brand will be able to rely on their name and their reputation as a way to keep the top talent in-house. The PlayStation generation will come in, scope out the situation, and see how the team works. If the reality turns out to be a bunch of fluff, they'll be out before the door has fully closed on their entry.

No company or industry can be competitive without bringing together the best know-how from around the world. I contend that the gravest challenge a manager faces today is leading an international expert network in a productive manner. One solution to the logistics versus productivity issue is to keep all key experts near by and the bulk of know-how a little farther out. Or, if this seems too obvious, a leader can build an international sub-level of trusted experts around the personal coffee machine to convey communications and assignments to far-off lands. Either way,

mistakes will probably be made along the way as in any long-distance relationship. Consequently, in 2020 there will not be a single internationally successful business leader who hasn't totally blown it and learned from this in a transparent and global arena.

Capital is flowing in different directions

Recent years have seen a dizzying array of tools and campaigns used to draw investment into developing economies. The first wave of investment consisted of industries such as construction, consumer goods and heavy machinery that were highly competitive in the home markets. This first wave was then followed by practically everything else – with the exception of some heavily regulated industries. The ongoing last wave consists of large, human-resource-intensive service organizations. A cynic would explain this by noting that in the service business the margins are low or the services always local. The truth is that the companies have lacked the resources and ability to transfer and localize their management practices and value creation to a different culture. This is, without a doubt, a difficult process. It's infinitely easier to produce an identical product, or maybe one with a different electrical connection or a modified user manual, for a faraway market than to enter into a profound dialogue with it.

Since this book deals primarily with leadership, we need to look at globalization from a different perspective. The one who has the capital can, if they so desire, steer companies, their investments and sometimes a considerable part of society. This is now widely accepted, and governments as well as companies are engaged in all manner of snappy protective manoeuvres to prevent this or that company from being acquired by the Russians or the Chinese. Only yesterday everyone was talking about the free movement of capital and free trade. Now, after the recession, everyone wants to curl up in their own country and build a wall around it – when it serves their interests.

The changing flow of capital will cause massive changes in Western economies. It's unavoidable that the big car, IT, food, energy and even consulting companies will soon have their first Asian major owners and executives. If the next major investments and corporate deals in the USA – or in the Nordic countries – were made by Russians, I wouldn't be all that surprised. At the same time, investments in Asia will increase. Even though this activity will become more of a two-way street, the direction in which *investment* capital flows will very likely change.

The fact that capital is starting to flow in a different direction will have a major impact on the future of corporate leaders. We're in a hurry to figure out how to acquire the top talent in these countries and integrate them into our companies as equal and motivated colleagues. A lot less attention has been given to what happens when the chat around your "coffee machine" takes place in Russian or when the chairman of the board is no longer the gentleman you once knew, but a colleague with a strong Korean accent doling out orders via a small webcam screen. In this situation you, as a manager, need to have a strong sense of your own vision and the way your value creation network functions. It doesn't matter what language your boss speaks. He will be interested in your ability to lead, influence and innovate as a part of an international team.

Changes in consumer behaviour

Changes in consumer behaviour will force executives to look for lessons beyond familiar industries and cultures. As recently as in the nineties, it was normal for customers to stay with the same insurance company, bank or grocery store all their lives. Not out of any particular sense of loyalty, but because they had no conception of anything better "out there" or any real motivation to change. This applied to the way companies and corporate leaders worked, too. We went to different seminars, saw new processes and technology and observed best practices from all over the world. We definitely looked as if we were looking for something new. Naturally, new things made us excited and sometimes afraid and we jotted down ideas in our seminar booklets. In reality, the most concrete thing we walked away with was an opinion on the taste in ties of the guy sitting next to us. We went home and there was nothing there to force a change. We were safe.

I know people laugh at me and think me an old fogey when I tell them this story. I was buying a present for my godson. I knew which product I wanted and knew I had a deadline (Christmas right around the corner, but all other attributes of the transaction were still open). By pure chance, I came across a website that compared 120 stores that offered the product in question by price, delivery times and customer service. Customer service response times were reported in seconds. The comparison included stores in twelve different countries, and the service could tell me how long it took each store to get the product to my door in Helsinki. It was an eye-opening experience. Technology now lets us find and share information in a matter of milliseconds. Shopping has never been the same.

I've kept a very close eye on changing consumer behaviour in different industries and the speed and depth of change still surprise me. I'm a Finn. Finns – especially the men – are considered to be both withdrawn and quiet. We don't talk, kiss or dance in public. According to the stereotype, a Finn sticks to the facts, avoids social contact and never shares failures or successes – even with the people closest to him. In other words, we're not exactly predisposed to providing good customer service.

The PlayStation generation, however, demands perfect and individual service. In 1995, it took a bank five days to reply to an e-mail query and that was considered perfectly acceptable, if not excellent, service. Now, the best banks offer 24/7 personal service with wait times that are less than a minute, no matter what day of the year. By 2020, contact will be made via video and the terminal or access device of your choice. This development will have an impact on all industries and business models.

Changes in consumer behaviour have already had an impact on the internal workings of companies. Bosses, pay cheques and companies have always been mercilessly compared in company cafeterias and around the coffee machine and water coolers. Now, potential employees can compare prospective employers and rate them based on how interesting a workplace they offer and compare salary levels on a local, regional, national or international scale.

About a year ago, I encountered a job seeker, and a very good one at that, who wanted a tightly defined trial period. Trial periods are par for the course so I didn't think it was a big deal. However, this 26-year-old Russian MBA was the one taking the company for a test drive before committing – not the other way around. I was confused – and a bit offended – at first. I can still remember feeling a little agitated when he asked, "Do your company's internal systems work like iTunes?" Then I realized he just wanted to verify that all he'd heard about the lack of bureaucracy, investment in personal development and efficient management was actually true. He wanted to give the company his all but not until we proved that we were going to live up to our reputation.

Access to information has radically changed consumer behaviour. And companies have embraced this fact. However, to me, the most disturbing aspect of the change in consumer behaviour is *speed*. Over the years I've had numerous discussions, some more philosophical than others, about where all this is leading. Wherever it leads, the one thing that's certain is that organizations and their leaders need to react faster. When consumers and employees change direction like a school of fish, the organization has to be at least one step ahead.

You can't retain credibility if you're telling employees or customers that you have heard their feedback and will make the necessary adjustments – *over the coming quarters*. Yes, consumers do understand that big changes take a lot of time. However, many of them have grown up in a Googlesque world where there is a new version of an application every week without advance warning. Feedback is collected and analysed every second and an application or service is in constant beta development yet is stable and dependable enough for people to use every day. This kind of speed requires not only systematic processes, but extraordinarily quick, front-line decisions.

The network economy is now the mainstream

The impact of the network economy on leadership and leaders is usually dealt with in a superficial manner. Everyone sees the impact when paper invoices are replaced by electronic ones that travel between information systems and online banks. Everyone knows that many information systems and even commercial services are combinations of services procured from different sources. Instead of making something or acquiring it, companies rely on services on which they superimpose their brand and business logic. It's easy. It's logical. And if this is all you are seeing, you have missed the boat.

As far as management and leadership go, the network economy will be one of the most important phenomena in the coming decades. No company can afford to develop business models based solely on their own resources. Small companies don't have adequate resources and for large companies it doesn't make financial sense.

When I entered the job market in the late eighties, people talked about the core activities of a company and those "other things" that could be outsourced. I made the same mistake many others did in thinking of core activities as big functions and outsourcing as a procurement issue. Technology developments and the way people think about getting information and how to get things done mean this distinction between "core" and "other things" is becoming blurred. Tasks and processes – either partial or whole – are being "networkified" in a serious way. Large outsourcing projects taking in entire company departments or functions will still happen in 2020, but it's the aggressive and active networkification of *partial* processes and services which will have the major impact on how a company works.

The heartbeat of this networked economy not only allows executives to occupy front-line positions; it forces them into the front line. An experienced TV producer and CEO, Saku Tuominen, said it well: "The CEO's position is changing from being an administrative, supervisory and recruiting figurehead to that of a producer." This, to me, describes the network economy leader to a T. Naturally, a CEO will be responsible for the company's profitability and its ability to serve its customers well. However, in order to be able to take responsibility for the operation and pattern sensitivity of the whole company and its services, he or she will have to act like a producer.

The producer is responsible for seeing that a team of virtuosos actually completes the production. The producer – not the talent on the team – is responsible for understanding the production from at all levels. A producer may be simultaneously responsible for several productions, in which case he or she has to hire associate producers, but can't blindly delegate responsibilities or decision-making powers down the chain of command. In addition, it's rare for a producer to produce a TV show or movie identical to one he or she has made before, and this is no different for our executive producers. As a result, a producer is forced to rely on *people* and his or her *network* instead of an *old, documented script.* People in the movie business probably don't spend a lot of time talking about a network economy, but it's obvious that a producer no longer worries about audio post-production or where to press the DVD covers. In fact, the producer may not even be present at the casting, because he or she relies on the casting network to find the best candidates.

What does the producer get with pure outsourcing? They get bulk. But bulk is not enough because every production is different and every schedule is tight. The producer has to start training his or her own network and its central players. You have to get to know each other, communicate shared goals to everyone and learn about how the people in your network function. In other words, you have to create a trusted relationship that isn't based on just performance and agreements.

For an executive, the network is essential for success. Utilizing the network requires high-quality leadership, sensitivity in the front line and vision. The members of the network must be equal players and all the members must learn faster than the competition. Learning can't happen if knowledge is concealed and communication is overshadowed by a constant fear of failure.

Competitive factors and *hidden* knowledge are often the reasons why an executive chooses to make autocratic decisions without tapping into the network. Or maybe fear will cause him or her to neglect this whole inconvenient and bothersome network thing. Yes, all companies have secrets and secret knowledge. But I also believe that if 90 per cent of all the information companies have appeared on the Internet tomorrow, nothing bad would happen – the very opposite actually. And the speed of change means the information that was relevant yesterday is probably useless tomorrow.

As a producer, you assume direct responsibility for a business or project's overall quality and continued viability. Your role is to oversee the process – to combine the best talents and information that you can find. Digital Cowboys, your most important source of strength, are adept at utilizing the network economy. You would be hard pressed to explain to them why a service you can easily buy off the Net is worth making yourself. It's better for the executive to just get on that train.

A change in the innovation race

The last external factor that has an impact on leadership is the change in the innovation race. Almost all the business success stories of the twentieth century involved an invention. For one it was the internal combustion engine, for another it was making a paper machine that ran faster than anyone else's. A third involved figuring out a way to industrialize the making of potato chips and a fourth led to a fortune in the development of anti-virus software for personal computers.

The next 100 years will bring more inventions, more knowledge creation and patents than ever before. Scientists and inventors around the world will come up with new inventions that will shake the very foundations of the societies we live in. More and more successful business ventures will be based on the rapid adaptation of service innovations and technology licensing. Many major corporations will establish small micro-units consisting of 20 to 50 people that can furiously combine different concepts and quickly test them with consumers. These units will draw the entrepreneurial doers and seers who excel at networking. The controlled, inspiring leadership of these individuals will be of key importance to innovation development.

Another key factor in winning the innovation race is the ability to scale new services to meet the needs of a whole continent or the entire world. Service development will require investment on such a scale that one country will never provide enough of a growth platform for any one service. Time-to-market or time-to-volume will be the moment of truth for many companies' new innovations.

On a practical level, this scaling involves understanding the culture, payment systems, legislation and technical facility of a wide variety of different environments. It's often assumed that growing your business is primarily a question of resources. The reason why large organizations hardly ever manage to pull it off leads us to the conclusion that something is not quite right with the management principles and steering systems. I often wondered why investments in the hundreds of billions didn't help the world's major banks get their online services off the ground in the late nineties, while at the same time small banks in Finland and Korea managed to create high-quality services quickly and scale them to serve millions of consumers.

Even today, the end-user experience for online banking in Finland or the USA compared to Germany is vastly different. Why can it be possible for one individual to manage multiple accounts from their smartphone in one country but need lots of papers and security numbers to manage only one or two accounts in another country? Yesterday, not many end-user were aware of those differences, but sooner or later they will know what they can expect and will either demand it from their current partner or simply take their business somewhere else – somewhere faster and more convenient.

The recipe for success in the innovation race combines all three of the themes we've touched upon: globalization, changing consumer behaviour and utilizing the network economy. Executives must understand that almost all innovations consist of a *nearly real-time* combination of knowledge, services, leadership and the fearless ability to make quick decisions to invest, kill or cannibalize projects. The old and slow annual design cycle combined with a leader who abhors social and financial risks doesn't work in this race.

Summary: The thousand-year traditions and wisdom of leadership still apply. A leader must, however, take note of the planets that surround him or her. From afar, they may look like stable, unchanging juggernauts. When you look more closely, you'll see they are constantly changing. It's of paramount importance to note that these planets will affect your decisions, prioritization and who are the best/worst people you could hire tomorrow morning.

Left side of the brain versus the right side of the brain

I've always had a hard time with abstract models and concepts. The problem wasn't in understanding them as much as it was with applying them to practical decisions and use of time. I remember sitting in class where a professor took a week and 200 pages to talk about strategic context. I failed to see how it was supposed to change me as a leader. It was even harder to see how I was going to take what I'd supposedly learned and convey it to my colleagues in an earthy and inspiring manner. Years later an older colleague explained it to me: "Strategy is the choices we make to make money. Changing your strategy means reallocating resources in such a way that everyone in your organization notices it in their everyday work. If the strategy doesn't tell you what you're selling and the new strategy doesn't bring any changes to everyday work, it's all just corporate fluff." That's all it took.

Leadership and management are equally as intangible. In fact, this pair can be even more abstract than strategy. Discussions and training sessions surrounding these concepts can result in very odd debates – for example, "Can I be a poor leader if I produce twenty per cent earnings before interest and taxes" or "Is there room for development in my leadership if the members of my team pretty much worship me according to personnel surveys?"

To discuss leadership and management successfully, I wanted to try to create an uncomplicated visual. My goal was to use a model so simple it worked in all countries, in all languages and for all industries. I decided to take a picture of my brain and split it into two parts – leadership and management – to illustrate a thought model that explained my insights and thinking on leadership.

In my simplification, the left hemisphere (management, control) is the one that, like a machine, deals with tasks *you've been assigned by someone else*. The left hemisphere doesn't really work that hard to create anything new, much less develop or challenge existing ideas. People with a strong left hemisphere are stunning in the same way a diesel engine is. They get things done and they'll go through a wall to do it – as long as there's a clear assignment and process to follow. Typically, no resources will be spent on trying to find a door or even understand why the wall is there in the first place.

A strong emphasis on the left hemisphere is typical of a management model I like to call industrial management. The goal of industrial management is to produce an extended series of quality goods with great efficiency. There's no room for soloists in any part of the process. Even if a limited series or a minor deviation would bring a considerable one-time profit, the industrial model doesn't allow for it.

As far as making money is concerned, the left hemisphere is

definitely more important. It's important that a company's daily processes keep churning on and people know what they are supposed to do. A company can't function if invoices aren't sent out, salaries aren't paid and agreements are not honoured. Herein lies one of the greatest risks of management, as well as illusions regarding what a leader's role is and what constitutes good use of his or her time. It is easy for an executive to put all the leadership and management eggs in the left-hemisphere basket.

All these left-brain administrative tasks can drag down any executive. A colleague complained some years ago, "I'm a conscientious executive and I've worked hard. Last night I felt I was staggering under my workload and printed off the intranet all the processes and tasks that a line supervisor is supposed to handle. I read the instructions carefully and marked each process in the project management tool, along with the recommended time they take. I then distributed the tasks over half a year. The end result is that if I do everything the corporate rules tell me to do, it'll take me on average 36.5 hours a week. Should I do what the company rules dictate? If I do, when do I have time to meet with my employees and, more importantly, my customers?" This conversation was a wake-up call.

My colleague was widely respected among customers for his ability to handle big projects. With fifteen years of experience behind him, he was being groomed for the next level of executive training and tasks. It was clear he felt a great sense of responsibility and even pressure to go by the book. He was a responsible executive and these tasks were within his comfort zone. At the same time, he wanted to show it wasn't his fault if he couldn't meet with employees or have direct contact with customers. For me, this raises the question: if all front-line leaders fill their days with left-hemisphere tasks, when do we create the new, learn and take our competitiveness to the next level.

The right hemisphere (leadership) concentrates on creation. The right hemisphere was born to question and develop. Value

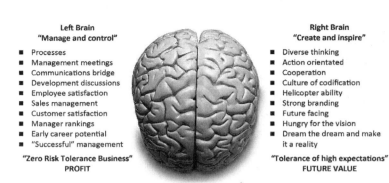

Figure 1 Left and right brain functions

creation and the competitive edge that guarantees a rosy future for the company (or executive) flow out of the right hemisphere. The competitive advantage generated by the right hemisphere can consist of cutting-edge human resource management, the most innovative product in its field or the world's most advanced distribution chain. These are not results gained by the mechanical repetition of a proven process. These innovations are the results of taking risks and creating something new.

Helicopter ability is often associated with people who use their right hemisphere. I am an amateur pilot so let me explain how I see this concept. Often helicopter ability is seen as the leader flying high in the sky, above his or her minions and problems. It's said that a leader needs to be able to look at things from a bit of a distance to get the big picture. Every young executive is warned about the dangers of micromanagement. Apparently you're not supposed to worry about details. Let's correct some errors here. Aeroplanes fly at 30,000 feet, high above the weather. It's all very nice, floating above the storms and unpleasant details. But this has got nothing to do with helicopters. Helicopters can't fly that high.

Now try seeing through the cloud cover from 10 kilometres. You have absolutely no chance of seeing your own troops engaged in their daily battles. An executive this high above the fray can always radio in – maybe send some e-mails – and ask how things are going down there. When it comes time to set targets, the local flight control will complain of difficult weather in the markets and wish for lower targets. A local manager in the middle of a leadership crisis can, with the help of the cloud cover, blow proverbial sunshine up your arse and say that the weather and other factors are solid – even if they are in the middle of a hurricane. Your cruising altitude will prevent you from making your own appraisal of the situation and you'll thus keep flying the (falsely) friendly skies.

No. The heart of helicopter ability lies in a combination of two things: the ability to *see things from high enough* and the ability to *land quickly at any time* should some detail or situation require your intervention. Naturally, you can land a plane but it's a lot slower and more complex than landing a helicopter. You can briefly fly a plane very low over places, but you certainly won't have time for dialogue as you fly over your troops.

As far as functioning in the right hemisphere of the brain is concerned, helicopter ability will be one of the central success factors over the next few decades. An executive can't outsource core decisions related to people or products. These are decisions that can't be based solely on information provided by messengers. Not because you don't trust your team. Of course you do. It's because you need to have a solid feel for the front line of your business. In the past, front-line analysis consisted of reading market and customer research and statistics. It might even include a few nice lunches and casual chats with customers – very aeroplane ability. Now, the accelerating development cycle and the consumers' unwillingness to conform to dictated trends are forcing executives to engage in more rapid changes of altitude than ever before. That's helicopter ability.

Hockey legend Wayne Gretzky defined the secret to his success: "I skate to where the puck is going to be, not where it has been." Even though I come from hockey-mad Finland, I can't really analyse Gretzky's style, but I think his philosophy describes how someone who actively utilizes the right hemisphere of his or her brain works. Hockey is disciplined and often even boring work. In fact, it's a little like a process. All coaches fall into similar patterns when they're short handed. They put in a couple of big guys with a particular player profile to protect the goalie. When it's time for a power play, they play the big shooters on the line and staff the sides with the virtuosos. The centre forward is busy masking at the goal.

Despite the formulaic way the game is played, the most successful scorers are the ones who can see where the puck will be next and can quickly organize how the situation will be played by the team. When the puck ends up where it was going to end up, everything is ready to go. However, no hockey team can ever win consistently with the world's best player – or even the three best players – if the quality of the rest of the team is a couple of divisions lower. Nor can you win by running through the same safe plays in an average manner or by adding a couple of prima donnas into the mix at the last minute. For an executive, it's not enough to be where everything is happening right now. You must be willing to take risks to go where the puck might go next.

Once, before a strategy meeting, I asked all supervisors in the room to write down on a blank piece of paper one mistake they had made. I gave them fifteen minutes to list just one single mistake from the past quarter. The point was not to laugh at the mistakes we'd all made. The aim was to be open about mistakes in order to help others avoid making the same mistakes. Once the time was up, we started going through our mistakes. I got to my colleague – someone who filled 36.5 hours of his week with processes and rules – and he was sitting in front of a blank piece of paper. "Pekka, no matter how hard I try, I can't think of a mistake I made last quarter," was his reply, verbatim. And I believed him.

I was convinced, and still am, that he had not made a single mistake over the preceding quarter. I'm also sure that if there was some small mistake or oversight he might have committed, he would've found it incredibly hard to mention it. In learning, and especially in innovation, appreciating how to quickly process your mistakes is of paramount importance. The left hemisphere sees mistakes as a sin and a violation of the all-important rules. The right hemisphere sees a treasure to be documented and shared. Just realizing this fact will often elevate an organization's ability to learn to a whole new level.

Summary: The larger the organization you work in, the easier it'll be to fill your day with left-hemisphere activities. People who crave that kind of security tend to be drawn towards large organizations. It's their safe harbour. In business, you often hear people talking about government organizations disparagingly for this very reason. However, both public and private sectors are equally guilty of left-brain stagnation. Instead of joking about the ineptness of other organizations, we would be better off spending our energy identifying, nurturing and developing those among us who have steady, ever-changing vibrations in the right hemisphere of their brains.

One of the most important tasks an executive has is choosing and training the supervisors one level down. And since the job you were hired to do was to innovate and help your company grow to the next level, you have to go where the puck is going to be. Who do you need to do this? Like it or not, you need an army of right-brained Digital Cowboys. They are your rising talent. If you're not ready to change and let the right hemisphere of your brain drive your strategy, you'll fail. You'll choose the wrong individuals, start repeating your old recipe for success and fail to engage the guys with the entrepreneurial spirit you so sorely need. NO FEAR is a plunge into the activation of the right hemisphere and accelerated personal growth.

The Digital Cowboys of the PlayStation generation

Before we go any further, we need to have a common understanding of what we, the authors, mean when we refer to Digital Cowboys. For most of us, the cowboy is an iconic image. He – almost always a he – hears there is trouble in town/on the ranch/on the prairie. The news usually comes from someone who has been shot/robbed/swindled by a bad guy. Our cowboy saddles up his horse and packs his rifle, pistol, lasso and a Bowie knife because he never knows what tool he will need for the situation. He heads into danger, ready to hand out justice. His tactics are wily and unorthodox, yet highly successful.

But Digital Cowboys are not the cowboys in a Hollywood movie or spaghetti western. They are the superstars of the PlayStation generation. They – men and women – have grown up being wired, connected and inquisitive. They eat and breathe the tools and the digital know-how. And they expect a different kind of leader in the corporate world.

According to Thomas Malone of MIT's Sloan School of Management, cowboys, by definition, are *independent*, decentralized decision-makers who have relatively low needs for communication. They must make independent decisions based only on what they see and hear in their immediate environment. Digital Cowboys, on the other hand, are *connected*, decentralized decision-makers. They make autonomous decisions but based on vast amounts of remote information available through electronic and other networks.[4] Instead of looking at the immediate landscape in a solitary world for information, the Digital Cowboy mines data from a multitude of sources. Traditional executives with an old-school management style of sitting in the middle of the organization, pushing orders out to their troops, will struggle to show their value to their team if they cling to their centralized decision-making processes.

Information is their currency. In the "old world", information was precious and you needed to hold it close. Otherwise, it lost its value

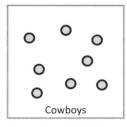  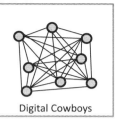

Figure 2 Decision-making models from Malone

and you lost some of your power. In the "new world", information becomes valuable knowledge when shared. Their views on information aren't the only changes and leadership challenges Digital Cowboys bring to your organization:

- They understand and seek the value of networks. Networks are both local and global in scope.
- They are problem- and solution-oriented. Once they find a solution, they are on to the next problem. And since the next solution requires different information, the composition of their "team" must be fluid enough to tap into new talents and new data.
- They are independent, yet connected. They work both inside and outside of your organizational walls.
- They are unorthodox in how and where they look for information.
- Company tools must interface and work with their tools – not the other way around.
- They need the latest and greatest digital toys. Old technology might block their access to new information and this is just plain unacceptable.
- They demand a flexible environment and hours – anywhere and any-time computing.
- They have fuzzy boundaries between work and life outside of work – their brain is on 24/7.

- They thrive in and demand organizations that let them build on their strengths and require leaders who are mentally and technologically equipped to inspire and empower them to do their best work.
- They are quick to shun those who don't bring something to the team.

While all of these traits are both challenges and opportunities, it is the last one – they are quick to shun those who don't bring something to the team regardless of that person's title – which will have the greatest impact on your leadership style. You need to realize that the chain of command isn't what it used to be. You need to prove your value to the team. You need to earn the right to be their leader. These Digital Cowboys either see you as someone who can help them get things done or an idiot who stands in their way.

You need to get them the tools they need and be willing to let them run with the project. You need to be open to unorthodox ways of getting things done. You need to be willing to provide access to information, people and plans to get the most out of your cowboy. Hoarding information and access ranks up there with not providing value to the team. In other words, you have to have NO FEAR as a leader. You have to be visible at the front line, as a translator, a go-between and someone who takes responsibility. You need to be open – from new ways to communicate to new levels of transparency. You need to be willing to engage and challenge your team and your colleagues, at the same time as you are being challenged by them.

▶ Your contribution to www.nofear-community.com

While Mark and I are fully immersed in the latest gadgets and both leverage technology to the fullest, we are too old to be actual Digital Cowboys. I have a lot of respect for the way they work, as well as for their values. There's a lot to learn there, and for the next 30 years I won't be able to avoid them. I'll have to lead them until

I retire from the job market around 2040. At home, we each have Digital Cowboys to raise so they are ready to hit the job market starting around 2020. Watching Mark's teenager and his friends leverage technology for schoolwork is – even for us – simply mind-blowing. If these kids with their virtual networks are able to create professional-quality videos for a simple school history project – something we would have read from paper in front of the class – what will they be capable of in ten or twenty years?

The PlayStation generation and their Digital Cowboys live and breathe digital communication and networked teams – they are true digital natives. Our generation is more like a band of digital immigrants – we didn't grow up wired and connected but we have been embracing the tools as they evolved. For example, after the millennium, I started to systematically transfer my teams online. I was leading a small, home market-based unit and the tools weren't very developed at the time, so I didn't really get much out of the exercise. Now the situation is totally different.

Eight people living in different countries and on different continents contributed to the writing of this book. Close to one hundred friends helped this book become what it is by commenting on the content and offering us their views. Although the core team has known each other for years, we only met face to face once when working on this book. Even then, instead of working on the book in any concrete manner, we spent our time together having discussions, debating, arguing and building our trusted relationships. Despite the time-zone challenges, I felt we were, as a team, very close to each other. Daily video conferences, document-handling in an electronic environment and shared task lists made us one. The team was assembled for a three-month production period and was instantly ready to go. This kind of collaboration is becoming more and more accepted and valued.

The soul of NO FEAR is to provoke discussion and growth in your team. For this purpose, we created the **www.nofear-community. com** Web community. The community is organized around the themes in the book. It's there to offer you the chance to comment on various subjects, theses and content. The service includes video interviews with the writers and other experts to supplement the contents of the book. We are looking forward to engaging with you, the reader, in a lively and insightful discussion.

We also hope you'll share videos, as well as links to other materials, with members of the community. Joining the community is totally free and you can invite any of your colleagues who are interested in leadership and innovation to join, too. No registration is necessary. You can use your Facebook or LinkedIn account to access our service.

We hope you'll find new ideas and perspectives for your everyday leadership. Challenge us and our thoughts as a part of the community.

Digitally yours,

The Authors

Beijing, Joroinen, Lausanne, Moscow, Munich, Palo Alto and Seattle

2 Why should I follow you? The challenge from the leader's perspective

I DROPPED OUT OF UNIVERSITY when I heard the siren call of business. Well, to be honest, the siren didn't call. I was just out of money. Regardless of the reason, I started down the rocky road of learning the ropes of leadership. I didn't have scientific or pedagogical learning to make the road very smooth. Over the next few years, I tried to imitate and apply the behaviour models and communication methods of the executives with whom I worked. I felt no shame in reusing anything I perceived as having value for my team and my company. I learned a lot from the heroic stories of success these leaders told. But I learned even more when things didn't work out the way they were supposed to.

Leaders from the Industrial Revolution enter the digital age

My image of a leader was someone whose central task was to create stability, hold the reins and take responsibility. It wasn't until much later that I began to understand a leader's real task is to build both goal-scoring opportunities and safety nets for the players on the team. As an egocentric person, I still find this concept challenging. It's not that the task is unpleasant or repulsive in any way – I just continue to have lots of room for improvement. I have to remind myself continually that I'm here to create scoring opportunities for others. It's a constant battle to fight the urge to grab the reins. Intellectually, I fully understand that doing everything myself is not practical, but letting trusted people accomplish a task in their own way is hard for me. Instead, it would be easy to fall into a micromanagement mode.

We have worked with and observed all forms of leaders. At one extreme are the CEOs who see their primary task as creating value for stockholders, plain and simple. For them, managing the job of "moneymaking" is an internal and invisible – almost secret – thing within the organization. To be successful, the leader just has to choose good workers, do the monthly round of reporting and pay out dividends to stockholders after the general meeting. These leaders take "creating opportunities for their team" to a whole new level. It is a very hands-off approach. They can just wait in the big meeting room at the head office and wait for their messengers to arrive with the glad tidings of success. A number of characterizations are often applied to this type of leader. They are clear, result-oriented, give others room to move, strive to become redundant, and so on. They are very good leaders, really – as long as the battle goes their way and the circumstances don't change.

The other extreme is the micromanager. Despite the big title and extensive responsibilities, he or she spends every day on the ground with the troops. And I mean, really on the ground – very, very deep in the trenches. This person's helicopter ability is severely limited. The skids never leave the ground – not even when seeing the big picture is of vital importance. This type of leader is often prevalent in middle or upper-middle management and they are often the people leading critical development projects. Projects instigated and run by this type of leader create growth or increased efficiency. Familiarity with details and the troops enables quick reactions and corrections. Recruitment under this leader is extremely natural and he or she probably grew into a leadership role after working as an expert.

However, this type of leader is so focused on the minutiae of a project, he or she is unable or unwilling to engage in a war on several fronts. When an innovation or new solution has been created, it needs to be scaled up fast – in weeks, not years. Put more simply, they lack the tools needed to lead an extensive network and lack an eye for the big picture. They often have trouble communicating with upper management. They want to present the details of an issue and engage in a deep discussion of all the possibilities, while their upper management just wants to know whether the light is green or red. Conversations are often a bit awkward and unproductive.

These are the two extremes of corporate leadership. One leadership model is based on strong processes and the other is based on deep expertise. These traditional ways of management had a place in the Industrial Revolution.

Good leaders know that they need to adapt their leadership style to the situation, their position in the organizational hierarchy and, most importantly, the needs of their employees at all stages of their development, expertise and maturity. In the 1960s and 1970s, concepts like "Situational Leadership", driven by Dr Paul Hersey and Dr Kenneth Blanchard and based on their "Life Cycle Theory of Leadership", became more and more accepted and eventually popular in many organizations. This employee-focused leadership was a big step forward. The digital revolution has changed the landscape dramatically, empowering the individual and challenging the organization to leverage an individual's increased potential for contribution far more efficiently and satisfyingly for the employee herself.

Leaders need to look in their toolboxes for strategies that inspire the Digital Cowboys who are riding into the business arena. Leading in this new era means more than "kind of" embracing the new tools and ideas. I bet we all know an executive or two who has written one or two blogs to his or her troops to show how open they are to communication and new ways of doing business. And we have seen the ubiquitous embracing of the iPad as the new, critical tool for CEOs. It's a nifty piece of equipment but I find it incredible that even the smartest people will justify getting one – saying they need one to keep up with the times. Sorry, guys. Buying or even using a new toy doesn't instantaneously make an executive a modern leader. The change required to lead the army of Digital Cowboys must reach much farther and go far deeper.

I have a very successful colleague in the retail trade. He liked to talk about *continuity* and *discontinuity*. These are all the things that would either survive or be terminated at the end of the current accounting period. In this chapter, we will illustrate the challenges of leadership using the same theme – keeping and fostering what is working today and tomorrow and retiring or terminating organizational and individual behaviour that doesn't serve our organization in the future.

What is the fundamental change when leading an enterprise?

Knowledge is power. Over five thousand years ago Sun Tzu in *The Art of War* was already extremely clear on that point. However, the *way you use knowledge as a leader has greatly changed.* The very heart of the change is that the way information is gathered, processed and exploited. Information combined with insights and expertise becomes knowledge. Information technology has fundamentally changed the amount and availability of information, as well as access to expertise and institutional or network knowledge. Social and expert networks, near-universal Internet access and Business Intelligence (BI) solutions are a few of the underlying technologies that fundamentally impact on an organization's effectiveness.

In the past, management systems were based on a model in which the organization gathers information from the bottom up, consolidates it and then returns the resulting conclusions back to the front line. Often this system is hard-wired into a company's management system and tasks – from financial management to

HR to R&D to production. Exceptions to the system are strongly discouraged.

In this "old world" system, information was precious and you needed to hold it close. Knowledge was concentrated with a few insiders, often in the higher levels of management. If the information and the knowledge were distributed too far inside the organization or even outside, the knowledge – limited as it may have been – lost its value and you, as the "leader", lost some of your power. With the vast increase in the amount of information and the ability of a network of expertise to transform this information into valuable insights and knowledge, information becomes far more valuable when it is shared.

In the age of the Digital Cowboy, information is fungible. It goes up, down and across an organization to fuel business knowledge and organizational agility. At the same time, strategic decisions require reacting to signals – which can be quite vague or weak – across multiple levels inside and outside of your organization. In strong *consumer* industries, such as retail, the most successful business leaders have worked systematically to understand these signals – no matter how strong or weak they seem. On the other hand, many companies that work in *business-to-business* markets make absolutely no effort to understand these signals. They based their market on a specific product or service and rely heavily on their long-term provider relationships to stabilize their bottom line. Well, consumer behaviour is hitting the value chains of business-to-business organizations hard. Their management is now yelling for quick reactions and wondering why the hell you couldn't read the signals. Who cares if they were weak or not – our competitors figured it out.

▶ What makes you authentic?

Equality is an important value in Nordic culture. The politics and taxation of the societies are set up to provide everyone with nearly equal access to education and the potential for success and influence. On the whole, personally I think this is positive. However, the downside of this kind of society is that all exceptions to the norm are suspect to the group to some degree. Being "different", "unique" or "outstanding" is not expressly forbidden, but the unwritten cultural code favours a certain amount of conformism. If you are nearly invisible and somewhat bland, in general you'll be taken more seriously as an expert.

We see this trend in the corporate world. It's rare to run into a corporate leader with a different and personal way of doing things. The stereotypical leadership environment drives you towards leaning on the left hemisphere of your brain to minimize the potential for errors. By sticking with "the code", you secure your own position until the next quarter. Deviating from the norm creates a chink in your defences. At the first sign of trouble, you're open to attack.

Spending plenty of time in the business traveller lounges at various airports, I find it interesting to observe fellow passengers from all over the world. And it is truly obvious that the "leaders" of many organizations with the mandate to innovate are slaves to trends, as well. The clothing brands we wear, the briefcases we carry, our all-important accessory – the phone – glued to our ear and the books we read display a great degree of uniformity. Even though fashion and vanity have always been and will continue to be important to people, I think it's safe to say that deviating from the norm is generally not considered an important attribute in a corporate leader.

Years ago, when I told a colleague about my ambitions regarding this book, his first reaction was positive. After a moment came the crushing "but". "But writing a book like this is a good thing for a retired executive to do. Then you can say what you really think without putting your career in harm's way." I was, and remain, truly shocked by this comment and wonder how many people are holding back and how much, not allowing the rest of the world to meet the interesting and insightful person beneath the façade.

Standing out from the crowd is not a virtue in and of itself. Good leaders do not join companies because they have an opportunity to stand out. Many top professionals choose the organization based on its stability. The leaders are a visible manifestation of this stability. The new generation of workers is looking for stability, too, but in their own way. For a growing number of experts, the *values* and the true *soul* of the company mean more and more. Companies know this and, until now, talking about values has been good enough. But it's no longer so.

Now employees are starting to ask questions. Does my boss have the guts to act based on these values, even if it sets us apart from the crowd? Can I identify with the leadership of the company to the degree that I am ready to ride into battle with them? Do I know what they really – and I mean *really* – think? Is the substance of these "company values" messages empty jargon generated by corporate communications? With the Internet, more information becomes available to these employees much faster. How an American executive behaves on a business trip to a foreign country – such as only eating at the American fast-food chain and avoiding local foods – can become a joke shared by the entire company. Or how many shares did the CEO sell last quarter while touting publicly "strong confidence in the future" of his corporation.

These employees are asking plenty of questions and they look to the company leadership for answers. The top people often flock to the charismatic leader who visibly "walks the walk" of corporate values. Now, people have always talked about "charismatic leaders". Charisma is generally considered a trait a person is born with – or isn't. Some leaders have gone through the same schools and training sessions as everyone else, but there is something about their personality which captivates. Something that makes people want to follow, no matter what happens.

What a stroke of luck it is for all leaders that this mystical charisma is not really a critical trait when it comes to leading Digital Cowboys. Your future as one of the top leaders of 2020 is not dependent on a natural-born ability to charm large masses. A leader's *brand* and *authenticity* will be important success factors. In the networked, global economy, all members of the value creation network are equal. Your list of Facebook or LinkedIn connections is a brilliant simplification of this. One of the people on the list may supervise a workforce of 20,000 people but they look like all your other "friends". We all use a 200x200-pixel photo of ourselves. So what brings people to your postings? What makes you interesting to follow?

When you provide your network with information that is thoroughly up to date and relevant, people can't wait to read your messages. If you apply yourself to this task systematically and creatively, you become a unique voice that is *believable and reliable*. You become an authentic player in your network. Authenticity doesn't happen overnight and you aren't going to build it by using cheap stunts or mashing someone else's words and claiming they are yours. Our colleagues from the new generation are keenly aware and highly networked. They broadcast knowledge within seconds and plagiarism is not acceptable. Authenticity is something you build over time – through consistent, genuine and credible actions. And if you think authenticity is a form of charisma – something you're born with – you are patently wrong. Period.

A word of warning: as a corporate leader, you start this task from a challenging position. What you say and do will come under much greater scrutiny. You will face more criticism than others. Regardless of criticism and the fact that people may ridicule you, do not give up. If you work hard at it and manage to stay as true to yourself as possible, you'll build up your own authenticity – think of it as your personal brand.

Uncovering, supporting and developing the authenticity of leaders in your own organization can be hard. When developing people, it's fairly easy to talk about left-brain matters. Financial results, cost structures and feedback from employees are straightforward stuff when it comes to evaluations. Using numerical values to create an evaluation and defining goals for the coming year are fairly simple. But how do we evaluate the authenticity of a leader? There is no numerical value or evaluation matrix. Do we just take each other's word for it? It's easy to say that I am what I am – a unique and authentic individual. Is that enough? What does an authentic leader look like in action? And who is the ultimate judge of our authenticity?

Team-building in a networked world is a fast and furious process. There is no time for "romancing" and getting to know each other. There is no time to slowly build your reputation as an authentic leader. A leader must quickly and efficiently locate the tools for creating and consolidating a virtuoso team. A leader's openness, transparency and willingness to put 100 per cent into the game – the hallmarks of an authentic leader – play an important role. These, coupled with the team's areas of expertise and internal chemistry, can create something unique –and maybe do it faster than the competitor.

Business schools are filled with fantastic case studies showing exceptional team examples. One of the more significant stories revolves around Leonard Bernstein – one of the twentieth century's premier orchestral conductors. Even though the case has nothing to do with business per se, it is perfect for illustrating the challenges a leader faces in an expert organization. The story is about the birth of Bernstein's *West Side Story* musical in the 1950s and the challenge of creating a new recording of it in the 1980s. The process was fascinating, unconventional and, for many, even revolutionary. Experts from a variety of fields were needed – dance, classical music, choreography and popular music. Bernstein put his team together using fairly unknown artists. The aim was to create a groundbreaking musical that brought together elements of classical music and jazz. The dance numbers bore the influence of both ballet and the underground dance movement of the time.

Bernstein was, without a doubt, a charismatic leader driven by an inner fire and a vision of what he wanted his team to accomplish.

Figure 3 Leonard Bernstein (© The Leonard Bernstein Office, Inc. Used with permission)

He threw himself into the game. He did not hide behind an old process or a familiar model. He had an incredible ability to make a team of people with very different backgrounds and areas of expertise work together towards a single, common goal.

People charged with leading Digital Cowboys wrestle with similar problems. Imagine you are leading a typical development project. You sit down in front of a video screen and face experts spread across various countries. You've never worked together before this moment. The members of your team are prima donnas in their own areas of expertise and you have a couple of weeks to make them work as one. Your company has numerous different process descriptions and models in various databases, but nothing that fully applies to your current situation. You're keenly aware of the fact that you're creating a new service and solution with no guarantees of success. Your team could work day and night for the next twelve months and the consumer might still reject the offered product or service. Will you and your team be considered losers as a result? Are you prepared to lead your troops into a situation of such uncertainty, and are they prepared and ready to follow you?

Leonard Bernstein was a genius of his time – a one-in-a-million occurrence. How can an ordinary team leader aspire to reach that level? How can you compare the creation of a world-class musical and the development of a next-generation accounting system?

The first advice is to take off your leadership armour. Be upfront and honest about your ideas and fears for the project. Set the example. Show the team that it's recommended – if not compulsory – that your weaknesses and leadership style be put on the table. If a leader does it first, it's easier for others to follow. Once those issues are out there, the things that are left are your strengths and the reasons you were put in charge of the project. Your authenticity is right there.

Unfortunately, this conversation and analysis are often conducted too late. They come at a point when the team has already hit the rocks. I've been guilty of neglecting all-important preparatory work when assigning tasks. I've barrelled ahead into problem-solving and trusted that the assembled team will bring the project to the home stretch, no matter what happens. Many projects like this have been successful, but many have also been failures. After the failure has become apparent, we've gone in and dug the team members out of their foxholes and limped across the finish line. Surely Leonard Bernstein had his own demons and fears to contend with. Despite this, he was able to psych his team into a state of systematic development. Had he not done the same himself – as the leader – getting the rest of the team there would've been impossible. By the way, his 1984 recording went on to win a Grammy and recordings of the artistic process became an award-winning documentary.

Where do we go from here? As far as *left-brain* functions go, nothing really sets us "Leaders of Digital Cowboys" apart from the mass of leaders on this planet. It's important that the more systematic and routine aspects of leadership are handled with care. Without them, nothing else is possible. However, a leader's authenticity resides in the *right side* of the brain. This is where you lead the development of new and creative processes without being saddled with a pre-chewed, left-brain model.

Years ago, at a low moment, a respected colleague offered these words of encouragement: "Pekka, a leader is paid to think and come to conclusions – not to learn the rule book by heart." This advice now needs to be updated to reflect the desire and willingness to make the Digital Cowboy a part of this creative process. When this is realized, a new generation of leader has been given the basic ability to function as an authentic leader.

▶ What is your value creation? I mean, really?

Many organizations offer their managers a "360 assessment" these days. The assessment is built on the responses to a set of questions, answered by the individual employee, his peers, his manager and some of his manager's peers. The goal is to achieve a full 360-degree view of the network surrounding the individual and uncover blind spots in his or her self-perception, as well as highlight strengths to build on further.

I always eagerly await the results of the biannual leaders' 360 analysis. For me, it best indicates how well I've done as a member of the network. Since the evaluating group in the 360 analysis is sufficiently extensive and spread out around the organization, it eliminates the "arse-kissing" that can sometimes plague linear leadership analyses. By analysing the results of this quick Web survey, it's easy to define areas where you need to develop yourself, as well the next steps the people working for you need to take.

Back when I first started working in an international organization, the results of these analyses were at times brutal and depressing. My subordinates felt I didn't give them enough time. People on the same level or above me in the organization complained of my meddling and offering advice in areas that were none of my business. My supervisor was pained by the complaints he received about me when I (and my troops) acted against this, that or the other traditional hierarchical model. People rarely complained that we were wrong when it came to matters of substance. The way we liked to communicate – openly and loudly – rubbed people up the wrong way. The fact that I advised my troops to solve the organization's internal squabbles in the client's court was frowned upon. If two sections of our organization were in conflict, I'd often decide in favour of one or the other based on which option provided the end-customer with the greatest benefit. If this model cannibalized the short-term profits of our company, so be it. Thus, we were accused of neglecting to look out for ourselves and the company.

After those early reviews, I was usually up against the wall – until the financial results and customer feedback started rolling in. It was hard to argue with the positive results. External factors also began to change perceptions. The IT sector got tighter and the economic bubble burst. The 360 analyses started to look quite different. In 2002, many colleagues found me to be immature, arrogant, difficult to understand and self-centred as a leader. Many leaders in my team received similar feedback. After five years of keeping our noses to the grindstone, the same analysis was performed on a group that was 70 per cent the same. In 2007, the same people who'd found us so difficult in 2002 thought we were inspiring, international, cooperative and adept at communicating with others.

Did we leaders change? Or did the organization? To a degree, both went through a transformation, but looking at my personal feedback and journey, certainly the bigger change occurred to me personally. An esteemed colleague participated in one of my 360 analyses. While he admired my "bulldozer" approach to moving projects forward, he also pointed out that my strong points were also becoming my weaknesses as a leader. This was my starting point as a leader. These were a few of his comments:

Pekka lives in the future – in his powerful imagination. When he wakes up in the morning, he sees himself in the front line, doing battle as a warrior, with the people he chose. For him the future isn't plans as much as it is a vision ... So strong is the living in the future that it gets others excited – excited about the picture of a future and their ability to reach it as a team.

But a big corporation means that chains of influence and command become longer and this gives rise to complications. His interactions [instead of being personal] became more like corporate communication. Some implementations of change turned into impulses that just got the change started and they brought with them the risk of too strong a forward momentum and personal superficiality.

A rude collision with the limits of his own influence forced a painful change. He had to choose which battles were worth winning. And it was even more painful to learn what kind of fighters he needs around him and who to trust when most of the work is being done out of his sight. As the chain of command grows good questions and sensitive ears become as important as a clear vision.

And so started my never-ending trek into the world of management systems. As the chain of command grows, how does a leader decide what is useful and what is just stultifying bureaucracy. How do you simplify systems so even the most literal-minded warriors/leaders can figure out what is needed and when – instead of delivering everything and the kitchen sink, just in case? And how do you link leaders and teams to form a value creation network?

I must admit I was incapable of parsing my role as a leader. Luckily, I had found some wonderful colleagues to work with and together we were able to start defining our roles and our model for value creation, mentoring each other. *Value creation network* is a concise way to describe the modern enterprise. The change in a leader's role, when compared to an old-fashioned organization, can be defined as follows: *from the head of a linear organization to an activator and developer of a value creation network.*

Stability, trustworthiness and decision-making were important factors in the previous leadership role. It was also said that a leader's job was to conduct a dialogue and a supervisory relationship with three to eight people. Some consultant or other stated that no one could lead and develop over ten employees. Not well, anyway. If this value creation maxim enjoys widespread acceptance, we must ask the following questions:

- What are the leader's desire and ability to create value as a leader of a new network?
- Is it possible to have an impact on the personal development of dozens, maybe hundreds, of people?
- How do you solve problems in projects that run through the whole organization without stepping on the toes of your colleagues?

Four main themes continued to come to the forefront, and these form the foundation for creating, maintaining and growing a value creation network.

1. Maintaining an entrepreneurial culture in all we do.
2. A single-minded monitoring of end-customer benefits.
3. A continuous activation of the value creation network using signals from customers.
4. Constant performance analysis and re-engineering of one's own parts of the network.

One thing that really had an impact on me was understanding the leader's dual role as fighter (warrior) and coach (healer).[1] In the morning when I woke up, I always felt like putting on my armour, sharpening my sword and heading off to battle. Battle alone wasn't enough. I'd find it frustrating to wake up to a day that couldn't offer a formidable enough foe. Some people laughed at my warrior attitude and probably thought it was a little nuts. Others found it refreshing and energizing. Still others found it too tiring and consuming a way to live.

As my company grew I realized that moving large crowds required something more than my role as a warrior. Energizing people was easy, but in addition to talk of battle and the example I set, the troops needed real tools to survive the battle. When I neglected this, the greatest degree of enthusiasm still resulted in defeat and disappointment. These mistakes and miscalculations resulted in me having the greatest respect for the kind of leadership where the fighter role and the coach role are equally visible. A coach alone without any nicks in his or her armour is not credible. A battle-hardened fighter who lacks the desire to develop people will soon find him or herself alone in the front line. A balance has to be found.

Maintaining an entrepreneurial culture in all we do

The practical implementation of an entrepreneurial culture in a large organization can be painful. This means that any part of my network – let's use a business unit as an example – should be able to start acting like an independent and separate company at the drop of a hat. In practice, it means that everyone in charge of a particular function ought to be prepared to take complete responsibility for their entire group. And not just as a group of professionals, but as a company responsible for sending its own invoices, paying its own salaries and maintaining customer satisfaction. In this model, recruitment plays an extremely critical role.

The people I look for first are those that want to be entrepreneurs. But they have to be more than just entrepreneurs. I look for people who have a desire to be entrepreneurs in their own field, but who also want to get their hands on big projects – fast. In other words, I want to create an environment where these experts can flex their own muscles, but also see and do new things more quickly than they would in a typical start-up environment. Investing our limitless resources training people demonstrated in practice the power of the network in a large organization.

Nothing is easier than standing at a flip chart and using your marker to list all the inflexibility, slowness and structural faults inherent in a large enterprise. It's much harder to sit down with a young, aspiring leader and list the ten concrete things that make his or her day more interesting, fun and financially rewarding. They are entrepreneurs and they want to know how you can speak to that entrepreneurial spirit. If a leader can offer nothing but platitudes on this matter, the Digital Cowboys will ride into the sunset, post-haste.

A single-minded monitoring of end-customer benefits

From a leadership perspective, the unceasing evaluation of customer needs is of critical importance. In many fields, including the IT service business, the practice is to produce large batches of a product based on megatrends. In many ways, the change in technology trends was foreseeable right after the millennium. It was obvious that the process automation cycle described by Forrester Research[2] was ending and a network economy phase was starting. The change to the network economy meant shorter product and service cycles. For example, early phases or cycles in personal computing had lasted at least ten to fifteen years. A cycle that long now is incomprehensible. Many other fields were experiencing similar cycle changes around the same time. These included the media's change to a fully digital production and distribution model, the total overhaul of the energy industry and the violent transformation of the forest industry. Regardless of the industry, the surest way for a leader to miss the boat is to avoid direct conversations with clients. Hanging out and being a friend with your clients is one thing. Committing yourself to understanding the profound changes they are going through is another.

In my experience, new opportunities in business have always arisen out of conversations with customers. Sometimes they involve technological innovations or best practices from other industries. Collecting and systematically developing these morsels of knowledge was one of the most central, as well as challenging, things I did as a leader. Truly understanding the needs of the customer and, most importantly, the end-customer takes a lot of time. Reading a superficial executive summary was never enough. I had to dig deep into the details to really understand the problems. In practice this required a lot of conversing with my own customer team and the customers' key experts.

A continuous activation of the value creation network

Keeping your value creation network bubbling with creativity, dynamic connections and vibrant members rests solely on your ability to communicate. A fire hose of information doesn't activate your network; it floods it. A slow trickle of communications and information will cause the network to run dry. Let's forget communication methods for a moment. Let's first talk about *what* you should communicate. I want to focus on communications in relation to the Digital Cowboys who form the foundation of your network.

A central purpose of a leader's communiqués is to tell people about new strategies and opportunities. Looking back and providing your organization with feedback about how things are going is allowed. The sharing of personal experiences and bringing your persona to the forefront via social media are important. They are good ways to increase openness and emphasize your own authenticity. Do, however, give a thought to whether posting a Facebook status about feeding your firstborn or driving to your mother-in-law's birthday brings any added value to your cowboys and their ravenous hunger for ideas.

It's obvious that the leader (as well as any other respected member of the network) is expected to have a nose for new business opportunities. Communications can be related to the opportunity itself, its exploitation, training or an analysis of what the competition is doing in that area. But you have to do more than just provide numbers and rehashes of someone else's analysis. It really doesn't take much to forward a bunch of research and clippings by McKinsey, Forrester Research and Gartner to your friends and colleagues. Your value in the network rests on your perspectives and opinions about the new strategies and opportunities.

The only thing worse than not providing an opinion – in terms of leadership – is sending messages when you do not understand the substantive core. And it is catastrophic if the information in the message turns out to be wrong and the leader's panicked reaction is to try to blame a subordinate for the mistake. A leader is expected to be able to engage in further conversation, and if you are an emperor with no clothes, you'll be found out in seconds. The value creation network shows no mercy to pretenders of this sort. I speak from experience.

As a young and inexperienced leader, I publicized a matter that I did not fully understand. I did not fact-check it before I opened my mouth. I was caught and it was a humiliating experience. Since then, I have made a point of writing all my own messages myself. It is important that I am responsible for every letter in a communiqué – typos and clumsy expressions included. I know people laughed at them. I'm not exactly a native speaker of English, but what the hell. Let them laugh. I know bland and odourless communication does not work when leading a bunch of Digital Cowboys.

Activating the value creation network with new business opportunities and thoughts is a leader's job. It's a task that no one can delegate, not even the person sitting in the biggest corner office on the top floor. If you delegate, you abdicate your role as a leader and you are no longer an equal member of the network. If you screw this up, your organization will become dependent on constant activation and flow of ideas from above. In a worst-case scenario, individual units in the organization can become totally passive. New members recruited into the organization will think this is the way things work and fall into line. Before you know it, a great network will have turned into a one-way tube of information, flowing from above to the waiting masses below.

I've made this mistake. I was developing my own communication and shared every single thought and idea I'd found, copied, created and/or further developed. However, I didn't encourage my supervisors and experts to do the same. Instead of ideas circulating through the network, I was flooding it with my thoughts and opinions. I believe this would fall into my 360 category of *"too strong a forward momentum and personal superficiality"*. Correcting the problem was as complex as my network. For some members, there were strong cultural differences at play. For others, a simple mention that they might do the same was enough. For a few, I had to resort to using financial incentives, monitoring and sometimes even what could be construed as force to make them active members of the network. Despite my mistakes, I'm not saying a leader should keep all personal thoughts and opinions secret. Fear of dominating the group must not limit your interaction with the group. You just have to make sure that all supervisors, all countries, various expert levels and all of the organization's functions are equally engaged in the activation process.

Throughout the book I keep emphasizing the role of *all* functions in an organization – both core and support. If a company's *support functions* isolate themselves from the *core functions* of the business, HR experts, CFOs, IT departments, security and all other support functions of the corporation can easily be outsourced. It is a curse for a support function to drift away from the front line of the business and be relegated to the background. Traditionally, a solution to the problem is to make the top managers of the support functions hang around at strategy and board meetings – without actually contributing much. I suppose the idea is that they should learn about the business and then apply what they have learned in order to correctly define their resource levels and activities for the coming year. But this is the wrong approach. A passive presence is not enough to make you a member of the company's value creation network. A leader must make sure that the support functions participate in the value creation network and its activation, and follow the same rules as the business functions. A leader of a "support function" needs to understand how their organization can make a difference to the core business. Mueller-Eberstein described in *Agility* the business impact the IT function of an organization can have on the business results. The same is true for HR, facilities, security, finance, procurement, etc. If a function or its chief has nothing but mechanical execution to offer – without any of its own input – the supervisor or the complete function must be ejected as soon as possible. The network does not tolerate freeloaders.

Constant performance analysis and re-engineering of one's own parts of the network

The fourth value creation tool involves the constant improvement of network members' – experts, teams and external parties – performance. In principle, this can be seen as a regular and systematic analysis of goals and results. Now and in the future, monitoring the efficiency of networked production machinery is probably the most natural task for a leader. I'm equally sure that the methods of value creation will deviate from an Excel macro-style support offered by the executive branch to the team members. More and more companies will find that their growth and profitability are dependent on a combination of extreme production efficiency and the efficiency and agility of generation of new service and product concepts. There are a shrinking number of companies, especially in the West, whose recipe for success is based on production ability alone. Since production efficiency and the creation of new service concepts must be managed concurrently, a leader must rethink his or her conception of efficiency.

I often look at my old calendars. Since 2002, I've had an alert in my calendar for every Monday at 5 a.m. which says, "REMEMBER TO EJECT UNDERPERFORMERS". It sounds pretty brutal, but it means what it says. You have to go through your organization and start the necessary measures to remove any underperforming individuals. This may involve further training or reassignment to more suitable duties. Firing the person is the last resort. By doing this, I was able to work for eighteen years without having to engage in any large-scale staff reduction consultations or waves of firings. The process was ongoing and no large-scale corrections were necessary at any point. This is a very straightforward aspect of performance monitoring. From a staff policy perspective it might sound a bit rough, but surprisingly enough it is precisely the sort of human resource leadership that experts respect. No one wants stowaways on board a ship that has to be in constant readiness to weather potentially violent storms. Conversely, all who create value for the network are always more than welcome on board.

Welcome to the Professional Service Firm culture

In the late nineties, I was searching for frameworks and models that could offer high-precision solutions to leadership challenges. After looking into what industrial management and production efficiency had to offer, I came to the conclusion that all companies should be led like expert organizations. I can hear the avalanche of disagreement even as I write these words. How could such a crass generalization apply to production-centric fields? What is so special and different about leading an expert organization?

A friend worked for a security services company. After working at his new job for a week, he told me it's enough if just one out of 100 employees actually thinks. The other 99 can just show up, guard the property and follow the appropriate processes. Six years later, he had a complete change of heart on the matter. In his business, brute strength, processes and the number of available security guards are important success factors. However, since the competition in the field has tightened, the importance of the company's customer interface layer has increased. Six years after our first chat, he told me that in his best units over half the growth comes through sales made by security guards. It is their front-line security expertise which makes them authentic – and successful – representatives of the company. If my friend is unable to show his appreciation of the security expertise of his employees, he will be unable to attract and retain the next generation into the security business. Without experts, growth stalls. The connection between leading experts and a company's future growth is indisputable.

So if the security industry – which has been around for at least ten thousand years – has so much to gain from the leadership lessons learned by expert organizations, I imagine the same applies to just about any other business.

At the International Institute for Management Development (IMD) in Lausanne, Tom Peters's *Professional Service Firm*[3] book was a revelation to me. At this writing, I am on my fourth copy of this book, and my current copy travels with me everywhere. His book on Professional Service Firms (PSFs) was first published in 1999. It was relevant then. It is relevant now. And it will continue to be relevant ten years from now because the culture it promotes speaks directly to your Digital Cowboys. Long before I read Peters's book, I'd admired the slick consultants of McKinsey and the Boston Consulting Group – the core examples Peters uses in his book. At that time, their entire operational philosophy seemed inspiring and energetic, but not particularly creative. At first glance, it seemed like well-educated and smart people taking care to listen to the customer, browsing their intranets for slides made for other customers, synergizing various presentations, changing the colour scheme and then presenting the result to the customer company's management.

Tom Peters's definition of a professional service firm is:

- A PSF ... Does work worth paying for **every time**. This ain't a game!

- A PSF ... Is well-known for **something**. It has a recognizable Signature and a Distinguishing Approach to Problems.
- A PSF ... Leaves a legacy. It Does Work That Matters and passes the test of time.
- A PSF ... Isn't afraid of the word "sell". It is proud of its capabilities and wants the world to know about them.
- A PSF ... Has a client list ... to die for! It is not afraid to dump dud clients ... internal or external.
- A PSF ... Is the "Place to Be". It is a Magnet for Hot Talent. It hires COOL and pays accordingly.
- A PSF ... Provides stunning growth opportunities for energetic individuals.
- A PSF ... Is exciting/vibrant/rockin'/cool.

As I transposed that list into the real world, I started thinking about the challenges facing my own organization. Even though I worked in the IT industry – which can be pretty dull in many ways – I saw exciting potential in the PSF philosophy. Instead of short and intensive *customerships*, our business consisted of *relationships* that lasted for years, if not decades. Our projects consisted of processing deep customer expertise into IT solutions. Instead of being able to rotate staff between different customers and exciting projects annually, many of my customers insisted on stable teams for three to five years. I had a very hard time seeing how an "exciting/vibrant/rockin'/cool" culture had any practical use for me. I was supposed to make a "rockin'/cool" culture for an expert who'd spent the last ten years working on algorithms for insurance systems? As a leader I figured I would make a total ass of myself if I tried to bring the themes from Peters's book straight off the page into my everyday work.

It didn't seem that building up PSF culture – especially the necessary leadership abilities – would generate any more cash in the short term. So some of my more experienced colleagues hemmed and hawed a bit as we went through the relevant IMD case. Relating the PSF list to their workaday world was impossible – at least on a practical level. This was 2003. The IT industry was still getting over the bursting of the bubble. The first members of the PlayStation generation had just turned eighteen and hadn't yet started making trouble in our working life. Five years later, when the first batch graduated from various universities and started looking for places to work, the situation started to change. As I said, the themes in Peters's book – and, in fact, his whole philosophy of learning and doing – go right to the heart of the values of the PlayStation generation. And, amazingly, many young people had somehow come to think that this is how big companies already work – regardless of industry. They were certainly in for a shock over their first weeks of employment.

At least four important themes stand out from the PSF definition list:

1. Develop a culture of cooperation
2. Invest in capability management
3. Be truly customer-centric
4. Embrace social networks

All four are age-old themes familiar from the strategies of countless companies. They are also things that a leader must commit to with all his or her soul.

Develop a culture of cooperation

A culture of cooperation is easy to see as just "normal good manners combined with the company's values". This is stuff you learned as a child. As a leader, I can't really be sure that if I recruit only civilized people, the culture will naturally grow – on its own. PSF philosophy posits, "All are responsible for helping all". The world of consulting companies – the original PSFs – operates around strong, independent partner organizations. These mini-networks are selfish and self-sufficient by nature. To offset this independence, PSF organizations created a culture where everyone was obligated to respond to any request for help originating from any part of the organization. The help request had to be precise and clear – not some convoluted spam e-mail sent to a big mailing list. If a recipient of the message repeatedly didn't react in any way to the enquiry, the rest of the network assumed that this person didn't really want to be a part of it. No reason was good enough to justify repeated neglect of the needs of your network, including high-flying title, time pressure, vacation, overtime, mother's birthday, etc.

People who – upon receiving a request for help – first consult their job description to see whether the request is any of their business are useless to me. They are not welcome in my network. It sounds rough, but this is one rule I had to follow ruthlessly. Conversely, if I saw a cardboard box sitting in a place where it might get in the way of a customer, I moved it away myself. Those who jumped over the box or kicked it out of the way, I disciplined immediately. I can't believe an intelligent human being would want to hide behind a job description. Or that someone would feel they are so high and mighty that a common cardboard box isn't their problem.

Putting this philosophy into practice is not easy, and I certainly wasn't able to respond to all requests for help. When you're dealing with an organization of 17,000 employees, the number of requests you receive every week runs into the hundreds. And that's not counting the ones from the customers. If I had ten different requests waiting for me in my e-mail and voicemail inboxes, it'd be natural to take care just of the easy ones first. These would include all requests in my own language, country and time zone. I decided the bigger challenge would be to start tackling the most confusing requests from the farthest distance away and written in a foreign language first. People who were physically close to me could often receive help from other parties. At some point even the people closest to me thought I was being a bit of a diva. Employee satisfaction surveys started reflecting this immediately. My closest colleagues said I was hard to reach and seemed to be on the road all the time. As with all new systems, I had a few kinks to work out as we strived to create a culture of cooperation.

Working through these growing pains is easier when an organization implements a 2:1 philosophy. Basically this means that everyone has two ears and one mouth. It's the art of listening. And it is the art of participating. As far as the culture of an expert organization is concerned, 2:1 refers to a much broader issue of leadership development. It's easy to say a leader has two ears for listening. But we're not talking about passive, one-way listening. As important as it is for a leader to carefully listen to signals in the network, it's also vital that everyone is encouraged to use their mouth to add value to the network. The personal opinions of all parties, including executives, are an integral part of PSF culture and in leading Digital Cowboys. All members of the network have the right to voice their opinions, but for the leaders it's also an obligation. When you remember that each message must have value for the network, it is possible to avoid most platitudes and clichés, as well as substance-free chatter.

The most important thing I got from building a culture of PSF cooperation is on my wall. It's a print-out from 2004 – a souvenir from a training session dealing with PSF culture. It states:

Hire people who like people
PROMOTE people who like people
Don't hire people who don't like people
God knows – don't promote people who don't like people

For me this has been the cornerstone for creating the kind of culture I want. I have copied this threadbare piece of paper countless times when heading out to instruct headhunters on the kind of executives I want.

"People who like people" can seem a bit vague, but look at the issue of building a culture of cooperation through this lens. What happens if the *know-how* of individuals is given significant precedence over *social abilities in building the organization*? Is it OK for the world's top technical expert, who brings oodles of cash and fame to the company, to isolate him or herself from the culture of cooperation? Even one recruitment choice like this can compromise your culture. On the other hand, cowboys tend to be individualists and I've taken on some real prima donnas for my team. I've let them play fast and loose with personal hygiene, office hours and time cards. I've allowed them to neglect company processes and, when push came to shove, organized all sorts of cover for people so they could make it in the maelstrom of a large organization. The one thing I've never allowed – and will not allow – is breaking the *rules of cooperation*. Making even one exception in the "all help all" philosophy degrades the entire community. That last sentence in the Tom Peters print-out makes a lot of sense: "God knows – don't promote people who don't like people". If you accept even one top talent who will not, doesn't want to or just doesn't have the common sense to work within your shared rules, you can never celebrate his or her accomplishments in front of the others. Your credibility as a leader is shot the second you do that.

The culture of cooperation is surprisingly challenging to work into the system. As a concept, it's incredibly simple but in practice differences between individuals and cultures can make it quite awkward for some people. The only way to make it concrete is to use crisis situations for this training. When it becomes apparent that there are people in your organization who refuse to work together, it's time for a brutally honest chat. The people in question and all supervisors involved need to meet the matter head on. These situations are as painful and awkward as the worst argument you'll ever have with an unsatisfied customer. But as far as the culture of your company goes, they're extremely important and educational. It's easy to pass the buck to HR or some other party. Don't do it. You and your organization need the learning experience.

Invest in capability management

Capability management balances meeting your current operational requirements with the development of future capabilities as the needs of your company change. At one extreme, you have managers who define the capabilities they need, write a list and send it to HR so their recruiters can find someone who fits the description. You provide a clear number telling them how many pairs of hands with a particular talent you need delivered, where and when. This is how many managers view capability management. In PSF, leaders see their company as a magnet for top talent. The magnetism isn't based on a steady pay cheque and a prominent brand, but on the fact that by working there, people can learn truly unique things.

Jim Collins – author of *Good to Great*[4] – succinctly described capability management. He describes it as getting "the right people on the bus, the wrong people off the bus". As business cycles accelerate, the process of getting the right people on and the wrong people off your bus needs to be assessed weekly. This is your job as a leader. The HR department can't – and shouldn't – do it, even though they can provide valuable information and opinions.

Once you have the right people on the bus, the great challenge of capability management is replication. A leader is always in possession of some sort of business portfolio. Your agenda is to ram through great production efficiencies, as well as fast service development and rapid growth – simultaneously across your portfolio. In modern business, Apple is a model example of this philosophy. As a part of the portfolio, the leader has business ideas – small-scale business starters. If the leader is lucky, one of these is a golden egg of such awesome power that the next phase involves aggressive operational growth. The leader's job is to constantly uncover capabilities that enable this sort of scaling and then train, develop and recruit the necessary talent as efficiently as possible. It's very rare that HR alone is capable of making resource and development decisions months or even years in advance. If the operative management doesn't make and manage these proactive decisions early in the cycle, growth will be stymied. Growth is rarely held back by a lack of potential paying customers. The limits to growth almost always involve a lack of expert resources for implementation.

There is rarely time to clone expertise in your own unit. Instead of cloning, you need to identify the Digital Cowboys in your network who ride into battle when needed – they may be in different units, countries or even outside your organization. You need to proactively find these cowboys before you need them, which means adopting a systematic practice of classifying people's backgrounds and areas of expertise. It sounds boring, but if you want to find the best bunch of cowboys in a large organization, it has to be done. In this situation it is the leader's responsibility to make sure that the disparate group of cowboys arriving on the scene all function based on the same code of conduct. The different areas of expertise must complement each other and a common, equal language must be found as soon as possible.

In Chapter 1, we talked about executives – and even CEOs – as producers. The producer has gone through the casting list, picked the right actors and approved the script. The director (project manager or leader) makes sure the production moves along at pace and that all the necessary expertise is present and accounted for, ready to work together to achieve a common goal. Eventually, when the production has been completed, it's the leader's task to evaluate the expertise of the individuals involved, as well as their continuing personal development – the leader is creating a capabilities bank. This often provides the most accurate picture of what kind of know-how is required and available for the next production. One metric used to measure the success of capability management is a successful production. In more and more organizations, the leadership's view of the necessary know-how is taking on a central role. We used to live in a world that offered a limitless number of production tools. This is not the case in the expert business. The changes people go through and their adaptation into life in a new corporate culture take time, and as a result many rewarding moments of growth are missed owing to bad decisions by the leadership.

Be truly customer-centric

Real customer-centricity is worthy of great admiration in the best PSF companies. It comes in a range of flavours and isn't necessarily a goal based on lofty principles. It is often driven by more craven factors – financial rewards and annual bonuses. Nonetheless, in a PSF firm, it goes deeper than this. From this broad subject, I like to pick two factors that have influenced the way I lead people. First of all, what's good for the customer should be the deciding factor if there is a conflict situation in your organization. If you have two different options to choose from, what's best for the client (or even better, the end-customer) should be the deciding factor.

In his book, Peters talks about a "client list to die for". As he states, "You're as good – or as bad – as the character of your Client List." As a matter of fact, he begins his book by focusing on the *client* instead of the *organization* or your in-house talent – *everything comes after the client*. You should be so passionate about your clients that you're ready to do anything for them. The flipside of this coin is that if you have clients to whom you aren't ready to give 110 per cent, you should remove them from the list. You shouldn't have any grey, humdrum cash-cow clients. Is this possible? Is it even sensible from a business point of view? My own response to this has become quite clear over the years. Yes, clients go through cycles and some projects are more interesting than others. It's the leader's job to ensure that the team is motivated and ready to serve the client with their best efforts, under all circumstances. No client should suffer the indignity of being a cash cow subjected to a soulless implementation process by the service provider

The other cornerstone of customer-centricity in the PSF theory is the team's (and its leaders) authentic desire to lead the client in their area of expertise. To Peters, PSFs are the primary sources of Intellectual Capital Creation. You have created a truly valuable network of experts but are you using it to lead your clients to new ideas, services, products or markets? Peters talks about why this doesn't happen:[5]

*Because, more often than not, we have **not** accepted the Leadership Challenge. We are **not** integral to our Client's lives. We are **not** Freaky-Cool-Provocative to be around. We are **not** on a Holy Mission. We are **not** providing wildly innovative, proactive solutions. We are **not***

*staying five steps ahead of the game. We are **not** routinely taking risks. We are **not** inventing the future ... and then leading our Clients to it.*

This requires great skill, experience and subtlety. No all-encompassing operational models exist to guide you. A strict case-by-case approach is required, even with different companies in the same industry. For example, I built a homespun system, whereby three times a year I'd collect all IT-related knowledge. Using this knowledge, I created a visual model over time that I believed to be meaningful. I sent the model to about a dozen trusted clients and a hundred colleagues. I used their feedback to make corrections and alterations. I took my model with me as I made the rounds – hat in hand – from customer to customer, team to team, to see how I could apply this knowledge model on a case-by-case basis. When four months had passed, I once again adjusted my views, corrected the parts that had undergone some sort of change and started the "process" all over again.

The model fitted some cases but unfortunately not others. The mistake I made was that I never used a large enough group to really test the model. In other words, I was unable to systematically bring together experts and scale up this vision we'd created, so that more clients and development projects could use it. It dawned on me that I was missing a key sector – the PlayStation generation. Their social capabilities and open-mindedness made them the perfect utilizers of this information. When I shared the materials in various internal blogs and wikis, I always received the most input and critiques from these experts.

In business literature this sort of operational development is referred to as knowledge management. As a term, knowledge management tends to lead us into the left brain – into the management sphere. Systematically documenting and passing on best practices and ideas is, naturally, an important part of the model. There are organizations where you can't climb the next step on the career ladder unless you've had, for example, ten articles in big papers or a couple of books on the bestseller list.

The most interesting and challenging part of knowledge management is how a leader contributes added value to all this activity. What precisely can you do to make sure that experiences in the customer interface can be translated into new services? How do you develop your own network to the point where each nugget of information can be turned into ideas? What do you need to do to make sure there is a common language between different countries and competence areas? In the PSF model, this is handled through top-quality experts, a master–apprentice operating model, as well as incentives and sanctions attached to data management.

Getting my own company to embrace the customer-centric model was an intense and often contentious process. Experts had a wide variety of roles and backgrounds. In many cases the key to a common language was a shared technology (SAP, Java, etc.). Our projects were also often quite extensive, both in terms of manpower and duration. To create client-centric solutions, we gathered our top experts from different countries, customer segments and technological backgrounds and threw them together. The team spent from a few months up to a year in sparring sessions where each participant described their own very customer-centric projects and explained how they would bring added value to others in the group. At the end of this, the groups were reshuffled and the team was given a large client project with a concrete value promise on which to work. At the end of the process, the client was able to evaluate how satisfactorily we were able to solve the problem. Conversely all the team members were able to evaluate which facets of the documented cases could be used as object lessons in success or failure in other projects.

The heart of the exercise was about *pushing* experts – and executives – into customer-centricity as part of an international team. It's not enough that the customer is satisfied now. The organization has to be able to further process and combine this know-how in a creative way. Digital Cowboys are incredibly well suited to this process. Conventional leadership can't start the desired change. Neither Digital Cowboys nor the rest of us are particularly keen on sharing our valuable little nuggets of information with colleagues. After all, we're all competing with each other in a way. A leader needs to be at the very core of this activity and learn constantly. A leader needs the ability to get the process started, lead it from his or her own corner and invite all relevant parties at his or her disposal to take part in it. The words *relevant parties* play a very important role here. The invitation process already shows whether the leader has any idea what kind of know-how and experts are needed for the task. This is the best kind of groundwork for managing a client and the customer relationship. Think of it another way – when you take an active part in the personal improvement of your key experts and the tools you use are real client problems, you are definitely leading the client and your team to "where the puck is going to be".

Embrace social networks

Social networks are one of the cornerstones of PSF culture's management system. Consultants who are networked with the client, their own international organization and often with policy-makers, too, are at the heart of successful consulting companies. As recently as the beginning of the new millennium, there really wasn't much to learn or copy in this area. It was so blindingly obvious it didn't seem worth emphasizing. Global markets and the entry of the PlayStation generation into the business world make social networking a critical success factor. A successful production operation is no longer the achievement of one closed group working off a limited, incestuous pool of ideas. For new services to be competitive, the views and expertise of a large and heterogeneous group are needed. This is where social networks come to the fore to play a central role in leadership. It is in everyone's best interest to find the best possible resources to throw at a problem.

At first glance, a leader's use of time in using social networks and social media looks like contact management – which is very much in the category of left-brain drudgery. One would think that if you know all the key people, your network is complete and your job as a leader is done. Can it be this easy? The PSF philosophy drove me to research the models of all the best consulting companies. Even though knowing and linking people via calling-card folders and later with digital tools is a part of all this, it's time to return to the root of it all once again. How do you create value for your network? How do you train your supervisors to work for the good of the network? How do you demonstrate the direct connection between this work and the money the company makes?

The development, utilization and measuring of the social network *must* be tied to the company's management system, as well as compensation in all enterprises. This is far too important a cornerstone of the operation to be left to volunteerism or happenstance. Even though it kind of contradicts the ideas of volunteering and spontaneity, you must be prepared to constantly single out those employees who create value for their network. It's hard to imagine an employee in any PSF company creating added value for his or her clients, year after year, without taking advantage of the resources offered by his or her network. On the other hand, if a company consisted solely of such one-man or -woman armies, nobody would ever want to network with anyone. The question would be: Why are they working in the same company? You can take this same approach to analysing and developing any enterprise or industry. If the network has no value in developing your business, you should give up using the word "company" and just turn it into a portfolio. This will allow you to skip all time-consuming celebratory speeches and honestly state that your company/enterprise offers no added value to its individual parts. Your experts are all so good they don't need each other for anything. As long as the results are good, everyone can live as they please.

To Digital Cowboys, this portfolio model looks like an absurd and nonsensical way to work. It's also the reason why top talent often doesn't want to work in large companies. It's an area where the leadership and management systems play a crucial role. The leadership has to, by setting an example through their actions, show how things ought to be. Are we going to get serious about working together and believe in the power of our network or are we going to make do with the portfolio model? In some cases, the portfolio model can be quite good. It's getting stuck in the middle which leads to damnation for all.

We *challenge* anyone working on developing leadership, business and management systems to acquaint themselves with PSF culture and its achievements. The growth spurts of practically all enterprises will be based on the leadership and efficient utilization of a new generation. Many of the principles and methods of a PSF enterprise appeal to the PlayStation generation. As a leader, you can't afford not to tap into the PSF model and principles.

Clarity, risk management and simplification

A very successful business leader once told me that the secret of his leadership was clarity – nothing more and nothing less. His office was the epitome of clarity. It contained a big oak table, a bookcase and a plastic plant. There were no papers on the vast expanse of the desk. The only items on the desk were a single pencil, a notepad, the day's paper and a PC. Nothing else cluttered the space. The bookcase held two books – a book on Finnish public law and the company's latest quarterly report – and a photo of his wife. When you went to see him, you always made it a point to arrive early. The meetings were quite relaxed – yet clear. Questions always related to results and their development. The meetings always concluded on time and he provided incredibly clear instructions on how to proceed.

I met him recently at the home of some mutual friends. I asked him, as is my habit, to give me some hints about how to approach my middle-age years as a leader. I was not surprised when he instructed me to remove from my agenda all things, countries and people that I found somewhat opaque or unclear. His instruction was to concentrate on those things that were clear and predictable. He thought the right time to get into growth markets like Russia and China was once their growing pains had subsided. "Invest as soon as things are on a steady and even keel" was his immortal wisdom on the matter.

I interviewed ten or so retired executives from different industries and countries while writing this book. I have the greatest respect for their views and I have learned a lot, in both my work and in writing this book, from these dear friends.

Every single leader I interviewed underlined the importance of *clarity*, *simplification* and *security*. It really got me thinking about the values and agenda I should drive as a leader. Was I hurting my troops by taking them from battle to battle, from foreign country to foreign country? Would invading just enough territory to put up a farm be sufficient? Then, we could engage in some clear, efficient risk-averse farming. Why does a leader or a company need more farmland if the current acreage is enough to feed us all into the foreseeable future?

I've thought about the clarity of Leonard Bernstein's working methods. Maybe he, too, had nothing but a Steinway Grand piano, a packet of cigarettes, a pencil and a strict schedule for creation. What I can say with some certainty is that he did not create *West Side Story* according to a project plan. His task as a leader was to put together a dream team of the top talent in many fields and lead them bravely into the unknown.

In PSF enterprises, corporate culture and management systems support the top talent and teams through a combination of clear operating models, a customer-centric focus and knowledge. We are talking about a very limited amount of creativity in order to focus on collective and network power. Digital Cowboys are going to be integral to creating a true PSF enterprise. Anyone tasked with leading them must be able to balance clarity, risk management and simplification with their individualistic, connected creativity.

Any and all of these challenges can create pressures from above and sometimes from the sides. Understanding how to manage these three issues involves wisdom gained from experience and – believe me – it can be a true learning experience. If you let the pressure from above influence your steering too much, you will have a hard time motivating your own experts to attack when necessary. If you are too risk averse to follow new business ventures, you will lose your credibility as a visionary leader. Look at it this way: the slow and predictable annual cycle of planting a field of wheat is not enough to engage the best brains in your business. The excitement that Tom Peters talks about is not there and this type of task will not allow your experts to develop skills or acquire new knowledge. The leader must have the courage to leave the wheat growing in the hands of trusted personnel and venture out into foreign territories with a network of like-minded individuals.

This is where your closeness to the customers and your ability to develop your top talent are worth their weight in gold. There's a third dimension, though. A leader must take the important lessons learned along the way and incorporate them into the management systems at every possible juncture. Your network is full of experts and knowledge. Dynamic management systems support these individuals and their teams. Strong leadership brings more like-minded troops into your network. Not only do clarity, risk management and simplification let you keep up with the pace of development, you are in the position to quickly scale up when it becomes necessary.

Summary: Network economies, globalization and consumers who move like a school of fish make life complicated. We are not living in a planned economy. With an infinite number of moving parts to content with, it is the leader's task to filter and simplify the data stream. Your own troops should always know in which direction to attack and who is in charge. For a leader, opening new avenues of growth and leading an international network of experts brings a new list of demands. Often it'll necessitate a re-evaluation of daily routines, the company's annual cycle and time expenditure. Sometimes value-based leadership with even the lightest of touches takes on a new, very precise meaning. What is your value as a leader and someone who develops his or her experts?

Bill Fischer was the lead professor in our IMD class ten years ago. His true straight talk and cut-the-crap attitude inspired us, but also forced us to face brutal facts. In other words, he made a team of 30 executives really think, contribute and think again. During those sessions, I really saw what world class facilitation and leadership means. It is not about giving answers to followers. It is about harnessing diversity to create action. There are a lot of professors, educators and leaders who try to do this trick with an ivory tower approach. Bill stands way above that league, so it was natural to ask him to contribute to the NO FEAR team. He is the Leonard Bernstein of facilitators.

Bill Fischer

Leaders and cowboys: unleashing talent in the digital age

"*Cowboys*". What an apt metaphor for today's millennial digital professional – or maybe not professional; maybe it's better to use the French word *amateur* for "enthusiast"? Think energetic, knowledgeable, fearless, risk-taking, "this gun for hire", ready to ride wherever fortune may take them – as well as being competent, capable and fiercely independent. These are people on which you can build a future. They are, in fact, building our digital future right now; they do not flinch in the face of big challenges. They are people who you can count on to be resourceful and inventive when encountering surprises and challenges. After all, didn't cowboys win the West? Yet, if you think about it for a moment, there are never really any "managers" appearing in the western movies. Cowboys never appear to have had real "bosses", in the professional sense of the term. And the notion of "leadership" is something that more often than not was settled out in the main street of town – *mano-a-mano* – with six-shooters. How well, then, does the cowboy metaphor fit our purposes in this book, and what does it help us understand? My impression is that the metaphor is more than merely "appropriate". It is "perfect". Understanding, of course, that it is "the winning of the West" which is really what was important and that the cowboys were a good means of getting us there.

Cowboys may have won the West, but they didn't do it alone. No serious portrayal of cowboys – be it analogue or digital – would be complete without "John Wayne" or "the cavalry" coming to their rescue. When faced with significant challenges, be it nature, bad guys or greed, the lone cowpoke was usually not up to the task by himself. In fact, one could argue that cowboys didn't really conquer the West at all, although they provided "the brand".[6] What really won the West was a combination of organizational innovations, such as the development of the "long drives" which moved cattle from the prairies to the markets, "trains" (or caravans) of covered wagons carrying hordes of settlers, and maybe even such short-lived organizational experiments as the Pony Express; technologies, in the form of the intercontinental railroad, telegraph and barbed wire; and, yes, certainly, the entrepreneurial spirit so vividly personified by the cowboy. This is not trivial stuff. The economic and social forces that these agents of change unleashed have truly redefined our world in the last century. Charles Francis Adams, Jr, an American military officer, railroad executive and great-grandson of America's second president – who "was there" in 1868 – observed:

Here is an enormous, an incalculable force ... let loose suddenly upon mankind; exercising all sorts of influences, social, moral, and political; precipitating upon us novel problems which demand immediate solution; banishing the old, before the new is half matured to replace it; bringing the nations into close contact before yet the antipathies of race have begun to be eradicated; giving us a history full of changing fortunes and rich in dramatic episodes. Yet, with the curious hardness of a material age, we rarely regard this new power otherwise than as a money-getting and time-saving machine ... [rather than] the most tremendous and far-reaching engine of social change which has ever either blessed or cursed mankind.[7]

In the nineteenth century, the force that Adams was describing was the liberating power of abundant geographical space – *the West* – which has subsequently powered the world economy for the better part of 150 years or more. In the early 21st century, these same sentiments could well be applied to the liberating power of abundant digital space – *the Web* – and its potential to open up limitless opportunities for the world economy of the future. But that, of course, will require a similar combination of "success factors" – organization, technology and cowboys. In fact, in many ways what Adams is talking about is being realized today through organizations such as Apple, Pixar, Google or even Nokia, and possibly Huawei. All of these are organizations that have harnessed the energy and ambitiousness of the cowboy mentality and have tied it to the leveraging power of organizations that take full account of this potential. They added an appropriate dose of leadership acumen and the result is an innovation track record that is writing, or rewriting, our futures.

We start, then, with the abundant human talent so typically associated with the "PlayStation generation"[8] – that is, the raw material on which the organizations of the knowledge economy of the 21st century will be built. Out of this raw matter comes ideas – if we are skilled in our management of talent – and from these ideas comes the stuff that makes our future. From everything we know, the availability of human talent is well above what is presently being utilized. When we survey participants at IMD, which is probably a much-biased sample because of the corporate seniority of the people involved, the numbers are shocking. Most senior executives, on average, see their organizations as receiving less than 50 per cent of the talent that they themselves are prepared to offer. One can only imagine how "junior" people feel. Too often, such organizations have become "prisons of the soul" – truly deadening places in which to spend precious time. This is unfortunate, maybe scandalous, not because such talent is so available, but because it is so underutilized. And it is a global problem of some magnitude.

The dilemma that makes this disconnect between "talent available" and "talent at work" so particularly alarming – no matter where you live – is that at a time when our planet's climate is imperilled, our ecosphere polluted, scarce resources being squandered, biological diversity at risk of decimation or more; when the existing economic order is being redefined, when emerging market opportunities abound, when there are still vast chasms between super-rich and super-poor; when technology appears to be the most vibrant it has ever been and when open-source communities have revealed the power of collaborative thinking – the need for human ingenuity and energy has never been greater, but our organizations are failing us. The most sophisticated social platforms built in human history, with resources and global reach that are unprecedented, appear to share one common attribute: the ability to diminish the human talent to which they have been entrusted. What an indictment of modern management and the practice of corporate leadership! We are hiring "cowboys" and turning them into "bureaucrats". What the modern complex multinational organization has mastered is the ability to take superbly talented individuals and convert them into average performers. How can this be? To my mind, this is indeed a failure of leadership.

What will it take to overhaul our existing approaches to organizations so that we can fully appreciate, and utilize, the promise of the Digital Cowboy? Let's begin with a simple premise that the future – uneven as it will likely be, but for the employers of Digital Cowboys, at least – will be led by firms for which "knowing things will be more important than actually making things". What would this mean? Well, for example, the real value of a Google, an Amazon, a Facebook or a CapitalOne, or even the gambling casino chain Harrah's, is not so much what they do – although all do perform their service delivery in an exemplary fashion – but what they know about their customers: you and I, and the great swaths of their market that they can "read", often as individuals, better than anyone else. And it's not just such "ethereal" producers as the ones I've mentioned, but also real product and service providers such as Frito-Lay, the salty snack-food producer; Ritz-Carlton, the luxury hotel chain; Holcim, the cement company; and Skanska, the construction company.

All of these firms are market leaders because they know more about their business and their customers than do their competitors; and because they've made that knowledge an explicit part of how they go to market – it is a fundamental part of their offerings. The acquisition, movement and harvesting of ideas are seen as being at the very essence of what they are and how they work, and key to such efforts is a sort of *"Idea Hunter"*[9] mentality that recognizes that the classical industrial engineering approach to organization – where scale and efficiency were the ultimate determinants of competitive success – is about to be replaced by the notion that knowledge is the ultimate competitive differentiator and that conversations – not tasks – are the basic building blocks of organizational performance in the new knowledge-intensive era. In fact, it's not too far a stretch to suggest that each of the firms mentioned above, as unlikely as this sounds, have adopted what Pekka has referred to in this chapter as a Professional Service Firm mentality. Think of it – working in cement or potato chips, but thinking professional services. That's exactly what is going on! And the beauty of all of this is that not only is the organization more "thoughtful" in the way it goes about engaging and satisfying its customers, but the talent employed is, typically, more fully appreciated and utilized as well. This is the power of thoughtful leadership at its very best.

Pekka speaks of what he learned from Tom Peters, and his book *Professional Services Firm*. I'm delighted because I taught the very class that he describes; I remember it well, and it was not easy. In fact, it was a "desperation move" on my part, as I was struggling to find a way to stretch the mindsets of his colleagues. I have always been an admirer of Tom Peters and thought that some of his bravado would do the trick ... maybe. It was abundantly clear, at the time, that Pekka's organization was being nibbled away from below by lower-wage/lower-price competitors selling commodity services, and that the only solution was to "move up the value chain into solutions". Yet, despite the frequency with which we hear such a strategy being bandied about, it is, in truth, much more difficult to do than it sounds. In fact, the "solutions segment" of his and many similar markets is already occupied by big-branded, consulting-type firms. Succeeding against them meant avoiding being deluded into thinking that just "average services" would be sufficient.

Tom Peters's message is far, far stronger. For me, the essence of that book is: thinking of ourselves as running a small (extraordinarily ambitious and well-run) business instead of running "a function" – and then, as we would hope any artisan would do, taking control of our lives and doing something that we'll be proud of for ever after. No more merely "serving our time". It's making the move from mere "overhead" to being part of the value proposition. I saw my wife Marie do this with her own small (20–30 full-time people), but extremely successful, catering business, and it was not easy; but for so many denizens of big organizations, with all their legacy encumbrances, thinking of running their own business is a stretch that is extremely difficult to make. Where do you even start? And who gives you permission to be different? To be ambitious? To be branded – even if only internally? To dismiss difficult or even enervating (internal) customers? Yet, without such freedoms, can it really be said that we're running our own business?

Here is where *Leadership* – with a capital "L" – comes in. Leadership is the absolutely essential ingredient needed to create a perfect balance between talent fulfilment and the achievement of business objectives. In our study of Virtuoso Teams,[10] what we discovered was that "Virtuoso Leadership" was always present. So much so, in fact, that one editor recommended that that be the title of the book: *Virtuoso Leaders*. Why? For several reasons:

- All of our teams were composed of incredibly young people, who needed a senior person for security – to provide "air cover" for their adventures.

- Their "adventures" were often rough-and-tumble. We discovered that polite teams get polite results and teams that "changed their world" work in far from polite ways. Leadership was necessary to "navigate" the balance between "abrasiveness" without slipping into "abusiveness".

- The adventures of these teams, incidentally, also remind the observer of "a culture of cooperation" such as you would find in a platoon – rather than a divisional headquarters – where, in the face of enormous personal risk, you are forged into a "band of brothers" and what you want from "leadership" at that point is more supportability than reportability.

- In fact, it turned out that the single most "effective" moment of innovative team performance, and individual contribution, in the teams that we studied inevitably came at the moment when the team believed that it had "absolute freedom" to contribute its talent in any way it could, and top management believed that it retained "complete control" – at the same time! This is a true leadership accomplishment. It requires both absolute clarity of purpose and absolute clarity of communication, plus sufficient self-confidence on the part of the leader to get out of the way and allow a team to "get on with the work".

- In all of this, "leadership" needs to be a contact sport. This is not about remote leaders sending e-mails praising accomplishment; this is about continuous and direct contact which keeps both sides sure of the intentions of the others. This, incidentally, sounds very much like Pekka's reference to Alexander the Great, and his penchant for being "on the front lines".

In fact, borrowing from the world of entertainment, increasingly what I think we will see for the leadership of "Digital Cowboys" will be *incandescent* teams – teams that burn brightly and then burn out, led by an "impresario" style of leadership. Teams in which great talent is discovered, nurtured and then carefully put into situations where they can both "star" and return the investment made in their careers. The great marine scientist Jacques-Yves Cousteau exemplified this approach when he admitted: "I am not a scientist, I am an impresario of scientists." And their impact is lasting, as one of America's most venerable broadcasting journalists, Eric Sevareid, admitted when he spoke of his mentor, the great Edward R. Murrow, who was not only a broadcasting legend, but an impresario to nearly an entire generation of radio and television journalists. According to Sevareid: "Murrow not only had an influence on my career, he practically invented my career." Similarly, in 2009, when the New York Philharmonic Orchestra made its groundbreaking debut in Pyongyang, DPRK, then music director Lorin Maazel stepped off the stage at the penultimate moment of the evening's performance in homage to his mentor – Leonard Bernstein, then dead for nineteen years – so that the orchestra could be led by the memory of Bernstein. To achieve such a legacy goes well beyond the traditional leadership model; even well beyond "leader as servant". This is about participating in unleashing talent and the making of legends, and passing down that gift from one generation to another, as a primary responsibility of leadership.

Finding leaders who can fulfil such roles is far from easy. When Pekka mentions the leadership role of Leonard Bernstein with the team that created *West Side Story* and changed the business of Broadway in the effort, he points almost naturally to someone who was one of the great innovators of modern music, but also an impresario who nurtured the succeeding generation of leaders in his field. Nonetheless, at Pekka's mention of him, I think immediately of his collaborator, the great ballet choreographer Jerome Robbins. Who was "the leader" of this team? Sometimes it was Bernstein, and yet it was always Robbins's dream that they were working with. Such fluidity of command, and generosity of rank, is another aspect of leadership in the future that has to be acknowledged and appreciated. Increasingly, it is not just "us" that we are leading, but rather an entire value chain that we are finessing. The innovation race of the future will be all about combining value-chain partners' assets into new business models – such as we have seen with the value-chain reconfiguring that was characteristic of both the iPod and the Nespresso Pod – and this makes leadership that much more difficult.

What, then, might be the personal qualities of such leaders? In a rather remarkable study of exceptional leadership in both old and new economies – the latter probably best thought of as the generation that preceded the Digital Cowboy – Warren Bennis and Robert J. Thomas compared the "leadership competencies" that they felt were shared by the set of successful leaders under 30 years of age and over 70-plus years that they studied. What they found were four shared leadership competencies that had been forged in the crucible of a high-risk professional challenge:[11]

- **Adaptive Capacity.** Hardiness; First-class noticer; Learner; Proactive in search of opportunities; and creativity
- **Engaging Others by Creating Shared Meaning.** Encouraging of dissent; Empathetic; Obsessive communicator
- **Voice.** Purposeful; Self-aware & self-confident; Possessing EQ
- **Integrity.** Ambitious, Competent; and having a Moral Compass

Are Steve Jobs or Bill Gates, then, possible "role models" for the Digital Cowboy's leader? I think so. They combine the visionary aspects so necessary to inspire the best in our cowboys' imagination with the personal self-confidence that is necessary to dream a big dream, make a big bet, license a virtuoso team, and then move out of the way. Are they "nice guys"? Does this matter? What does matter is that they have instilled a personal loyalty among the many talented individuals who have worked with them, and they have played a major role in developing the next generation of industry leaders who will succeed them. While neither has been necessarily the easiest person to work for, their ambition has never been questioned – nor has their integrity. Not bad for leaders who have also changed our daily lives with the products that they both conceived and then produced!

But there are many others who like Jobs or Gates can unleash the power of Digital Cowboys and still harness their energy. I was recently struck by the insights of a chronicle of the war against malaria, entitled, appropriately enough for our "cowboy metaphor", *The Imaginations of Unreasonable Men*.[12] According to author Bill Shore, himself the CEO of Save Our Strengths, an organization dedicated to ending childhood hunger, great movements are launched as a result of:

- An *unwillingness* to accept that the failure of imagination has become routine: "… imagination cannot be bought and installed like the latest software, or taught in an MBA program. … [quoting Sun Tzu] every battle is won or lost before it is fought. … Not believing that malaria could actually be eradicated was a failure of imagination that distorted and undermined the way the malaria community went about its work until someone leaped unreasonably over the hurdle." This failure of imagination is big; it is endemic in the modern world of organizations! Recently, while working with a group of Marketing High-Potentials at a world-class fast moving consumer goods company, I was startled by their conclusion that firms with brands and products that delight and energize us dream bigger dreams than do others. They felt it with their company and were quite envious of companies that they admired, and which they felt had "bigger" dreams than their company had.

- An *unwillingness* to accept the status quo: "The status quo yields not an inch of ground without a fight". Yet, as Shore observes: "In a field that has suffered a stalemate for many years, eccentricity might be just what is needed."

- An *unwillingness* to accept "average performance": Shore quotes Victoria Hale, founder of OneWorld Health, as observing that: "Five years at the FDA taught me one thing in particular: the success of a product depends primarily on the product team. The drug and its qualities are often secondary." This gels with everything that we've seen in the *Virtuoso Team* experience: great teams have the power to unleash great talent!

- An *unwillingness* to be an average leader: great teams also require, we think, great leadership in order to function at their peak. Shore points out the criticality of leadership: "... someone with persistence bordering on stubbornness, confidence bordering on arrogance, the long-term persistence of a cathedral builder, and the immediate impulses of an emergency-room doc. ... And it takes a boxer's willingness to take a punch and come up off the canvas."

- An *unwillingness* to be "reasonable" in the face of great challenge. Effective innovation leadership requires the self-confidence to be unreasonable. I recall the CEO of one of Europe's largest cellular network operators telling me that the single most difficult requirement of his job was "being unreasonable". But, as Shore illustrates, big change is often impossible without an unreasonable person leading the charge. Another "Shore" – George Bernard Shaw – opined in 1903 that "all progress begins with the unreasonable man", and one need only recall the unreasonableness of Prime Minister Winston Churchill during England's darkest hours in the Second World War to appreciate the power that one unreasonable individual can possess.

- *An unwillingness* to "fit in": the sorts of people who are required to drive ambitious and unreasonable change possess personalities such that their names ".... invariably left an invisible but palpable tension in the air, like one of those high-energy transmission towers that can be valuable or dangerous, depending upon your point of view". This would be entirely appropriate for most of the leaders that we chronicled in our *Virtuoso Teams* book: Miles Davis, Sid

Caesar, Robert Oppenheimer, Leonard Bernstein and Jerry Robbins, Roald Amundsen, even Thomas Edison, would all fit this description. So would Steve Jobs or Bill Gates!

The great opportunity of our lifetime is that nearly anything is possible, if only we have the courage to allow human ingenuity to flourish and to provide the right type of leadership to point it in the right direction, keep it headed on the right path, and keep it revitalized. Digital Cowboys are very much a part of our hope for the future – but only if they have the leadership that they deserve!

To learn more about Bill and to continue the conversation, go to **www.nofear-community.com**.

3 Hey, Old Man. What do you know about me? The challenge from the Digital Cowboy's perspective

DIGITAL COWBOYS ARE THE SUPERSTARS of the PlayStation generation. They – men and women – have grown up being wired, connected and inquisitive. They eat and breathe the tools and the digital know-how. Since their teens, the Internet and mobile phones have been part of their lives. If they have a question, they "Google" it. Finding a (most likely) outdated encyclopedia and checking there wouldn't even cross their mind. Digital Cowboys will be the experts in your value creation network, driving the teams' productivity by leveraging collaboration tools and best practices. They will be the catalyst and driver for the high-performing teams of today and tomorrow.

As I've stated before, my definition of the PlayStation generation includes all people born after 1985. Of this age group, I separate into the Digital Cowboy group those who wish to work as experts – those who leverage the digital tools, networks and mindset. This applies to all industries and educational backgrounds and both genders. Before we can look at ways to solve current and future leadership challenges, we have to understand how our future troops view the challenges. As we know, the cowboys generally have no problem letting you know how they feel. So, it's time to throw the ball into their half of the court.

Hey, Boss – this is not just work

The following is a direct quote written by an expert born in 1988 for a document related to a development discussion. Please keep in mind, this is a high-performing individual we would like to see in a leadership position. Somebody born in 1968 or 1978 would likely have pushed for a management position as the "logical next step on the career ladder". But when this member of the PlayStation generation was asked about taking on a management position, he replied like this:

As far as working as a supervisor is concerned, the only thing that interests me is the company car. Our child will be born this spring and I want to spend as much time at home as possible. I may take six months off to take care of the tyke. I might consider supervisory duties after that, but I'd have to be able to pick my own team. I don't want to do the same thing over and over again. On second thoughts forget the car. I'd rather use public transportation. So forget about the leadership duties. It'll help save the planet.

How do we get someone who thinks like this to embrace the challenges of leadership? Is this group of experts a bunch of self-centred SOBs? A McKinsey report[1] looked at the expectations that the PlayStation generation has as consumers. I find these consumer expectations apply to job expectations, as well:

Desire for individual self-determination: having control over what matters, having one's voice heard, and having social connections on one's own terms. The leading edge of consumption is now moving from products and services to tools and relationships enabled by interactive technologies.

The list describes a combination of self-centred thought and highly developed social networks. Our cowboys are used to taking care of their own business. They are creative problem-solvers, desire a high level of autonomy and are experienced networkers. In other words, *they want to spend their time on things that matter to them*. Dan Pink, who has authored several books on the changing world of work, based his book *Drive* on 50 years of behavioural science. His findings looked as if they were ripped from some Digital Cowboy manifesto. He states: "The secret to high performance and satisfaction – at work, at school, and at home – is the deeply human need to direct our own lives, to learn and create new things, and to do better by ourselves and our world."[2]

So, we have the theory and data to support it, but how do we use this information? What does it look like in a business setting?

As part of a training exercise, I worked with a development team to collect a list of common issues that came up in development discussions. Our aim was to find a list of attributes and targets for improvement that would help build a *bridge* and a *common language* between leaders and top experts within the organization. More importantly, we wanted to get a handle on what Digital Cowboys thought about leadership in its everyday manifestations. The quote about the company car that started this section came from those discussions. At the end of the exercise, it was obvious that we were not on the same page – maybe not even in the same book. Something wasn't right and we wanted to understand how far from each other we actually were. It was becoming increasingly clear to everyone that changes had to take place in the brain hemispheres of the leaders and those being led. The list included many differences that were attributable to normal variation between the experiences of generations. I won't go into those in any detail. It's more important to try to understand the expectations and factors that will help us get the most out of the Digital Cowboys. It's also important to get our young talent excited about leading and leadership. All of a sudden it seems that the path to leadership is not the most desirable road to take.

What do you expect from a leader?

The PlayStation generation expects those who work in leadership positions to provide *enthusiasm*, *responsibility* and *cohesion*. Of these, enthusiasm – the honest energy and optimism with regard to the business of living – is generally considered a trait of children and the young. Its opposite is the experience, restrained caution and distance brought by age – at least up here in the cold North.

Enthusiasm

The enthusiasm that the PlayStation generation represents and also expects from its leaders goes a little deeper than just being childlike. Their enthusiasm is more a passionate interest or eagerness to do something. This enthusiasm won't go away when you "grow up".

This generation knows only this post-industrial age. Its members in the *industrialized* countries have not known war, hunger or any massive social upheaval. In the *emerging markets*, where change over the last twenty years has been much more pronounced, this energy has been directed down different avenues. When the environment has been more extreme, more extreme methods have been called for. At one end of the spectrum you have movements prone to radicalism and even terrorism. At the other end, among people of a similar mindset, you have a retreat from large, traditional organizations and the drive to establish exceptional start-ups. Many of these start-ups were born as a result of rebellion rather than systematic analysis. Fortunes have been created without manufacturing a tangible product or by hijacking or robbing someone else's fortune. The fact that a company worth billions can be created from the vision and capabilities of a couple of individuals is extremely inspiring and happened far more often in the decades of the digital revolution than at any time before. You no longer have to be born with a silver spoon in your mouth or live atop a huge pool of oil. *Enthusiasm is born when you hold the reins and control your destiny.* A leader must understand and use this enthusiasm as he or she builds a leadership strategy.

It's not about preaching a baseless gospel of positivity. It's about creating an atmosphere of continuous analysis and new opportunities. And considering that, in an era of social media, just about any products or services being developed can be distributed just about anywhere, opportunities have no bounds. Albert Einstein said, "There are only two ways to live your life. One is as though nothing is a miracle. The other is as though everything is a miracle." Those are words to live by for someone leading members of the PlayStation generation. I often change the words "a miracle" to "possible" because I do believe that everything is possible. And since everything is possible, the role of the leader truly becomes the role of the producer – sitting on a wagon full of innovation, holding the reins.

Practical leadership, then, requires a strong personal vision and the ability to take risks. When this works out, the result is an energy that moves the team forward. We've all been in meetings that leave us energized and ready to tackle any challenge. On the other hand, at least I'm willing to admit that I've sat in meetings after which I had no intention of going out and conquering anything. The same logic applies on an organizational level. To create energy, you need a good deal of rationality and exacting deployment methods – great work for the left hemisphere of the brain. But the team also needs that emotion and excitement brought by a shared mission – right-hemisphere stuff. A deep energizing enthusiasm is produced by a diverse team with a variety of perspectives on the matter at hand and the ability to process these into a workable synthesis. Victory or death!

If the producer is still developing an industry-specific vision and know-how, he or she must – at the very least – be able to *steer* discussions and critical decisions. A leader who tries to retreat behind a wall of standard-issue management jargon, away from the actual substance and decision-making, is an abomination. So-called professional leadership will not be the hottest property around in the coming decades. Even though leadership in different industries shares many traits, it's impossible to be an inspiring leader if you have no knowledge of or interest in the industry you're dealing with. It is easy to see the value created by professional management as just massaging the left brain. It's the kind of value creation an

Excel macro is capable of. A leader must – by his or her own example – be able to get the team to engage in a continuous, pulsating and completely transparent process of ideation and productification. My own experiences as a producer indicate that trotting out my own views on things – even when they're wrong – is a lot more inspiring to the team than retreating from the front line. What could make a person happier than being able to correct the boss? I love giving people that opportunity.

Responsibility

The PlayStation generation has a reputation for self-centredness. However, people fail to appreciate this generation's high degree of accountability, coupled with a strong focus on community. Their sense of responsibility regarding the environment and surroundings is pronounced when compared to, say, members of my generation as a whole.

Don Tapscott wrote in his book *Grown Up Digital*[3] on expectations of authenticity and moral integrity:

So it's not surprising that they [Net Geners] care about honesty. Among other things, they have seen the giants of corporate corruption, the CEOs of Enron and other major companies, being led away in handcuffs, convicted, and sent to jail. It's far easier for Net Geners than it was for boomers to tell whether a company president is doing one thing and saying another. They can use the Internet to find out, and then use social communities like Facebook to tell all their friends. [...] They do not want to work for an organization that is dishonest.

Members of the PlayStation generation constantly follow societal phenomena around the world. The college campus meetings of the 1960s have given way to worldwide groups on Facebook, where debate about the pros and cons of any given question spreads extremely fast. There's the media truth, your own truth and the truth of the people in your network. These are weighed against each other.

Responsibility is expected of the leader in business, too, albeit in a different sense. The philosophy of the Digital Cowboy dictates that you own and take responsibility for your own horse, carry your own gear and show up when you're summoned. A cowboy is a pro who takes responsibility for himself and his actions. He never quits and leaves the herd to the predators – slinking off to safety when the going gets hard. In the same way, it's assumed that leaders are accountable for their actions and opinions. No freeloaders are allowed and every leader must have a clear, public opinion on where it is that they are headed. Making adjustments to this opinion is right and acceptable if something is wrong. In addition to accepting mistakes, understanding that all rapidly developing plans have holes in them big enough "to ride a horse through" is an important facet of honesty and transparency. Not everything has a solution to it – yet. For a Digital Cowboy, this situation is not an abomination. It just means that you have to get moving on the stuff that has been thought through and decided upon.

If things are going wrong, there is nothing more absurd than a leader hiding behind corporate guidelines or home office strategy. Instead, a leader pushes through the challenge, using whatever is at hand. That's what a proper cowboy would do. I keep coming back to this theme for a purpose. It is for this reason that large organizations have such a hard time keeping the best cowboys in their troops. A real pro isn't fazed by a bit of bureaucracy or some slowness. However, when he or she realizes, over the course of daily interaction, that no one is prepared to be held accountable for anything – well, that makes the cowboy mad. Add to this a rigid hierarchy and an atmosphere suffused with fear and you've basically murdered creativity and enthusiasm. The best way to combat this kind of catatonic corporate culture is to choose and train the kind of experts and supervisors who take responsibility for decisions. Erring on the side of excess – i.e. being too accountable – is always easier to recover from than dealing with a mess made by someone who doesn't take any responsibility.

Honesty is an integral part of a sense of responsibility. Politics and plotting have always been a part of life in an organization. Many of us used to hear from older colleagues that business is a game measured in quarters. "Life is a game and business is a big game using someone else's money," a star executive succinctly put it in the nineties. Even if we, the elder colleagues, are prepared to put up with organizational shenanigans, our Digital Cowboys are exceptionally allergic to it all. The PlayStation generation expects total honesty and directness from its leaders and their organization. If you've grown up in a communications culture where everyone has access to their own worldwide media outlet and has no problem with airing grievances publicly, we're talking about a whole other level of transparency compared to that of ten years ago. A leader had better look at the way he or she operates. The old model might be OK, familiar and acceptable for you, but your idealistic cowboys will not want to participate. Not even if you pay them more.

Cohesion

The third expectation is cohesion – working together to form a united whole. A leader's primary tool for creating cohesion is producing a *shared reason and purpose* for working together. People go to work to earn their daily bread and their kids' daycare fees, but that isn't inspiring. Social glue – parties, trips and playing basketball with your workmates – is not enough. These social tools are important for getting to know the people and for lowering the barriers between different age and expert groups but they won't motivate them to work together. It takes way more work than a few get-togethers.

Let's look at the typical progression. We achieve the first phase leading towards cohesion by getting the team aglow with enthusiasm for tackling a particular mission. Months later, the team has succeeded, failed, rebuilt and redirected *together* enough times to consider themselves a team. They don't need T-shirts or coffee mugs to show who belongs to the team. This is real cohesion. As a leader, seeing the team you lead turn into a cohesive unit continually focusing on innovating and creating is an incredible moment.

But you never get to step back and bask in glory. Quick and constant changes in your network make it challenging for a leader to function as a creator of cohesion. Cohesion takes a hit as production cycles speed up and members of the network change positions more rapidly. This is particularly challenging in situations where the team is in five different countries and who knows how many time zones. The Digital Cowboy has every right to expect that the leader has this situation in hand and can keep the team together.

Harnessing social media and technology is essential. They can be used for a lot more than just updating a Facebook group of the team members. The heartbeat of the network is felt through social media. Production development happens via social media. Left-brain activities, such as processes, timetables, memos and quality control, are on the Net and often shared. New ideas created by the right hemisphere – such as innovative services or even the tiniest development ideas – are there on the Net along with user and end-user feedback. Most important of all, the communications between team members – your value creation network – happen in social media. You don't need a crystal ball to realize that many of your leadership activities will take place via this medium. The PlayStation generation expects leadership to be a motor for cohesion. Cohesion is built around a shared mission and purpose. In real life, value creation happens where the battle rages – on the front line.

Digital Cowboys as part of an organization

When the Digital Cowboys enter the labour market, they will become a driving force in society. Some are already here, changing the way we think and lead. For me, one of the biggest surprises has been that Digital Cowboys expect their boss to demand more! It is weird when an employee wants more and more challenging tasks. We're not talking about stuff said in the heat of salary negotiations. The message is that I, as a boss, must be able to challenge these experts to give their best – and then some. But this cuts both ways – if you demand their best, they expect your best in return.

As I said earlier, the PlayStation generation expects those who work in leadership positions to provide *enthusiasm, responsibility* and *cohesion*. I've asked the same of the cowboys I lead. I have said to employees you can't be a cowboy of mine if you're not interested in bringing these same attributes to the team. Since work groups consist of very different experts, enthusiasm, responsibility and cohesion manifest themselves differently with each individual. Regardless of their form, these are the three basic pillars on which everything else is built, so they have to be somehow recognizable and in a position where they can be developed. It's not just about good manners. I consider going through this basic stuff so important, I tend to do it face to face with my experts. The hard core of team-building, i.e. direct communication and leading by example, starts during recruitment and continues subsequently. The same model must be repeated when choosing people for supervisory positions and during training. There are a number of things that are non-negotiable. In contrast to what many people think, members of the PlayStation generation are more than prepared for this.

As you can tell by this book, I am fascinated by Digital Cowboys and what makes them tick. When I started researching and showing interest in Digital Cowboys, it made some people worried. Would this result in ageism at work? Would my ruminations result in a situation where leadership is geared towards the young and agile alone? Was it my aim to drive all the old fogeys born in the forties up to the seventies out of our workplaces? The fear of discrimination hung around in the shadows.

I think the opposite happens with Digital Cowboys in your workforce. In my experience, Digital Cowboys tend to lessen all brands of discrimination on the job. If the leadership does its job and everyone is on an equal footing, there is no room for discrimination. A team member is only part of the team as long as he or she creates value for the other members. If a member can't create value, then he or she is useless to the team – regardless of age, gender or nationality. When old merits and academic degrees count for little, everyone has to do their best. I have not encountered ageism. I've seen quite the opposite. I love seeing someone who's been doing the same job for twenty years come alive when put on a team with some enthusiastic Digital Cowboys. Life experience and perspective combined with the creation of something new makes people come alive.

Digital Cowboys learn from their older colleagues and openly put their own knowledge and perspectives on the digital table. The greatest mistake a leader can make is to try to separate the generations and expert groups from each other. Leaders often set a very compatible group of like-minded people to work on a task. And then assign a supervisor cut from the same cloth. It's normal. It's easy and it avoids interpersonal conflict. Digital Cowboys would like to use the same model – they get to ride into the sunset with a group of people their own age, who listen to the same music and are pretty smart. STOP. For the sake of *learning* and *personal development*, this scenario must not happen in any way, shape or form anywhere in the organization. The way to do this is to avoid all quotas at all costs. No group should by default have x number of Finns, Americans, Indians or Russians just because it "looks good". Quotas kill equality. Everyone in the team is entitled to one vote, one task and one value he or she creates for others.

So you have created an enthusiastic, responsible and cohesive team to tackle your latest development challenge. The team is working hard when all of a sudden your lead engineer decides the snow is great in the mountains and doesn't bother to show up for the weekly team meeting. Her Facebook page shows slopes of fresh powder and she has even posted a video of her "first tracks on the mountain" exploits. You are about ready to blow up when you also notice that she has posted her latest batch of code that resolves a sticky customer issue. Here is one of the biggest conundrums of the PlayStation generation.

Unlike previous generations, Digital Cowboys are much more finicky about their use of time. Fewer people are willing to work fifteen-hour days and traditional careers or titles are not their top priority. Hobbies, friends and their social life are equally as important. Maybe it's my background as an entrepreneur, but sometimes this really annoys the hell out of me. There have been numerous occasions when I've lost my temper with my Digital Cowboys. To my sense of morality, it's unbelievable that at a time when the whole team is on fire and deadlines are approaching, a key member might decide to go and be a ski bum for four weeks. And no amount of money or promises will convince them to postpone it by a couple of weeks. It's as if the operator of Alexander the Great's biggest catapult decided he needed to go see a masseuse right in the middle of a bloody battle. Their decision probably involved some life–work balance philosophy but I feel it displays an unforgivable degree of indifference to and disrespect for the rest of the team.

Finding the balance between these two opposing views has not been easy. A Swedish Digital Cowboy once told me, "We don't have working hours, just hours." Digital tools have done away with the idea that you have to sit in an office from nine to five in order to create value. You carry your know-how with you and your tools are in your laptop, smartphone or iPad. Work is best done when the drive is there – when you're inspired. If a brilliant idea hits you as you are dropping off to sleep, your laptop is there and you capture the idea – creativity and productivity 24/7. The downside of this model – which makes it more demanding for the boss – is that it requires constant reachability, connectivity and access to resources and data. If you worked on Saturday and Sunday, you don't get to take off Monday and Tuesday. Someone on your team might need input from you or a customer needs to reach you. The demands of accessibility increase when the team is dispersed over several time zones.

The only way to get past the challenge is to openly agree on specific rules of engagement. These rules are sacrosanct and apply to *everyone* – from the boss on down. There are no free rides. If you break a rule, you lessen your value to the team. Accountability and open feedback are part of the agreement. A word about "open feedback" – the PlayStation generation has grown up with shows like *American Idol* as their example of open and helpful criticism. This method of giving and receiving feedback seems cruel and unusual to previous generations but is totally OK with our *American Idol* followers. Honest and – at times – brutal feedback in the work environment can often be the only way to make things work. And I, as a leader, must be prepared to listen and learn when my team critiques my performance. I also have to be ready to face the "Idol" panel every day.

What do we mean by a truly international experience?

I was born in the early 1970s. The first time I spoke to someone who wasn't Finnish, I was thirteen – I was on a class trip to Stockholm. The first time I conducted business in a foreign language, I was eighteen. It still took years after that before I was able to have a true friendship with someone who wasn't Finnish. I'm sure many readers are smirking to themselves and wondering what rock this troll crawled out from under. The picture and international experience that Digital Cowboys have of the world are vastly more developed than mine or those of the more "experienced" members of the workforce.

Today, it's totally normal that a ten-year-old has 40 gaming friends in 40 different countries. A kid in Finland can attack and take a virtual bunker with his virtual back against a virtual wall alongside a kid in Brazil. They are both battling with a couple of kids in India and a kid in California who form the opposing team in this virtual game. Meanwhile, a thirteen-year-old kid in the UK just added a new Facebook friend in the USA and some kids in Switzerland who happen to be at the same boarding school as the kid's friend, sharing pictures, music, games, news and political insights and opinions. This is totally normal for the PlayStation generation.

I talked about the impact of globalization on competitiveness and leadership in Chapter 1. The PlayStation generation – experts in global networking – has infiltrated the ranks of most international organizations already – and they keep coming. From a young age, they have played online games and communicated via international networks and social media. While language and cultural barriers haven't disappeared, these experts do not have a problem cooperating in an international network – unlike your average executive in a Western company. Most of us understand and agree, no company or industry can be competitive without bringing together the best know-how from *around the world*. I contend that the gravest challenge a manager faces today is leading an international expert network in a productive manner.

Leading an international expert network is always hard, even under the best circumstances – whether the core of the team consists of Digital Cowboys or not. I've observed that there is no correlation between the ability to work in an international team and the number of countries you've visited. This applies to both tourism and corporate tourism, where, in the name of internationalization, you pay brief visits to corporate offices around the world. You have to put in the time to learn about the cultural differences and how to make the team push for a shared goal.

The difference between a "nice" international team and a good one is in the ability to create value. Productive international teams evaluate their results on a continuous basis and have a fearless attitude. While they are aware of differences, every member of the team is able to function without prejudice or fear within his or her group as the ground rule of "striving for joint success" is clear and contributing and leveraging the maximum value from each other dictates collaboration and communication. Experts know their own area backwards and forwards.

A direct way to develop internationality is to make all education investments part of shared customer projects. The best international teams consist of team members who can quickly create and maintain the necessary trusted relationships between themselves and with the customer. For a leader, being directly involved in your international team's work on customer projects is a sterling opportunity. In addition to coaching opportunities, you have the chance to constantly evaluate your people. You'd be hard pressed to find a more valuable source of information than people's conduct during customer projects.

Individuals who are unable to work in an international group owing to fear of ignorance need help. Fear and caution are among our most primal feelings. This is where you have to switch from autopilot to manual control. By manual control, I am referring to conversations, models and encouragement offered by the boss. We all have fears and prejudices related to other people and nationalities. However, I am particularly optimistic about the Digital Cowboys' ability to adopt and understand the capabilities needed to develop an "internationality" mindset.

By focusing on developing the work done for the customer, direct communication and awakening the enthusiasm sometimes dormant in each and every member of the team, the Digital Cowboy is a valuable resource for the whole team. We older generations are the ones who can present a problem and some extra work for the leader. We have the necessary experience and language skills, but we are often strangely averse to creating trusted relationships. Some call this wisdom, but I call it a waste of resources and just plain stupid. Our caution not only limits our own development, it also stalls the internationalization of the work environment. How do you spot the problem? If you have a list of five calls to make and you assign the farthest location to the last call spot, it means you find this interaction challenging and are working to avoid it. If you let the support functions take care of home base first and then pay attention to the outfield, you need to look at your motivations. Are you sending your cowboys the wrong message about international teams? Are you providing a faulty model? The answer is simple. If you can trust your colleague 2,000 kilometres away as implicitly as you trust the one sitting next to you, you are ready to compete on the global stage. Your task as a leader is to coach your team and choose those people who are able to overcome their fears and have a burning desire to learn.

Let's look at the real moment of truth for all teams – the moment they meet the customer. The complete team should be part of that moment. Everything on the project up until that point has been a cakewalk in a safe environment. Meeting the customer is never safe and predictable for all members of your team. If not everyone can meet the customer face to face, it's important that all feedback be shared with the whole team – no matter where the individual members are based. Nothing eliminates a sense of equality more efficiently than working on something for weeks before a customer meeting and then not hearing anything about it for months afterwards. Feedback is always important, but in an international environment it is vital. The supervisor needs not only to communicate what the customer said, but also to provide team members with an accurate description of the customer's current situation. This model must be hard-coded into the DNA of every leader – from the front-line supervisor to the CEO.

Summary: The PlayStation generation has transparency, an emphasis on personal value creation, good communication skills and a natural ease when functioning as a part of an international network in its genes. They are exceptional members of any international team. From a leadership perspective, your central task is to create a work environment where everyone is on an equal footing and the key mantra relates to value created for the members of the team. Do not sweep problems under the rug. The team is steered by the development sparked by constant feedback and critiques. While the PlayStation generation evokes fear in older colleagues, you must hold steady. No one is kowtowed to or pulls rank because of age, gender or title. Look at it this way – there is a new team meeting tomorrow and it's like going to a new *American*

Idol audition. If you put yourself out there 100 per cent, you'll make the cut. The "voters" are not distant leaders in their ivory towers, but the members of your team. The best part for the new team is that one of the people auditioning alongside them is you – THE BOSS.

In order to increase my own understanding of the Digital Cowboy world, I asked Philipp Rosenthal from Germany to tell us what the world looks like from his cowboy boots. Philipp was a young and eager expert when he ended up in one of my coaching groups. Together, we've thought a lot about how to decrease the amount of anguish large and inflexible organizations tend to generate in Digital Cowboys.

Philipp Rosenthal

Avoiding negative organizational gravity while becoming a Digital Cowboy

Recently, a French company announced their plan to eliminate e-mail as their main communication channel. The well-established way of keeping people away from productive work will be replaced by so-called "social business tools" – an awesome plan and a thrilling prospect.

My only concern with corporations making such strong statements is that it's not about turning the digital workplace into an activity stream mimicking Facebook with the company's own logo on top. It's bigger than that. We have to acknowledge that we are facing a significant shift towards something that I'd like to call the "social corporation", where value is derived from the network – not a hierarchy.

Imagining a conversation between an established and experienced talent and an information worker of the new generation makes me chuckle:

"Hey! Could you please consolidate your ideas together with sources into a PowerPoint and mail it to the evaluation board? And please ask them to return their comments in their copy of the file by tomorrow EOD."

"Can do. But why don't I just kick off a group space and post an invite on the participant's profile walls? I could dropbox the background information and add some TinyURLs to the workspace next to the exec summary for further reference. Then we could openly discuss and develop the feedback? And let's publish a short blog post inviting the rest of the corporation to contribute comments and perspectives if they can relate to our subject?"

"Uh ... I guess."

Yes, the conversation might be slightly exaggerated. However, it could happen if we don't understand that the world around us has already started to change.

Hierarchy and managerial thinking as the key to economic success are outdated. The established communication channels and (un)productivity techniques cannot cater to the requirements of internationalized and decentralized operations any more. The social web is teaching us how to connect millions of people, keep them engaged and motivate our information worker or talent to tap into their innate abilities – sharing, contributing, inspiring others, networking and relationship-building. Add to this a new spectrum of converging communication channels which enable an individual to manage information based on situation, preference and urgency. The changes are profound, as are some of the challenges.

I experience the challenge myself every single day despite the fact that I work in an environment that is generations ahead of the corporate average. Our work tools and culture already reflect the emerging revolution of information work. However, encouraging the – I apologize – old-school talent and leaders to adapt to social work tools and refrain from e-mail as the "let's stay in touch" channel is, to put it mildly, not easy. Change in established behaviour is one of the most difficult goals to achieve, despite there being willingness to leave old paths and move on to the next level.

Looking at the big picture, it's not just about altering communication behaviour. It's about changing the fundamentals most of our corporations are built upon – organizational hierarchies and the power of an individual, which increases as he or she steps up the hierarchal ladder.

The move to a networked organization that allows the Digital Cowboys to unleash their talents to their full potential requires the ability to *let go*. *Let go* of careers that are based solely on individual success and knowledge. *Let go* of hoarding information to create power. *Let go* of the impression that a ship can be steered only by a captain. The leadership challenge is to create an environment where talent, experience and expertise are measured and rewarded. It is an environment in which helping others and seeking help are considered benefits, not indicators of weakness.

Some companies have already addressed that cultural change by introducing career paths that are not linked with managerial power. Instead, the paths are linked to the ability to lead by example and to establish a footprint through a network. Supporting thought leadership (see also Forrester Research blog: http://goo.gl/5cxwZ) could have a significant impact on an organization. When paired with enterprise 2.0 inspired thought, behavioural leaders' change can be driven into the organization subversively.

In a recent conversation, a very experienced CEO identified the cultural framework of his corporation as the potential bottleneck for change. The CEO – with his affinity to IT innovation and the will to move forward – promised to be a sponsor of the enterprise 2.0 approach. He was sure that the majority of managerial lines below him would probably fear the loss of control and, therefore, create boundaries and hurdles for any initiatives in that field. Network-oriented work and open exchange across hierarchical levels would not be compatible with the corporate DNA – yet.

Part of the conversation with this valued sparring partner was about the KPI- (key performance indicator) driven decision-making around enterprise 2.0. There is no question that major investments have to deliver a reasonable ROI. I am simply challenging the "wisdom" of trying to find a metric for each single move and initiative. Measuring certain effects such as productivity increase, e-mail and attachment volume, travel and communication cost, the CO_2 footprint or project efficiency makes total sense – provided one is willing to go through the pain of a zero measurement revealing the current nightmare of inefficiency.

The effect of a connected organization and the power to accelerate business development, as well as the speed of innovation and market impact, are hard to quantify but definitely not out of reach. The positive effect of activated knowledge and the power of crowd sourcing are already widely acknowledged. Even major consultancies have adapted this perspective. I don't want to create the impression that I am against the quantification of results. However, an approach that is too numbers-driven will most likely override and suck up any momentum. In the same way, a solely technology-driven approach won't create any significant impact in an organization either.

To return to the dialogue at the beginning, if corporations want to build a bridge between their current talent and future employees, they have to start with a common denominator. They have to take a first step. They have to be aware that change and adjustment need time to sink in. And moving away from well-established paths requires hand-holding – not a handbook. A new platform or new communication and collaboration rules are never the solution in themselves. Only if users/employees are comfortable with the new way of working will they be able to unlock their potential and create value beyond the initial work instructions and goals.

Now is the time to launch programmes and initiatives that introduce alternative ways of communication and collaboration within corporations. Now is the time to change the culture from solely hierarchical to a more balanced model where boxes provide framework and orientation and where results and agility are derived from powerful networks. Now is the time to build the (not so) Wild West where Digital Cowboys can roam the corporate landscape.

Maybe the previous conversation between the established and incoming talents might end up more like this:

"Of course. That makes total sense. But please don't forget to back-link to the forum that we've created in the innovation space. And poke Stephen to RSS the group stream in case his moderation is needed."

In addition to my thoughts on the social enterprise, I would also like to share a personal experience, for two reasons. First, the NO FEAR community and this book are built on sharing experience to initiate discussion and ultimately to learn from each other. Secondly, I have experienced what "new leadership" is and how much power Digital Cowboy energy can create – for me personally as a leader, as well as for the corporation in which I work.

As someone who's driven by passion and the will to leave a footprint, I am always seeking new challenges. In an organization that is organized solely by boxes and clings to hierarchical thinking, pursuing personal challenges usually isn't an option. You have an assignment based on your role and your position and goals that have to be achieved. Creating new roles and initiating new business opportunities aren't part of everyone's job description.

In contrast, my own employer considers thought leadership to be a major driver for success – individually and for the corporation. Thinking out of the box – leaving the personal comfort zone and challenging the established – is not only allowed, it is proactively requested. It is an environment in which the new generation of workers can excel – where networking and sharing knowledge are *essential*.

Thought leaders themselves aren't managers. They aren't even leaders in a classic sense since in most organizations a certain managerial level is required before you join the "leadership" ranks.

Thought leaders have to attract their followers and be "nominated" as a virtual leader – without reporting lines or shared business goals. Together with a powerful and committed network, thought leaders initiate change within a company and in close interaction with markets and customers. They are corporate change agents and marketing activists. Nominated leaders typically hold existing roles which serve as a launching pad for the new role, the new challenge and, ideally, a new business opportunity.

This is when the challenge for the formal organization comes into play. If the potential of the new approach is recognized and acknowledged, an environment has to be created for the "informal" leader to really drive change. Control and formal management procedures have to slowly fade away in order to create freedom and flexibility. The thought leader is put in charge, equipped with the trust that the slowly forming virtual organization around him or her will not lose focus on the existing goals but will allocate enough time and energy to investigate the new opportunities and understand whether it's a go or a no-go.

To be frank, there's no way that any formal process or managerial practice would be able to balance this without driving people crazy with conflicting goals. If the organization can "feel" that the new thing is supported, they might look into it out of pure curiosity. If there's something in it for them (always the best motivation), the organization will support it proactively without a formal mandate – especially if it's substantial enough to become a profitable new initiative or even a new source of business.

For managers, this process means letting go. It is the only way to break the existing hierarchy silos to enable virtual organizations to form. It is the only way to unleash the talents of your Digital Cowboys. At the end of the day, it means losing power. This is probably the most frightening thing for managers in today's companies – not being in control any more.

Sounds like a lot of theory? Nope. Been there. Done that. Got the T-shirt.

I have fully experienced the process myself. Nominated by a small group of colleagues (including parts of our management), I became an evangelist for the modern digital workplace – the enterprise 2.0 ideology that I talk about in the beginning of this article. With continually increasing freedom and support from my direct management, I was able to form a virtual team overcoming functional, geographical and hierarchical borders. Together, we were able to improve an existing business and lead a successful new market entry in less than a year. It was like finding an old piece of silver in the attic and polishing it until it's shiny and sellable. It created so much impact that the offering was named as one of the strategic corporate initiatives for the following years and the "go to market" process became a blueprint for similar activities.

Throughout the entire process there were no direct reports in sales, marketing, consulting or solution design. There was no big team that I was able to order around. There was a network – driven by *excitement*, *passion* and, most importantly, *belief*. The organization's belief in my virtual team of "followers" and in me to identify and initiate the right moves was essential.

The belief in this "joint mission" was essential. But so was the knowledge that the cross-functional virtual team – which consisted of multiple disciplines – could compensate for individual weak spots. I'll be brutally honest. I am a marketing professional for sales and business development. When it comes to methodological consulting practice, large-scale project management or software architecture, I would *not* consider myself the best fit. That, however, didn't really matter because the team could provide the professionals needed to really drive all the various stages of pre-sales, sales and initial project work. Leading the development of this "best of breed" team across borders without a real shared business goal was an amazing experience.

I once said to a coach, "I want to be a leader. But all leadership positions are taken." That was before I had learned that leaders don't have to sit in structural boxes. Business leaders can and, in the future, will be nominated by their followers just as followers nominate their leaders in social media by following their tweets or blogs.

Eventually, I was nominated to be the global leader for the offering that my virtual team and I had created. Even in this new position, the operational business is not reporting to me. Digital Cowboy leadership can and does work on a larger scale as well. *I truly believe that this is the future of leadership in businesses.*

Corporations will have to develop a new balance between formal structure and networked operations in order to unleash the potential that lies within their core asset – the people. They will have to create an environment in which the momentum of social media can kick in and accelerate business in a way that we have never experienced. They will have to craft a culture in which people are motivated and incentivized to share and collaborate and in which managers have NO FEAR of setting the stage for something which isn't fully in their control.

This is the stage for the new business leadership. This is the stage for Digital Cowboys.

Yeehaw!

To learn more about Philipp and to continue the conversation, go to **www.nofear-community.com**.

4 How does an industrial-age relic turn into an authentic leader?

CONVERSATION ABOUT LEADING DIGITAL COWBOYS and the changes they bring to business creates two leadership camps: those who think changes aren't needed and those who see the opportunity for cataclysmic change. Those who think the status quo is fine think there is too much importance being invested in a single generation. Leadership, they say, is a tradition going back thousands of years – one generation can't change the tradition *that much*. At the other extreme, we have the leaders who see an opening and embrace the opportunity to create extensive change. They spend a lot of time thinking about how to lead an organization and how to develop oneself as a leader – how to generate change, what to define as its aims and how to verify that change actually took place. They spend time uncovering the new skills and behaviours the Digital Cowboys bring to the table to determine how these changes can drive change. I – and my co-authors – fall into the second camp. There is no doubt that the PlayStation generation is a recognized phenomenon – both politically and socially. The question is how do you successfully lead this group of cowboys without getting trampled or pushed to one side.

Before we venture down the road of change, I want to address one fundamental shift that all leaders need to make. Industrial management models have hammered the phrase "human resources" deep into our psyches. The thought of an employee as a production tool is so ingrained in the leader's mind that the difference between the hatchet and the man carrying the hatchet has blurred. Thinking of humans merely as resources drives the leader away from "expertise development". It leads us down the value chain in the exact opposite direction of where terms like Digital Cowboy and moneymaking should take us. A deep change needs to happen in the thinking and language of the leader. Human resources are human beings. The human resource must become the individual. This thinking is the basic foundation that you must have in place before you try your hand at leading your Digital Cowboys.

Quantitative estimates concerning resources shift from tabulating number of bodies, titles, ages, salary levels, etc., to focus on an individual's *value creation ability* – their expertise. This is easiest understood through networking. A value creation network is not born out of an interconnected mass. If this were the case, every company could be called a network. For the network to function, we must understand that individuals who use their own exceptionality and specialization to add value for the other members create the network. A person tasked with leading Digital Cowboys does not have human resources. He or she only has individuals. Period.

In 2009, I was lucky enough to catch an excellent presentation given by IMD professor Martha Maznevski. It opened my eyes. The presentation was called "Leading in turbulent times: how to manage uncertainty". It had nothing to do with Digital Cowboys per se, but the simple models described were an excellent fit for this subject area. Her main point is that change is here to stay and we must all learn to embrace uncertainty and make it part of our operating reality. The changes necessary for developing a new style of leadership are going to push us out of our comfort zones. Business complexity does not cause discomfort. The discomfort comes when you – as the leader – are responsible for making precise decisions on what to simplify and what to amplify in the organization and then helping your experts understand the changes. If you choose wisely, you free up your experts to use the new simplicity to tackle complex issues.

The model has two interconnected sides to it: simplification and amplification – actions a leader can take to make an organization both agile and capable of assimilating new things. Simplification focuses on clarifying the organizational structure, vision, basic processes and culture. Maznevski wrote, "Go back to 'What is our business? What are our values? Our vision? Our mission?' These are the most important things to simplify. Who is our customer? What is our value proposition? What is our business model? And importantly, what are our values? – because values guide people's decision-making in uncertainty."[1]

Every successful leader will recognize and accept this list. Real simplification will not, however, happen voluntarily. Executive staff and middle management have always loved structures that are created incrementally over several years. The first law defined by Northcote Parkinson, born in 1909, says it all: "Work expands so as to fill the time available for its completion". Basically, the number of office staff will grow at a defined pace regardless of the amount of work available. Other words of wisdom from Parkinson are "An official wants to multiply subordinates, not rivals" and "Officials make work for each other".

The amplification agenda emphasizes an organization's ability to confront change and errors. Every change and error must be seen as an opportunity. This happens by strengthening all informal and cross-organization networking and knowledge creation. Curiosity, immediacy and courage are at the heart of the change. The leader's own behaviour is central to this activity. It must be given space and the leader must set an example in the front line – as the head of the customers and Digital Cowboys.

This really highlights the fact that we aren't dealing with a conflict situation. Instead, it is a situation where both parameters must be stretched as far as they'll go. *Simplification* forces us to reduce the value placed on the "traditional" leadership abilities and the company's leadership systems. *Amplification* demands space for renewal and variety. The leader's central task is to balance these extremes by intertwining them. They are the backbone of all change. The intertwining is largely an ability to display flexible leadership in live situations – on a case-by-case basis.

The call to simplify – or distil – structures and ideas is part of our journey. Along the way you need to question what is worth reorganizing in your sphere. Is the model I set for others suffused with curiosity, immediacy and courage? Am I willing to admit this is a journey and mistakes will be made along the way?

How many mistakes did I share today?

The Maznevski model attempts to shepherd the leader and organization into a state of constant dynamic action. This requires a rough, almost daily battle between production efficiency and exploratory development. Personally, I like to talk about an offensive, aggressive mindset. By aggressive I do not mean ramming opinions through violently. To me, aggressiveness refers to the fact that everyone in the organization must be ready for "full body contact" in a game situation. It means that we embrace change with all our strength and the strengths of each player. We have – as a team – an unshakeable faith and energy. When you use force and inject your emotions into the game, you take risks and make mistakes. You can't know how the situation is going to turn out, but you trust that your team will come through, somehow. For some leaders, full body contact is pretty hard to handle.

Full body contact is not typically part of the picture of leaders and leadership. Strength, clarity and determination are good leadership attributes the world over – we like the picture of a square-jawed, confident executive, leading the troops. We also know it is hard to keep it up 24 hours a day. Leaders are ordinary mortals – well, most of them are. Everyone has moments of weakness. We have all made plans that make absolutely no sense. Or at least no one could really call them clear. And we have all experienced moments of uncertainty when standing up in front of the troops. What if instead of "hoorays", my troops start laughing after I've said my piece? What if someone in the front row yells that I have the wrong slides? What if I can't recall last year's EBIT percentage? What if the paper says I'm the proverbial 90-pound weakling? Do I have the courage to drive into the golf club car park in an unwashed, ten-year-old Toyota? What if my fellow leaders think I'm *weak*?

All these fears make it very easy for a leader to succumb to the temptation to sit on the next hilltop over to observe the battle. The troops see your handsome silhouette, framed against the gorgeous setting sun. Luckily for you, they aren't within shouting distance. At dinner with your management pals, you can regale them with stories of high daring in battle. Why spoil a good story by telling your pals you spent the battle crouched behind the left flank? Full body contact requires that this leadership model change. And making the changes stick on a *permanent* basis will be very hard.

Digital Cowboys simply don't believe in leadership hype. Healthy scepticism is one of the expert's built-in sources of strength. The PlayStation generation tends to view scepticism as a primary human right. I think the *right to doubt* and the *right to screw up* are the most important building blocks of a healthy organization. They have, in my opinion, surpassed in importance the right to privacy and traditional decorum. The fact that your opinion is being challenged is not impolite. It's not a personal attack against you. Organizations that thrive on doubt and uncertainty are born to create new things. The opposite is a model where operational models, organizational particles and human resources all know their roles to the nth degree and keep repeating the same performance. These are not expert organizations. They are machines.

As a young leader, I was totally full of myself. I'd sold my company and my current account was bulging. I got stellar marks as a leader and I was proud to lead a loyal elite team. All the people important to me were within shouting distance. As I mentioned earlier, this leadership model stopped working as soon as I stepped into an international team full of PlayStation generation experts. Even my nickname – the Bulldozer – which had started as a friendly joke, came under closer scrutiny and took on new meanings. What used to signify that I was strong and surefooted now meant I was arrogant and pushed everyone else aside. Getting in front of the Bulldozer was a bad idea. I was genuinely mystified.

I figured the only way to change was to throw myself in front of the coming punches – no matter how basic the situation. In practice, it meant that even for the most insignificant matters, I'd tell everyone just how I screwed up. I started this while gathered around the familiar coffee machine, but social media and digital channels offered me a way to reach all my colleagues. I never let anyone proofread my blog or correct the grammatical errors in the mass internal mailings I sent. I wanted the messages I sent to come directly from me – straight from the front line to the front line, unfiltered. Not surprisingly, nothing brightens a day like telling someone you blew it. Of course, your subordinates are twice as happy, as they see that their leader is an ordinary mortal who screws up. I started educating my own reluctant managers and experts in this same principle.

How many mistakes did you document and pass on to your team last week, last month or this year? If your disclosure is met with comments like "things like this happen, you'll do better next time", your organization has reached a sufficient level of transparency. However, if the same message has generated a series of suggestions for fixing the situation or other similar ones, you and your organization are only lurking on the outer boundaries of transparency.

Documenting, sharing and quickly receiving suggestions for corrections with an open mind may sound like a simple change in the grand scheme of all this change. However, I place these at the top of the list. These changes, if anything, will strengthen the organization and give the whole team licence to try new things and to fix old problems. Digital Cowboys, as a group, are no easier than any other group to coach in this area. Like everyone else, they often want to sweep their mistakes under the rug. For this reason the leader must set an unequivocal and absolute example with his or her own behaviour. The core outcome of this change is the ability to learn more quickly and to avoid repeating the same mistakes at different organizational levels. The constant hunt for mistakes, problems and targets for improvement adds spice to organizational life. Transparency is important, but an even more important part of leadership is the ability to learn quickly.

To the reader: If after reading this, you come to the conclusion that your problem is a *lack* of errors, seek help. Either your judgement is severely impaired or you have died and just didn't notice it.

Anatomy, physiology, psychology or psychology, physiology, anatomy?

I was once given a slide that listed what's required for change in an organization to be successful:

1. Anatomy (structure, organization, technical solutions ...)
2. Physiology (roles, decision-making processes, information channels ...)
3. Psychology (the mindset maturity of key personnel, beliefs, holistic vision, inner determination, timing ...)

At the time, I breezed through the list and thought it made a lot of sense. I also read the note below the list: these things had to be implemented in the right order – 1 then 2 then 3. That made a lot of sense, too. To be honest, I wondered why they'd bothered mentioning something so patently obvious.

The order of execution is the key to the changes leadership needs to make *but the list is in the WRONG order*. I believe the correct order is 3 then 2 and then 1. Traditional industrial leadership is based on the idea that you first build an organization – the anatomy. Once the upper level of the organization is defined, the leader calls the other people in to form the lower ranks. At this point, leadership defines people's roles, as well as decision-making forums and processes – the physiology. Once this work is done, we move on to psychology. Leaders invite the troops to kick off jamborees and workshops. A number of walls are covered in a storm of Post-it notes in an attempt to build a shared vision and mission. Afterwards, everyone goes back to their desks. After the hoopla, the time is ripe to meet the experts and – most important of all – the customer. Psychology is undertaken as soon as the next level of supervisors synthesizes the results of the Post-it-note marathon for their teams. The end result is a complete solution presented to the experts for their execution. Anatomy, physiology and psychology – and it's all *wrong*.

Working with Digital Cowboys has turned this model upside down – psychology, physiology and then anatomy is the correct order. For me, leading the cowboys has emphasized the importance of the psychology phase. Defining direction requires fast and fresh minds. Experience is important, but so is the ability to analyse increasingly weak market signals and to quickly take actions when it comes to planning and implementing change. It's not about driving people into a mass frenzy or public displays of commitment to a project. It's about giving all parties involved an equal opportunity to participate in the process. There is no complete solution presented for execution.

As a leader, your role in this model is to encourage and ensure a systematic dialogue. This does not mean you just delegate tasks and innovation to Digital Cowboys. You ask tough questions but avoid unnecessary decrees. You must be insistent in challenging people and getting them to deepen their opinions. You must be brutally honest with yourself if you're going to demand the same of others. These interactions must always be followed by a DECISION. It's capitalized because deciding is scary. A decision has to be made and a leader is responsible for it. You can have endless collective discussions and conclusions, but in the end the buck stops with the leader. The decision is binding for all – until a new decision is made.

Once a direction has been chosen, it's time to define the physiology – the roles, implementation-planning and communications mechanisms. After that, we can move on to the easiest part – organizational structure and metrics. It is at this juncture that we truly measure the leader's own power. It's easy for traditional organizations to start a new process by defining parameters. I like to use an aerospace term and call this the negative inertia of an organization. A brave leader – who turns the model upside down – will run into great scepticism, hearing shouts of "Without clear specifications people can't do their jobs. Without definitions the organization will panic and people will just step on each other's toes. It'll be chaos!"

Chaos is a relative concept. A Digital Cowboy can live with the chaos produced by the push and shove of people, ideas and actions during this phase. A situation like this will motivate a smart cookie to really put everything into the game. If, on the other hand, weeks and months are spent sketching out structures and systems without anyone actually telling the team how they will win the next fight – the "business as usual" model – the efforts are all for naught.

Psychology is the most important development area for every leader of Digital Cowboys. You can never be too good at evaluating people's capabilities or motivating your team to search and dream. You not only have to develop this expertise in yourself, you also have to develop it in your experts and subordinates. This entire demanding field of leadership was summed up in two sentences by the French author and pilot Antoine de Saint-Exupéry: "If you want to build a ship, don't drum up the men to go to the forest to gather wood, saw it, and nail the planks together. Instead, teach them the desire for the sea." If the leader doesn't love the sea, the men won't make good boats. Psychology first, physiology second, and anatomy last.

A leader is a producer, not an invisible delegator

A leader's job is to be on the front line with the troops, providing guidance, clarity and motivation – leading and directing by example. A good leader is really a producer. When leadership changes from wielding influence through structures to steering critical change projects, the producer metaphor is absolutely right.

In Chapter 1, we talked about executives as producers. The producer goes through the casting list, picks the right actors and approves the script. He or she makes sure the production moves along at pace and that all the necessary expertise is present and accounted for, working together to achieve a common goal. Your role on the front line mirrors this type of interaction. To truly take on the role of producer, a traditional leader must develop two different capabilities:

1. An understanding of the end-customer's issues and opportunities from everyone on your team. The team is you and *anyone* who touches the product or service.
2. Helicopter ability

The first and most demanding part is getting deep into any issue or process from the perspective of the end-user. For example, the producer of a children's movie has to evaluate every actor, script, set and advertising image through the eyes of a child – not through the eyes of management. Walt Disney's *Toy Story* could not have been produced by a team that had never read a fairy tale or played with children. Looking through the eyes of a child made the movie believable and authentic. In the same way, it is impossible for a leader to make technology and personnel choices related to developing a product or addressing an issue unless he or she fully understands the end-user – who actually may be the end-user of your end-user.

In a network economy model, a deep understanding of the end-customers' issues is critical. Products are no longer produced in-house. They are assembled via global subcontractor networks.

The subcontractors do not work for your end-customer. They work for you. They understand your needs, which are very different from your end-customers' needs. You see my point. Your company's metrics, production lines and the functions of the product can be close to "perfect" – but if they don't meet the customers' needs, the customer can't use it. Service quality and functionality can be at the top of your company's agenda but if you have failed at your role of understanding the end-customers' issues – one of the key capabilities of a good producer – the best quality and services are moot.

I've read several memoirs by different movie producers to better understand how they think and manage. It's obvious that there are big differences in the models the successful ones employ. However, regardless of the model, they have all developed and use the helicopter ability – our second key capability. In Chapter 1, I talked about the importance of helicopter abilities. The heart of helicopter ability lies in a combination of two things: the ability to *see things from high enough* and the ability to *land quickly at any time* should some detail or situation require your intervention. It is the ability to comprehend the big picture, as well as the critical details – the agility to move quickly between the big picture and the details, from one situation to another. The ability to see the whole field from a thousand feet gives the leader an understanding of what is happening in the company and the market. A light touch on the helicopter's stick with the left hand and we are immediately close to ground level. Here it's easy to provide immediate problem-solving support, or at least evaluate what kind of support actions might be needed next. For leaders, using helicopter ability is usually more a question of willingness than ability. Being afraid of stepping on someone's toes or fragmenting time usage into too many bits or underestimating subordinates often keeps us too high for too long. So we wait until the forest is on fire before we go and take a look. That's a curious sort of leadership.

The central task of the producer is to keep a constant eye on the progress of the whole. But a movie requires vast amounts of talent and resources. They are too big for a single person to manage. Successful movie producers rely on a team of capable assistant producers. In the world of the producer, a major production has multiple sub-productions that are their own discrete projects. The producer owns all of these areas. But while the movie producer is the only one on the movie truly using his or her helicopter ability, the leader of an organization must encourage every member of the organization to tap into their own helicopter abilities. In fact, it needs to be everyone's right. When the agreed-upon metrics and processes tell us that things are OK, it requires a lot of confidence for a sub-producer or expert to bring up something they perceive as an issue or problem.

There is a difference between helicopter ability and random acts of interruption. A leader's penchant for getting involved in details – which used to be considered potentially awkward – becomes a leader's right. The helicopter model encourages transparency, challenge and shared learning agendas. Expressing an opinion and meeting the team on the front line does not make you a meddling micromanager – unless you are using your ability for PR stunts. Leaders who fly in as tourists to gawk and offer "help" are a joke. Those who familiarize themselves with the problem quickly and create real value for the team in the middle of their pain are living legends and known as great producers for years afterwards.

If the acceptance of helicopter ability is not hard-wired into your corporate culture, cross-organization intervention by a producer-leader is clearly going to be a sin. This takes us back to the issue – and problem – of company anatomy being the driving factor in an organization. If the leadership started with defining structures and positions, there will be a firmly entrenched hierarchy. Offices on the higher floors are filled with award-winning executives, but owing to the leadership code, not many are going to question their neighbour's approach. Many books deal with this subject. *How the Mighty Fall*,[2] by Jim Collins, for example, charts the different phases a company goes through, from its rise to its fall, with fantastic attention to detail. Without going deep into the book's findings, many of the global examples lend themselves to an obvious conclusion: success feeds existing structures, making them more self-satisfied. Inevitable future problems, such as changes in markets, technology or competitive environment, are registered in time, but the method palette employed for change leadership has become a prisoner to the structures and the anatomy. Everyone has their own well-defined row to hoe. There is no producer culture and childless engineers are writing *Toy Story*. And no one is encouraged to stand up and mention that the emperor has no clothes.

Is Facebook my value creation network?

These changes in leadership affect the way you act *within* your value creation network. But they haven't addressed the way you physically create and maintain your value creation network. This involves defining your privacy limits and your own daily system for managing the network. A traditional line organization keeps the resources a leader needs below his or her position in the hierarchy. It's a network of sorts, but as far as helicopter ability goes, it is too narrow and defined to be a true value creation network. You have to be able to create different levels of interactive and trusted relationships throughout the whole organization. Increasingly, you also need to create these relationships directly with customers, your customers' customers, the media and other experts. You have to be able to breathe the same air as your target market, and your network provides the necessary access and feedback required to make decisions that can be quickly tested and refined. In a leader's life, it means simplifying down to the essential, asking the right questions and creating a network of sufficient breadth that draws some of its breath from outside the organization.

At its simplest, the mechanical construction of the network involves gathering and classifying the contact information of important people. Whether you use Facebook, LinkedIn or a systematic Excel chart is irrelevant. It's all left-brain stuff for now. However, as soon as we start talking about creating value for the network, we're dealing with profound change for the leader. We've already gone through personal vulnerability, change leadership priorities and the leader's role as a producer. We will now add something I like to call *hypercommunication*. Finding and sharing nuggets of information and opinion that might be of value to others, as a journalist would do, is at the heart of hypercommunication. The value can be directly financial or societal, or it can be something that brings joy or optimism to people's lives. Unlike with a media outlet, a leader's communication should never be about just passing on information. A leader has to connect the information with the team's mission and psychology. The interaction can help uncover new perspectives or network connections. Interaction creates value, which is why I use the word *hyper*. We're dealing with intense intellectual activity.

As we specified earlier, a network consists of exceptional individuals. Having access to the right parts of the network on a case-by-case basis is more important than "owning" a vast network. Especially since the network can't be owned. In and of itself the network doesn't have the power to effect change. It just produces noise. The network combined with leadership is a whole different story.

For me, learning this was a long road. It's hard to find and reduce the things that really interest your network. I'm a lazy person, so slipping into just passing on information without making the ties to business explicit is easy for me. I thought my biggest problem was getting my network to work on my messages. In the first few years, I felt that of all my thousands of colleagues, it was always the same 20 to 50 active individuals who reacted. I knew the stuff I wrote was being read, but the opinions of a few dozen key people were not enough to give me a comprehensive picture of everyone following the messages. One reason was that the messages were not sharp and accurate enough for most people to grab on to. Customers and other partners were in immediately. Thoughts and, at times, sharp criticism from that quarter were actually the main reason I started to believe in the potential of this sort of communication.

We built various communications channels and events that were 100 per cent uncensored; everyone was free to say anything they wanted. Every second and every cent invested in this activity paid back fivefold. However, as far as my own organization was concerned, I was still mystified by my inability to engage people right from the start. Part of the problem was surely related to my bulldozing style, but there was also the fact that people just weren't used to direct communication. That was understandable. The people in the organization had a picture in their heads about who was allowed to say what sorts of things. What was truly shocking was the unwillingness of people at my level or above in the organization to engage in interaction. Some feared for their image, others wanted to maintain a distance from the troops or were afraid of the amount of work involved. What if someone had further questions or comments?!

Back then, in the early 2000s, I was so self-sufficient that I didn't actually care about this attitude. Now, working with Digital Cowboys, I no longer have this option. Were I to decide to sideline myself from this kind of communication and networking, I'd be sidelining my own personal development and that of my experts, as well. By making this decision, I would essentially forfeit my right to lead anyone and should immediately try to find a job where I had no subordinates around to ruin my day.

My strength in network creation turned out to be the ability and the desire to build trusted relationships quickly. This surprised me since I never thought of myself as a particularly social type. At school, the other kids had more friends and at university I was probably considered a bit of a stodgy bore. Within the company, and especially in situations where I was able to breathe the same air as the customer, everything was different. It was and is the coolest thing about being a leader. For one coaching session, I was asked to describe the most important facets of my "method". The thing is, it's not a method. It's my way of working with the customer. Its most important facet could be described as social risk-taking. At its core is a true identification with the customer's challenges, situational awareness and the guts to express your own strong, well-thought-out opinions and solutions. There's a thin line between good social risk-taking and arrogant directness. That thin line is authenticity – an authentic desire to understand the customer's challenge and an authentic commitment to your own solutions.

The same model that created positive energy in face-to-face meetings with the customers turned out to be a key element in managing my network, too. And the same basic model is valid whether the individual works for your own company, that of a partner or the customer's.

I have tried to describe my way of managing my networks at many training and coaching sessions. Basically, I've split the task into three parts. I try to be as systematic and brutal as possible with the basic element of the network – the contact information of its members. Brutal in that I am quick to remove any members who have been unwilling to participate in the discussions. Secondly, in addition to the spontaneous information searches I engage in, I have a weekly schedule for publishing information. Like an "editor-in-chief", I keep a list of things I think will be valuable. On Sunday evenings, I look through what the week has brought. The stories and ideas that make it through the Sunday cull go into blogs and other internal media. Some things I transfer to other parties for additional information and perspectives, others I elaborate on myself. As the editor-in-chief, I am responsible for all content, regardless of who processes it. The third and by far most pleasurable part of my system is to get little groups of people, consisting of members of my teams or employees of the customer's organization, together for gatherings. The preparations forced me to think and face-to-face is a great way to really get to the root of the matter at hand. We would do this with bigger groups three to seven times a year. It virtually became pretty much a weekly occurrence. We summarized these dialogues for consumption by larger groups. It's worth noting that all the time we invested in this produced something far more valuable than just communications: it defined the change targets and actions that the organization needed and for which I had to act as a producer. I was constantly refining my helicopter ability.

After the training sessions, people created their own ways of gathering and managing their network, depending on the industry, country and position. It was easy for all to see that the network was of importance and profitable when dealing with Digital Cowboys. It took a lot of time to develop but the benefits were indisputable. As far as developing members of older generations, like mine, the biggest problem is still the "knowledge is power" myth. Fear of failure is always present, of course. The far bigger fear was that someone else might benefit from your particular titbit of data. That's why each morsel was weighed and analysed with such care and then, in the end, often not passed on at all.

Choosing people: old world versus new world

Just as you choose your network carefully, choices about who you work with, who you believe and who you influence are central to a leader. For the most part, these people will be from outside the leader's own organization. Inside the organization, the leader's immediate subordinates are an important subgroup, but still just one subgroup. And as with your network, you have to make choices that involve building, pruning and fixing – always the most painful part – relationships.

Over the years, I've learned to love the human resources department. My love is not based on the department's ability to create three-character codes for every human eventuality – M03 equals good leader, deserving of leasing a car, knows how to develop people, and so on. Instead of the inspirational world of the HR code handbook, my deep feelings are based on the department's undisputed impact on the earning ability of the expert organization. The HR department is the first point of contact – they are the gatekeepers that prospective employees first meet. From a recruitment point of view, they determine whether the applicant has the prerequisites to fit in with the organization's mindset, culture and psychology. If these prerequisites can't be created with a fair amount of training, then all the rest of the stuff, such as knowledge and skills, is a waste of time. So what happens if they pass the first hurdle?

The applicant's versatility and changeability in the social network are key facets of a successful personnel choice. The idea of a Renaissance man naturally sounds great to a leader. It'd be awesome to have an army of 1,000, each and every one of whom knew how to do everything real well. The absurdity is pretty obvious, right? This da Vinci stuff doesn't work when you're trying to find individuals with whom to work. Someone with real expertise and the desire to combine it with the expertise of his or her colleagues is a treasure in an expert network. With the best, this desire turns into a way to influence others and learn more. Many have this desire, but unfortunately fear or the golden handbook on how to behave in an organization prevents them from having an impact on their colleagues. From a leader's perspective, personnel choice involves a focused expertise profile, a path for developing it and, in the world of the Digital Cowboy, evaluating the applicant's social capabilities.

The age of the Digital Cowboy will emphasize these three areas in personnel choices:

1. Leading international diversity
2. Choosing and supporting focused expertise and know-how
3. Leading an end-customer-centred production efficiency

This list is very different from the traditional personnel-choice criteria. Speeches tend to glorify results, good people skills and quality leadership. The real world is a different matter altogether, for very human reasons. These attributes, which are traditionally trotted out, are vague enough to mean almost anything. In choosing leaders in the real world, the attributes that matter to many, myself included, are: easy to communicate with, a similar cultural background and similar values. A leader may unconsciously (or very consciously) choose key personnel whose training will not eat up massive amounts of expensive time. As a result, deployment and the early phases of the project will probably go a lot faster and more smoothly, but the leader will not have created a team that can conquer a market – never mind the world.

Repositioning the entire recruitment and training processes is of vital importance in the personal development of the leader. All successful leaders always bring up the importance of personnel choices, but few talk about their own role and results as coaches. The fact that you can remember to rattle off what percentage of turnover is used for employee training does not mean you are interested in developing people. Having subordinates who are more experienced and intelligent than you doesn't mean your contributions as a coach are not incredibly important.

Digital Cowboys are, generally, more well rounded and international than their supervisors, but as leaders and supervisors, they are total rookies. As a result of their individualism and the unbearable lightness of their being, their models of leading and leadership are incomplete. If you try to solve this problem with the traditional team leaders' supervisor training sessions, I can tell you the results won't be much to brag about. The reason is that twenty years ago structure – anatomy – provided an unequivocal justification for leadership. In the world of the Digital Cowboy, leadership flows out of psychology, the ability to provide a shared direction and mindset, as well as willingness to live on the front line with the troops, facing the customer. So teaching a model of leadership has to start with something very different from interpreting the rule book and daily missives. You can do it only inside the organization and in a way that involves the entire leadership of the organization.

Of these three areas, leading international diversity will rise to the top when choosing personnel. All talk of globalization will cease over the next twenty years as the subject matter comes to seem increasingly ridiculous and old fashioned. The fact that we can all use Visa or Amex to pay for things securely all over the world would have seemed incredible in the 1980s. In the 2030s, an American teen opening an account in a Swiss bank or the CFO of a Norwegian company being Chinese won't raise any eyebrows. This development will increase the facility of leadership at an accelerating pace. It's a question of who can build a two-way, open trust relationship with anyone on the planet.

As open season is declared on many commodities, price and availability in the world markets are no longer driving concerns. We are entering a Renaissance age of expertise and focused know-how. A leader must have the ability to pick, develop and support the experts and competencies necessary for the company's next growth spurt. Trusting the HR department to find the necessary experts on the labour market after the executives draw up a strategy is too slow and too tardy. The central decisions involved in the strategy process are nearly impossible to make without the direct influence of the focused competencies HR is going to be looking for – soon. Discussions centred on geographic strategy will undergo a change in many organizations. Ruminations about countries will be replaced by discussions about which competencies and services the organization can best deliver. For these discussions to achieve any real success, a leader must be acutely aware of the know-how of the front-line troops.

Yes, the third area is a monster of a term: leading end-customer-centred production efficiency. The roots of this monster are in the pressures exerted upon unit prices and quality by global competition. Leading in this dimension takes our thoughts to the leader's ability to monitor the cost structure, squeeze the last drop out of every subcontractor and take care of logistics in all its manifestations. Isn't this a basic profile for a result-oriented leader? In my opinion, this definition must be broadened with the term *end-customer-centred*. In the context of fine-tuning production efficiency, end-customer-centredness refers to evaluation of the product or service throughout its life cycle. Every leader responsible for production efficiency must understand how the end product is used. When the services and products are a part of the network, a brilliant product or service or the company that made it can quickly drift into dire straits without actually making any visible errors if they are not focused on the end-customer.

All these three basic capabilities demand that you combine brutal production efficiency with leading a fast expert network. Looking at this from the point of view of the change the leader must effect in him or herself requires that time and profound thinking be applied to the problem of helping people develop. You have to love HR more fiercely than ever before. Their task of supporting line management will become even more fascinating in the age of the Digital Cowboy. To start off the new age, send HR the wonderful PSF culture motto I mentioned in Chapter 2. It's worth repeating here:

Hire people who like people
PROMOTE people who like people
Don't hire people who don't like people
God knows – don't promote people who don't like people

Me and my chairman

We have looked down and across our value creation network. It is time to look up to see what hovers above every CEO – the company board. It is my blunt estimation that every Monday you have to sit down and evaluate which parts of the company, as well as individuals in it, create value for stockholders. You gauge this with the additional dollars and euros in the company's net worth – be they through growth or profit. After this evaluation, you have to decide what to keep, what needs to be developed and what has to be jettisoned as fast as possible. Digital Cowboys accept this analysis and model. Nobody wants to ride into battle if only half of a platoon is really attacking and the rest is just thinking about it.

The corporate board is a platoon in your army. They are not on the outside. I have tried very hard to understand the reasoning behind the desire to keep the roles of the board and the operative leadership clearly separate. I think this approach is widely used as an excuse to lurk in the bushes while the battle rages on. When we're dealing with the stockholders' money, no one should have the right to stand off to the side of the action.

For years, I've listened to CEOs complain about having to do things a certain way because of the board. In coaching situations, I've heard colleagues verbalize the more or less unrealistic wishes and expectations they have for the board. Even if my experiences with this are limited to European companies, I still think something odd is going on. Company boards are generally excluded from the value creation network. Yet owners, employees and especially our top expert Digital Cowboys believe the board has *a comprehensive strategic vision* that can be used to bless the suggestions of the leadership. Any information provided will have been cleaned up and given a nice spit shine. You won't find the CEO waiting to tell the board about all the mistakes that were made – quite the opposite, in fact. They are there in a monitoring capacity; they have to make sure things are above board legally. They also have to manage risks – probably based on the spit-shine information they have been fed. The board's one very powerful tool is it can pick and fire the CEO. Add to this a faceless, institutional ownership and you've got a dangerous combination. It means that reactions to changes are too slow and too herky-jerky.

There are opposite examples, too. I've had the pleasure of working with truly fantastic corporate boards and owners that obviously breathe the same air as the employees and customers. To a Digital Cowboy, this speaks of commitment and a desire to move things forward. It often leads to a real professional fostering of ownership value, as happened in Sweden, the USA and Germany last century, not to mention the real power of the owners in emerging markets in Russia, China and India in the last ten years. I don't think the family companies of these countries would be successful if the owners and the corporate board let the operative leadership go about its business on its own.

To rectify this situation, I suggest CEOs take members of the board into the company's networks and let them wield influence. It's hard to see how the fact that the weak signals of the company make it all the way to the owners could undermine the CEO's position. As we've said, in the world of the Digital Cowboy any station based on position rather than value creation is doomed to fall. Approving this model makes the CEO look strong and it sets an excellent example for the next level of leaders.

Summary: When you are surrounded by highly skilled people on the front line in real situations, you gain a high degree of affinity with your team and will develop your leadership skills. Change begins with attitudes. The attitude to have is that creating affinity with the Digital Cowboys is fun and will create stupendous value. I am a living example of the fact that a leader can learn and change continually. You need some good advisers, some Digital Cowboys as merciless sparring partners and a generous dollop of guts.

In the following section, Kari Hakola talks about the inner vistas of the CEO in the age of the Digital Cowboy. Even though companies are becoming networks, the model provided by the CEO is still important. All investments in the organization and training supervisors will be watered down if Digital Cowboys perceive these things as being off-balance with each other.

Kari Hakola

The CEO as the number-one change agent

I have been in the IT business for 40 years. I have seen – and been part of – multiple cycles of technology hype and have followed philosophies championed by management gurus. Through these 40 years, I have seen the profound and sustainable impact technology has had on business models, practices and competencies. IT has had a tremendous impact not only on our own industry and organization, but also on the business of our clients – from retail, telecommunications, banking and insurance industries. For the last seven years, I've had the luxury of working in an environment where many of my colleagues are Digital Cowboys. They are well educated and very ambitious. I've seen the power of diverse and virtual teams. I've seen the momentum created when you get the right combination of ambitious business challenge, collaboration and personal growth. I've also seen how it can go terribly wrong. The success – and failure – usually rests on leadership – or the lack thereof.

It's easy to set challenges

At its most basic level, leadership is about making decisions and choices based on sketchy information. There are always two sides to a choice: the actual goal and your implementation capability – the "what" and "how". These are the day-to-day decisions of balancing goals and resources. A wise leader occasionally makes choices that push the team and stretch implementation capability to its limits. People will say that the goals are impossible to achieve. But most often, a strong and well-led team reaches the goals – even if only just.

Solid change management supports the team in meeting your given challenge. The array of change management methods is practically limitless, and much good advice is available in books, from consultants and from those who have been there before. Regardless of the method, leadership and deep cognitive know-how form the core of change management. This reliance on know-how is why leaders tend to employ those change management methods that have a proven track record. People like to engage in demanding tasks using familiar methods and with people they know. This is fine when sailing in charted waters, but when you sail off the map, you have to question your old selection of methods and be ready to develop and embrace new ones.

The CEO has powerful tools in the toolbox

The instruments a CEO uses to do his or her job are *strategy*, *organizational structure* and *personnel*. Strategy provides the long-term thinking and framework for high-performance execution against demanding goals in the here and now. Strategy is a manifestation of the skill – not the plans – of a tactical leader. Vision for future opportunities and implementation capabilities combined with the guts for some radical resource allocations – these are the kinds of decisions for our best leadership brains.

The organizational structure is an outline or reflection of businesses, business processes and functions – a company's choices regarding what the primary dimensions of its actions are. And structure is ecosystem, the choice of the prime business partners and one's own role in the ecosystem. From a CEO's perspective, structure defines the people closest to him or her, what can and must be demanded of them and what each individual is accountable for.

The clarity of the organizational structure is essential for short-term performance. Demands and responsibilities must be *clearly* defined. In a world of increasing complexity, organizational structures will tend towards greater complexity, too. But as organizational structures become more and more complex, in the name of simplicity some drastic choices have to be made. Whatever the structure is, the profound changes will cut through the units of a line organization. There must be other ways to increase agility and performance than complexity of structures.

As early as the 1970s, Henry Mintzberg proposed that – in addition to business units, management hierarchy and support functions – a "technostructure" was an integral part of the organizational structure. A technostructure is a team of executive staff who are charged with standardizing internal activities across businesses and adapting to external changes. Over the last 30 years, IT assisted business process and infrastructure standardization through businesses and industries. Process automation has produced incredible results for the *internal* workings of an organization, supply and demand chain efficiencies through industries and service automation. IT's work shored up the position of the technostructure as almost all operational development is done through processes.[3]

But in most organizations I'm personally familiar with, the technostructure is rather ineffective at adapting to *external* changes. The mindset and implementation methods of internal standardization have taken root so thoroughly that they are used to meet challenges posed by external changes. The result is a never-ending stream of projects that have the right goals but lack the firepower to enact real change.

A healthy technostructure is critical from the CEO's point of view. In most organizational structures, the executive staff is physically and mentally closer to the top leadership than the front-line units engaged in "real" business activities. It means their ideas and implementation methods tend to have a more immediate effect on the leadership than the opinions and suggestions percolating up from the real front line. If the technostructure is healthy, this is a creative conflict. But if the technostructure is internally oriented and too strong, it will be the start of a cancer.

The last of the three tools is personnel. Most of the *operational* work is done by automation and systems – machines and computers. People take care of complexity, the complexity of renewal. Like the proverbial three-legged stool, strategy and structure fail without the right people in place. The people you choose impact the dynamics of the work. The challenge with picking the right personnel is the fact that only history is certain. It's hard to evaluate someone's ability to solve challenges before you get to know them. And it's even harder to evaluate how they will react to other individuals, diverse groups and personal growth forced by unavoidable change. For most leaders, the mistakes you make in recruitment are the hardest to correct mentally. How long do you have to remain loyal and supportive? When is the right moment to make that final decision?

How do you make the most of and balance these three areas? A CEO has access to one very special option – the broadest choice of whom to listen to, believe in and work with. Picking the right network and the right mentors will make the difference. While the organizational structure and personnel choices are necessary, the people the CEO chooses from both within and outside the organization have a strong impact on the dynamics of the work that is ultimately performed. Do not let the structure imprison you. It's only one of your instruments.

Why do organizations and CEOs get stuck?

I have a long career working as an executive and consultant with many large corporations behind me. I've seen leaders use one or all of the tools in their leadership toolbox with varying degrees of success. Something which happens over and over again still surprises me.

- The endless planning and the seriousness with which plans are created. People still tend to think that making a good plan takes you halfway to the goal. In reality, a lot of planning is often a manager's comfort zone. There is by far less serious steering and demanding than planning. And by far too few brutally honest and open "lessons to be learned" discussions after each fiscal period and change-step – especially in cases that ended in failure.
- The strong focus on content – i.e. on "what" rather than profound execution capability and capability development. This is OK in "business as usual" situations, repeating what is already known. But it's not OK in the real changes. Profound execution capabilities and stretching those capabilities are as important as making the choices on "what".
- The fear of doing things differently or trying something new. Building something unique requires unique capabilities.
- The amount of broadcasting-type communication compared with good and critical questions and true dialogues. The

ability to create something new is entirely dependent on the quality of the dialogue. A real dialogue starts with openness, a good question and the courage to disagree.

- Delegating the personal development into the line organization and HR function. When the CEO does this, the levels below him or her will do the same. Once the chain of delegation reaches the lowest supervisory rung, there is very little room to manoeuvre. The focus is only on acute problems and short-term competence development. This creates a lack of networking ability. We know that real personnel development takes place in the social context – in diverse teams with people who are probably more talented than you, all working on challenging tasks with serious attention given to personal growth.

Simple enough and joyful – "social" networking

Professional communities and networks are vital in all professions. They are the quickest avenues to learning and are a way to measure your own competence level in relation to others. At nearly every organizational level, an individual can find "peers" with similar scope and challenges. The CEO is in a unique and often lonely position. He or she lacks the natural and close peer groups that make immediate feedback – the measuring of one's personal performance, as well as continuous, intense and quick learning from one's peers – possible. This increases the risk of a CEO remaining a prisoner of his or her choices, promises, beliefs and plans, alone atop the organizational structure.

To avoid isolation and imprisonment in his or her role, a CEO must actively build both *external* peer groups and *internal* key relationships of trust and feedback. This is especially critical during changes that cause upheaval within a company. The CEO must be intensely present in the change. The important factor is not the CEO's title but his or her actual role in the change. The role – which can change from project to project or from team to team – is based on a clear and solid insight into one's own strengths, both as a front-line warrior and a coach for other warriors. Everyone on the team must play in their strongest position to create the strongest team possible.

The organizational structure must be simple and clear. Simple and clear especially from the short-term performance point of view. Then the organization structure and the ecosystem structure leave a lot of room open for those who are responsible for strategic changes. This gives the CEO a natural space to manoeuvre in – the space cuts across the line organization and breaks down hierarchies that create stumbling blocks. When a leader has a clear picture of his or her own role and the benefit of a few decades of intense cooperation with hundreds of people, he or she has the punching power and dynamics to enact change. This is the best part of being the CEO!

On the front line, the CEO is a warrior and a coach. As the warrior, he has the sharpest impact, but the coaching role is the one that reverberates through the organization and is many times greater than that of the warrior. However, in both roles it's critical that attitudes, behaviours and capabilities are dealt with and discussed openly. This includes openness about one's own capabilities – strengths and weaknesses – as well as totally honest discussions about issues and experiences. Ask why did we succeed or fail and how do we pass on the lessons learned as quickly and efficiently as possible.

Between the CEO and the front line is middle management, and this layer is always conservative. It needs to fulfil this role in order to sustain the day-to-day life of an organization. However, it should never become a layer that stifles the organization. By spending time in the front line, the CEO experiences the evolving dynamics and fast-paced decision-making and understands the challenges the front line faces. The CEO not only gathers invaluable knowledge, he or she can personally offset middle management's conservatism.

There are always heavy time restrictions on how much a CEO can personally work on the front line. But any time the CEO can spend with the front-line troops has a drastic and positive effect. It creates an affirmative role model, makes it natural to demand more of others and can change the criteria used to evaluate and choose personnel. As many young people of the PlayStation generation are still in the lower ranks of organizations, working on the front line gives the CEO the opportunity to directly interact with them.

From my own years in the trenches, I can assure you that experiencing the capabilities and enthusiasm of the Digital Cowboys and their powerful snowball effects across the organization is a joy. My first experiences with the Digital Cowboys were "culture shocks". I was not used to their very direct habit of testing my value to them. My position and my experience did not have any meaning. The only important issue was my value to them in the specific situation they were in at the moment. It forced me to sharpen my value. In some cases it was my more profound understanding of customers' pains and how to influence the customer, in other cases it was my contact networks, and in still other cases the insights on how to achieve international business growth and so on. On the other hand, the Digital Cowboys' behaviour made it very easy for me to challenge them – to ask the hard questions and to push them to dig deeper and evaluate the options. The joy was to see how rapidly the valuable insights were applied and shared. "Stolen with pride" is great.

The painful learning for me with the Digital Cowboys was their communication tools and habits and being part of their communication networks. Communicating becomes a time issue as the number of channels increases. But even more it's the rhythm of communication, the expected immediate interaction. It made me learn how to "listen to the channels", to carefully think when to interact and how to interact, to develop my own style to use for one-to-one communication and group communication. It's my own path to "hyper communication", as Pekka calls it. It's probably a never-ending path but it provides powerful tools.

Most importantly for the CEO, looking at things from the front line forces one to focus on what's essential and to simplify things. Simplify the message. Simplify leadership and management systems. Simplify the "technostructure" and support services. This will *never* happen by making generalizations across the organization and transferring the implementation to the executive staffs. It's possible only when the top executive *personally* engages across borders and hierarchy levels and brings change and example to a line organization from the outside.

To learn more about Kari and to continue the conversation, go to **www.nofear-community.com.**

5 What I as an executive should change in my company, in practice

THERE IS ALWAYS AN ENDLESS DEVELOPMENT wish list on a leader's desk. The management's goals are *growth* and *profitability*. The leader's task is to find a balance between *stability* and *change*. Digital Cowboys as consumers, customers, partners, employees, subordinates and leaders will change everyone's agenda. Both the list of high-priority matters and the methods for implementing change must be modified.

The development of an organization's capabilities, operating methods and structures involves the same themes as a leader's personal development. Simplifying and amplifying procedures and the sensitive balancing of the two form the starting point for a leader's readiness for change. Unlike in industrial portfolio management, it's no longer just about clearly delineated investment and divestment decisions. Successful fundamental changes are so much more than just tools of production.

Theoretical versus practical models

When thinking about methods of change management, the model published by MIT's Sloan School of Management in 2003[1] is a good place to start. It divides changes into three categories: dramatic, systematic and organic. Dramatic change descends from the top (from senior management), systematic change is generated

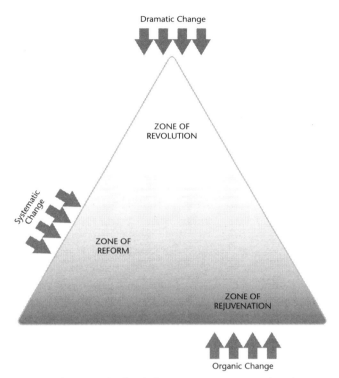

Figure 4 MIT Sloan methods of change management triangle

laterally and organic change emerges from the grass roots. These three forces interact dynamically, each providing the primary, but not sole, thrust for a key transformation process. Dramatic change incites revolution, which provides impetus; systematic change orchestrates reform, which instils order; and organic change nurtures rejuvenation, which spurs initiative.

Dramatic, top-down change requires systematic leadership and the organization's commitment to change. Systematic change requires leadership and, once again, success depends on people's level of commitment. Organic change is the most natural, but still needs systematic specification and leadership in order to spread throughout the organization. In other words, no change direction works alone. The message of the Sloan article is that, for success, these three dimensions need to be combined.

The model is theoretical. It does not tell you *how* to achieve this change in practice. For me, that is a problem. My own background was in small-scale entrepreneurship – working with customers and world-class experts on the front line. Suddenly, I found myself in a large corporation and needed to find ways to ram through some changes when my company was acquired. Unfortunately, neither MIT Sloan nor any number of other archives of wisdom could provide me with a formula for making this work in practice – quickly. I focused all my energy on finding a way to systematically blend the *theory of change* with *real, front-line experiences*.

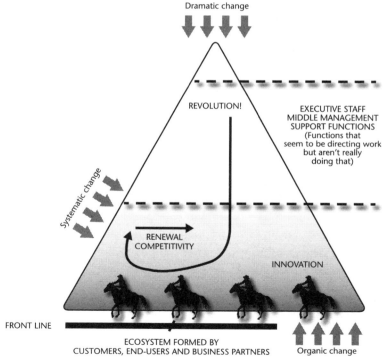

Figure 5 NO FEAR methods of change management triangle

At the core of our model is the simple fact that dramatic, top-down changes are not the most important tool when it comes to leading Digital Cowboys. Yes, this generation is prepared for major changes. But change is not driven by portentous pronouncements and revolution from on high. The strong, Pattonesque clarity of industrial leadership doesn't really work in the world of the Digital Cowboy. Only matters of extreme simplicity – which directly matter to or impact the Digital Cowboys – work as top-down changes. Compared to the list in traditional industrial leadership, this list of changes is pretty brief.

In today's world, placing emphasis on the top-down direction – the "dramatic" change portion of the triangle as shown in Figure 5 opposite – drives an organization into a spiral of paralysis with surprising ease. The spiral starts when higher management places pressure for renewal and innovation on middle management. If the other change directions – "systematic" and "organic" change – haven't developed knowledge of the markets, leadership qualities and organizational courage, the whole machinery actually starts to slow down. And middle management will do what it always does when confronted with something new. They put their trust in constant management, oversight, results evaluation and reporting.

Instead of getting closer to the front line and the customer, middle management spends their time interpreting the daily orders emanating from executive staff and drawing up new instructions. To a Digital Cowboy, this looks like indecisiveness, too much process and annoying red tape. At the same time, middle management becomes frustrated because unremarkable results beget a new series of major changes without an open analysis of past mistakes. Many top-down changes tend to turn people's thinking in the organization inwards and towards the preservation of the current status quo, instead of outwards to the customer or market. With this inward focus, even customer-centric matters such as increasing sales power cause months of internal turmoil. The inward-focused problem is only compounded when applied to "non-customer-facing" functions like product development, reforming production processes, human resources management or other parts of a business.

In the industrialized world, almost all major organizational changes emanate from *within* the organization. In his book *The Innovator's Dilemma*,[2] Clayton Christensen outlines convincingly what happens when companies strip the disruptive potential from new ideas, often to protect the currently successful models. The economic age we are currently living in will show that all major organizational changes and epiphanies – especially ones related to competitiveness – *are born or die on the front line.*

In the Sloan model, these are the "organic" changes. But the changes can't affect business until all three directions of change work together. We never know which country or corporate function will push a change to the surface. The only thing we do know is that the change happens on the front line, where there is real customer interaction. If the organization's psychology is healthy, a cluster of experts will immediately nab the idea. The experts then have the opportunity to quickly combine this information or idea with ideas or information across the organization which amplify it. At this point, leaders all the way up the organizational ladder are alert and evaluating the idea. This happens, though, only when the organization is healthy. In far too many organizations, those ideas never leave the front line or are stifled on their way up.

If the organization's psychology is dysfunctional, the leaders – in their grand rhetoric – tend to emphasize their interest in big-time matters only. In this scenario, you would never bother a leader who runs a billion-dollar business with an idea that has made or saved a mere $10,000. You achieve a complete slowdown when you add a rigid hierarchy and a layer of organizational fat charged with relaying messages from the front line to the leaders.

What was one of the first things Larry Page implemented after he returned as CEO to Google to break the "corporate hiatus" and get back to the innovative start-up culture? He broke down the hierarchical and logistic barriers. Page made some subtle but interesting changes to how the company operates right away. He told his "Googlers" (Google employees) to change how they run meetings to make them more focused and productive and directed executives to meet in public areas of the Googleplex so they would be more accessible to hearing ideas from rank-and-file Googlers.

The front line as part of an ecosystem

In 2003, when I first started thinking about this model, the front line was facing the customer. Today, in 2011, the front line is part of a whole ecosystem – customers, sales, services, partners, suppliers, and even opinion-making groups – most of the time. In 2020, it'll be part of a whole ecosystem all the time.

In the networked world, pushing through change is impossible if the leader does not live within and understand the ecosystem and is not truly part of the network. In many cases, understanding the borders and limits of the ecosystem is much more important than an accurate recall of organizational resources.

The ecosystem brings the front-line experts some new and confusing opportunities and roles. It's not enough to give the customer what they want any more. When you have your own resources *and* access to a powerful network, the ecosystem provides you with opportunities to surpass customers' expectations. Instead of delivering what was ordered, you create something completely new. This often results in something that changes or creates markets.

The leader's endless development list now has a new column – modern methods for accomplishing change. Top-down revolution may still work when the matter belongs in the simplify basket. The clarity of the strategic direction and radical changes in how strategy is implemented belong here. Smaller-scale simplification is achieved by having business functions draw their own conclusions regarding what is beneficial to them in the organization and shutting all other noise out. Sadly, *this dispersed model tends to guarantee that the forces of change remain weak*. When top-down change has been ordered, it's time for a reality check before going public with it. The change has to be tested in relation to the company's front-line fighters and the rest of the ecosystem first. Once this analysis is done, get on the horse, blow your horn and get the roadshow rolling!

If the leader wants to *profoundly* strengthen or transform something, the change must start from a different direction. I've been through this – and screwed up – with the development of my own expert organization. The organization experienced many heady, rousing moments during the 2000s. Our small team often thought we'd come up with something revolutionary. The ideas often came from the front line after meeting with a customer. Bursting with energy and creativity, we created some nifty PowerPoint presentations and even White Papers. We recorded some on-demand videos for our staff, where I, rosy cheeks aglow, told a group of 1,000 experts which way we were headed as of *now*. This resulted in a slew of excited e-mails and phone calls, all echoing the same sentiment: "Wow! What an exciting direction to be heading in!"

Fast-forward a year and all we really got out of it all was a hangover. The experts took the materials and studied them well. They went to the customer and tested the materials and ideas. We experienced some success – mostly thanks to brilliant individual performances – but not for the broad organization. What happened? We had professionally covered all the basics. All the incentives, the capabilities of individual experts, monitoring processes, support resources and a lot more were in place. Nevertheless, a change that would impact the entire organization never got off the ground.

Things started happening organization-wide only when we realized that instead of making grand pronouncements, we needed to put together strong clusters of experts, train them and let them innovate and execute. Through intense debates and the combining of different areas of expertise, we started to develop services that were suitable for different target markets. We could see what we should do ourselves and what we should buy from the ecosystem.

For us, this was the first step towards the systematic change described in Sloan's model. Not everyone believed in the process and some experts departed. There were individuals who didn't want to "demean" themselves by partaking in this sort of blended interaction. They saw their own expertise as being something so unique that they could discern no value in trying to influence others – never mind learning something themselves. But the ones who stayed became the central resource for the deep change the organization would go through. These experts formed the solid cores of the change clusters. The successful change clusters help culture develop and act as change agent for the broader organization. In my experience, their strength is vital for developing business and their impact is one hundred times stronger and more influential than that of a sermon delivered from above.

A company can't buy or import true expertise and experts. It must acquire quality raw materials – people who can work as a part of the front line and ecosystem and develop them. Not to take anything away from the importance of making good personnel choices, *helping experts develop is really the key for managers and organizations.* It's a question of profound cooperation between very different experts with their different capabilities and the results that can be generated this way. We see, over and over, that the success with this process has a defining impact on the company's ability to grow and make money.

The same type of process brought democracy and social equality to Arab countries in a way nobody in 2010 would have thought possible. And it has revolutionized leadership and political structures in the Arab region in less than six months. It took the PlayStation generation and a bit of courage, sociability and communications technology to kick-start the revolution.

Many companies are going to undergo a similar kind of turmoil in their "social and leadership structure". The tool palette will not include Molotov cocktails and tanks, fortunately. But you will be faced with the brutal reality of your best experts in the organization asking, "What kind of organization do I want to learn and work in?" The other parties in the ecosystem – your customers and partners – will do the same. Top experts within your organization and the ability to lead them are a prerequisite for value creation.

Organizational change in the age of the Digital Cowboy

How should an organization change in the age of the Digital Cowboy? The following lists five targets for change. It's not a comprehensive or unequivocal list by any means. These are, however, the things that come up most often in conversations with leaders and Digital Cowboys. They are:

- Strategy and strategic leadership
- Innovation, productification and scaling
- Support functions
- Culture, strategic capabilities and training
- Technology

Our greatest expectations for dialogue with the digital NO FEAR community are related to this very chapter. It is of vital importance and interest to hear about your experiences and opinions of these and additional change targets and the order of implementation. We are interested in hearing what worked and what failed big-time. As you read this section, if we trigger a thought, argument or question, join us at **www.nofear-community.com**.

1. The high and mighty strategy

Digital Cowboys have high expectations of corporate strategy. When we interviewed young experts around the world, as a part of our coaching sessions, these expectations kept coming up. For some, the way in which strategy is formulated generates expectations of deep interaction and the ability to have an impact. The colleagues who'd built a successful 30-year career were a counterweight for this barrel of hopeful thinking. The only thing they could recall about strategy was the term "customer-centredness". The most cynical comment I ever heard came early in my career when a very cultured retail store leader said at a meeting that "strategy always comes up when your subordinates are in trouble and the money's just not coming in. Otherwise you only need this fluff once a year for the corporate government."

In order for a strategy to succeed, a Digital Cowboy *needs both a task and a purpose*. A task is not enough. You cannot work towards a goal if you do not understand the purpose. You can't actively or successfully network for a task if the purpose can't be used to discern which network is most suitable for reaching the goal. But purpose alone also isn't enough. A Digital Cowboy is very proud of his or her own expertise and wants to develop and contribute the expertise for a particular and valued task.

In school, I was taught that *strategy* is the choices you make regarding customers, products and production methods. *Strategy implementation* is directing resources. A strategy's success is gauged by measuring how many resources are transferred during the project. If the redirection of resources was unsuccessful, the management needs to be honest and admit that the strategy goal may be right, but the implementation is not.

The expectations Digital Cowboys have for the strategy work done in an organization are comprised of four simple things:

1. **Purpose and vision:** A Digital Cowboy wants to be proud of his or her community. Traditionally, profitability signalled which companies were worth belonging to. Nearly everyone loves the rich and beautiful. However, Digital Cowboys can see financial profit as a negative in some cases, especially if the management tries to use it to illustrate the company's purpose or use it as the primary yardstick for success. A product or service that was created together and collaboratively and is unique, pioneering, ecological or incredibly well designed seems to instil more pride in

this bunch. In a world of plenty, everything is possible on a material level. Thus, mechanically repeating something that has already been created will not turn anyone on – not even if every euro that went into it meant three more for stockholders and employees.

For *purpose* to be understood, the strategy and the attendant discussion must be taken to the end-user level. If the purpose isn't to create superlative services and products that generate excellent customer feedback, no one will want to put their career and valuable time on the line and devote valuable working years to it.

When describing the purpose, it's important to include a timeline. This is often neglected since management has more freedom to schedule tasks based on resources and progress if a timeline isn't in place. In reality, purpose without a schedule is pretty frustrating. Strategy creation and communication are tests of the maturity of both the management and the organization. They show whether strategy is taken seriously or whether it's just a loose pronouncement. Strategy turned into concrete services and schedules is the Digital Cowboy's purpose.

2. **Task and volition:** Assuming the organization has a clear purpose, the Digital Cowboy will expect to receive – as a part of the strategy – a task. In defining the task, precise borders separating products, geographical regions or industries are not pertinent. Turning strategy into a vision of what competencies are needed tomorrow or after a year *is* important. In contrast to the way working life is currently structured, a Digital Cowboy will not expect tasks and expertise to remain the same until retirement. This results in flexibility but it also forces the best organizations and leaders to continually analyse how to further develop their experts. The most visible sign of poor leadership is the leader who postpones expertise planning and redirection until a financial crisis. However, if a leader can cut the strategy up

into tasks and gather the Digital Cowboys together with a clear understanding of the areas of expertise required by the implementation, the individual and the team develop a strong volition to surpass expectations.

3. **Strategic change:** Strategy is a long-term pursuit. For a Digital Cowboy, "long-term" does not mean "written in stone". Even though a decision regarding direction has been made and tasks for all parties have been defined, it is the "civic duty" of all parties – not just the leaders – to evaluate both the purpose and the task continually. Strategy, and especially its implementation, may and can be adjusted. In traditional leadership models and tradition, the front-line evaluating strategy would be considered impertinent questioning or disloyalty. But this is not how a Digital Cowboy thinks. The purpose – the justification really – for superlative services can't be seen anywhere other than on the front line. Changing the strategy based on signals from the front line is not just important; it's essential.

4. **Success and failure metrics:** The biggest changes in corporate strategy processes are related to the monitoring, transparency and communication of successes and failures. Since Digital Cowboys live a real-time and transparent life on the Web, it's natural that all members of the network expect to see how the strategy process progresses – ideally in real time. Limiting strategy monitoring to the upper levels or making it dependent on financial reporting alone dooms the whole process.

   Everyone knows there'll be delays and things don't always work out the way they were originally planned – it's the nature of developing something new. Face it, each half-truth or unpublished miscalculation by the leader will end up on the network anyway. Every mistake, failure and delay is something everyone will want to learn from. This is so important that measuring information distribution and utilization is becoming a central leadership tool. We're not talking about how many Facebook friends an expert has, but about how many people in the organization have found help and advice in the expert's messages and views. When we're discussing the birth of strategy, evaluating its success or redirecting it, nothing is more important than getting these little nuggets of information together as fast as possible.

It's rare to be able to use the same metrics throughout the change leadership process. Often at the start of a strategic change process, knowing the outcome is impossible so defining the metrics at the beginning of the change is artificial. It's more important to deeply understand and measure the actions and tasks used to drive the change. If the goal is to create a new generation product, it's a little weird to measure the product's sales and costs. Sales will, for a long time, be a big fat zero and costs cannot be forecast. Maybe a better metric would be the creation of the capabilities, expertise and ecosystem that are expected to work for the organization in general or the particular task. In other words, what we did right, what we did wrong and what we'll do differently. A traditional business leader may shy away from these metrics because they're "soft". To the Digital Cowboy, they prove that management wants to understand how the change itself can be implemented.

When trying to convey your strategy, the result is often a nightmarish bloat. On the one hand, investors and analysts want you to summarize your strategy in three clear slides. The board expects a 40-slide presentation with a heavy emphasis on numbers that can be monitored on a monthly basis. Then there are the customers who skip the slides and just want to keep nagging you with questions like "Which of your products are better for me and my specific situation than anyone else's?" or "Where is the world headed?" or "Which products and services do you believe in three years from now?" With all this piling on the pain, how are you going to tell people which competencies are needed for all this? Aren't the expectations we place on the strategy process a little on the heavy side?

Creating the strategy

When changes originate from the outside, real vision can't be created in the minds of a small management team. It's nearly impossible for the management alone to have a sufficient understanding of the required capabilities – what's needed and which parts need to be strengthened. In preparation for writing this book, I read through a huge number of strategy papers from the last twenty years, and it is precisely this part that often seems a little light. If the parties that have an impact on strategy are not willing or able to think about execution, the odds are they will never go to the front line for any purpose other than to preach about their plans. The strategy needs to produce an accurate picture of new services, offerings and customer engagement models – ones that will really create value.

Unlike with industrial management, the know-how required for this process is rarely found inside the organization. The only areas with the potential to draw these conclusions are the expert clusters working on the front line. We can get pretty close to a solution by comparing this information with other information emanating from the ecosystem and adding a little of the leader's own vision. Inviting representatives of these clusters to upper management strategy meetings can only alleviate the pain, not treat the sickness. Dropping some new faces into a strategy meeting might produce some good ideas, but the resistance put up by old operational models is usually too strong for the necessary fundamental change to take place.

In my experience, the only way to create a strategy is to start a continuous and systematic strategic dialogue with the expert groups. Vision can be created only as a result of intense interaction. It is management's job to take these discussions and the competing ideas generated by different work groups, and find the seeds that they'll let the wider network germinate. But to be clear, I am not advocating letting the organization-at-large make the decisions. This is not a singing competition to be decided via a text message vote. The whole point of the involvement of the front-line clusters is to avoid that sort of superficiality. Rough, analytic choices are always involved in the creation of a strategy, and the whole network can never be in possession of all the information needed to make these choices in an informed manner.

It is only after this process that the preconditions for defining purpose, the tasks necessary for success and financial goals, have been met. Now we can say that a strategy has been born. As we approach the moment when the strategy comes into existence, it's important to make sure that all the central service units and support functions are staffed with people to whom the direction is as clear as it can be on the eve of battle.

Traditionally, the strategy management process can take two paths within the organization. One is the top-down model, where the strategy is born "somewhere" and handed down to the experts – on an FYI basis. The other model is to distribute the financial goals from up high to the units/countries and expect the units' strategic choices and actions in the return mail. As far as leading Digital Cowboys goes, both have their good points. When done right, a top-down model can strengthen leadership and provide a shared direction. A distributed model helps people commit to the goals, but unfortunately there is a lack of strategic vision on a corporate level in relation to the ecosystem. For the Digital Cowboy, more important than the chosen model is that the purpose, its related schedule and, to a certain degree, task definition are concrete enough. They enter working life from a strongly visual world, so choices that are made must be visualized and productified into a prototype on some level before they can be used as tools for generating discussion and development.

A Norwegian Digital Cowboy colleague of mine summarized his expectation of the strategy in the following questions:

1. Which products and services will we be the best in the world at?

2. Who are we going to sell them to? How do we add to our know-how of the new target market?
3. What expertise and resources are we going to invest in the development?
4. Which partners (i.e. ecosystems) are we going to work with?
5. When will the prototype, test version and first deliverable version be ready? A month's accuracy is sufficient. When do we have our millionth end-user?
6. How much money are we going to make with these products and what are the metrics for the project's success over the following months/quarters?
7. What are we going to relinquish for the new strategy?

Is this level of detail too granular for a "strategy"? This list is certainly different from those produced by most strategy processes. In fact, upon seeing this list, our CFO declared that it confused strategy and business planning with each other. A board professional I interviewed for NO FEAR also said the list was too detailed for strategy work done by the board.

Personally, I understand this Digital Cowboy's list. It aims to understand – on a concrete level – what we are about to do. Without answers to these simple questions, there is nothing concrete with which to identify.

This series of questions is a healthy way to put the boss on the spot. If I am not able to answer them, I'd be presenting a strategy without having given a single thought to its front-line implementation. It doesn't matter whether this list is complete, wrong or right. It works very well for me. After seeing it, I have asked these questions of every strategy presentation I've seen. For the sake of clarity, it'd be great to create a strategy where one arrow said "growth in China M 10 => M 500!" That's clear and everyone understands what it says. However, from a Digital Cowboy's perspective, this proclamation is hollow and doesn't build confidence that the organization knows how to reach the goal or that it eventually will happen. Strategy is not a bunch of empty phrases handed down from on high. Its purpose is to define the purpose of the company and the tasks of the individual. Strategic leadership requires an understanding of both these areas. In my experience, the disciples of industrial management can somehow still grasp purpose but the understanding of the tasks is incomplete.

2. Incubating and scaling innovation

The front-line clusters are the organization's most important feelers. A "cluster" sounds a little unprofessional and earthy, but there's a reason for using this word. I certainly do not want – under any circumstances – to use the word "unit" here. A cluster typically congregates around a couple of top experts and, most importantly, a task. The best clusters are never created as a result of an edict by the management and by recruiting the personnel resources they've defined in their eternal wisdom. A working cluster does not usually consist of a homogeneous competence or age group.

The strategic purpose of the cluster is to quickly produce development ideas, as well as products. Its birth always aims at preparing the idea and the work group for rapid growth and reproduction. Its fuel is the excellent dynamic between the experts and a strong desire to build something great – as a group of independent entrepreneurs would do. For a cluster to succeed, it's vital that it breathes the same air as the front line and the whole ecosystem.

The opposite of a cluster is the executive staff development unit, which is supplied with a task, money and a group of very smart people. The thought that this group will retreat into its chamber for a defined time period to develop a product for front-line business is just deplorable. All development done by the company, including improving support functions, is equally deplorable. The idea behind a cluster is that regardless of the product development tasks, members of the cluster must spend a part of their time in dialogue with the customers. I know there's a school of thought that wants to provide these groups with the peace and quiet supposedly required for concentration. Working with the customer is seen as having a disruptive impact on focus and professional productification. What I've learned from the Digital Cowboys is that far more innovations are created during open interaction than in isolation.

From an organization's point of view, the birth, lifespan and measurement of clusters are complicated. Because a large organization has to live within the scope of certain structures, chains of command and responsibility, a cluster that comes into being quickly is difficult to place in the organization. It would be easiest for a leader to decide that extracting the top teams out of the organization is the most efficient path. Drumming up support for this decision is easy, but it is problematic in many ways. It goes against the grain of the Digital Cowboys' basic thinking. The presence of A- and B-class citizens in the same organization is not conducive to bringing together necessary competencies – nothing good can come of it. Even the best cluster will eventually reach the end of its road and the limits of its growth if its experts isolate themselves from others. I've witnessed how great ideas and top experts are driven into a situation where insularity and looking out for your own team have resulted in a brilliant idea dying on the vine.

I think a cluster needs an opportunity or problem originating in the customer interface as its life force. In many cases, the expert network will become interested in the subject, and it is at this juncture that organizational culture plays an important role. If the culture allows and supports networking across unit and country boundaries, the germination of clusters will pick up speed. It's a challenge for the organization and its leadership to constantly analyse these fragile sprouts. The normal mechanisms of a line organization are very bad at it. Normal life in an organization tends to focus on caring for the big, growing trees. Sprouts grown by a couple of experts are too small to be interesting. A leader's job is not limited to caring for a fragile beginning. Equally important is asking the right questions and making the right comparisons to verify that, in addition to a brilliant customer solution, something truly ahead of the rest of the pack is being created. The experts in the cluster tend to fall in love with what they're doing, so challenging their thinking is no longer possible without outside help.

If everything works out, an idea that is born organically will find strong experts and a sponsor who has power across the organization. Now comes the critical moment – the decision whether or not to move forward and how to manage the risks. If the decision is GO-GO-GO, the expert cluster must organize. This has to happen regardless of which phase of its annual cycle the company is in. It has to happen even if the line organization is screaming for better planning and if this new development threatens to cannibalize current plans. In the service business, which I'm personally most familiar with, these teams tend to number fewer than fifty individuals. The entrepreneurial spirit must course through the soul of each member, as well as the desire and ability to have an impact on the rest of the organization.

Assuming a cluster born in this manner achieves its goals, the following challenge is more serious by an order of magnitude. How to scale? Of course, it matters that the cluster has achieved something unique, but on an organizational level the effort will be a waste if it can't be scaled up. Something with real financial impact, where a cluster consisting of first five and then 50 people grows into a business of 5,000, is rarely achieved intrinsically. For this reason, the thinking and actions of the experts and the management must be geared towards influencing things in the entire organization. Preparation for this must be started on the day the cluster organizes itself around a task. This tends to be forgotten. I have had many bitter experiences with this.

It's rather like how an entrepreneur starts a small business. No matter how great the business idea, people seldom start the business by developing a generation of leaders, support functions or recruitment geared for rapid growth. In my experience, large, stable organizations are in the same state. New innovations and interesting strategic options are generated from front-line experiences. The vision and energy of the Digital Cowboys are needed to further develop the ideas. For the organization to utilize this power, the whole organization – down to every little support function – must be prepared to act. In addition to front-line action by the management, it means starting the first clusters, as well as the supporting network. It also means the organization must support any and all models that encourage entrepreneurial attitudes and courage. No school will teach you how to be a successful entrepreneur. Growth cluster leaders and experts don't come ready formed from anywhere, either.

3. Coaching people and executive education

In the past, I've asked the people I coach, "What are your methods for coaching your team?" I have to admit, the stupidest response I ever got was "I don't have to. I only hire the best people. If they can't cut it, I fire them." I thought it was a joke at first – typical of an alpha male's sense of humour. Unfortunately, it wasn't.

Leadership, like all capabilities, is forged under fire. Without the development of capabilities, a company can't grow. I'm a simple man; once upon a time, I took this to mean that I had to take my team to where the fire was hottest. At the same time, I had to start thinking about systematic coaching. I will pass on the advice I got from my helicopter security trainer during a complicated training exercise. He told me, "Tomorrow I'll drop you head first into a seawater pool six metres deep, inside a wrecked helicopter. Your task is to get yourself and four unconscious mates out of the helicopter and to the surface. *I suggest you prepare for it tonight.* It's a lot nicer to go through the basics of seat belts, doors and the different phases of passenger rescue on dry land for the first time than underwater." The same applies to coaching the experts in your organization.

I tend to split coaching people into three areas: professional coaching, knowledge refinement and support of expert clusters. Professional coaching and education involve the conscious coaching of every expert and leader, too. Coaching is largely the purview of specialized subcontractors and trainers. But making sure their experts have an understanding of the surrounding ecosystem is the job of the organization's leadership. Leadership must also make sure experts can easily network with other experts in different areas of the organization and ecosystem. The core of professional coaching involves constantly feeding this network with successes and failures.

Knowledge refinement

General knowledge implies drinking from a fire hose of information – it's messy, unorganized and most of the water goes everywhere except in your mouth. Knowledge refinement implies you are gathering and distilling large amounts of data into usable and valuable knowledge. The general knowledge of the company and leadership is the code that the organization's psychology is built upon. Granted, general organizational knowledge can be quite broad, but it is important that your experts have a firm understanding of it. However, knowledge refinement is what advances the implementation of the strategy. First and foremost, knowledge refinement is a question of understanding signals originating from outside the organization.

When coaching Digital Cowboys, general knowledge takes on a different meaning from how it's understood by older generations. Information retrieval and a continuous flood of data are an everyday fact thanks to the development of technology and networks. More time is spent on retrieving and processing information than ever before.

In my own career, the best business schools have imparted a lot of knowledge to me. I count this as a part of general management knowledge. In the early nineties, when I was accepted into IMD, I had no idea where to find real-life cases and world-class sparring partners to support my work. Now, in 2011, I can feed a few words into a search engine and find a limitless supply of perspectives from inspiring thinkers and speakers. Back then, we read management literature and the biographies of great business leaders, but finding these kinds of books wasn't always easy. Before the age of Amazon, finding business literature in the middle of the forest was no picnic. Now the cowboys let loose at our workplaces have the tools and networks to acquire breathtaking volumes of information.

Business schools like MIT, IMD and INSEAD, as well as corporate trainers, are extremely interesting partners when dealing with general knowledge and knowledge refinement. Their role lies in refining information and making its impact more profound. At their best, they can help alleviate information bloat. Owing to their extensive feelers into the business world and society at large, they can really improve the quality of information. An outside sparring partner has often helped get to the heart of the matter and carefully pick the right sources. When there are a lot of sources, a leader has to choose whom to listen to and believe. It's part of developing the skill of knowledge refinement.

The entire organization needs outsiders to regularly take it to task. The best customers and other parts of the ecosystem do it all the time, naturally. Even though this is the most important "source" of knowledge from the outside, I've noticed that an analytic and objective questioning of the status quo by a good business school professor is always refreshing. With Digital Cowboys, the added value is created through analysis and substance. It's no coincidence that the best educational institutions are brilliant models of a PSF culture. For leaders, increasing their knowledge in a business school class, surrounded by 30 cowboys from their own organization, can be pretty tough and humbling. Some are so intimidated by it that they never indulge again. They're afraid. This is an environment where all participants will be stripped of the medals they previously earned. I've experienced some very uncomfortable moments in a lecture hall that, at the time, reminded me of the Colosseum in Rome. There's no place to run and the young lions are tearing you apart with their questions. In the front row, an experienced professor keeps throwing tasty morsels to the lions, as I do my best to keep my thoughts and dignity together. As a leader, this was one of the best and most educational moments of my career!

Expert clusters

Supporting expert clusters is the most organization- and industry-specific of the three forms of coaching. For this reason, it's highly unlikely that you'll find an off-the-shelf coaching solution that fits your needs. In our organization, we built 90 per cent of our coaching guidance and training curriculum ourselves – with the help of a couple of outside partners. The conclusion I've drawn from many years of coaching is that management has to take personal responsibility in this area. You can and should leverage expertise to keep you sharp, point you in the right direction and show you your blind spots. But you have to help strengthen important issues and projects in the organization – you have to be a part of the change. If I hadn't gone into the lions' cage of business school myself, I would have reflexively raised myself above the level of my experts. And I would have been wrong.

In Sloan's model, this is about creating the "spiral" – this is when you productify front-line experiences and then transfer it to the whole organization. This is not a question of coaching individuals. It is about amplifying – as per IMD Professor Maznevski's model – the best thoughts and visions. A systematic change in attitudes, behaviour and operational culture can't be outsourced.

I don't want to get too bogged down in the details of the programmes. The general formula is pretty simple: we nominated a work group from inside the organization to build all the content for the programme. The members of the preparatory team were picked from the top individuals in business and from human resources management. We wanted to achieve diversity straight out of the gate so the preparatory team included members from several countries. We didn't pick any line supervisors for these programmes. Instead, we picked top experts from around the world. This was a source of controversy since the supervisor traditionally always knew things first. This was the paradigm that we wanted to blow apart. The coaching, which stretched out over a year, was divided into five modules and numerous exercises and preparatory tasks to be performed alongside regular work. For the duration of the training, unit leaders – yours truly included – engaged in interaction and were physically present. Receiving feedback for the coaching itself was of paramount importance. Receiving feedback about the real customer cases we were dealing with was even more so.

Even though the model we used was labour and time intensive, it was a success in developing the people and changing the organization from within. For the participants, it opened up expert networks and created trusted relationships. Instead of knowing each other just as work acquaintances, the members of the group were forced to really learn to trust each other. The work done here supplied us – the coaches – with a lot of information, not only about issues on the front line, but also opportunities in the customer sphere. It was great to receive immediate feedback concerning several countries and customer situations. The coaching produced product ideas and services that we were able to monetize in different units. The deepest change occurred in the way the organization worked. Up until this time, the experts had felt that their only common denominators were our brand, the CEO, reporting and the financial management system. Changing the process allowed everyone in the expert network the same opportunities to look for information on the Web, read shared strategies and review service descriptions. The coaching combined with the process of doing things together strengthened both the individuals and teams.

Why am I telling you this? Isn't this very basic team training? Well, I think that the kind of interaction I've described would not have been possible on this scale ten years ago. The social and technical skills of the PlayStation generation were needed. We needed advanced methods of operating in a virtual environment and a deep desire to function in an international environment. It's worth noting that the group being coached was not all youngsters. Our age range was 21 to 64. But it was the energy and dedication of the Digital Cowboys which inspired and catalyzed the entire team to bring our top game into play. Nobody asked for any special privileges and none would have been forthcoming. Everyone had to prove their own worth and development as a part of a customer team.

The entry of the Digital Cowboys into the world of work does not decrease the need for any of these three forms of coaching. I think every organization needs to take a long hard look at the ways in which coaching is conducted and the points that are emphasized. Digital Cowboys need more knowledge in areas related to wielding influence in the organization, leading people and analysing information. In addition to general knowledge, many organizations have to develop their own models for creating and strengthening expert clusters. When the success of all organizations depends on constantly feeling out the front line, there really is no alternative to embracing this change. The best method the leader can use to get a sufficiently deep and fast view of the world outside the organization is to engage the Digital Cowboys in a shared arena – without fear.

4. Support functions for Digital Cowboys

I've made a conscious choice to use a simplification where an organization is split into business and support functions. In my thinking, all parts that earn money are part of business and all those that don't are support functions. Thus, management, supervisors and all the traditional support functions are part of the latter group, regardless of whether the salary is being paid by a business unit or some other party. The purpose of this division is not to downplay anyone's importance to the success of the company. I'll use an old and wise adage – to win a battle, you always need more Indians and fewer chiefs.

Organizations often have a tendency to grow their support functions faster than their business functions, frequently creating complexity in the process. We talked about the importance of simplification and amplification earlier. Support functions definitely belong in the simplify basket.

The support functions, roles and structures of organizations were born in the age of industrial leadership. Even though many companies are, by nature, expert organizations, their operational culture is something altogether different to the PSF culture we explored in Chapter 2. These organizations are very hierarchical and personnel choices never stress customer-centredness, the sharing of information or learning from mistakes. In fact, in many cases quite the opposite. The soul of support functions is controlled by tightly defined processes and service descriptions.

As we've proved earlier, all central forces for change are born outside the home organization, in the ecosystem consisting of customers and partners. The same applies to support functions. In earlier times, it was possible to define the development agenda for personnel management within its own function. No more. For example, an IT unit used to build its solutions based primarily on standardization, security and cost limits. Now this is no longer advantageous as far as the earning potential of the organization goes. Generally, managing support functions is not getting any easier. It's getting harder. The demands that the customers – the employees of the organization – place on the rate of change and the user experience of the services conflict with the standardization and slow-moving planned economy thinking inherited from industrial management.

Digital Cowboys' expectations of support functions are clear and simple:

1. **Self-service like iTunes.** Employees must be able to take care of all basic processes via a self-service channel. In service development, changes in strategy have to be moved from *technical* solutions to *usability* solutions. Service functionality should not be developed remotely by a support function. It should be done by the best minds in the front-line clusters. The paradigm where we first define business needs and

then implement them over years is too monolithic. Service development needs to test its thinking and prototypes with the front-line clusters on a monthly basis. In practice, this means that any support function must be able to release a new version of the services every week.

2. **24/7 personal customer services.** For every support function implemented as a self-service, there must a service that provides personal expert support. Since Digital Cowboys rarely observe office hours, the personal service must work around the clock. In complex processes, self-services result in a drop in productivity instead of an increase in time savings. So help needs to be available, and too many calls are a good indicator that the self-service process needs additional simplification.

3. **Perpetual learning and transparency.** Support functions are saddled with the ghosts of industrial leadership – control and monitoring. In this scenario, all processes must work flawlessly and people who make mistakes must immediately be punished! For support functions to keep up with the users, this must change. Instead of fearing mistakes and trying to cover them up, support functions must create a totally open PSF culture, too. A feedback channel for the end-users must be attached to every support function service. The leaders in support functions must tell us about their mistakes like everyone else. The philosophy must be that the odds on someone who makes a mistake being fired are of an order of magnitude smaller than those for someone who hides a mistake.

4. **Trust.** If the previous part really gets under way, we can cut down on the amount of energy expended by control roles. Control is often justified by referring to laws and statutes. In my experience, these do not generally hold water. I like the idea that we can lessen control in order to increase productivity. How about if we take something like the approval of travel expense invoices and make it possible to process them without a supervisor's approval? If people

knew that every tenth one was manually checked, it'd be enough to keep everyone honest. The knowledge that any discrepancies would result in an official reprimand would be sufficient to make employees toe the line. Digital Cowboys and everyone else would immediately accept a model where people are responsible for their own actions. There are, of course, situations that arise while we exist in the no man's land between manual and digital business management for the time being. For example, there are many documents that require a legal signature. While there are electronic and services-based solutions available, lawyers always claim that, in a contested legal situation, the lack of signatures may prove contentious. So what? If thousands of people save an hour a month, the company can use the savings to hire an army of lawyers just to sort through the legal issues. Conflict is a sign that we are alive. Lack of conflict is a sign of impending coma and eventual death.

Trimming the organization into a series of dynamic front-line clusters is impossible without development in the management and operation of traditional support functions. If this doesn't take place, the best Digital Cowboys will simply not want to work in the organization. They love their own time and themselves far too much to devote several hours every week to red tape. Because efficiency is an important gauge, it's important to combine extremely high-quality self-services with personal service. At the same time, thoughts about the human resources in support functions must be changed into deep thinking about what kind of individuals the service needs as experts. Nobody wants to work as a part of a bulk mass, least of all a Digital Cowboy!

Wait a minute! Who's going to pay for all this? Usually when businessmen and -women discuss support functions, the subject of price comes up. Does your IT function cost $50 per month per person? How much does your human resource management cost per thousand people? You have to understand this discussion from a left-brain perspective. From the vantage point of understanding customer-centredness and the big picture, discussions about unit price are of secondary importance. *The support function's primary job is to remove administrative tasks from the front line.* A 5 per cent decrease in administrative tasks in the front line is more important for the organization's ability to make money than a 10 per cent drop in unit price.

5. Technology: my smartphone is my access to the value creation network

Mark is providing an entire chapter dedicated to technology. But I'm a nerd and I feel that I just have to comment on leading Digital Cowboys from a technological standpoint. Here, elementary strategic matters are connected with the development of the corporate ecosystem, as well as the social and value creation network. In both of these cases, the issue is service availability. The individual experts must have access to all the services using any device at all times. Consequently, the trusted members of the ecosystem, which includes different companies, must have access to intra-corporate networks. In practice, this means that you have to re-evaluate the experts from your IT units. Many have probably ended up being "purchase agents" for different support functions. For this critical job of delivering value through IT, you don't need administrative know-how and purchasing agents – you need integration, security and usability experts.

One of the biggest technology challenges an organization faces is its sea of basic and often proprietary or highly customized applications. An application environment created over a number of decades generally isn't supporting rapid networking and the use of a social network and media tools. A Digital Cowboy will have a hard time understanding the continued presence of such a cumbersome mess of applications that has driven people crazy for years. Why is it allowed to exist? Why doesn't the IT or corporate management do anything about it? Why didn't the fact that I submitted an improvement suggestion or pointed out an error have any impact? There's always a pat answer to these questions: "Let's not be so hasty. These things take time in a large organization."

An old application environment has a lot of organizational history "hard-wired" into it. It is not designed for the future. Before you can move IT forward, you have to dig out the antiquated core of these massive systems. The creation of tools for rapid networking and interaction atop this hard core is an area that needs to be strengthened. It is also an area that is undergoing rapid development.

I've worked on large IT ventures for the whole of my career. I have humbly learned to understand the slowness with which big systems change and how hard they are to manage. A small change in one place triggers a crisis in another. Very clear areas of potential development are neglected because of fear and laziness. When nothing is changed, nothing can break. I encourage organizations to evaluate each IT system through the eyes of the Digital Cowboy and the front-line clusters. An annual process where everything is questioned is healthy. If every piece of technology is under the watchful eye of your own expert network and their ideas for improvement, IT will take a quantum leap forward.

Additional motivation for taking a closer look at the technology strategy is the fact that organizations are opening up and becoming a part of the ecosystem. It wasn't that long ago that people outside the organization couldn't see how the processes and technology worked. Now one firewall after another is being eliminated so members of your network can see inside your organization. In the traditional world, partners and top experts built up the image of the organization with the architecture of the corporate HQ. Now the use of technology, working methods and support for expert work are increasingly visible outside the organization. Technology is turning companies into real glasshouses. Technology use and development targets need the eyes of the Digital Cowboys and other players in the ecosystem to look at issues from an external perspective. In the job market, this already means more to the organization's brand than the traditional high-rise HQ.

If this chapter up to this point hasn't stimulated thoughts and comments then it is time to read the following section by Sberbank's CTO, Victor Orlovski. Victor analyses organizational and technological change from a ringside seat. Sberbank is Russia's largest bank, with roots stretching back a century. Today, the bank employs 240,000 people and has 19,367 offices in ten time zones. In the last few years the bank has undergone a rather extensive modernization process involving, among other things, rapid implementation of new technology and process renewal. When you read about Victor's experiences, take note of how the Digital Cowboys impact the bank's customers.

In addition to an organizational modernization process, we are dealing with a major concurrent change in the market. Historically, every Soviet citizen was once a Sberbank customer. As recently as twenty years ago, the bank had no need for a competitive strategy – *everyone* had an account with Sber. But this is all changing quickly. Russia's young generation expects a totally different order of speed and service. Digital banks are the benchmark. iTunes and the best online shops define the level demanded of customer service. Russia is going digital at a breathtaking pace and a big bank must react via their structures, leadership and technology. To be clear, it's not just about purchasing and installing new technology. We are dealing with issues of leadership and the raising of a new generation of experts and leaders.

Victor Orlovski

What will the future be? Is the world vertical?

We live in an exponential and scalable world, and I think this should be the beginning and the end of this chapter. It is a world where practically any event can be a "black swan". Life in this world with NO FEAR may be the only true way to survive: climbing up exponentially and looking down each time is fraught with the danger of falling and only those who are really thrilled by "climbing the rock" against time can be well on the way to success. Imagination and passion for change are the major drivers of today's capitalization. Do you need any examples? You will find them in this chapter. All-knowing Russian and Chinese businessmen invest in the Silicon Valley start-ups so that practically every slightly significant new venture immediately gets about $100,000 in exchange for a stake in the company. In the scalable world, where every company can become a Google, an Apple or a Groupon, investing in everything is the most appropriate investment strategy. And if you – imagine that you have such money and are not risking your last dollar – invested $1 billion in 10,000 start-ups ($100,000 thousand in each one), even if you managed to guess right only once, you can hit a jackpot of several tens – possibly hundreds – of billion of dollars and recoup all your investments with interest.

Someone may think the Internet boom of the last two to three years is the next bubble. Wait and see. But, from my point of view, the end of the twentieth century, with the boom and then the dot.com bust, was rather a false start. At that moment, investors had the correct perspective of the Internet role's in society, but then – at the most crucial moment – the investor's hand faltered and pressed the "escape" key instead of "enter". Their fear threw the investors from the heights they had achieved. The rule that has worked is that you are at the exponent and that is why you should never look back. But "mankind", especially those born between 1990 and 2005, keeps pressing *all* the keys, strongly and sweepingly. Those who are called digital natives in the pseudo-scientific environment are already starting to teach us how to confound the virtual and the real world.

A friend of mine, whose son recently won an international online game together with his team of friends from Latin America, China, Malaysia and Australia, offered an opportunity for the team of young winners to meet – at the expense of the amiable father

– and sip Cokes at some place where the team could talk "face to face". Over more than five years of joint Internet battles, the team had never – in the opinion of the father, who is definitely a digital immigrant – got together. The kids only saw each other and communicated via the Internet. The son, quite reasonably (from his own outlook), replied to his father: "Why? I see my friends every day." Father could not understand his son and the son could not understand his father. It's simple – for the children born during the digital revolution, there is no difference between the virtual and real environments. And we will never be able to understand this.

I watch how easily my children – eleven, eight, five and three years old – master the digital world. Their brains are programmed to use various remote controls, keyboards, mice, video, and many other things, that our parents would have had to apply every effort in order to learn and master. And just see how they use the iPad! This device is designed for children's brains, and they already "beat" Steve Jobs himself in their imagination, saying to me: "Dad, why can't I do this or that on the iPad?" or "What a pity that the cat is unable to do so". My son plays Angry Birds with both hands, pulling the string with birdies on two iPads simultaneously. All of his friends sit in their own rooms and use their devices to communicate with each other via Skype because it is more "fun" for them.

We're trying to improve ourselves – we touch-type and check e-mails from our phones. But we are not digital in the full sense because we haven't embraced the idea that we do not need reality any more. Look in its face, raise a glass of champagne, touch a cold faceted glass with your lips, feel the notes of a fine rose scent of bursting bubbles ... all of this is in the past. Forget it. You could use the Internet for this. Do you say this is not what you need? Wait and you will get everything – bubbles, smells and tactile sensations – as all these will be available in the digital reality very soon. But there will be something you will miss and it is clear what it is – your youth gone for good. But for those who are twenty now, there will be no memory of phonographs or vinyl records, and it is unlikely that our kids will recall the good old three-inch disks and 2 MB flash drives.

The world has never lived in an era when each year was marked with the emergence of something fundamentally new, radically changing our lives. Now, every quarter there is a new product. Every month there are new discoveries. Every day we can access volumes of information that – at best – took our grandmothers several years to process. A weekly edition of the *New York Times* contains more information than a person received in the seventeenth century during his entire life.

Is the world flat?

In this respect, none of us knows where this exponential growth leads us. We certainly do know and can analyse where it has already led. First of all, it is the deepest integration of many different areas and spheres of our lives: capital markets flowing from one industry into another or from country to country at lightning speed, travel which has become simple enough for practically a billion people, human communications – the world has not fully appreciated the capacities of social networks and the change in mankind's paradigm in this connection – areas of expertise, data processing and storage. Gaps that existed in society – craft and professional differences, inequality in access to expertise and information, inequality in the field of finance – are rapidly disappearing. The world *is* becoming flat. The world is becoming integrated and accessible. Consumption is also becoming flat. This also applies to the "spiritual" world with the consumption, processing and accumulation of data. And we see it in the "material" world, where the gap between the living standards and the goods consumed by the "20 per cent of the richest people" and the "20 per cent of the poorest ones" keeps continuously narrowing. See for yourself where the fundamental difference between the Tata Nano and the S-class Mercedes lies.

Perhaps the largest and most impactful trend of the latest fifteen years has been the breakthrough of the real world into the virtual one. The principal and – one would think – impassable gap between the "non-existing" world of "imagination" and the real material world is rapidly decreasing. In the past, this seemed *impossible* to overcome. Think of it as the modern virtual world becoming the successor to the "book-painting-musical" world. Today, differences still exist and there are several unsolved technical problems. In particular, our brain potential still exceeds by a thousand times the functional ability of a modern personal computer, on which I am typing these lines at the moment. We are unable to transfer tactile or taste sensations or odours through the network. The computer is still poor at synthesizing and perceiving speech. That is why a part of our "basic" emotions demands we "disconnect" from the digital for communication and networking. In addition, the virtual world is still just an environment – a means for communication of "biological" people with each other – and to a lesser extent it is our permanent "companion" and "partner".

But the above-described problems will be solved in the very near future. By 2020, a personal computer, at a price not exceeding today's PC, will have comparable computational functions to the human brain. By 2050, the same PC will have comparable computational capabilities to the entire global population.

The virtual environment will cease to be a means of communication. It will no longer be an infrastructure "object". It will turn into a "subject" with which we will maintain an emotional personal and social relationship, and my grandson will be able to discuss this entire article with his computer in a serious manner. Computers themselves will no longer be personal. Today the simultaneously used resource of all personal computers in the world does not exceed 0.1 per cent. This is the fault of the archaic operating systems, databases and other software tools. The era of distributed computing is not far off. Today, much of the current resource of my computer is idle while I am typing this text. But tomorrow – hopefully others agree with me on this – it will be used for various calculations "assigned" from above. I'll be able to sell or give away my computer resource free of charge. My single resource adds to the multiplying effect the growing computing power of each PC adds to the increasing network utilization of such capacity. This will be a new virtual technological revolution with new opportunities and new challenges. Do not look down. We are riding the exponential and there is a vacuum below us. Cling to the new – it is less frightening than a jump from this dizzy height.

Even now, scientists are successfully experimenting with speech synthesis and analysis, and have managed to transfer smells and tactile sensations. Special systems learn to read signals directly from the brain.

In one of the MIT laboratories, researchers offered me a head frame to wear and told me to "imagine" the number of my home phone – they explained that I must imagine each digit of the phone number. The phone connected to the frame rang up another phone and after a certain number of beeps, my beloved spouse picked up the receiver.

I was, at first, convinced it was a fraud and thought about exposing the trick. But it was not. The frame was reading and interpreting my P300 brain signal. The "trick" was first tested on poor mice, which had electrodes implanted in their heads that could transmit brain impulses to a computer monitor that could display the silhouette of the neighbour's cat – and not just any cat, but precisely the one they had recently seen – at the moment when the familiar "meow" came from around the corner.

Modern researches in the holography application field will surely be able to generate a holographic image of ourselves that would be practically indistinguishable from our real self in the very near future. Think of it. You'll be able to pat your favourite spaniel Charlie while comfortably settled in your first-class seat, having created your home environment around yourself and having surrounded yourself with those household items dear to your heart – underlining the point that a flight may lose any sense of meaning for the "X" generation.

Thus, the fusion of the "virtual" content – which is already tightly integrated into our real life – and the full integration and interpenetration of the virtual with the real world will most likely find our generation – those who are 30–45 years old now – still "alive".

The corporate world

There is a problem of identity loss in today's virtual environment – you can be virtually anyone. No rules of identification and general rules of behaviour exist as such in the Internet environment. The environment was a big information dump from the beginning – if Google hadn't come along, it would still be a dump. Doesn't it remind you of the primitive-communal virtual system? This anarchy may never end. But another thing is more likely: the virtual environment – like any dialectically developing community – will need self-organization. This self-organization will be brought by users themselves. The future Internet environment will have communities with organized structures, strict authentication of users and the "binding" of real users to their virtual entities. Rules for using the network, moral and ethical standards, and laws will take effect. Someone will somehow regulate these laws, punish their non-compliance and show mercy for the sincerely remorseful. And this Internet environment, like any better-organized system, will banish the chaos to the outlying districts of history – creating digital ghettos and slums.

When the Internet has a relationship system in the image and likeness of modern society, this environment will be much more comfortable for doing target business. Judge for yourself how Internet companies make their living today – advertising on the sites, Internet trading with goods and services, communicating with users, and providing entertainment. The product itself is not typically created in the online environment. It is developed using technologies – but not using the network – outside the context of the virtual world and then placed in the virtual world to be promoted and sold.

Imagine that you wanted to establish a company, hire workers, design and make a product through the virtual network. It is almost impossible to do this now: the network lacks the basic rules, creating extremely high risks for an investor. I forecast that such rules will be created in the virtual environment primarily as a self-organization mechanism, and then they will be supported by both virtual and real communities.

In this context, the virtual environment will be both deeply integrated into our human, personal and social relationships and will become an excellent tool for business organization and development, not only in terms of the "Internet business" as an "alternative" channel for promoting a product or a service, but also as an adequate environment for the real business of both an independent enterprise and entire industries. Without leaving the network, one will be able to register a company, find the capital, employ the staff, create and develop a product or a service, list a company on the IPO, arrange mergers and acquisitions and much more. The virtual environment will supply business – including the financial ones – with tools that are difficult even to imagine at present.

I work in the banking industry and it is one of the most conservative industries in the world. Conservatism is associated with the fact that all activities of the banks are related to the analysis and taking of "reasonable" risks. Bank clients investing in the bank also assume the risks. In general, although the banking industry is the most over-regulated and controlled sphere in all countries without exception, the banking business is primarily based on trust and on the history of client–bank relations. In this context, banks per se cannot be innovative companies. However, banks as service companies – they release no products; they sell a service – that sell virtual services – what else can you call the redistribution of financial resources? – should be high-tech and rely on a solid IT foundation. One can, thus, summarize that a bank is a high-tech service company – or rather it should be such to successfully compete. However, a conflict arises. How does a bank become a high-tech service company *and* remain a conservative company by nature? And will banks manage to survive, if they fail to become high-tech service companies, preferring to remain conservative entities and assigning primary importance to their historic conservatism (i.e. reliability)?

I'll try to reply farther on, giving proof by contradiction.

War and peace of generations

I'd like to return to the question of how the exhibitor that previously led us to the complete and unconditional integration of the virtual and real world is now working for a "break", and what new threats are emerging for all of us.

A reduction in society gaps and globalization, in inequality, including the digital one, and crazy data flows that are inevitable under the dialectics laws that are still valid despite the collapse of the socialist empires, hold surprises for us in the form of new gaps and threats. The world is being inevitably compressed and integrating according to one measurement, and expanding, becoming more complex and differentiating according to another.

The vertical course of time during the exponential period – let's assume that it has already arrived – does not leave us any chance to live our lives on one planet with our children and grandchildren. Our minds and mentality do not have time to adapt in the new environment – changes occur too often and are so fundamental that it is almost impossible to adjust to them unless you are young. Our children easily adapt in the virtual environment, adopting the new rules and interacting in the new reality. The exponent at the point of their birth and personal enhancement is far ahead of the point at which we were born and evolved. The eternal conflict of parents and children – the conflict of generations that cannot understand each other – has always existed in history. But all previous conflicts were localized on the same "vertical" axis of consciousness, and in that respect had a localized nature. The conflict of generations that will be initiated by the virtualization and technologization of our world has no comparison to anything mankind has experienced before. We, our parents and our children now live in a completely different world. We, our children and grandchildren "will scatter" along the "vertical" axis of the consciousness and in our mentality even more. We – my generation – will never be able to absorb the virtual world, dropping into and dissolving in it. And the virtual world and its relationship with the material world will develop exponentially, thus estranging our children from our own grandchildren even more.

How will these changes affect us? The problem is not solved using the results of scientific and technological progress. Instead, we can look at two common, equally survivable and completely opposing philosophies. They are embraced by two of the richest men in the world. The first philosophy is embraced by Bill Gates, the technology guru who said that "... he who perfectly masters the electronic mail tools will become a millionaire in the 21st century". The second philosophy is embodied by Warren Buffett, a man who is not on friendly terms with the computer and avoids gadgets and computing devices. One can still get ahead in business and become rich without registering in the network. However, it will be impossible to become modern, and it will also be impossible to understand the younger generation, without being born in the digital era.

It is highly probable that children will not be taught to read and write in about twenty years. Why bother if it will be possible to use the interface reading your mind, and recording all required information in the clear head of your child. Imagine how the mentality of a person will change if he or she is not taught to read or write – he or she is taught only to think quickly and remember. And how that very person will look at his grandfather – it is incredible, and I know I would like to see it.

Another gap will occur in the sphere of accumulation and use of knowledge. Mankind will rapidly move towards specialization of knowledge. It appears that a limit slowing down a "chaotic" accumulation of knowledge in the virtual environment will arrive, and the time for "distributing" such knowledge over the shelves will begin. The euphoria and resources offered by the Internet in response to the cry of "I want to know everything" will certainly stay in place. But as our expertise grows, areas of expertise will become the lot of a very small number of people – each discipline will be largely unique. Everything new will emerge at the interface of these disciplines, with the "new" quickly transforming into a specific isolated area of expertise.

Until now, the major scientific discoveries happened and created value for mankind only thanks to the "crossing" and "cross-pollination" of the industries and areas of expertise – physicians involved in chemistry, mathematicians taking up economy, etc. Creating value through the synergy between the areas of expertise will become a great problem in science, research and manufacture for the next generation of explorers and pioneers.

Practical integration of the virtual and the real worlds considerably advanced the development of relationship in society, especially in the sphere that is regulated by the laws, rules, procedures and arrangements fixed in various agreements (corporate and governmental ones). The official world is still wearing ties and suits, which is old fashioned in itself. The real world as a mix of technologies and the everyday practice of human communication between people and their ideologies threw off the relics of the old world, where grandparents differed only slightly from their grandchildren. The hairstyles, manners and style of moustache would vary, but their expectations and experiences were still somewhat similar.

The first signs of such dramatic impact of the virtual world and technologies on our thread-stitched world of the mid-twentieth century have already been seen in the economic crisis of 2008–10 and revolutions in the Muslim world. All these events are a war between the old-world remnants and the brand-new coming world.

It is still unclear how the virtual world with its outlook will affect the human mind in terms of religion – the changes may become dramatic.

As I have already mentioned above, the impact of the virtual world on the world of economy is still minimum since today's investors can hardly define or understand the environment of the virtual world. But this environment will organize by itself, and it will not only create new attributes of old businesses, but will also drag the existing business models of the real world into the virtual one by breathing into it a completely new reality.

Fear is a bad partner when you are moving up. While the world was moving over the horizontal plane, fear of the new reality helped the conservative species to survive, but it also allowed the more aggressive innovators to move forward into the melting pot of history. One cannot look back when moving forward.

Poor rich banks

I'd like to go back to the banks. What will these behemoths do in this new world? They are remnants of the era of large, insurmountable barriers between the industries, members of the world's elite shop, the world keeping the guild rules in their original form. They are all of these things, yet they stand against the background of practically erased boundaries in the capital markets. Money in the modern world is a virtual commodity. How did it happen that the banks remain such conservative entities and, at the same time, manage to survive?

First, entry into this highly over-regulated market is played by the banks' rules. If you want into the game you play by the rules. Secondly, there is the historical conservatism of bank customers – when was the last time you took money to your neighbour and left it there at a high rate of interest?!

Does it mean that the banks have nothing to be concerned about? They most certainly do. The banks are at the centre of a potential explosion, and the threat to the banks as standalone players is coming from ...

The virtual high-tech world. And this world is not "climbing" through the barbed wire and the banks' defensive Mannerheim line composed of laws, prohibitions, licences and other restrictions. It is stealthily penetrating the bank, creeping into it and drawing users – bank customers – away through the back door, as it were.

The emergence of the virtual environment promoted the onset of payment tools in this environment – the so-called electronic money. The arrival of mobile operators and mobile phone accounts urged the mobile operators (note that the word "mobile" reflects not only the mobile nature of communication means, but also the philosophy of these companies) to look for new sources of income via payments for services and goods using the mobile operator account.

And while the network turnovers and balances on the mobile customers' accounts were insignificant, banks just did not see them and serious investors were in no hurry to invest in the virtual environments.

The virtual and the real worlds are no longer isolated from each other. The rules and regulations of the virtual world are not so strict – yet. But in the near future, rules will be normalized and standardized for the market players, and these rules will be free of the "remnants" of the past. The rules will be clear to the players and will guarantee the "neo-conservatism" to "neo-clients" from the era of digital natives. And this new client will see no difference between the real and the virtual environments.

The boundaries and barriers that have been protecting the banks from hostile penetration of new players in their financial Eden for such a long time will therefore be erased.

The question is what should be done? And I have only one possible answer: you need to be flexible in this variable world. Banks still have the above-mentioned, competitive advantages because the world still has digital immigrants like you and me, who invest our savings and get credits in the old good banks. This means that banks have some time during which they can invest in penetration into the virtual world, and before they say a non-returnable goodbye to its more than 500-year modern history, which raised banks to the top of today's business pyramid. Banks themselves should exit their bank and enter the virtual world through the back door, climb over the exponential vertical curve of the modern high-tech world and then pursue clients who have mentally gone far ahead.

Otherwise, in the case of banks, we will become history. Judge the facts yourself. The Internet and the mobile communications market did not exist 30 years ago. See how this market has grown over 30 years and then compare it to the growth of the banking market within the same period. Extrapolate the resulting model 30 years forward. Do you need any more proof?

I dare to assume that the business that fully integrates with the high-tech virtual environment and erases the boundaries between the real and the virtual worlds, in terms both of its services and products, and between its customers and itself, will be the only one to survive in the 21st century.

One can climb over the development exponent only by applying tremendous effort, and if we do not climb we will automatically fall under our own inertia.

Lessons learned and missed

What does the image I just described – even though it's maybe arguable – mean for the financial industry and Sberbank? There are no prescriptions for the financial industry, and the same goes for publishing companies, producers of copiers and office supply manufacturers. These are all industries of the present and past – but not of the future. It's hard to imagine, but even companies like Google will run into problems that have no ready answers. Why would you need a dedicated search engine when it can be integrated into a social network ...

But let's just imagine a bit, at least about the financial industry.

What should a large and successful bank do today to maintain its success tomorrow and the day after tomorrow? I'll use Sberbank as an example since I currently work there.

First of all, the bank has a set of unquestionable advantages compared with the aggressive Internet environment:

1. **A loyal customer.** Trust in banking means more than in any other industry. You don't lend to and borrow money from a guy up the street, do you? Granted, some peer-to-peer lending sites are thriving, but not everyone will trust their money to an Internet provider.

2. **Tough requirements for client authentication.** Banks require strict registration and authentication of account holders. It is almost impossible to open an account with fake documents or without any documents at all. It means that the personal data accumulated by the bank is much more reliable than similar types of information collected by a social network. You don't trust what I say? Type "Bill Gates" in the Facebook search engine and you'll see that the founder of the legendary corporation has many names in the famous social network – and most probably none of them belongs to the real genius of the technology of innovation. To conclude – banks know their clients, while the Internet world only thinks that it knows them.

3. **The legal aspect.** Most countries are providing an environment of trust for banks and their clients on a legislative level. Most countries are creating an environment of trust based on banks. And that's mainly why you offer your credit card in a restaurant or hotel with almost NO FEAR. This environment is so harmonized that the client can relax and allow the bank to automatically pay household costs upon the service provider's request. The client knows how to act in this environment and is used to trusting it to a larger extent. In the modern Internet, there's no anarchy – there is chaos. There are no rules and laws there – the Internet is

a world of Digital Cowboys, the Wild West of the middle of the eighteenth century. The Internet was created by young experimenters with no real management skills. They created a trash world without rules and systems. This world happened to take the hearts (eyes and hands, to be precise) of over two billion people, but it is still as wild as it was at the start. In contrast, the real world is used to existing according to rules and trusts these rules. There are obvious legal restrictions – what the banks can do and what they *cannot* do. These restrictions will, of course, be diluted over time. But while the state government and money as such exist, restrictions will remain and provide the banks with a slight competitive advantage through the banking licence and the existence of banking legislation.

4. **A special feature of the banking business is the selling of "risk".** Seventy per cent of the income of big non-investment banks is coming from the margin between investing and collecting resources. Banks know how to sell risks and make a good profit on it. It is not possible to acquire such know-how in an instant. One has to learn and try for a long time to understand it. Neither the Internet companies nor the mobile operators have this know-how and they won't possess it in the near future.

5. **The existence of cash.** Cashless payments and electronic money surrogates are playing a more and more significant role in our lives, but nothing can replace cash money for the time being. In Germany, cash payments make up over 50 per cent of all financial transactions, and in Europe overall it's 65 per cent. Small transactions are still mainly conducted with cash. It's partially a matter of convenience and technology. Advances in technology mean that cash flow will be reduced. But cash means not only convenience. Cash is confidentiality. Cash guarantees confidentiality and therefore allows individuals to experience more personal freedom. And if cash is not connected to its owner, the cash is out of his or her

control. Because of legislation peculiarities, the know-how of cash operations is "bread" for the banks.

These five points are enough to ensure that the financial industry has a chance of surviving and has a medium-term reserve of stability.

Prescriptions for surviving in the exponent

I will try to generalize the pros and cons and propose for Sberbank a development scenario that can create a successful, hi-tech service company in the present and the future. I see it as follows:

A. The Internet environment should become neither an alternative channel for selling and providing classic products and services nor a classic channel for providing "alternative" services and products. It should be an *extension* of the bank's products and services. I will try to explain. We already accept the Internet environment as an alternative reality. But the fact is that it is a real environment that has "no alternatives", and experiments there should be conducted as seriously as if they were conducted in reality. Imagine that you do not have several virtual lives on the Internet, but you have only one, real life. This is how the Internet will look tomorrow, and the financial community – mainly banks – can contribute to the Internet community and make it more correct from the legal perspective and more systematic. Internet offerings should be an organic extension of our front-office network, our products and services.

B. Banks should take their trusted environment and transfer this environment to the Internet. The unique knowledge of clients and the trusted environment in the offline world – the rules of trusted partnership and the strong authentication methods – should be transferred "as is" to the Internet space. Banks have the possibility of doing this, unlike many companies jumping to the Internet. While the banks can bring their trusted environment, they can't bring their rules, as they do not possess ownership of the online environment.

C. Clients should understand what their own benefits will be by using the Internet channel to communicate with the bank. This benefit can be offered by the banks by creating partnership with different non-financial services. The bank can act as a trusted authentication provider, and the other services can use the bank's authentication service to be sure that the user is really the person he claims to be. The client, using a trusted bank's Internet environment, gets much more than he loses. The client loses the possibility of calling himself Steve Jobs, but gains a chance to earn money, bonus points or whatever else the bank's marketing department dreams up. In the end, we are following the rule of the road and we do not rob banks because we understand the personal benefit of doing/not doing it. It is the same with the Internet – the fact of being inside the bank's legal environment should become a benefit in the minds of users, the bank's clients.

D. Like heals like. This means that offline and online products and services should be the same, and they should be developed and implemented by people who think like Digital Cowboys – people who were born in the digital era and are digital natives. Yes, such people are still too young to hold high positions in the banks. But if the managers of the "old era" don't free up space for younger and less experienced comrades with an "upside-down" mind, the banks in general will not be able to satisfy the requirements of their new clients. Managers in the banks should – to some, even if small, extent – be Digital Cowboys, understand something about social networks (I mean live there, not only know that they exist) and think in a "What is our client thinking?" way. But of course, they should not lose the know-how from point 4 above.

E. In the end, banks currently have a much bigger capital (money) and stability reserve to conquer the market than most modern technologies. To become Facebook, it's enough to just buy it and not destroy it afterwards (point "C" will help in this respect). Maybe such an asset can seem too expensive – but it isn't at all. It is just an antidote that guarantees that there will be porridge for us, the bankers, when we get old.

The five points that I listed above (A–E) combined with the preceding points 1–5, in my opinion, give banks good chances not only of survival, but also of growth and stability. The right moment should not be missed, because otherwise the price that the bank will have to pay will be too high.

Understanding this, we should move forward towards the unknown future, towards the threats and opportunities of the exponential and big-scale world that we are living in.

To learn more about Victor and to continue the conversation, go to **www.nofear-community.com.**

6 Places of magnificent growth and magnificent failure – emerging markets

AN EXECUTIVE'S BRAIN goes through cycles of three to seven years. First, the word pounding in his or her head is PROFIT, PROFIT, PROFIT, and then the next phase of the cycle is dominated by GROWTH, GROWTH, GROWTH. When economies switch into low gear, it's time to save, and the phrase you hear out of the executive's mouth is *profit improvement*. It's the only way to be a leader. Fast-forward a year, maybe two. A variety of left-brain activities have brought a lax attitude towards expenses back under control and profits are on the rise. At the same time, the markets are recovering so analysts no longer get excited about a well-tended, stable and conservative company. Suddenly, there's a spring to the executive's step and speeches are full of pronouncements about growth. Smart corporate leaders have attempted to combine these cycles by uttering the following magical phrase – *profitable growth*. WOW! What a surprising and revolutionary concept!

The purpose of this chapter is to demonstrate how the challenge of leading Digital Cowboys ties into future economic growth – *profitable growth*, that is. My understanding of the situation is that our success in the biggest growth regions of the future is dependent on whether we can integrate the Digital Cowboys of these regions into our organizations and networks – both in their home markets and in our global organization. In those emerging regions, more than half of the population was born after 1985. While Digital

Cowboys will be important to growth in established markets, they will be critical to growth in the emerging markets. This is why the importance of understanding the PlayStation generation and its impact on management is more important in these markets than anywhere else. A pool of young experts is not an alternative resource to be tapped a few years down the line. In developing markets, hiring people from the PlayStation generation and developing and promoting some of them into leadership positions is practically the only way to achieve consistent growth. To succeed in the emerging markets we have to learn and change quickly. And the lessons learned there will guide us in our broader organization in all markets, too.

The amazing leap from second to fourth generation

In their list of problems facing developing markets, every corporate executive and board member includes tiresome bureaucracy, corruption, a volatile political situation and, in some industries, protectionism. In Russia, you can add fluctuations in the price of oil to the list, seeing as how it pretty much dictates the rate of economic growth. No one will deny that these problems exist. What we can differ on is how serious these issues and their impact on the ability to generate business are. My experience is that these well-known and publicized issues are *all* surmountable. You just need patience, serious front-line leadership and the painstaking cultivation of trusted relationships.

I think the biggest obstacle to increased investment and business in developing markets is a lack of leadership capacity. These markets suffer no shortage of growth opportunities, foreign reserves or knowledgeable experts. The biggest problem is finding and cultivating leaders. Leadership cannot be transferred or imported into companies; leadership talent has to be home grown. For my own part, I can say that Tieto's investments in China and Russia would've been much more extensive had we been able to find a couple of hundred leaders to deploy to the front line and in various levels of our organization in these countries. For us, it was not solely a question of training local leaders. The other parts of the organization, including support functions, should be able to develop at the same pace. Leadership and cooperation capabilities should be developed in all countries in which we operate. The network can't function if the roles and competencies are not balanced.

When following developing markets, note how consumers, ordinary people, can leap over several generations of technology to find the tools they need. In the old days, if you wanted to see what was the newest of the new in tech, your best bet was to go to Palo Alto or a hip café in New York. Now you need to go to a Starbucks in Beijing, Seoul or Moscow. I'm always amazed by how open minded and fearless people can be about trying and using new things. Most people don't make a big deal out of it. It's just a normal part of life.

So, as far as exploiting growth in these markets is concerned, general knowledge and technological prejudices are not the problem. It's the slowness with which new generations of leaders are developed. There is no shortage of brilliant engineers and experts in various fields, but leaders interested and experienced in the development of their staff are rare. While there are many and varied reasons for this, the biggest influence is history – not in the sense of a long chronology of historical events; I mean history in the sense of learning from example and from mentoring – "this is what I learned from my boss and he learned it from his boss" and so on. Learning management and leadership are primarily about learning from role models and applying those lessons in practice.

In developing markets, these role models have been in short supply over the last 40 years; there is only limited precedence to guide leaders. The economic growth of China and Russia started only a few decades ago with political and economic reform. Entrepreneurs who struck it rich during those periods did so primarily thanks to the privatization of capital and government properties and traditional patriarchic and top-down management approaches. When the economies were growing at an annual rate of 8 to 20 per cent, there was no time to *develop* future generations of leaders. In many of these countries, a variety of organizations *copied* an old, strictly hierarchical structure and model. The guiding principle behind leading an organization like this is not so much NO FEAR or "Management by Objectives" (MBO) as its polar opposite, "Management by Fear" (MBF). The managers and leaders hired in the nineties were mostly familiar only with the leadership model of the command economy. Building the leadership structures of companies using the same model was natural for all involved parties. In the early days, it worked.

Today, the rapid increase in wealth has resulted in the kind of laziness that can prove fatal in business. Business is still seen as a tool for distributing budgets, both public and private. I've been involved in numerous discussions where nobody enquired about the substance or value of the actual product at hand. Only the project budget and how it was to be distributed among the interested parties were discussed. It's not even about corruption much of the time. It's more about short-term management of cash flows. By the way, this is a tendency found also in many mature Western corporations, and it is often driven by a senior middle management defending managing their "turf" for short-term success.

Finding leaders interested in long-term development isn't easy. Many still remember the banking crisis and how it made all the money we made disappear or end up in the pockets of a few. Why plan for the long term if something can come along and undo it all? Or why invest in developing an employee when tomorrow he or she could get a better job at the company next door? Why change a thing when it is working right now. It's no wonder that developing people, teams and the corporate leadership culture is not at the top of the list of actionable items – especially when business keeps growing and money keeps flowing into the corporate coffers and the executives' private treasuries.

In the future, the lack of leadership role models combined with the thought patterns and values of the PlayStation generation may create a tenuous situation. Society and politics are already feeling the rumblings. Companies in the emerging markets will be next. Leaders must think now about the challenges their companies will face – not tomorrow:

1. The biggest obstacle to growth and increased investment is limited availability of leaders. The few that are available are often inexperienced.
2. Owing to language and cultural barriers, the "leader shortage" can't be solved by foreign reinforcements.
3. Labour markets have overheated, so recruitment of the few "stars" is expensive and close to impossible.
4. Your only choice is to find, cultivate and train your own leaders.
5. Practically everyone you'll choose for that process will be a member of the PlayStation generation.

Since developing a generation of leaders takes years, you have a decision to make – develop a new generation and stay in the market, or don't and get the hell out fast.

Now here's the important part. Deciding that you do want to grow and help people develop is easy, but making it a reality is harder, slower and costlier than you probably imagine. It's not a question of sending people to expensive management courses or delegating the transformative stuff to the HR department. In my experience, making members of the PlayStation generation into future leaders in developing markets does not happen the way we're used to doing "management development" at home. It's a very different ballgame. For the kind of change we need to happen, leadership models must first take root in the organization.

It touches on all the themes related to transparency, front-line leadership and authenticity that we've dealt with in previous chapters. In practice, this means developing personnel across the whole organization and over national boundaries. *There is no way this can succeed unless the organization's leadership is fully committed to the process.* For the existing supervisors and executives, it is not enough to be a manager. They need to be coaches for the next generation of leaders. While the leadership model grows within a company, the coaches must constantly question their methods and actions. How do I create value for this alert but impatient group of people? How do I make sure that my message as a manager and coach is delivered to the next organizational level and the ones after that? How do I instil in the person I'm coaching a desire to really develop people? *How do I personally demonstrate to him or her that all future success of the organization and of the coach is directly dependent on his or her ability to lead and develop other people?*

How to add value for Digital Cowboys in emerging markets

In earlier chapters, we touched on a leader's role as both a warrior and a healer. The warrior takes a very active role in the front-line battles of the company. He or she is an important member of the team and is very visible – to the team and to the organization. The healer role is to personally help team members succeed – in short-term critical situations, in their own personal growth and in their efforts to lead and develop successful teams of their own. The healer – who makes sure to give credit where credit is due – is, more often than not, invisible, especially to those outside his or her team.[1] Regardless of the market or industry, Digital Cowboys *assume* their leaders will be warriors. But they *expect* leaders to be healers even more because, in this role, the leader can really assist the cowboy in his or her own growth. These are not black-and-white roles; a good leader can and should act both as the warrior and as the healer. However, we associate "strong leadership" with the warrior, and it is easy for that role to become the dominant one. This is especially true in emerging markets.

The leader occupies the top spot in the hierarchy – *alone*. He or she makes the decisions and lives or dies by them, with his or her boots on. The public image of the leader has no room for failure or doubt as to which battle the ship is setting sail for. The people closest to him or her may get to see the healer once in a while, but the healer is mostly kept under wraps. The healer's role is seen as a sign of weakness and not suitable for leaders high up in the food chain. A change is coming, though. The PlayStation generation always questions the authenticity of a leader's message. "The search for the truth", after receiving messages from above, results in a staggering amount of idling and inefficiency in an organization – energy better spent on productive implementation. The corporate response to the questioning is often of the "shut up and buck up" school of management, closing the door on any potential for a dialogue. This model is doomed to failure.

The traditional leadership doctrines and models in the developing markets tend to emphasize the left brain. Sometimes to the extent that – in some cases – right-brain activities, such as creativity, innovation and open communication, are seen as negative or even threatening.

I learned a lot about this chasm through correspondence and by reading the comments people left on my blog. Some of my troops were honestly worried about my public image and felt unnerved by my open and candid manner of communicating. For example, the fact that I might admit to being tired in a message I wrote on a plane late on a Friday night was interpreted as a sign of weakness. Many wondered how my "enemies" might use this kind of information. Some of the comments were so interesting, I flew into the offices from which they originated – I went from Moscow to Beijing to Pune, India. I wanted to meet the people who wrote the comments, face to face. At first, all this amused me. But then I realized it was a real problem. It was clear that the organizations were surprisingly politicized. My ruminations led to direct questions about what the PlayStation generation expected from me as a leader. I found there are four specific areas where you – as leader in the emerging markets – provide true value to your Digital Cowboys.

1. Providing "protection"
2. Sharing leadership knowledge – the type of knowledge that inspires and motivates your cowboys to strive for both personal development and improved customer relations.
3. Establishing a systematic plan or model that helps people and teams grow in both competence and leadership skills.
4. Creating diverse teams that combine talents from "home" and "abroad" – teams are not allowed to be separate islands. Don't let cultural barriers get in the way. Home troops – and other leaders – need to gain a global perspective.

Protection

The easy stuff had to do with getting protection – the kind of protection associated with the Mob. Using the term in a corporate context is clumsy, but that is the issue. Younger leaders and experts wanted my assistance in creating a public shell and network that would allow them to work in peace. They weren't lazy or trying to cadge a free ride. They simply needed to be protected from the politicking and noise of the organization so they could get on with their work. Supporting a request for help can be quite awkward in practice. You have to try to do it in a way that doesn't break or belittle the authority of the immediate supervisors. However, I can never leave an expert in need of help alone. My tactic was to make these experts a part of the development process. Take them into situations that involve line supervisors and representatives of support functions. In some cases this model worked very well. But then there were the times when the supervisor would fire the employee who had entered into this dialogue. Talking directly to the executive or across organizational boundaries was interpreted by the direct supervisors as being disloyal – talking with the "enemy". Every time I found out about this, I hired the person back at double the salary. The fact that this person had the sense to ask for help and was willing to enter into a dialogue was no reason to fire him or her. I wouldn't interpret it as disloyalty to the company.

Knowledge

Another way to create value is to constantly feed the expert with knowledge and hooks that can be used for both personal development and improved customer relations. I've observed that Digital Cowboys in developing markets are very active and eager to learn new things. Their desire to learn technical details and their ability to slog through vast amounts of information are fantastic attributes. What is much harder and of central importance, as far as personal development is concerned, was putting these experts in front of the customers. If there are things in PSF culture that supervisors in developing markets find difficult, they are usually related to just this.

The combination of the customers' hierarchical organization structure and politics within the hierarchy of their own units scares the Digital Cowboys. For this reason, many things that the customer really needed to hear were left unsaid. Not because the expert or team supervisor in question wasn't full of ideas, but because all the time and energy spent thinking about how to present the ideas without stepping on anyone's toes took the wind out of everyone's sails. As a result, it fell to me to show by example how to be brave and really look for the best possible solution in customer situations. For learning purposes, an open briefing by the client engagement team before a customer meeting was a valuable tool. An even more valuable tool was the feedback discussion immediately following the meeting. We'd discuss what we'd do better the next time. To instil some courage in my experts, I'd tell them about my successes and, more importantly, my failures. Not because they offered any valuable lessons or insights, but just so everyone could see that failing wasn't the worst thing in the world. If a leader is allowed to screw up and correct him or herself in front of a customer, anyone can! The idea that trying to do something was not a punishable offence was a huge step forward and it gave my people a lot of energy. In many cases, it's as if they finally let go of the handbrake when driving their car.

A model

The third important task is to establish a systematic plan or model that helps people and teams grow in both their own competence and leadership skills. I don't mean a model in terms of process or theory; this model or plan is more like committing to a diet or a long-term fitness programme to improve your health. And just as with a diet or health plan, you must have total commitment and a belief that it must and can be done. It takes discipline for it to become just a natural part of your day; it's hard to stay focused and committed and sometimes it's completely frustrating, but the longer you stick to the diet or plan, the better you feel. The foundation for your plan or your model is an intense commitment to coach and share knowledge – the lessons learned – with the leaders you are developing. It's more than providing "an education". It's putting yourself in the middle of the model so you and your team learn and grow together.

I wrote earlier that learning management and leadership is about learning from role models. In developing markets, these role models have been in short supply over the last 40 years; there is no precedent to guide leaders. As you create your model, think of it as creating your own leadership "history" for others to follow – a history that inspires and motivates your leaders to develop the next group of leaders. A major obstacle to creating your model is the conflict between how the warrior and the healer are viewed in these cultures. This is a job for the healer, yet the healer may be seen as being weak. Another obstacle is the idea that know-how is born out of education, which is strong in these countries. Many think that universities create leaders – not companies. Belief that good leadership comes from formal education and with the progressively better job titles that follow is strong. People who inherited wealth or became wealthy during the redistribution of billions of dollars' worth of assets are seen as being above even all this. They get to be good leaders because they made lots of money.

Before we could even create a model, we had to tell the Eastern cowboys that developing people is in the interests of us all – it's vital for a successful organization. Once this phase is over, it's on to the next phase and your role – establishing the plan. This can be a hit-or-miss operation. Your existing leadership development ideas may or may not work in a different culture. I should know. For years, I'd spend time thinking and talking things through with my line supervisors and then leave them on their own to develop their own style or take on an issue. Despite the best intentions and smart people, real change never took place. I kept repeating this mistake, over and over. It had worked in the past so I foolishly thought this model would transfer to my new teams. Lesson learned. Based on these sometimes painful lessons, we now embrace the four areas and are constantly improving. We are clear in our examples, communication and expectations of our managers

The cowboys in these developing markets are thirsty for knowledge. Once we rid them of their fear and get them to understand the significance of attaining that next level of development, nothing can prevent us from winning the battle together! Sharing knowledge, know-how and credit with employees – the foundation of a good development model – does not undermine a leader's authority. It sounds like a cliché, but when I worked with managers in developing markets, I had to repeat it daily. And you will likely have to repeat it too, whether you are building an organization in an emerging market or simply trying to get your existing organization ready for the ongoing change.

Eliminate the "home team" advantage

In international business, there are always "home teams". The reality in all global organizations is simple – the home team has more protection, knowledge and face time with you, the leader. They have a history with you and know all the ropes. Or at least, they think they do. The "away team" always feels that the home team has the advantage and wants to get their hands on that "advantage". To tap into the most extensive network of experts, you have to get the home troops out of their cosy existence – their little islands of competency – and get them to embrace the new team players and cultures. When I say "culture", I don't just mean national differences. Organizational and functional cultures can also create barriers. Your job as a leader is to lower the barriers – not only the barriers between the home team and the away teams, but the barriers between you and your teams, regardless of their location. The ability to do this is directly related to your own ability to read and adapt to new cultures. In truth, the real cultural barrier is your own mindset.

The cowboys of the developing markets are often a few steps ahead. They are interested in learning and really understanding the culture of other countries and the way they do business "over there". Based on my own experiences, I can't say the same about my colleagues in the USA and Europe. I don't think this is because of arrogance or a sense of superiority; we don't see this hunger for learning from others in the home troops' younger generation. I think it comes down to ignorance and groundless fears. But these issues can be overcome by getting those reluctant members engaged in some "corporate tourism".

This means taking a two- to four-day business trip into the developing world. I use the term "tourism" because it so neatly describes the characters that show up at the office – representatives from the home office, all tense, with their wallets and passports hidden in their inside breast pockets and a bewildered look in their eyes. Naturally, it's wonderful when they begin to realize that mobile broadband works better in downtown Moscow than in New York (they got Wimax in 2009). It's amazing to them that students in Beijing are more adept at using mobile services than their own sisters and brothers in London. They are surprised that the public transport system in Singapore is cleaner, safer and more punctual than the one in Berlin. Or that the selection at a sushi restaurant in St Petersburg is more extensive and fresher than at one in Paris. For many, these are the first real-world experiences with other countries and cultures, unfiltered by their own cultural bubble or their local media. All these experiences begin to increase understanding and more openness if the tourists get immersed in the local life, have lunch with the local colleagues instead of going to McDonald's and try coffee that is not from Starbucks.

Tourism is the first baby step. Value creation won't really happen until you and your employees become travellers. There is a huge difference between tourists and travellers. Tourists scratch the surface. Travellers want to learn, to adapt and take some risks. From a leader's perspective the training of the home troops needs to start quickly. Is it expensive? Yes. Does it take some time and coordination? Yes. Does it improve teamwork and team agility? Yes – it makes your global network function. True integration is best achieved by mixing up teams consisting of members of the PlayStation generation and making them work together. Each team member must be given an equal chance to influence and create value. The leader's role is to make sure this equality is fully realized, as well as evaluating the results and the performance of individual members.

When you are building a truly international organization, expectations and actions need to be clearly communicated. In the interests of developing a workable culture, removing all experts and supervisors who do not want to work in an environment like this is important. This applies to all levels of the organization. Telling the tourists apart from people who really want to commit to this shared development can be difficult and requires a leader's full attention. Nothing destroys a Digital Cowboy's motivation and excitement more quickly than direct or indirect doubts emanating from the home office. Pessimism based on ignorance and fear kills the culture of cooperation dead. The moment you hear "The Germans always do ..." or "Those Finns just can't ..." or "The team in China said ...", you know you have a problem. If a company has chosen growth in the developing markets as a strategy, it must have the courage – as Collins puts it in *Good to Great*[2] – to check who's on their "bus". Achieving real growth together will not tolerate a single supervisor, unit or function standing around and not pulling their weight. You either get rid of them or you change your strategy. This is a decision a leader must make and execute.

I am extremely optimistic about many developing markets, especially China and Russia. I'm a cautious and reserved Finn, so I shouldn't be saying things like "extremely optimistic", but it's how I see the situation. My optimism is based on the political and economic reforms I have had the opportunity to witness and which are unfolding in front of our eyes. These changes alone do not correct the structural faults created over decades, but the primary reasons behind my optimism are the values, attitudes and international outlook of the PlayStation generation. They are making a particularly visible entry in these countries. This involves the inevitable transparency in communications that follows digitalization – like the Twitter phenomenon seen in Iran or the Facebook-driven collaboration and self-organization in Tunisia and Egypt in 2011.

Put these things together and the modernization of society and business will be unstoppable. People's ability to look for information is limitless and they are not afraid to come to their own conclusions or share these conclusions very quickly and very broadly. In five to fifteen years, these young lions will start taking their places in the heart of society. It's of critical importance that we use this time to develop a new generation of leaders. The countries and companies that can get all this right in their structures, education systems and leadership coaching will grow and prosper. Those that rely on simply trying to adapt the inflexible Western models or hierarchical and functional structures will flounder. It's a new generation.

I grew up in Finland, lived in Germany and have spent a great deal of time working and studying in Russia and the USA. Mark, a native German, has lived in the USA now for over a decade. Between us, we have family on nearly all continents and have worked around the globe with customers, partners and colleagues. While we have our extensive experiences from the countries we live and have worked in, we have found it invaluable to learn from people who have their own perspectives. In order to evaluate the changes and development these emerging countries are undergoing and the challenges leaders are facing, I asked three thought leaders to give us their insights.

Arkady Dvorkovich, adviser to President Medvedev of Russia, has watched and influenced the planning of the Russian modernization programme from a ringside seat. He was personally involved in planning the substance of the Skolkova innovation project – the Russian "Silicon Valley". Arkady is also active in many Russian educational organizations.

An additional point of view on Russia is from a "Westerner" who built a very successful organization in the country. It is provided by Birger Steen, a Norwegian leader with experience in leading a team of experts in Russia. Birger managed Microsoft's Russian operations from 2004 to 2009 and under his stewardship it grew into an expert organization. As the current CEO and president of Parallels, he is now leading this Russian-founded technology company with headquarters in the USA.

The other "emerging" market in this section is now the second-largest economy in the world – China. With over 70 per cent of the population under thirty years old and tremendous, continuous growth, the development of leaders in this market is critical for every organization. Alex Lin, the CEO of ChinaValue, gives us his views on the vast armies of Digital Cowboys in China. Alex works in the very heart of Chinese social media and been a central influence in the development of Chinese leadership training and e-commerce.

Arkady Dvorkovich

New leadership for new leaders

Nobody has a clue how to manage the new reality. People who lived through the pre-Internet era are still at the top of most modern societies. And many of them are still hiding from both the opportunities and challenges of the digital world. Societies they lead are missing chances to achieve a better quality of life in a faster way. It is a big mistake. It is a pity.

And those who spend more hours on the Internet than in real life are not experienced enough to utilize the new skills for the benefit of the society as a whole. They either do not want to be the leaders, or they are not allowed to. So, the winner is ... A radical? A drug dealer? A spammer? Probably one of those – certainly one of those more often than the IT professional or an engineer, or even a venture capitalist looking for business opportunities that can radically change traditional markets.

All that is true, independent of where you live. But it is very much country-specific also, especially when countries like Russia are concerned. So, what is different about Russia?

It is the largest country in the world if measured in square metres. It is a member of the G8, G20 and BRICS, but not of the WTO. It has the lowest income per capita in the G8 and the highest in BRICS. Does it matter? Yes, it does. The consumption pattern of an average Russian is similar to that of the top 20 per cent in the rest of BRICS and to that of the bottom 20 per cent in the rest of the G8, while the level of education is higher than the OECD average. And, unlike in the other BRICS, Russia's young generation is rapidly shrinking. So, the question is who will take over the management of the mighty empire? The most successful from the smaller group of youngsters or the most aggressive coming from the outside? And what should be done to ensure the former?

Many people believe that the single thing needed to transform

Russia into a successful country is to start and win the war against corruption. True, corruption is the most important obstacle to the positive development of the Russian economy and society. But the question is who is going to win the fight if not the new leaders? And if there is no one else, how can we make sure they would commit to delivering?

There are three stories that make me confident that the mission is possible. Let me tell you those stories.

Story number one: The New Economic School

Once upon a time, back in 1991, one of the top economists of the Soviet Academy of Sciences met his good friend from the Jerusalem University and they decided that it was a good moment to establish a Western-style school of economics in Moscow. The Soviet Union had collapsed in just a few weeks and, with the support of George Soros, they had managed to open the school. I was one of the lucky ones to be among its first students.

There are many amazing stories about that school. One is about its name. The founding fathers of the school – Valery Makarov and Gur Ofer – thought that it should be clearly differentiated from the existing universities, while still being a Russian educational institution. After twenty years of its existence, the school still has two names: the New Economic School or NES (English version), and Rossiskaya Economicheskaya Shkola – Russian Economic School or RESCH. It is a real miracle that NES has survived both the disasters of the nineties and the aftermath of the 1998 economic crisis. And it has the top-rated master's programme in eastern Europe.

Is that because *"nes"* in Russian means "miracle"? Or is it because of the true leadership shown by the people who run the NES? Probably both. The key point is that the school cares about the future of its graduates, as everyone understands that the future of the school itself depends on the success of the graduates. And the graduates care about the school's future in return.

The story of NES is a true success story of the modern Russia. Its graduates teach in the best universities around the world and work for the Russian government and the private sector. No budget support has been provided to the school ever. And it is likely that the New Economic School will become a part of Skolkovo University, to be created in partnership with MIT.

Story number two: Local miracles

Being the biggest country in the world, Russia cannot be a place with the same lifestyle, traditions and level of success across the board. All 83 Russian regions, members of the Russian Federation, have their own governments and legislation that should not contradict the federal laws. And it is common knowledge that as an investor you would be better going to a limited number of regions with efficient governance rather than choosing your destination based on pure economic criteria.

One local miracle of Russia is Tatarstan, situated on the Volga and Kama rivers. Its capital, Kazan, was founded 1,000 years ago. It has been the focus of a huge investment project that ignited the transformation of the town of 1 million citizens into a third capital of Russia.

When asked about success stories, almost every Russian official or investor would mention Tatarstan. The most famous thing about the republic's leaders is that they started using e-government technologies before everyone else. All schoolteachers have recently been provided with notebooks. And, amazingly, already almost a quarter of all requests for the registration of marriages are being submitted via the Internet.

What is so different about Tatarstan? People. Besides using the local wealth (including oilfields and the KAMAZ factory, the famous Soviet heritage), they proved to be most efficient ones in implementing federal programmes. What is important is that they have NO FEAR, learning from the best international examples.

Teams of high-ranking officials are travelling around the world to find the best possible practices. And they have NO FEAR in applying these locally, with creative changes that make the application suitable for national traditions. And the world is going to become familiar with this success story in 2013 when the global Student Games (Universiada) will be hosted in Kazan.

Another similar example is the Kaluga region, just a two-hour drive to the south from Moscow. Having no natural resources besides land and no real competitive advantage besides proximity to Moscow forced the local leaders to find a way to become a jewel for foreign investors in the first decade of the 21st century. The recipe was a simple one to invent but apparently not the easiest one to turn into reality. Make an investor happy, give him your cellphone number, and he will return with money.

As a result of creating a friendly environment for investors, Kaluga has turned into the biggest car assembly cluster in Russia and has a rapidly growing industrial sector. Ask Volkswagen, Volvo, Siemens or Samsung. More jobs in the manufacturing sector are being created every month. Where do they find managers and industrial workers? They are raising them locally and bringing the best from other, less successful regions, including Moscow. Some people drive there from one of the largest global capitals every day. A true miracle.

There are smaller miracles in Russia as well. IT jewels like Kaspersky antivirus, David Yan's ABBYY Lingvo, Belousov's and Galitsky's Parallels, and Yuri Milner's mail.ru are already considered global brands. There are not many in the consumer sector, though. "Russian Standard" banking and the alcohol brand created by Rustam Tarikoare are among the few. In arts, the newly created Theatre of Nations and the Moscow Photography House are examples of the organizations where art directors (the amazing Evgeny Mironov and Olga Sviblova) should devote half their time to construction management ...

What puts all these stories into one basket of modern Russian miracles is the presence of unique talented leaders. There is one little problem, though: Russia is an extremely big country and the number of similar success stories is not far above zero.

Story number three: Smile, you are being followed

For the last two years, Russia has lived not only real life, but also a virtual one. First, my generation spent a few months searching for old school friends in "Odnoklassniki" (Classmates). Then the even more informal network of mostly current school students, "vKontakte" (inContact), took over leadership in virtual reality. More and more people have been registering on Facebook and LifeJournal. But those engines have changed real life only marginally.

Tweeting made a real change. Many were doing that long before June 2010. I created my account four months before putting the first tweet on the Web. It happened after Dmitry Medvedev, the president of Russia, visited Twitter headquarters in Silicon Valley. The Russian Twitter population has grown five times larger since then. It is still only around half a million (40 per cent of them following the president), but it is sufficient to bring messages about what is happening to the top of the political elite.

What is different about Russian Twitter is the fact that many of the politicians read, tweet and respond to messages of "ordinary" people without assistance – including the president. It is not easy. My experience of being followed by 25,000-plus people and communicating with a few hundred of them is a tough one. There are lots of attentive and creative people, lots (but fewer) of annoying guys, many specific well-grounded requests for help, and dozens of interesting ideas and projects.

I should make two confessions.

First, I do use Twitter to find new leaders. And I have found a few already. Some of them are already involved in projects for which I am responsible. I am sure that many of those guys are the future stars in the government, business or non-commercial activities.

Second, I do use Twitter to force people to start thinking in a different way. Thinking critically, but without quick, ungrounded assumptions. Thinking with respect for another person's opinion, but without idealizing anyone. Thinking realistically, but with NO FEAR. And I am confident that it works in many instances. Not in all and not immediately, but more and more.

I do hope that some of my friends in the government will join me and some of the regional governors in this endeavour. It is about connecting people to the decision-making process. It is about a new, unprecedented level of democracy. And it is about new opportunities for many of those who never thought they would have the chance to utilize their talents.

These stories are about people who love learning by doing, are ready to change and are not afraid of taking responsibility. Such people are in the minority, both in Russia and around the world. The difference is that Russia is not going to survive without leaders who will provide the change. So it needs at least a few hundred professionals with leadership qualities. Exactly as my friend Pekka argues in this book.

So, what else besides the New Economic School, a couple of decent regional governments and Twitter is needed to build-up the leadership capacity in Russia? Let me mention five key pillars on which I believe we should concentrate attention and resources.

Pillar 1 – a cross-cutting network to support talented kids and accompany them through the whole studying period. Such a system existed in Soviet Union, crashed in the nineties, cannot be recreated now in the same form, and is crucial for raising a new generation of professionals.

Pillar 2 – real progress in modernization projects and a commitment to deliver on related promises. Forty specific initiatives have been launched since President Medvedev came up with the modernization agenda. New national priorities such as radical improvements in energy efficiency and provision of broadband Internet access across Russia have gained broad public support but with much loudly voiced scepticism regarding the chances of success. This is particularly true of the flagship initiative to create an innovation centre in Skolkovo, near Moscow. Consistency and incisiveness can re-establish trust and the willingness of people to play an active role in the further development of Russia. Stepping back or changing those priorities will turn people away and minimize the chances of success for the new generation of Russians.

Pillar 3 – privatization. One can say, "Who cares about who owns Russian companies?" In fact, it matters a lot for the ability of young people to take real responsibility and develop leadership qualities. Those working for state companies largely do not care about competitiveness. Risks are minimal and cash flow is always there. Privatization and elimination of preferential treatment of state-controlled entities will force managers to change their mindsets, which is the most important change we should be seeking in the immediate future.

Pillar 4 – providing social mobility. We have to bring people from the local to the regional and from the regional to the federal level. This is true for the government, for law-enforcement agencies and for businesses. Also, we should make an effort to stimulate rotation of people among agencies, as well as between the government and the commercial sector. Everywhere, it is important to keep a rational balance between experienced staff and newcomers who will produce real change.

Pillar 5 – opening up to the rest of the world. This is not just about removing barriers for trade and investment. What is even more important is to make our young generation familiar with the best international practices. We should use for this purpose both companies and professionals coming to work in Russia. We should send our students, young engineers and managers to the best schools and companies abroad. We should allow new companies to start applying international and best national standards in domestic practices. That will inevitably change the approach that will be used by young people in daily life and work.

Summarizing all the above-mentioned pillars, I have to admit that there is no simple recipe to compensate for the lack of a new generation of leaders and professionals in today's Russia. It takes a multi-pillar routine to work. What makes me optimistic is that we do have success stories, and the key lessons are crystal clear. The most important one being we should have NO FEAR in meeting modern challenges as we have the instruments not available to anybody before. Information technologies enable us to reach out to millions of people within national borders and outside. And choose the best possible options based on all available information.

The most difficult part is not to educate people how to find the right piece of information, nor how to use it or even create it. It is how to create a new value amidst the mass of information. The most likely solution to the problem is that the new value will be a function of free, softly regulated interaction among talented people. If you prescribe exactly what people should do, the outcome will be as anticipated. But it is not going to be a new outcome. If you force people to go through multiple real and virtual social networks, you have a real chance that new ideas will be produced and brought into real life with NO FEAR.

Birger Steen

Leading a team of Russian super-professionals

The world is changing. Really

In the early evening of 9 November 1989, I was sitting in my college study hall in Trondheim, Norway, and watching the Berlin Wall coming down on live TV. It felt like someone was rearranging my mental model of the world by physically moving pieces of my frontal lobe around. One of the displaced bits locked into a new set of synapses – they must have been somehow primed for this moment – and I suddenly realized that I would learn to speak Russian and go to discover and work in that riddle, wrapped in a mystery, inside an enigma, that is Russia.

Fifteen years and a few jobs, schools and degrees later, I arrived in Moscow as the newly appointed general manager of Microsoft Russia, at the time Microsoft's twentieth-or-so-biggest subsidiary. I felt I had finally arrived. I also felt apprehension and not a little uncertainty. In the intervening years I had indeed learned to speak Russian. I had practised it for a short time as an oil trader in the disintegrating Soviet Union in the early nineties. I had also been a strategy consultant, a media executive and a dot.com CEO, before joining Microsoft as leader of the company's Norwegian business. But despite a strong feeling of finally arriving at my professional destination, nothing had really prepared me for leading the business of the world's biggest software company in the world's biggest country. Russia was, and is, immensely rich, in pure monetary terms, in natural resources and in terms of intellectual horsepower. At the same time, it was nearly invisible as an IT economy. Piracy kept the software market small. A mix of Cold War carry-over attitudes and more recent bad press in the West kept many of the global brand-name IT companies in holding patterns of limited investments in-country and occasional chief executive visits with vague statements of future plans. Microsoft was one such company, and that clearly needed to change for us to have any real impact.

A greater immediate priority was the challenge of leading some 160 young, bright and ambitious Microsoft employees in the country (a few short years later we would employ more than five times that). In previous leadership roles I had always put great stock in my own ability to assess, attract and retain strong performers with high potential. How would this work in Russia, where I had no network, no real in-depth understanding of what companies and institutions develop and educate the best? More importantly, would these smart and proud twenty-somethings – our average age was 28 – accept the leadership of a foreigner from a tiny country to the north-west?

Why should they follow me? My previous teams had been very much meritocratic groups of equals built on Scandinavian norms of non-hierarchy and collaboration. Such success as I had in prior jobs was more to do with what these teams had taught me than what I taught them. Meanwhile, I knew – from my own prior experience, from studying Russian history and literature and from plenty of anecdotal evidence from other expatriates – that the story of leadership in Russia is largely one of strong, autocratic, sometimes brutal *voievode*. Even my distant compatriot Rurik fitted that narrative. He is assumed to have been a Viking prince who was invited to KievanRus to rule over local factions who could not settle their differences alone, and he went on to father the dynasty that bears his name and hence the first 500 years of Russian tsars. The odd exception, like the brilliant and well-loved general Suvorov, seemed only to confirm the rule. How would I need to adapt my own leadership approach to be successful?

The answer was, in the end, very simple. I didn't attempt to change a thing, because there was nothing I could change, not at the core. I firmly believe that the only way to lead, particularly an army of twenty-somethings, is from the heart, from the centre of your own conviction, albeit with a profound willingness to listen and to learn. Leadership, quite simply, must be authentic and cannot be faked. So my model for leading in the Wild East of Russia stayed the same as it had been in the peaceful, egalitarian social democracies of Scandinavia. Here I will attempt to summarize and illustrate what I see as key elements of that model.

The individual is sacred. The company is a tool

Traditional industrial-age companies, well suited to the requirements of large-scale, centralized design, production and distribution, often see friction in assimilating young, "digital native" talent. Two very different sets of implicit assumptions cause this friction. Traditional companies are ends in themselves, and the individuals joining them are a means for their growth and perpetuation. Meanwhile, the cowgirls and -boys entering the workforce in the 21st century, and even more so those digital natives growing up in fast-changing emerging markets, see the company they join as the best temporary means to their own ends. This fundamental difference in perspective shows up in a lot of different guises – widely differing time horizons and expectations in career development, retention/attrition problems, low acceptance of formal authority and, in more serious cases, violations of company rules or applicable law. Each one of these can ruin your year or your entire business.

Conversely, team members, whether "native" or "immigrant", who understand that they are seen, treated and listened to as individuals with legitimate goals and purposes that may differ from that of producing maximum returns for shareholders do, by and large, reciprocate with above-average loyalty and productivity.

So what does it take to achieve this?

First of all, over-communicate. Attention span is a fundamental limitation in all human endeavours. All the more so for digital natives at work, who live in a real-time social network where few activities go on for more than a couple of minutes. The seven or so facts most people can keep in our short-term memory stack are continuously recycled by texts, tweets, e-mails, chatter, over-the-cubicle shout-outs and commercial breaks in the Spotify stream in their headphones. Dr Edward Hallowell, a specialist in Attention Deficit Disorder (ADD), names the resulting condition Attention Deficit Trait – a work-induced, temporary form of ADD. To break through the clutter as a leader, you need two things – *impact* and *frequency*.

At Microsoft Russia, we eventually had 35 offices across 11 time zones, from Kaliningrad in the EU to Vladivostok on the Pacific coast. We nevertheless started every Monday with a 30-minute town hall meeting for all employees, physically in Moscow, by video link in the other locations. The format – a continuous physical meeting, as opposed to an interruptible e-mail or podcast – as well as the frequency – every week – kept us focused. And although themes

> **What we've got here is a failure to communicate.**
>
> One of my biggest failures as a leader also involved communication – or lack thereof. In January 2007, Russian authorities charged the director of a local high school with violating intellectual property rights by purchasing and using computers with pirated software. Besides being the "injured party", Microsoft had no involvement in the case. This did not prevent Russian media from taking great interest in Microsoft's role and point of view. My mistake was in believing we could somehow stay out of a clearly negative media situation based on the facts. As it happened, the case attracted international attention and remained on top of the local agenda for several months. Lesson learned: The facts are meaningless unless they are communicated, transparently, clearly, repeatedly and with conviction.

would change, the core message would always be the same: Here is our plan to reach our goal, and here you can see how we are doing against it and ask questions.

Second, flatten your perceived organizational structure through transparency – not by restructuring to compress the formal organization. Make yourself and other leaders accessible – as accessible as everyone else. Tear down office walls and put your C-level execs in open seating plans like everyone else. If you think some of your folks need offices, be fanatical about "open door" accessibility and allocate offices to those who particularly need or want them, not by rank. Build skip-level relationships, in the form of regular one-on-one or one-on-many meetings. Show up for the New Year's party and stay until the end. In organizational terms, "flat" is not a geometric property. It is a collective state of mind where there are no unnecessary barriers separating your best people from each other, from their leaders or from the best facts, resources and opportunities to do a great job.

Third, be a ruthless meritocrat and do-o-crat (i.e. "the one who does gets to decide"). Your new twenty-something whizz-kid, your thirty-something star seller and your experienced PR manager have all chosen you for their purposes. They will leave you if you cannot deliver. That goes both ways. Praise, pay and promote your best, brightest and highest achievers and let them in on the most important, exciting work you have going on. Conversely, help your lesser achievers find their next job, inside (if the problem is role fit) or outside (if the problem is company fit). But note this: *even as employment ends, the individual in question is still sacred*, whether they are leaving you or you are helping them on. It will cost a bit extra to ensure that 100 per cent of your "company graduates" land well in their next step. It is worth it. In this case, the only thing more important than a first impression is a last impression. If you do departures well you will build a lasting community where the number of people rooting for you is many times the people you actually employ. Lifetime employment? Neither you nor your Digital Cowboy employees want that. But you will both benefit from a lifetime connection.

Your company needs profits. Your people need a long-term, inclusive and compelling purpose

So, your team came to you, and to your company, for their own ends. How effective they will be and how long they will stay depend on your success in aligning those objectives with those of the company. Most troubling for traditional company managers, these ends are almost never trivial or even the same for all your team members. Sun Tzu suggested that "the skilful employer of men will employ the wise man, the brave man, the covetous man, and the stupid man". Hold on a second. Did he actually go out and search for the covetous and the stupid? Probably not. As a general in the small kingdom of Wu, however, he was realistic enough to assume that he would have to play mostly with the cards he was dealt, and that he would therefore have all types of soldiers in his army. So will you. Hence, the challenge for "the skilful employer" is to motivate everyone on the team to move in the same direction, although for very different, individual reasons.

At Microsoft Russia, we invested significant time and resources in examining individual motivation, and using that to formulate a shared team mission. We collectively wanted to grow Microsoft's revenue in the country by a factor of ten, while at the same time having a real, positive impact on the local IT industry and society at large. Our mantra was "to make Russia more important for Microsoft and Microsoft much more important to Russia". This, while ambitious, was a direction that subsumed a number of individual ambitions. First, we had a significant population – and I count myself as a member – who were drawn to a version of national romanticism. While disillusioned with the stagnation of the seventies and eighties and the chaos of the nineties, these employees believed – and still believe, I think – that Russia has unfulfilled promise as a global IT leader. We set out to change that. Related, but not 100 per cent overlapping, another group of idealists believed that smarter and more comprehensive use of IT could improve lives and businesses in the country. Thirdly, we had a significant number of younger team members who saw Microsoft as a smart first or second step in a great business career. Fourth, of course, any real sales organization will always have a significant population of "covetous" team members for whom a growing pay cheque was paramount. In reality, we probably all have a bit of it in us.

The important point here is that our team mission was both big and bold enough to get noticed and wide enough to contain a number of individual, personal goals. In the end, we *did* grow the business by a factor of ten in four years. And in the process, we *did* help build the Russian domestic software industry as software piracy declined, we *did* help put modern IT into millions of homes and thousands of businesses and schools, we *did* help hundreds of young Russians along in their career while they were taking home a decent pay cheque.

We are all from Denmark – or somewhere similarly small – so even Russians need to learn how to be a minority

Traditional companies used to have home markets. Most of their employees were citizens of a country where they spent 99 per cent of their lives. So in the twentieth century, companies – and leaders – from smaller countries were at a disadvantage, as their home base limited the scope of their opportunities, at least initially. Managers and leaders from the largest countries had the best career outlook. Globalization has turned this upside down. Today, hailing from a small, sparsely populated part of the world such as Scandinavia is actually a *benefit* because businesses born in a small home market are forced to approach problems and opportunities as global from the get-go. So, you are Danish. Speak your own language, and no one (well, technically around 0.1 per cent of the world's population) will understand you. Speak English, Spanish or Mandarin well enough, and you just multiplied your addressable market by 100 or even 1,000. In the Nordic countries, all TV companies have acknowledged this – all their programming is aired with the original soundtrack and subtitles.

Digital natives, all over the world, also get it. They live in social networks that blanket the world and they see nothing strange about making friends or doing business twelve time zones away. To digitally native companies, and to their employees, even China and India are small countries, in the sense that they represent at most one sixth of the addressable market. And the USA, while still accounting for more than a third of the world economy, is home to less than 5 per cent of the potential customers for most consumer products – in the long term.

So, proficiency in global communication, in English, by e-mail, a CRM system or on Facebook, with colleagues, partners or customers in different countries, is a necessary condition for success. So is recognizing that the confines of your own country, language and culture limit your business potential. However, being capable of connecting with a global market (and motivated to do so) is not enough. "Nothing Sucks Like Electrolux" was a technically correct but not fully localized translation of a great vacuum cleaner ad. Sasha Baron-Cohen's Borat spoke, um, English, but failed to realize his potential, so to speak, during his US sojourn.

At Microsoft Russia, many of our most talented people had experienced similar kinds of cultural barriers first-hand during business trips, company training, reviews or other interaction with the global – in reality mostly North American with a dash of western European – culture of the company. It became a pattern. The young, bright-eyed IT-specialist-turned-first-line-manager would come back from her first company training session with a desolate look on her face. The training had been good. The other participants were impressive and sociable. The problems worked on had been both interesting and important. And "my" champion had offered up great solutions. However, some well-spoken Brit on her team had reformulated it in a more elegant way and a charming Italian had sold it first to the session leader. So when the time came for evaluation and feedback, my young colleague would get to hear that she "had some good ideas but insufficient interpersonal impact" and "should perhaps consider an individual contributor career".

This was a real morale problem – great people failed to get recognized and appropriately promoted and retained, not because of skill or potential, but because they failed to translate culturally. It was also a business problem. Great project proposals would get turned down because they weren't adequately "globalized" – not articulated in the right form or language.

So we invested significant time, resources and money in bridging the gap. We sponsored "reverse expats" – high-potential Russian employees who were posted to global HQ roles to learn and help us get involved in key projects at the right time and in the right way. For selected "network hub" jobs, we hired expatriates, most of them Russian-speaking, with a track record of success in connecting across global teams and time zones. We offered English-language classes and specific business school courses to high-potentials. We hosted 35 of Microsoft's top executives for a week in Moscow, St Petersburg and Nizhniy Novgorod to learn about Russia, its people and its technology potential. We brought together leading Russian software developers, investors and company decision-makers in Seattle to evangelize the potential to develop software in Russia. This bridge-building effort contributed to better investment decisions and more resources for our team as well as career opportunities for many of our best and brightest. It also helped us grow our business while expanding our company's impact across the country – making Russia more important to Microsoft and Microsoft more important to Russia.

In the end, the Russian subsidiary of Microsoft was recognized three times in five years with a "Platinum Award" – the highest achievement in the company. Two years in a row we contributed more to Microsoft's bottom-line growth than any other international subsidiary. Russia became a first-rate citizen of our company. Mission accomplished.

For five years, I had the opportunity to lead an ambitious, spirited, energetic group of digital natives from the largest country on earth. For those of you who are reading this book, I owe each one of you a depth of gratitude: For most of you, I realize, "unfollow" was literally just a click away. Thank you for your hard work, your passion and your trust. On a very personal level, most of all I thank you for how you received me as a friend and educated me as a leader and businessman in one of the world's most dynamic markets. As the old Russian adage goes: *Vekzhivy, vekuchis*.[3]

Alex Lin

The Internet is speeding up the integration between China and the world

New York Times columnist Thomas Friedman put forward the democratization of technology, capital and information as the key to making the world flat in his book *The World is Flat*. In Chinese, the character 平 (sound 'Ping') has two meanings: one is flat, the other is fair. And the most critical force for making the world flat is technology, symbolized by the Internet. As the head of a leading Internet technology company in China, I have the privilege to see and lead a part of this change and its impact on our country and people. Some of the insights gained in this role should be of interest to the current generation of leaders from within China, to those coming to China hoping to be successful, and to the young Digital Cowboys entering the workforce in China, as well as across the globe.

In April 1994, during the conference of the "US–China Technology Cooperation Alliance", Chinese representatives and the US National Science Foundation discussed access to the Internet for China. On 20 April 1994, 64k Internet access was initiated for educational use and research and development in Zhong Guan district, Beijing.

Over the past sixteen years, the growth of the Internet in China has been explosive. China has the biggest number of Internet users in the world. By the end of 2010, the number of Internet users in China had reached 457 million – a 73.3 million increase compared with 2009. The Internet Penetration Rate has jumped to 34.3 per cent – 5.4 per cent higher than in 2009. It has already surpassed the world's average of 30 per cent.[4]

In terms of the way users access the Internet, the percentage use of desktop computer, mobile phone and laptop computer was 78.4 per cent, 66.2 per cent and 45.7 per cent respectively in 2010. Laptop computer usage increased 15 per cent compared with 2009. The growth rate of mobile phone and desktop computer usage is 5.4 per cent and 5 per cent respectively. Mobile Internet users number 303 million, an increase of 69.3 per cent compared with 2009, with mobile phone Internet users constituting 66.2 per cent of overall Internet users – up from 60.8 per cent in 2009. Mobile users have become the main driver for the growth of the Internet in China.

Economic and political impact

The Internet boom has boosted growth in the Chinese economy. During the past sixteen years, the annual growth rate of the information technology industry in China has been 26.6 per cent – from less than 1 per cent of GDP to around 10 per cent. As of March 2010, over thirty Internet enterprises spanning multiple industries have been listed in the USA, Hong Kong and mainland China. The culture of the Internet industry has become the most significant component of Chinese industrial culture.

Recently, the most obvious characteristic of the Internet economy is reflected in the growth of electronic commerce. Over 92.7 per cent of small-to-medium companies and 100 per cent of large companies have Internet access. About 43 per cent of Chinese enterprises either own official websites or build their web store through the electronic commerce platform. Some 57.2 per cent of enterprises are communicating with clients and provide consultancy services by using the Internet. The average percentage of application of e-commerce or Internet marketing is 42.1, while 21.3 per cent of enterprises are using e-mail or e-mail direct mail (EDM) as the most common means of Internet marketing.

The transaction volume of e-commerce in China was over RMB3,600 billion in 2009. Online shoppers number 100 billion. Online shopping was the fastest-growing online application in 2010, expanding 48.6 per cent. Online payment and online banking have annual growth rates of 45.8 per cent and 48.2 per cent respectively and are steadily outgrowing other online applications. The standardization of professionalized services such

as digital certification, online payment and logistics is gradually becoming established. The Internet era is driving business activities in all areas of the economy, and this is still just the beginning of the growth.

Equally important is the Internet's impact on political democratization and information flow. In the 1990s, the Chinese government started the comprehensive project "Government Go Online". China has 45,000 state-owned portal sites, with 75 central government websites, 32 provincial government sites, and 33 prefecture-level city government sites. In addition, over 80 per cent of county government has established government websites to provide more convenient online services for domestic working and living.

Although government sites have expanded greatly, access to information and news is still restricted on traditional media. Social media give the "other side of the story" and provide information citizens would not be able to find anywhere else. The Internet has become the most important channel for the Chinese to receive information. In addition to providing news and information, the Internet is helping the Chinese people in their fight for freedom of expression, their right to know and their right to government administration. It allows people to voice their opinions about events such as the Wenchuan earthquake, the melamine incident and the Beijing Olympics. Even national president and premier Hu Jiantao and Zhu Rongji have used the Internet to communicate with the Chinese people about the introduction of political democratization in China.

While Internet growth in China is extensive, problems do exist. One of the major problems is uneven regional and countryside development. At the end of 2009, the Internet penetration rate of the eastern and western areas was 40 per cent and 21.5 per cent, respectively. Urban Internet users represent 72.2 per cent of the total number of Internet users, while rural Internet users represent only 27.8 per cent. In order to narrow the gap, the Chinese government needs to devote more effort to more even Internet introduction and access. The other ongoing issue is the tight control the Chinese government exercises on the Internet. The Chinese government should change its attitude to be more open and confident towards the Internet.

What these changes mean to business

The Internet is continually evolving, which means opportunities for businesses in Chinese markets. The following four development trends are of note:

Trend 1: Social computing or networking. According to forecasts, Facebook has become the driving integration force between the PC and the mobile phone in many areas of the world. While Facebook itself is still not a main player in China, similar sites in China are experiencing strong participation. People – especially students – like the model, and the leading sites for Chinese students – RenRen ("people") and Kaixin ("happy") – are strong and growing. Social networking is becoming the hub of our networked society. As of 2010, in China mobile devices have replaced the personal computer as the most common Internet access device. There are over 1.82 billion mobile devices worldwide, with China possessing the most mobile users in the world.

Trend 2: Sensor computing. The interlinking of mobile devices, social network websites and communication products by 2012 implies considerable market opportunity for sensor computing. This will provide more mature environmental perceptions and usable functions. It is favourable news for the Chinese e-commerce market.

Trend 3: High-end analytics. According to Gartner, data storage volume in 2012 will be five times greater than in 2008. The challenge of enterprise is to transform the data into information and use the information to make better decisions. Analysis tools in the past could only provide the analytics post-incident; however, high-end analytics will provide the vision from reason to results.

Rapid analysis of operational measures is highly valuable, meaning that high-end analytics may overcome the technical disadvantages faced by some Chinese enterprises.

Trend 4: Cloud computing. By 2012, roughly 20 per cent of enterprises will not own any IT assets.[5] Virtualization, cloud services and even the PC devices used by staff will encourage enterprises to cut IT assets. For China's industrial restructuring, cloud computing lowers operating costs while boosting the flexibility and responsiveness of growing businesses. Especially for organizations without a mature IT infrastructure today, cloud computing offerings will allow them to quickly adopt technology capabilities without costly and slow internal upgrades and deployments.

While development is moving quickly and covering more and more organizations and people, China still has a huge untapped potential today. The latest data by the Internet marketing research company comScore shows that Canada has the highest Internet penetration rate – 68 per cent of Canadians are regular Internet users. The percentage of French and British, Germans and Americans is 62 per cent, 60 per cent and 59 per cent respectively. The White Paper "Status of Chinese Internet" was released by the Chinese government recently. It outlines investment priorities and shows that the Chinese government will continue to promote the development and popularization of the Internet in order to increase the penetration rate to 45 per cent in the next five years.

You can steal my technology, but you can't steal my value

Technology is positively impacting our world, but technological advantage in itself is only a temporary advantage. Innovation and flexibility create the competitive advantage.

Yes, China has jumped on the Internet bandwagon but not necessarily in the same fashion as our business counterparts in other parts of the world. However, technology is not the driving force – value is. Companies like QQ.com – an instant messaging company with over

640 million users – don't tout their technology or features as the reason for their success. Their success comes from their loyal user base. Keeping these users happy is the key driver for their business. The founder of Alibaba.com, the world's largest online business-to-business trading platform for small businesses, once said, "… the reason why I have been so successful is that I have no idea about technology … it needs to be easy to use for people like me".

The truly innovative and successful companies do not rely on proprietary technology. They differentiate themselves with innovations in business models, systems, environments, marketing and so on. So it doesn't really matter if the competition can "steal my technology", they cannot "steal my value". My friends asked whether I was concerned about Intellectual Property (IP) protection at my own company. I had to tell them, "Not really." We have IP in the functions and features we have developed and are considered the leader in our market sector. But we continue to win because of the overall user experience and the user-base size.

Another challenge we face in China is an attitude in the young generation which I find more often in the older generation in other cultures. It is a resistance to adopting technologies or a belief that "knowing technology" is only for one functional group and other functions and their leaders should not get involved. This has led to a situation whereby young people in China who are very interested in the technologies are usually not as interested in business – and the other way around. Managers often don't like to be seen as being too technical for image reasons. Traditionally, China's leadership philosophy is more one of holism.

Young Chinese entrepreneurs are not using the Internet to drive business locally or globally, and very few of them are using the Internet for daily business – yet. Over 70 per cent or so of the Internet population in China is under 34 – the PlayStation generation you talk about in this book. Most young entrepreneurs are still rooted in "bricks and mortar", using the Internet only for e-mail, social networking and news.

E-commerce has been plagued by a lack of online trust and credit in China. Collaboration between industry, academics and government is addressing the challenges. In 2008, I proposed that the Chinese National People's Conference (NPC) introduce "Limited Real Name of Internet" regulations that would improve e-commerce and create a more reasonable Internet environment. In 2010, the minister of industry and information technology, Mr Liyizhong, announced, "China will implement Internet Real Name ..." Like all new legislation, it is not a complete answer to issues of fraud and illegal business dealings, but it is a good start.

For business success in a global economy, access to information and expertise is critical. I had a recent conversation with friends based in Seattle about the changes that have happened over the last decades. On Skype, we were discussing technology and social media and their impact on businesses here in China. I jokingly commented to my friends that we were living proof that Marshall McLuhan's "Global Village" was alive and thriving – some fifty years after he coined the phrase. Obviously, China is poised on the brink of an Internet "Golden Age". Only sixteen years ago, the only outlet for information was the *Xinhua News*, the *People's Daily*, CCTV and China National Radio – all highly regulated sources of information. Today, we have the Internet and access to global news and insights *and* an outlet through which to voice opinions. Sixteen years ago, we could not imagine online shopping, chatting with friends and playing games with people halfway around the world. When McLuhan predicted that the world would become a global village, people said he was crazy. Today, if you do not have access to the Internet, people will think you are crazy.

What are other challenges I see in China today? A different way of thinking, learning and educating is necessary to embrace, drive, invent and develop the next new wave of technologies and business models. Perhaps the biggest roadblock for our young entrepreneurs is China's education system and the "traditional culture" (a tricky term), which lacks elements of individualism and independent spirit. People like me, who have been educated in a Western environment, can easily embrace the concept that creative ideas and technologies are great business resources. Those educated in China's more traditional education system are usually very "functional" or "silo"-focused. Chinese people are very smart and well trained in their functions – look at the winners of major maths competitions – but the education system does not foster the critical and independent thinking that leading international universities provide. "Out of the box thinking" is discouraged and uniformity is highly valued. This leads to functional excellence but lack of creative solution-finding and cross-discipline inventions.

People care about *short-term results* – not how they are achieved. Our manufacturing sectors have benefited from collaborations with global companies and experts. This has not happened in our media or education sectors. To date, there has been limited international influence in these areas. But this is beginning to change. Accountability for results is increasing and the awareness of necessary change is rising. For example, in 2009 China's minister of education was forced to resign. During his six years in the office, China's education system continued to be plagued by academic dishonesty, corruption and arbitrary fees. This is not a system that develops the best and brightest minds. It is as if real talent has been accidental in China up to now – however, the talent base is huge in China, so this feeling is bound to change. The government understands the importance of the young generation and has made corrections to protect this critical asset.

Where do we find our best minds? Let me share my personal experience. One of my own Digital Cowboys is Mr Zhu Bo. Before he became ChinaValue's technical manager, he studied computer science at a local university. He dropped out – like Bill Gates and Pekka – to build his own start-up and eventually made his way to our company. Why did he join? Mr Zhu joined because the leadership of the organization envisioned building a fundamental new business model and technology. And he, as well as many of the other employees in our company, knows he is encouraged to pursue his own career goals in the company.

ChinaValue is similar to LinkedIn in the USA yet quite a bit bigger – we have 600,000 professionals providing content based on their areas of expertise and ChinaValue has become a site where high-profile people can communicate their thoughts and ideas to each other. We are a progressive and evolutionary company but not in the exact Western sense. When you think of the Apples and Googles of the world, you picture a relaxed and undisciplined working environment – the Wild West kind of feel. This is not the only environment that stimulates creativity and innovation. The philosophy of Confucius emphasizes personal morality, correctness of social relationships, justice and sincerity. While I have brought some of what I learned and experienced as a senior manager at Intel into ChinaValue, these principles have influenced my life and in turn influence the culture of our company. This discipline does not destroy creativity. In the same way, a relaxed environment does not guarantee creative results. As Mark Twain once said, "… there is more than one way to skin a cat".

The convergence of Eastern thought and Western technology is forcing change within China's massive economic powerhouse. From the outside, it probably looks more like the "Wild East". But progress is happening in many areas. The arrival of the Digital Cowboy in our workforce and in leadership positions in all areas of society will eventually force changes at all levels of organizations. It promises to be an exciting time. As the motto of one of my alma maters, the HEC Paris business school, states: "The more you know, the more you dare". What will our Digital Cowboys dare?

To learn more about Arkady, Birger and Alex and to continue the conversation, go to **www.nofear-community.com**.

7 Technology – your saviour or your nemesis

Mark Mueller-Eberstein

IN CHAPTER 4, we highlight the importance of "simplifying and amplifying" within an organization, and in Chapter 5 how to transform the organization itself. As a leader, you are responsible for making precise decisions on what to simplify and what to amplify and then helping your experts understand the changes. Simplification focuses on clarifying the organizational structure, vision, basic processes and culture. As we briefly touched on in Chapter 5, technology is one tool to help with this process. The decisions and actions you take with and about technology can make your organization more agile and capable of assimilating new things. Those organizations that are fortunate enough to have the diversity of talent and capabilities in their ranks *and* have a culture, infrastructure and simple and amplifying processes are the ones in which the Digital Cowboys will strive and excel.

Technologies and the technology department can be a key asset or a key blocker to the "NO FEAR organization". In this chapter, we will dive deeper into how technology can be an asset for you and your organization. We also look at how the support functions of the IT department and how this department understands and interacts with other business functions have to change.

In my fifteen years in the IT industry, I have seen how technology supports an organization or truly works against it. One such organization was a large South African services organization.

Sarah Mocke, a senior IT architecture consultant, got pulled into a situation at an organization where "putting out small fires" had turned into a fully fledged three-alarm fire. And how hot the fire was burning became very clear when she met with the IT leaders, who were in complete panic mode. Not because of a "security issue" or a system "being down" – no, they were in a panic because the HR department had made it very clear to the bank's leadership team that the IT infrastructure was getting in the way of the business and creating a massive HR problem.

The bank was experiencing a tremendous brain drain. Over a few months, the actuary workforce – most of them Digital Cowboys – left the bank to join their competitors across the street. During exit interviews, it became crystal clear that the actuaries were leaving because the company did not provide the IT tools they expected and needed to be successful. The most glaring hole was the lack of an internal social networking and knowledge management structure. The lack of a networking infrastructure meant actuaries could not efficiently create and share business knowledge. They felt unproductive, underutilized and disconnected from other experts and knowledge networks.

The bank's leadership team put the blame squarely on their IT. Up to that point, the IT team was completely unaware of the demand or need for "social networking". But even worse, they were also completely unprepared to address the challenge by building the infrastructure and tools necessary to enable the capability in a reasonable turnaround time. The Digital Cowboys took their toys and went to play in the competitor's sandbox.

As we have seen earlier in the book, the demands of the Digital Cowboys of their leaders make us as leaders either valuable contributors to the team or a burden to work around. The same holds true of corporate support functions and tools like an IT infrastructure. They can either be an enabler of the stars in the business or a hindrance. Look at it this way – IT and what is *not* possible in an organization's environment can become quickly a threat to the business itself.

When I wrote my book *Agility* in early 2010,[1] the story of the "South African bank" losing its actuaries was still considered an exception. Not even twelve months later, we hear from HR leaders *and* the people responsible for corporate IT infrastructure how employees of all ages demand that the technologies that they use at work have *to be at least as good as, if not better than* the technology they use in their private lives.

In late 2010, organizations were still having serious discussions about Facebook. These discussions are reminiscent of late-nineties arguments about "Internet access" for front-line employees who might waste valuable company time and surf the Net for news. Many executives – typically the same ones who thought "the Internet and e-mail" were distracting – think Facebook is a diversion that keeps their employees from delivering their "real" work. Only six months later, in 2011, the integration of social networks into business processes and the inclusion of "consumer electronic devices" into the corporate IT infrastructure are topics at every discussion we have with clients.

Information versus intelligence

Almost all current and definitely all potential employees have embraced social networking and mobile technologies outside of work. They have to wonder why they don't have access to the powerful information and networking tools at work. In addition, many expect – and are beginning to demand – that their corporate IT supports their personal electronic devices and that "business tools" are as easy to use as their iPhones.

A young lady had just finished her MBA and was looking for her first job. A bank in Melbourne offered her the absolute "dream job". The job included a very impressive compensation package – a generous salary and bonus, a nice company car, an outstanding healthcare plan, a health club membership and a few other nice tidbits.

Before she would seal the deal, she asked to see her workplace, or at least something close to it. The recruiter and the hiring manager took her to "her" cubicle – a typical office environment with a nice laptop. She switched on the computer and looked at the tools available. She would have access to the latest business, statistical and analytical applications.

But there was *no* communication environment – no access to Facebook, Twitter, Skype, E-zine articles, blogs or any of her other social networking sites. Surprised, she explained to the recruiters that those were the tools that had made her so successful in her studies and previous career. She asked, "Why don't you support them or use them?" The answer was "they could interfere with *our* business applications". So she explained again, those tools were part of her workday, necessary to her success – they made her more productive, creative *and* informed. The recruiters looked at her blankly. Our Digital Cowgirl quickly bid them "good day" and signed up instead at a technology company that provided the mature infrastructure and the tools she needed to do her job.

The recruiters probably saw the tools as an overload of information that distracted their employees. Our Digital Cowgirl saw the same tools as the path to better insights and greater expertise. She didn't see distraction or potential security and compatibility issues, she saw access to information and intelligence, the required knowledge, missing. Information is just the facts. Intelligence is the ability to gather and synthesize the facts into relevant and knowledgeable action.

In Chapter 5, we touched on the importance of *knowledge refinement*. It's about taking information and turning it into intelligence. It's about understanding signals originating from outside the organization and making these signals work for the organization. The young MBA graduate inherently understood the importance of technology as an enabling tool in this process.

Information technology has fundamentally changed the amount and availability of information, as well as access to expertise and institutional or network knowledge. Social and expert networks, near-universal Internet access and business intelligence (BI) solutions are a few of the underlying technologies that fundamentally impact an organization's effectiveness.

Our experts need to gather, synthesize and distribute the most up-to-date and relevant information and connections across company lines. They need access to front-line interactions and cross-functional relationships with the complete ecosystem. The organizations must have the tools that allow them to locate the right people and connect to useful knowledge banks. It's impossible to build clusters and to amplify if you can't connect your experts. The world is not going to become less connected, so understanding and managing these increasingly complex and interconnected systems will make the difference in how effectively and agilely your employees and your organization can manoeuvre.

But the technology is really just "the tip of the iceberg". Technology doesn't work if you do not understand how experts interact and work. Your technology solutions need to provide these knowledge workers with tools that let them work the way they want. One of the most enlightening ways to understand how your cowboys work is to look at their workspaces.

The typical workspace of a Digital Cowboy looks very different when compared to "traditional" offices. First, physical location doesn't really matter any more. Their "workstation" or the desk they use on a given day is only one of the places they work. Their "data" (information, contacts, sources, networks) is available from any of their devices and at (nearly) any place and time. Their main workplaces probably have multiple screens, and while a laptop is the norm, it is often complemented by an iPad and a smartphone with advanced computing and connectivity features. The current technologies the "connected" employee expects today are:

- Digital media access
- E-mail access at any place and any time (messaging functions on social network sites are replacing "traditional" e-mail)
- Instant messaging and texting
- VoIP (e.g. Skype)
- Blogs and a wide variety of information sources and distribution platforms
- Remote access to information wherever the experts are and wherever the information is stored
- Social networking
- Virtual community (interest groups) and infrastructure for clusters
- Mobile access
- Game technologies for entertainment but also learning and collaboration

The use of these technologies is as common among Digital Cowboys in emerging economies as in any other economy in the world.[2] These technologies provide *continuous* access to information and the cowboys' network. Technology is an accelerator and the level of technology capabilities is an indicator of organizational (and personal) agility. If you are going to be a successful producer, you not only need your network of experts, you need effective tools for them and yourself.

The leader's role in all this change

As a leader, you are charged with both creating stability and driving change. In the past, this meant stability and change within the corporate walls. Today, you play this role in a global ecosystem. Your experts and their knowledge are spread across multiple regions. How you connect these teams is crucial to their ability to perform. If your expert clusters are local, your challenges are limited – but so are your opportunities. Gathering the insights and amplifying them across the organization is far easier for local teams than trying to do the same with a widely dispersed expert cluster. The physical location of your clusters or teams is not your only challenge. The nature of how teams form, operate and disperse has changed.

Team membership has become much more volatile. Experts may join together quickly and then disperse just as quickly. Other teams work together for years developing a product. In this situation, experts will probably join and leave the team over the course of the project. You must ensure that the team can connect, build trust and share common goals – easily.

The PlayStation generation automatically understands the technologies that facilitate this type of teamwork. While you can and should provide wise leadership and coaching, you also need to supply technology that supports them in their work. With this in mind, there are three ways in which you can support your Digital Cowboys:

- You need to *enable and embrace* – not *block* – technology access for them *or* for you
- You need to get over your fear that most information must be kept "secure"
- You need to learn, enable, embrace and evangelize the tools your Digital Cowboys choose to use

Enabling the change

You are tasked with enabling – not blocking – access to tools and technology. One of the most efficient strategies to achieve this is to consciously forge alliances between business leaders and IT leaders.

Why is it so hard for many IT leaders, in charge of this critical support function, to have productive discussions with their business counterparts? Why do many business function leaders avoid including their IT counterpart in strategic discussions for their areas? Why do the Digital Cowboys set up their own "shadow IT infrastructure" that fits their needs instead of engaging with their corporate support colleagues? Often because red tape, lack of transparency, unclear expectations and the perceived high costs and pain of those interactions make avoidance such an appealing choice. But it is a choice neither you nor your organization can afford any more. Your improved alliance with IT is a symbiotic relationship – you both benefit from the alliance. More importantly, your Digital Cowboys benefit. You and your front line provide the vision, the resources and access. IT provides the tools, technical, self-services and process innovation. Everyone wins.

The role of IT professionals is changing dramatically. The actual management of the IT infrastructure will be outsourced more and more to be handled by service providers. This means the "in-house" IT department can become a valuable adviser to the other business functions on how IT can improve those functions and their processes. Your organization needs an IT team that is truly a partner for the business and not simply a "technology procurement and maintenance team". For example, IT can partner with HR to discuss remote and flexible work environments. They can partner with the facilities department to discuss space and energy savings. They can help the marketing department to leverage new customer management and the sales teams to impact more customers with more efficient value generation.

Your IT team can even be your first stop when you are trying to figure out if a new venture is "good for business" or an "innovative but completely crazy distraction". They should be able to identify which IT capabilities – remote work, order processing, CRM, advertisement optimization, etc. – drive your business forward while analysing costs – less office space, greater productivity, etc.

As the producer who enables, you need to let the team know that newer technologies can mean that "fast and easy" solutions win over "great and complete" overhauls. You can remind them that "off-the-shelf solutions" can often deliver what is really needed by the front line and are much faster and cheaper to implement. I find that the front line nearly always prefers simplicity and fast availability, while other support functions demand the highly sophisticated customization. If you feel you are in such a situation, ask yourself who is truly driving business forward and decide to prioritize for simplicity and speed. And look outside the traditional highly customizable offerings for enterprises. Leveraging consumer technology solutions and online services instead of building and maintaining your own IT infrastructure is often a superior strategy for agility and achieving your business priorities in a much shorter timeframe.

Getting over the security fear

For the past 30 and more years, leaders have acted like the magistrate of a medieval city when it comes to corporate information and computer infrastructure. A wall as high and as strong as possible surrounds the organization and only a very few are allowed to enter through the well-guarded gates. This sounds exactly like a corporate IT security strategy. But is this model still the right approach for the next decade? New weapons eventually made the city walls obsolete just as new technologies and a network economy are making old ways of operating IT seem antiquated. The city walls blocked trade and stifled growth. Inhabitants eventually tore the walls down when they began lowering the quality of life instead of improving it. You, as a leader, must also be brave and tear the walls down where they inhibit your organization and the positive impact the Digital Cowboys can have.

This doesn't mean throwing caution to the wind.

Instead, you and your IT experts need to take a critical look at the complete picture of your business *and* your IT infrastructure and make informed decisions. IT architecture consultant Jeff Johnson is one of the leading experts on designing and managing the access of devices and users to corporate networks and data. He explains the fundamental decisions an organization has to make when deciding on the security model that works for their business needs.

- Who has access to corporate resources and information based on:
    1. Who you are
    2. Which device you use
    3. Where you are

For example, the head of accounting doesn't need access to the new engine designs coming out of your R&D department. However, your marketing director might provide a different viewpoint and be able to suggest feature changes that R&D hasn't considered. So you give your marketing director access to the information 24/7. But then she leaves her laptop at the office and has to use her teenager's laptop to check on design changes. The kid loves to surf the Web but doesn't really see the need to update his security software. When the laptop tries to access the company intranet, security software on the corporate side identifies a nasty virus and boots the marketing director out.

- What are the classifications of information based on their impact on the business:
    1. Low
    2. Medium
    3. High

What is "high-security information" – really? What is really critical and worth the highest level of protection? What really needs to stay secret – nuclear missile launch codes, the recipe for Coca-Cola, or the new car colours for the coming year? What about a company's financial information or memos concerning international subsidiaries? There are plenty of compliance regulations that need to be considered, certainly. In the era of Wikileaks and financial disclosure obligations, more transparency than you ever imagined will be the norm tomorrow, and what we consider "high security" today will make sense – and be cost effective – only for a very limited amount of information. As organizations, we have to think about other ways of remaining "compliant" and staying ahead of our competition. And a very different approach to information security with higher transparency might be the way to go.

To be clear, we are not advocating giving up on IT or facility security. We are advising that it is important to make informed decisions. Overprotection can come with a high cost in actual spending on technology and it can reduce networking and knowledge creation. By forging alliances across business units and IT, as well as tapping into the power of your expert network, you – as the producer – can drive the discussion on what kind of information needs what level of protection in your organization.

Security is more like a sliding scale and making informed decisions is a key way to attain organizational agility while maintaining a level of security and compliance necessary for the specific organization and its needs. One thing is clear – just building higher walls and hiring more guards is probably not the right approach for the future.

Embracing and enabling your Digital Cowboys

The PlayStation generation is far more familiar with the latest technology tools and has discovered ways to use them to boost productivity, business intelligence and system efficiencies. They work from anywhere at any time and fully expect their company to support this mobility. You or your IT team no longer have the option to dictate device "standards". Employees will use the devices and services of their choice. If a simpler and more powerful offering becomes available, the Digital Cowboys will be the first to adopt it, leverage it and bring it back into the organization.

A true leader needs to embrace the technologies his or her team demands and then adopt the technologies themselves. A team cannot abandon effective ways of collaboration to accommodate a leader who does not embrace their technologies. The time when we as senior managers can look back at our own and our predecessors' experience to determine which tools and processes work best is over. Seventy per cent of the productivity improvements in the last ten years have been driven by IT. And that was mostly on mundane tasks. In the coming years, we will see similar improvements for knowledge creation and innovation – again driven by technology innovations coming from the PlayStation generation.

The good news is this embracing of new technology is getting easier and easier. It is getting simpler to use for individuals. Many executives are discovering the ease of use of the latest smartphones and tablet computers and can't imagine a life without them any more. Just like the younger generation, we simply expect them to work in our corporate environment, too.

The consumerization of IT

Before we go any further in our discussion of technology, we have to fully understand the impact that consumer products have had on corporate IT. The phenomena of bringing mobile devices like a smartphone or iPad to work or searching social media sites for consumer trends are not limited to the PlayStation generation. We have all slowly begun to bring mobile devices and services that were designed for the consumer market to work. However, the PlayStation generation is the first to have grown up with them and cannot imagine a world without them. For them, "on the go" access to the Internet and their social networks is as "normal" for them as wearing a wristwatch or having a telephone is for us older folks. The Digital Cowboys fully expect and demand that all of their toys and tools are going to work just as well at and with the office environment and as they do at home or on the road.

Simplicity, the exponential increase in computing power, reduced device costs, availability of services, expanding infrastructure and new tools all drive this trend. A 2011 Apple iPad commercial ends with the tagline "Technology is not enough. It needs to get out of the way." We don't want "technology"; we want stuff to just work.

We expect that when we turn on a new device, it is ready to go. We don't want complicated operating instructions. If the phone claims it will automatically sync with our other devices, we expect it to just that – automatically. And why would we spend hundreds of dollars on expensive software when we can get an app to do the same thing for a couple of dollars? If my laptop is stolen or lost on a trip, do I want to go through the time-consuming process of ordering a new one from my company's approved vendor and be without tools and access for weeks? Or do I want to go to the nearest local electronics store and just pick up another one the same day and be up and running again?

The iPad and other tablet or slate devices are a great example of how quickly "simplicity" takes hold in the business environment. Consumers are eating up the iPad and its popularity means it is quickly impacting the professional organization as we write this book. Based on analyst estimates, the original iPad, launched in June 2010, sold 15 million units in 2010 alone. Including Android-powered slates, in the second half of the year over eighteen million tablet devices flew off store shelves and many made it into the business environment. In early 2011, we see these tablets in nearly every large business meeting. This is just the tip of a formidable trend. A 2011 Gartner report[3] on iPad states:

Media tablets seem to be everywhere. The iPad brought to life a new model of computing centered around Web browsing, applications and media consumption, which is a smash success. It makes computing practical in many new locations where a laptop or a smartphone just wouldn't cut it ... It has spawned a huge variety of applications designed for the unique environment ...

The iPad, and an anticipated larger wave of media tablets, has captured the imagination of business leaders. Some companies have issued them to business and IT leaders in the spirit of exploration. Others see areas in which they can use media tablets to bring computing into settings that were not practical or were too cumbersome to use traditional approaches. For the consumer, the iPad brought a casual but rich experience onto the living room couch, or the train, or while waiting in line at the bank. In turn, IT organizations are finding new places where tablets can deliver information and media in ways that were not practical, too cumbersome or just too unwieldy. Tablets remove the burden of computing and let the user merely act – and get useful work done.

Will the iPad replace all the other devices? Probably not – at least not right away. There are better solutions or devices than the iPad for a specific capability or use scenario – the iPad doesn't compete with a PC for the creation of complex documents or a TV for at-home entertainment. Instead, it adds opportunities, from watching a movie when on the road to video conferencing when sitting in the park to reading a magazine while reviewing pictures from a photo shoot. However, in conversations with other executives, I continually hear, "I can get fifty to ninety per cent of my work done on the iPad. It doesn't do everything but I'm beginning to think it does enough for me and my organization most of the time."

The "new devices" like the iPad are not "replacing" the PC. But because these devices can also do some of what only PCs could do in the past, the need for PCs (and the frequency of updating them with new hardware) is decreasing – not going away, but definitely decreasing. At this time, there are no "standard devices" that meet all the various current and future needs of all users.

It will probably be a long time before there is such a device, but Victor Orlovski, in his earlier contribution, alludes to the potential of "singularity" through information technology becoming first as powerful as a human brain (in about 2025) and later maybe even a physical part or extension of the human body (around 2045, based on Raymond Kurzweil's predictions[4]). Independent of how integrated people will be with computers in fifteen or twenty years, what we do know for sure is that these changes in the broad consumerization of IT and mobility directly impact and determine the IT capabilities needed in an organization today. The capabilities and functions of the end-user device and their simplicity will exert a great deal of influence over infrastructure development, flexibility and maintenance, and their simplicity as well.

The cellphone changes everything

Recently, the CIO of a large European chemical company was with his peers at a roundtable discussion I was hosting. He bemoaned the fact that his teenage daughter never answers either his e-mails or her cellphone when he calls. For her, it's texting (either phone-to-phone or more often from Facebook message to phone) or nothing. "Doesn't anyone ever call anyone any more?" he moaned. For teens messaging has become the primary communication vehicle, replacing phone or e-mail. And, Simon Silvester would probably point out that the habits of today's teens are guaranteed to shape our business and social future.[5] His group interviewed a large group of teens and, while some of their observations are somewhat light hearted, they do provide insight into the importance of the cellphone as a networking tool:

So, from our observations of teens, it's clear that in the future:

1. Everyone will be socially networked. Because otherwise they'll never get invited to anything.

2. No one will ever lose touch with anyone.

3. Social networks will be valued for the security they offer. You don't want to date a psycho. Psychos don't have many friends on Facebook.

4. Celebrity culture will grow and grow. The Internet allowed people further into celebrities' lives than ever before. The mobile Internet allows them to follow those lives 24/7.

5. It will be normal to let other people monitor where you are through GPS. If your signal switches off, your friends will think you have been kidnapped. And your boy/girlfriend will know you are up to no good.

6. *Linking the cellphone to the cloud will be important. Lose your cell and you lose your life.*

7. *You don't pay for air. Why pay for music and film?*

We have now over seven billion people on our planet and over 5.3 billion mobile phone devices. Compare those numbers with a mere 1.4 billion TVs – a technology that has been around for over sixty years – or the 1 billion PCs – which have been around for 30 years. Smartphones, the mobile phones that can also access the Internet (and the social networks) and provide e-mail and messaging capability, are becoming the norm, indicating that not only teens but most people want to do more than just talk and text with their mobile devices. These devices are more powerful than the PCs of only half a decade ago and far outperform the data centres we used to send people to the moon.

Moore's Law, reflecting and predicting computing power accurately over the last decades, also holds true for those small mobile computers, the smartphones. It is hard to imagine, but the smartphone devices of today are just the beginning. The devices for mobile use will become 1,000 times more powerful in the next ten years alone. People will continue upgrading their handset far more often than other components of their private IT environment. The two-year cycles prescribed by the mobile companies in the USA today will soon be unacceptable for more and more consumers. In Thailand today, middle-class youths are already updating their mobile device every six months.

It is increasingly clear for many that the USA has not been at the forefront of the mobile revolution. The USA is, by most measures, an "undeveloped" mobile (and also Internet access) nation. Japan has been, for a long time, the most "mobile" society; partly because during long commute times in public transport systems the mobile device allows entertainment, connectivity and communication in an otherwise "isolated" environment. "Texting" was popular in Europe and Asia five years earlier than in the USA; even the German chancellor, Gerhard Schroeder, and his secretaries were texting each other while sitting in parliament in the early 2000s. Looking across technology use in different regions, and especially at the PlayStation generation, we can spot early trends and deduce how they might impact not only our own region, but also those where we are going to be active and those from which many of our customers and future leaders will come.

Are there speed bumps in the way of the mobile revolution we can already identify? Is there anything that might slow the revolution down a little? Providing the devices and their users with access to the network and information will be the biggest challenge in the next decade. Mobile network bandwidth is limited and not growing at the same exponential rate as mobile device computing power. How much will this slow us down? On current trajectory, the limited mobile bandwidth will have an impact, especially on the "usefulness" of mobile devices in populated areas. By leveraging both Wi-Fi access and mobile networks, some of this traffic will be absorbed and the slowdown will be less noticeable. It remains to be seen how this potential bottleneck for mobile collaboration will develop. Certainly it might slow the Digital Cowboys down, but not stop them. And let's not forget a quote from Bill Buxton, the Canadian computer scientist and visionary: "It's not the devices that are mobile; it's the people." People, and especially our Digital Cowboys, are like water. They will find a way.

What does consumerization mean for the leader of the Digital Cowboys and the IT department of an organization?

With these new mobile devices and services – which were primarily designed for the consumer market – entering the corporate world, a fundamental shift is happening. While employees are enjoying the simplicity and ease consumer devices bring to their work environment, the IT department faces huge challenges in terms of security and manageability. Leaders find they have to fundamentally rethink these IT mainstays. IT also faces a tremendous increase in expectations of simplicity and the service levels they are supposed to provide. Corporate standards for PCs and phones are out, and all content and data must be available on the Web and in web formats that can run as standardized "apps". On the other hand, IT can also benefit from the lower costs and flexibility of these devices and the potential standardization and certainly simplification of how data and services can be provided.

For example, more and more companies are implementing a "Bring Your Own Computer" policy. Instead of fighting a decidedly uphill battle against corporate-issued PCs and new consumer devices, organizations provide employees with a "digital allowance" and computer, tablet or smartphone specifications. Individuals choose their device – some choose high-powered options while others go for versions that suit the type of work they perform. The analytics department and the PR department have very different needs. From your expert's perspective, they choose the tools that best suit their needs and allow them to collaborate in their clusters. Since they are not tied to corporate procurement, they can easily change apps and hardware when newer and better devices become available. Consumer devices are nearly always less expensive than devices and tools "designed" for corporate customers. Many workers are fine with a $700 desktop or iPad and don't need the standardized and guaranteed hardware configuration of a $2,000 laptop any more. It doesn't matter whether the device lasts for three years or not – newer, better and faster devices come to market quickly and can replace the existing device easily as more and more data and functionality reside on the Web or in the cloud, and less configuration, application installation and data transfer is necessary. The bottom line is that your experts probably get "more for less". And IT doesn't need to oversee who gets what (or procure it). IT still needs to make sure the corporate data and infrastructure assets that have been deemed "of high business value" are kept secure – but without building new "city walls" that get in the way of simplicity, connection and collaboration.

▶ Simplify, amplify and the IT infrastructure

As we stated at the beginning of this book – fear is the greatest contributing factor keeping us from growing as business leaders and empowering our experts. Information technology has often been mystified and left to the "geeks". The lack of knowledge and the mystification have created a fear of technology discussion for many business function leaders. As we outlined earlier, it is high time to overcome that fear of technology and dive into IT infrastructure for a while. Trust me; it is not as hard as many of you might fear. Before we start, it helps to have some of the basic terminology surrounding IT infrastructure and how to "measure" (and compare) the level of advancement or maturity of an organization's IT capabilities. Researchers at Gartner, MIT and Boston University pioneered a model that showed how organizations and IT mature, which I have used as the models for research and discussions with hundreds (or by now maybe thousands) of leaders from IT and business. The key is that the more mature the IT infrastructure supporting an organization is, the more agile that organization can be and often is. The simplified description of IT maturity looks something like this:

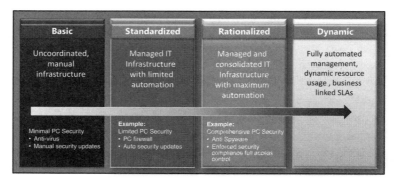

Figure 6 IT infrastructure model

The levels of maturity are Basic, Standardized, Rationalized and Dynamic. Obviously the goal of a company is to move quickly out of IT survival mode and work their way to a point where IT is considered a valuable business partner within a company. For example, security improves from highly vulnerable in a Basic infrastructure to highly proactive in a more mature infrastructure. As a company moves away from Basic IT infrastructure to a more dynamic and rationalized infrastructure, business benefits *increase* and IT costs *decrease*. IT rapidly becomes a key strategic asset to support and *drive innovation*, profitability and customer satisfaction. The model is most valuable when utilized not against the "IT infrastructure" as one unit of observation, but broken down into core IT capabilities (such as messaging, analytics, collaboration, mobile access, etc.) and when those capabilities are evaluated and prioritized against the business needs of the organization. While it is not necessary for an organization to be "Dynamic" in every area, it is critical to have the maturity to manage and embrace change efficiently and securely to meet the requirements of the PlayStation generation.

The high cost of "spaghetti architecture"

As Pekka mentioned earlier, many organizations have an IT infrastructure that has grown over decades and improved through add-ons and customization. We recommend you radically question that kind of IT infrastructure. It is important to understand the cost and downsides of the "organically grown" IT infrastructure many organizations still have today. When we try to visualize their IT infrastructure, many of the diagrams look like a bowl of spaghetti. And while Italian pasta is very tasty, in the IT environment a complex "spaghetti architecture" comes with significant downsides and costs:

- It is hard to manage and to keep secure
- It is fragile and often unreliable
- Any change requires extensive testing
- Reusability of a developed functionality is rarely possible

- There is a lack of Web or mobile accessibility of the business data
- There is a lack of extensibility for leveraging business or technology opportunities

No wonder many IT organizations spend over 80 per cent of their budget and time just on maintaining that existing and inflexible infrastructure. It is the exact opposite of simple and agile both for the IT team itself and also for the business they are supposed to support and enable.

Clearly, it is often complicated and costly to rebuild a grown and complex IT infrastructure and supporting business applications from scratch. However, many organizations find a way to move their infrastructure and their end-user experience forward, improving the services they offer to businesses at lower cost. One area of optimization is the move away from a silo approach and end-to-end proprietary customized IT functionality – such as applications specifically for HR, sales, inventory management and finance. Instead, companies are moving to a model that provides data, independent of the "system" in which it is stored, to users who actually need it for knowledge generation via mobile and "off-the-shelf" applications. Instead of providing access to a specific business application through a dedicated interface – which often needs to be individually programmed and maintained – these organizations create a layer of software that collects the data from the various individual applications and provides this data to standard interface applications like a web browser or Excel. Then, for the IT team, it doesn't really matter whether the CFO looks at the latest accounts receivable information through an Excel spreadsheet on her computer or an application running on her iPhone.

Information and software capabilities appear to the users in a way that is simple to understand and use *without* the users needing to know how it all works. This simplified approach to providing data to users is called "abstraction". The graphic in Figure 7 (which you can use or "whiteboard" when talking with your IT colleagues)

Technology – your saviour or your nemesis

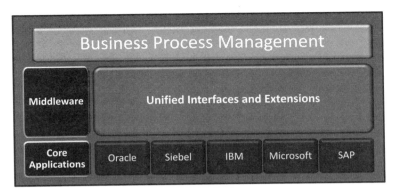

Figure 7 Abstraction layers in an IT infrastructure

shows how these layers interact. The middleware is responsible for much of the "abstraction" experience.

IT organizations changing their application architecture through the systematic use of such an interface layer or "middleware" achieve tremendous gains in their flexibility in terms of answering business requirements, and you will hopefully find that many of your CIOs are already thinking about the concept and its benefits. By supporting them in that endeavour of simplification, and explaining to other business function leaders why simplification and giving up on some familiar but very costly customization for a niche of users is the way to go, you can help make the changes happen faster. By abstracting the business applications with a unified middleware layer, a lot of the complexity can be removed from the system. After such a transition to a modern middleware layer, simple configuration changes can be made in a move away from very complex and manual custom coding.

IT infrastructure and "abstraction" may not be your cup of tea but it is important for the organization to understand the impact these areas have on your efforts to simplify and amplify.

Oracle has published the outcomes of some of their "abstraction projects".[6] British Telecom was able to retire 46 legacy applications that had to be maintained prior to their abstraction push. Oracle quotes Norm Fjeldheim, the CIO at Qualcomm, saying, "We are always looking for ways to simplify what we do and manage"; he estimates that the abstraction clean-up saved them $5 million that otherwise would have been spent on "custom development" to provide a secure knowledge management system to their employees. ING was able to reduce manual IT administration by 75 per cent and calculated annual cost reductions of 3.7 million euros. And Eaton Steel was able to reduce their invoice resolution time by 28 days, which had a considerable positive impact on their cash-flow situation. Making IT a business asset and unlocking the potential often hidden in "inaccessible" databases is driving tremendous business upside. Top-performing organizations are almost three times more likely to use analytics in their business activities than lower performers.[7]

Key for the business leader is to understand that an outdated IT infrastructure not only gets in the way of providing the Digital Cowboys with the infrastructure they need, but it has a direct and costly impact on the broad organization.

Simplification versus customization

Making data available for a variety of end-users' devices and applications to leverage creates another challenge or opportunity. How do you give the users the "best possible experience"? Traditionally, we standardized on a very limited number of applications that could retrieve the data from back-end systems. We then developed a custom-coded application that gave exact functionality to our users. We – IT and management – were in full control of making changes and determining information access. This approach is no longer viable in our world of different devices with a myriad of potential applications. Why write a custom application for the CFO and his team for their smartphones? Isn't there "an app for that"? As mobile devices become a standard for information access, IT teams and leaders need to fully think through the ramifications. Do they create custom applications or look for simpler solutions?

If an organization chooses customization, the IT department would need to create a separate application for each computing platform they support or which their users demand.[8] The cost of development of a mobile app is usually between $30,000 and $130,000 per app and per platform. Analysts predict that by 2015 customization will include platforms for the iPad, Android, Chrome, Windows, Windows Phone and BlackBerry. Of course, this list includes only current platforms for 2011. Another often overlooked challenge is the distribution of the "app" to the intended users. While software distribution in a corporate IT environment on to Windows PCs is easy, no "private" distributor of apps on to mobile devices exists today. "Apps" have to be installed from the smartphone manufacturers' software distribution centre, the "appstores", iTunes store or the "marketplace", and therefore are available to every user. Not just the people in "my organization" or "my CFO and direct reports".

There is actually an upside to this situation for the organization and the IT team (excluding the custom code developers). Because the customization is not feasible, most users will install the apps provided by traditional providers, consulting organizations and new organizations. For example, SAP and Oracle offer apps that provide BI capabilities. And because these "off-the-shelf" (or better, "off-the-appstore") applications are requiring standardized data formats and inputs, the only service the IT department is going to provide to the users is simply to furnish the corporate data in the appropriate format. And development and support of the apps (and the organization's end-users) will be provided by the third parties who developed the apps. Life for the user and for IT becomes actually simpler that it is today.

The advantages of cloud computing

Cloud computing is becoming mainstream and is making a tremendous difference to how our organization manages IT capabilities. Over the last decade, consumers have learned to rely on much of the IT capabilities they use being available "on the Net" or "in the cloud". Like consumers, organizations can use cloud computing services – both hosting and delivery models – that are offered by a variety of vendors for their IT and business needs. The people entering the workforce have grown up with the understanding that many of the tools and services they use daily are already "in the cloud". They wonder why a proprietary implementation or "build-it-yourself" strategy is a first choice or even considered when they can get a needed capability "now" via the cloud.

People, especially from the PlayStation generation, expect to be able to access and interact with all their digital "stuff" (music, financial records, personal and work e-mail/calendars, social networks and contacts, personal photos, medical records, games) no matter where they are. So more and more of the digital data and the applications that make the data "useful" is stored and managed in the computing cloud. And what holds true for consumers is becoming the norm for companies, as well. For the organization, "cloud computing" might mean that their data is in a remote data centre or "public cloud". The data and application could be stored and managed in an environment provided by another vendor, called a "private cloud". The end-user wouldn't know the difference; but IT and the business organization could take advantage of the additional level of control and lower costs.

The computer itself – specifically, the PC – isn't a destination any more. As fewer people use only one computer or device to manage their life and their work, the PC becomes just one of many ways of accessing your digital "stuff". Amazon's "music cloud" illustrates this idea perfectly. When you store your music in their cloud, it is available wherever you go and whichever device you choose.

The offerings and technologies are available today and more are coming to market all the time. At a deeper level, the challenge leaders thinking about IT face isn't "the computing cloud" as a technology, to reduce costs or as a way to acquire IT capabilities; rather, it's a very broad change in people's expectations of the timeliness of capability availability, the simplicity of acquisition and use and the service level provided to the end-users. The benchmark for the performance of your own IT infrastructure for your people is the simple and easy consumer experience they are living with every day.

Learning from previous technology cycles

If you are a baby boomer, you remember rotary-dial phones and electric typewriters. If you are a Gen X, you grew up on MTV and fell in love with your first PC. If you are a Millennial or Gen Y and born after 1985, you are probably a true digital native – connected and wired wherever you go – and others are just digital immigrants trying to keep up with the latest technology evolution.

Many people leading organizations and deciding which technologies to deploy grew up in a very different world to that of the PlayStation generation. They have experienced several new technologies coming into the workplace but often haven't embraced all of them fully. IT people are often systematic and critical thinkers, not necessarily embracing every fashion or trend just because it is "new" or "cool".

For many IT experts, there must be clear and well-communicated reasons to make the changes we are now requiring both in the way they deliver the services and in the way we want them to engage with the front line on business priorities. It helps if they have their own lived experiences of the "new world". Also, leaders must support them as they develop and implement a well-thought-out roadmap of the changes. Sure, there is fear of change, but often there is even more denial that changes are really necessary. They may lack a clear understanding of the business reasons or fail to see the dramatic shift in expectations and ecosystems. More often than we would like to see, some lack the profound and necessary professional knowledge and skills to enable them to engage, prioritize and execute. While none of obstacles is "new" in human history, changes in work style, organizational collaboration and technology were far slower in the past. People often retired before they really needed to embrace the "new" for themselves or their organization. As long as you are not planning on retiring all your "slow movers" in the next few years, this gradual evolution is not an option any more.

Of course, there have been previous technologies that revolutionized the workplace but not at the pace we are experiencing now, a pace that has taken on a completely different dimension and requires multidimensional adaptation.

Technology changes faster than any of us ever believed it could. It took about a hundred years for the mail system to go from initial commercial application to a ubiquitous, worldwide business communication tool. It took the phone seventy or so years to become a mainstay in many households. The fax machine worked its way into every business office in about twenty years. It took only fifteen years for the mobile phone to become a universal business tool, and it is hard to remember that previously you needed to carry coins with you in case you wanted to use a phone booth while outside the home or the office. It was only in 1995 that Bill Gates wrote his famous e-mail on the "Internet tidal wave". And what a wave it was; it fundamentally changed our societies and our economies in only fifteen years. E-mail breezed in in under five years and instant messaging took only three years to become a mainstay. Social networking? Over six hundred million users joined Facebook in a two-year period with a growth rate of 3–5 per cent every month. We don't know what the next communication innovation will be and who is going to offer it first, but we do know that it will probably be here in about eighteen months.

Future trends

Predicting the future in a world of exponential growth in the multitude of networking effects is a pretty hard business. While we do not know what will be next, we do know that some trends and developments are fairly predictable:

Computers will be tremendously more powerful. As we outlined in the section on mobile devices, Moore's Law still holds true for them and other computer-chip-powered devices. His theory describes a long-term trend in computing hardware and basically states that the quantity of circuits that can be placed inexpensively on an integrated circuit will double approximately every two years. Intel co-founder Gordon E. Moore described the trend in a 1965 research paper. He predicted that the trend would last about ten years. Almost fifty years later, the trend is still going strong and is not expected to decline until 2015 or 2020, or even later. The capabilities of digital electronic devices are strongly linked to this law. Processing speed and memory capacity are improving at (roughly) exponential rates and are not slowing down. In *Mobile Mania*, Simon Silvester predicts that mobile devices will be 1,000 times more powerful by 2020,[9] compared to those entering the market in 2010.

Wireless access to information and people any time and anywhere continues to expand. Owing to today's limits on available mobile bandwidth, data access is not necessarily unlimited. However, everyone will want to have HD video for conferencing on mobile devices at some point in time, and we will need to have the technology in place to support this trend.

The Internet itself is transforming. Facebook nicely summarizes this trend: "The web is shifting from a vast encyclopedia of information to a social environment that reflects our real identities and the relationships and information we care about."[10] It is too early to tell for sure, yet, but there is a high likelihood that for lower- or medium-security concerns, identification by association (through an individual's friends and network) might become a formalized option in the online world.

Consumerization of IT, driven by simplification, lower cost and cloud computing offerings, will continue to raise the bar for corporate IT services. Devices and services will transform users' expectations and drive down delivery costs. For most IT capabilities that had to be implemented "on premise" in the past, there will be a cloud-service alternative to those "on premise"-created capabilities. Mobile devices will access those cloud-based capabilities as do "traditional PCs" today.

And then there is the **network effect**. As more people adopt a technology, the value for all participants goes up while the incremental costs for the individual, organizations and the providers of the technology go down. As the cost and other barriers to entry diminish, further and broader adoption increases faster and faster, amplifying the impact.

While the technology continues to evolve rapidly, your existing processes, culture, infrastructure and past investments can, and often will, get in the way of adopting the new technologies, slowing down the effort to simplify and to amplify the change you as a leader want to drive.

Managing technology change

The lack of a technology legacy actually makes new organizations – or even societies – much more adaptive to embracing the new. Underground copper cables were cutting-edge until fibre optics took over. Many developing countries, especially in Africa or other previously remote areas, didn't have landlines but that didn't hold them back. They simply went wireless.

So what do we do – those of us who have a legacy, existing infrastructure and cultures? How do we deliver a level of services that might not be quite perfect or ready to our business partner today and still keep the current business running?

While the speed of the technical transformation and broad adoption has changed, we can still learn a lot from the previous cycles. We can especially learn how to manage the change and the implementation itself. We just have to ensure we do it a little faster. There is additional good news. As IT becomes simpler to manage and IT capabilities simpler to acquire through a variety of options, the quicker change is possible.

Some of the key learning from previous technology cycles that comes to mind:

- The PC revolution started both in companies and in homes around the same time. The early adopters within the employee base brought the experience they gained at home into the workplace. Education offerings at work, which helped them master the new tools, were leveraged both in the work and the home environment. The expertise derived from practising at home in turn made them experts in their social environment. These people often became the change agents who then pulled the majority along with them. The synergies between private and business use of IT are even stronger today than just a few years ago. People – not only the PlayStation generation – bring their expertise of the social networking environment to work and can use that gained expertise in clusters and cross-functional teams.

- Internet usage, especially as a research and information acquisition tool, came of age when many households had a PC and AOL made it very easy to join the Internet community. Using "search" engines to find and access information was a skill that had to be "learned". Now, explicit training in using the computer and "the Net" as a research tool in the corporate environment isn't even necessary any more.

The trend is the same with mobile technologies. Most people learn to use the new technologies and services at home and bring that "expertise" back into the work environment. The simplification of consumer devices has made the shift happen at lightning speed. In the workplace, members of the younger generation are usually the earliest adopters. When encouraged, they can and will share this knowledge with their colleagues, mentor the "willing majority" and act as the change agents we need in the organization to amplify our strategies. As new technologies, services and processes become available to consumers, you and your organization need to foster, encourage and support the valuable change agents in your organization.

While users enjoy the simplicity of newer technologies and the value they add to daily life, these bring specific challenges for the IT team and the business itself.

- Connectivity is expected anywhere and at any time.
- The PlayStation generation expects that their consumer tools from home will work at the office.
- Consumer devices and experiences are simple and IT needs to provide the same simple experience with their tools and experiences.
- New capabilities are available nearly immediately on consumer devices; employees expect the same speed from your organization's IT infrastructure.
- The time of "eighteen-month IT projects" is over. Even if a project is designed to take that long, the spec will have changed three or more times ... completely.
- Cloud computing is becoming a de facto standard option for consumers and they want to see this technology at work.

Again, technology is often simply an indicator, enabler and catalyst. For the PlayStation generation, technology changes go far deeper – affecting their quality of life. Their work life has a very different level of importance to their self-image and life priorities compared to the baby boomers. Technology needs to support their desire to "work to live ... not live to work".

For example, what happens when an employee's child gets sick? Many countries have sick-leave policies that allow employees to take care of their dependants. Traditionally, the employees called in sick, stayed home and took the full day off work. But in a work environment where physical presence in a certain location is optional, a sick child probably means moving a few phone conferences to accommodate a doctor's appointment. The parent can still be a productive member of the team by working remotely for most of the day. Employees are happy that they do not have to compromise on their duties, and the business impact can be staggering. A study from the University of Rotterdam documents that organizations with flexible work hours and remote work capabilities reduce sick-leave time by 2–20 per cent.

The same study shows that for organizations that have processes, policies, management examples and IT infrastructure to support flexible work models, facilities costs drop by 15–50 per cent and travel expenses by 5–30 per cent. Of course, simply providing technology does not guarantee that employees will adapt to the changes easily. Leaders needs to communicate *and* model acceptance themselves. For example, people often complain about video conferences and virtual meetings not being effective, especially for people who call in remotely into a conference room full of people who are deeply engaged in their discussion. The leadership at Shell instituted a simple and very effective solution. They set the policy that if one person is "remote" for a meeting, all participants have to "dial in" individually – even if they are at the same location. This quickly ensured that documents were shared in advance, people were on time and everybody had a chance to contribute.

The next technology cycle and new leadership

We looked earlier at the expectations of the PlayStation generation, and especially the Digital Cowboys, of their work environment. By now, it should be clear that the next-generation IT, services and process infrastructure needs to evolve to support much richer and more mobile environments. The device that an individual uses will become the "personal knowledge access portal" that presents what is relevant to the individual in a device-specific way and is fully customizable based on the needs and preferences of that particular individual and his or her needs, role and tasks. It can be a webpage, an app, "tiles", a vendor-specific "mobile interface" – but we might call it something completely different altogether. Whatever we will call this "knowledge presentation interface", it will do and show exactly what I want. It will be easy and simple to customize, to use and to adapt – for me – and fully integrated to support my personal and business knowledge and data management, as well as my collaboration and communication needs.

For security or compliance reasons, the information and data available to me through "my personal portal" while inside the corporate environment might be different to when I am travelling to a foreign country. But transitioning from one environment to the other will be seamless, and the expectation of the end-user is that the corporate infrastructure works as well as the rest of the Internet and cloud infrastructure the individual is using.

Key characteristics of the next-generation infrastructure for organizations will be:

- Enabling the next-generation portal framework and presentation applications
- Allowing full user personalization of their environment, including devices and data sources
- Providing a business dictionary
- Showing activity graphs
- Supporting and fostering enterprise mash-ups

A "mash-up" is a technique for building applications that combine data from multiple sources to create an integrated experience. Many mash-ups available today are hosted as sites on the Internet, providing visual representations of publicly available data.[11] Enterprise mash-ups are lightweight applications that combine data from two or more sources to create something more valuable than the sum of their parts. They are often developed to solve a particular problem, implemented in days rather than months, and use standards like POX (plain old XML), Atom and RSS to make sharing and subscribing to them easy. They often combine internal and external sources of data.[12]

Laptop, PC, tablet or phone?

What will the devices our Digital Cowboys use actually look like? For a while – maybe for a long, long time – there will be a multitude of devices for the end-user, who will probably own several at once. Certainly, the time of the desktop PC and the mobile phone that only was a phone is over for most of us.

For the end-user, squeezing more information and capabilities on to a device means usability becomes a huge factor. Is the screen big enough to really read? Is there too much glare when I go outside? Does my tablet weigh too much? Does my battery life support my travel schedule? How many charger cords do I *really* need?

Will knowledge consumption or creation be the deciding factor in what to choose? It is unlikely that devices will only be used to consume information. Probably some form of input will continue to be necessary. From note-taking to quick communication via e-mail to Facebook updates or Twitter, input will have to be simple, fast and easy. Observing the Digital Cowboys of today, we know that the majority of our future communications will be short and timely. Also, many people using the new tablets such as the iPad for their work feel the devices are already sufficient for much or most of their work. For some, there will still be a need for "specialized" devices or extensions for these otherwise optimized devices. Authors of long texts will still want a keyboard or a mouse and designers of any kind want a large external screen to fully see creations.

For the IT strategist and architect, this trend towards multiple and mobile devices means that information must be available on and through the Web. And it needs to be in web formats that can be consumed by a multitude of devices and applications without the need to develop "custom interfaces". Also, because these devices can now do some of what only PCs could do in the past, the need for PCs is decreasing. Not going away, but decreasing.

What do the trends mean to the business leader?

As the business leader of an organization, what do these key trends in mobility, simplicity and IT availability mean to you? How do you ensure that your IT team is providing the services and the infrastructure that are empowering your Digital Cowboys and your ecosystem? Does your IT help you to simplify the organization and amplify changes? How do you ensure that your IT infrastructure becomes a key driver of the organization's success?

First, think from the position of an end-user. You are one of the most valuable end-users in your organization and you should raise the highest bar on your expectations for yourself and for business flexibility delivered by your IT team. If you can see a clear business case in any business function – from facilities to HR to your core business – that could be improved by IT, it nearly always can be done. And it can be done much faster and cheaper than a few years ago, especially if you are also flexible in accepting a simple "80 per cent solution" instead of demanding a "110 per cent" implementation.

But you also have the responsibility to articulate the business case and the business need to your IT colleagues. You owe it to them to support the changes and simplification, often against other stakeholders in your organization. The methods that work for the development of your front-line experts will also help you in identifying and delivering the IT projects that benefit your organization the most. You must build and empower expert clusters that include the front-line and IT services people if you want to accomplish amazing innovations in your services and processes.

You personally need to lead the charge of simplifying and updating your IT infrastructure. You need to over-communicate – in all directions and at all levels – the goals the organization is aiming at and the roadmap it is following. You have to be up front and transparent in your communication that some of the changes will require the organization to say "goodbye" to familiar niche solutions. But you also have to communicate that the resulting new way of providing the service to the business is the way to keep the organization successful. Let me walk you through the process and some key questions that will help you to make the most of your IT discussions with the IT experts and experts from other business functions.[13]

As we outlined earlier, many people, especially in the IT profession, need to understand the rational and the business case before embracing and supporting a significant change. As the leader, it is your responsibility to understand and to communicate the answers to their questions. Before diving into in-depth analysis with your CIO, it can be helpful to do a very informal "back of the envelope" calculation and leverage available tools for an initial "gut-check". This way, you can at least have the right numbers in place to help focus the discussion before you begin to engage the broader teams. Also, schedule time with the other leaders in the organization for the initial assessment and get them and your CIO at least philosophically on board.

There is one key question that you need to answer quickly. Is your CIO – and his or her organization – embracing the ongoing changes and is he/she getting your organization ready to use IT to drive the organization's mandate to simplify and amplify? If not, it is probably time to make significant changes. But if you have a true partner in IT, these discussions will help your organization unleash a tremendous potential for acceleration. And don't forget, this is a two-way street. Driving the fundamental change in the IT services requires your ongoing involvement, your availability for discussion, questions and prioritization and your support for the new IT strategy and their leaders.

Once you have answered the first question, you are ready for a broader analysis of your organization. Very few companies have all of the data readily available in detail but they often have a pretty good idea of the ballpark numbers. Here are a few of the questions you should be asking before you plunge into real planning:

Productivity

- Does my company calculate productivity of knowledge workers?
  - If so, how? And what is the amount per year and employees?
  - If not, can I use an industry-comparable number for productivity?
- How do we measure innovation?
- What is the age range of my workforce?
  - As older workers retire, many companies see an increase in productivity as younger workers are more willing and able to embrace new productivity tools.
  - And how do I manage and value the existence and also the exit of long-term expertise and knowledge?
- If neither of these numbers is available, what are my fully burdened costs per knowledge worker?

Sick-leave cost

- Per employee, per year, or total for the organization.
  - Sick leave reflects the corporate culture. Bad morale and bad management lead to high sick-leave percentages.
  - Companies that support work anywhere at any time typically see lower sick-leave rates.

Attraction and attrition

- Recruiting costs; including headhunters, relocation, sign-on grants, and so forth.
- Rate of attrition. Bad company morale definitely impacts attrition rates of the people we would like to keep.
- What are the expansion plans in emerging countries with an even greater need to recruit and develop leaders from the PlayStation generation?

Total cost of ownership of workplace

- This is the total of direct capital investment in hardware and software plus indirect costs of installation, training, repairs, downtime, technical support, and upgrading.[14]

Communication and collaboration

- Where do your employees work – from a desk, in the hall, in meeting rooms, on-site with clients, from the airport, from home? What technology does your organization provide today?
- Which technologies and capabilities do your employees already use that are not provided by "your IT"?
- If employees are spread across geographic areas, how do they communicate and conference – in person, over the phone, or not at all? Are they using the Web, video and audio conferencing capabilities?

These are the "technical" questions that start your conversation. From a business leadership stance, there are additional areas to tackle. These are the top business questions to ask your CIO – or, if you are the CIO, be ready to answer them:

On business

- How are you partnering with your business colleagues to help meet their and my key priorities?
- How can I help you to think like a CEO so you can focus on the organization's top priorities?
- Your first 60 days on the job – what do you need to do for the organization?
- How are you attracting and keeping superior talent both in IT functions *and* in the broader organization?
- How can we use technology to enrich our customer experience?
- Are my people engaged in creating services innovation with our front-line employees?

On IT

- How much of our IT budget is allocated to maintenance of the current environment and how much for new capability delivery?
- What are your key IT metrics (admin per server, uptime, etc.) and who do you benchmark against? Are you benchmarking against the industry leader or, better yet, against "capability leaders?" Are you benchmarking against global standards?
- What is your plan to make data from one functional silo (e.g. sales data) available to other business functions (e.g. HR) to improve business intelligence?
- How do you improve simplicity and self-service for all users – development, finance, management, legal, etc. – of our IT infrastructure?

Rethinking how things get done

After the deep dive into technology, we will look at how new technology capabilities have changed business itself, and specifically at what the Digital Cowboys can achieve when leveraging the new technologies. We pointed out earlier in the book that one area of radical improvement is the way people collaborate and self-organize within an organization across organizational, geographical or even technological boundaries. We have discussed the new type of leadership necessary for the future within an organization. But technology has changed the role of the leader in collaboration scenarios even further.

When Jimmy Guterman asks "Do we need leaders?" he raises a very valid question as to what kind of leadership is necessary in our connected world.[15] His *Harvard Business Review* blog concerns the rapid changes in Egypt which – unlike other recent instances of social unrest – seemingly lack a clear leader. He asks where Egypt's Nelson Mandela or Aung San Suu Kyi is. The resounding response to the blog was "Yes, we need leaders", but not so much to lead. New leaders become leaders because they are able to influence through contribution, network and knowledge – which takes us right back to the leader as a producer. Technology doesn't change the role and importance of influence but it certainly increases the number of people a leader can reach and reduces the time it takes to reach them.

The new technologies and the way the Digital Cowboys collaborate to change societies and their personal environment will also have a fundamental impact on the way organizations work in general. For example, let's look at "collaborative product development". In the past, an organization pulled teams together and gave the team top-down direction. Today, technologies support a self-organization model that works surprisingly well – not only within an existing organization, but beyond organizational, national or time borders. The open-source projects (such as Linux), knowledge aggregation through Wikipedia, knowledge sharing through TED, awareness and fund-raising campaigns on YouTube or expertise sharing and education on LinkedIn discussion groups are examples. We had not even imagined these possibilities at the end of the last millennium. And we didn't comprehend that the primary motivation of the PlayStation generation for contributing is not about getting a pay cheque, but to do and to be part of something "cool". The use of technology and the shift in motivation therefore raise questions about the type of organizational structure and leadership necessary in the new world of collaboration.

Collaboration technologies, as well as a completely different attitude to collaboration for a common goal, are already changing fundamentally how "things" are created. Who would have thought that a bunch of computer science enthusiasts – supported by a few corporations – would be capable of developing and marketing an operating system (Linux) that obliterated a complete business sector surrounding Unix? Dominating products such as HP-UX, IBM's AIX, Solaris or Siemens' SINIX are "non-existent" as market forces only ten years later.

The *Encyclopaedia Britannica*, written and edited carefully by experts, was the "gold standard" of knowledge documentation. Then Wikipedia came along and outperformed the *Encyclopaedia Britannica* on nearly all objective measurements – including number of references, topics covered and "being correct". How did this happen? How could a loosely and unstructured group of people outperform an organization and a scientific approach that has been driven to perfection for over a century? And how could that happen in only a few years?

The way Wikipedia has been "developed" into the largest repository of answers is a fantastic example of how knowledge is being created and leveraged in the new world, mostly by Digital Cowboys. Drake Bennett, in *Bloomsberg Business Week*, reflected back on Wikipedia's first ten years:[16]

Then in 1999 came a simple idea from a search engine entrepreneur named Jimmy Wales: a free encyclopedia written by experts donating their time to draft articles and conduct peer review. The content would be copyable and ad-supported. It would be called Nupedia.

The site's team of PhDs moved at a pace that was indeed scholarly: In its first year, they created barely a dozen entries. Wales realized Nupedia needed a feeder site, something less exclusive that would attract writers. On this new site, any visitor would be able to write about whatever he wanted.

The experiment was unpopular with Nupedia's editors and advisory board – the fact that the uncredentialed were encouraged to contribute could, they reasoned, threaten the credibility of the encyclopedia. After less than a week, the new site was spun off into a separate project called Wikipedia – a "wiki" (Hawaiian for quick) is software that allows users to write, edit, and link Web pages on a single shared document. Within a year the new site had 20,000 articles. When Nupedia shut down in early 2003, it was stuck at 24 articles.

Wikipedia is an example of a massive social networking group contributing voluntarily and creating a repository of peer sharing and knowledge. Imagine the impact on your business if you could support the same type of community for your organization and your innovations. Or imagine the impact if you leverage the community to validate or even develop your next product offering?

Product development teams can use an internal knowledge base *and* external intelligence. Ask an executive from the automotive industry what it takes to develop a car these days and you will find an answer that is based on the continuously improved experience of over one hundred years of corporate design and benchmarking against the traditional competitors. The answer you hear is probably similar to what an executive or the chief editor of the *Encyclopaedia Britannica* would have said and believed in 2001. But as we now know, the real competition might be somewhere else – somewhere unexpected.

Joe Justice, the "Team Lead" of Wikispeed, founded the company to change the way both manufacturers and consumers think about cars.[17] Wikispeed has a distributed development and production environment with facilities in Seattle, Washington, and in Hillsdale, Michigan. Their CAD, Web, PR and legal help – completely volunteer at this point – are located in Baltimore, San Francisco, Cincinnati and Detroit, with other contributors located around the world. The team's first product, the SGT01, placed tenth out of hundreds of competitors in the 2010 Progressive Insurance Automotive X Prize. The company attended the 2011 Detroit Auto Show and is taking and receiving orders. While they are still in their infancy, it will be interesting to see whether the business model is going to work and whether the product will succeed in the market. Nevertheless, the team has demonstrated that the style of collaboration combined with the right tools and processes could create a complex product – a result "unthinkable" only five years earlier.

Beyond product development, organizations are tapping into this type of knowledge base and social network on an organizational level. Voices.com is one of the largest agencies for "voice talent" – the voices you hear in commercials, documentaries, etc. It is a nice example how tapping the network of social media can transform a complete aspect of the business. Often their clients need a "local" talent for a recording to keep production costs down to the minimum. In the past, Voices.com used a costly broad awareness programme to recruit talent on a global scale. Areas strong in acting talent had lots of choices, but these areas weren't necessarily where their paying clients were located. By switching nearly all their recruitment funds into Facebook advertisement, the CEO, David Ciccarelli, explained they reduced the time it takes to respond to customer requests and can now allow their clients to search a database of more than sixty thousand voices representing over one hundred languages.

How does recruitment via Facebook work for them? When they have job openings, they run ads on Facebook targeted only at people who fall into these categories based on their Facebook profiles. For a specific client's job, only people complying with the criteria below would see the advertisement:

- They live in Canada
- They live in London, ON
- They are exactly between the ages of 24 and 40
- They speak English (UK or US)

Facebook then displays these ads to approximately thirty thousand people daily. While there are only maybe twenty clicks per day, the costs to get the "right people" are minuscule. Targeted advertisement saves the organization from the bombardment of résumés that come through traditional recruitment platforms.

The recruitment success worked so well that Voices.com made social media a critical part of their overall marketing strategy. For them, "fans and followers" mean traffic, traffic means customers and customers mean sales. They have developed a corporate social media strategy and employ a full-time Social Media Manager to pursue these efforts.

Was it hard to make this switch? Ciccarelli reported, "The implementation was a matter of weeks and was more the implementation of a full-blown social media strategy rather than treating online engagement as a pilot project or series of experiments. We decided to be industry leaders by embracing our community through social networks. We also built our reports, dashboards, posting schedules and guidelines in the same timeframe using the online tools." Now think about your own recruiting costs or advertisement targeting ability. What could your organization achieve by rethinking the "traditional" approach?

We are just at the beginning of seeing how technology and its use in different ways are changing our work environment and making historical "best practices" obsolete. Technology supports greater collaboration, and the collaboration model and communication flow of the Digital Cowboys will become the new norm, especially in regions were people under 30 make up 60 to 70 per cent of the population. We wrote about the communication model of Thomas Malone in Chapter 1, and it is worth looking at it again through the lens of the changes we have discussed in this section. From Malone's summary:

But in the knowledge-based economy that is emerging, globally connected, decentralized decision makers will play increasingly important roles. Figuring out how to design effective decentralized systems and how to manage the continually shifting balance between empowerment and control will not be easy. But I believe that mastering this challenge will be one of the most important differences between organizations that succeed in the next century and those that fail.

Malone was right that – in 1997 – the design of those effective decentralized systems wouldn't be easy. However, consumer technologies for mobile communication and social networks have simplified the process. And so Malone is correct – today the Digital Cowboys are *connected*, decentralized decision-makers. They make autonomous decisions based on vast amounts of remote information, knowledge and expertise that are all available through electronic and other networks.[18] The task for you, together with your IT organization, is to enable and support this new collaboration model through tools, data access, technologies and processes.

Let's look one last time from a cost perspective at all this communication and collaboration traffic the Digital Cowboys' work-style creates and the cost of building and maintaining the infrastructure necessary to support it. Malone was correct when he identified the fundamental shift in communication structure the Internet – at that time with only 30 million users – would enable and drive: "… But in addition to being a technological enabler of other organizations, the Internet's technical architecture and governance structure themselves provide models for structuring highly decentralized organizations."

By 2011, the cost factor of the communication itself for organizations and for the Digital Cowboys is tending towards zero. Many Digital Cowboys today are permanently connected through "flat rates". So after a basic high-level communication infrastructure is available and procured, additional "units" of communication become basically free.

As we have outlined from multiple angles, the cost argument of complex IT projects that forced many in the past to choose between different IT capabilities improvements is far less of a factor today. Already, large improvements can be gained by relatively small investments and scalable capability availability through cloud offerings.

Summary: Traditional executives with an old-school management style of sitting in the middle of the organization, pushing orders out to their troops, will not be able to tap into the network of information, innovation, knowledge refinement and experts. Wikipedia and Wikispeed are prime examples of this new way of thinking and organizing.

As a business leader, it is your responsibility to ensure that your IT organization understands how critical this change is for the success of the business and to support them in making the changes needed quickly.

The opportunities are phenomenal. The PlayStation generation can and should be the change agents for our organizations on all levels. By leveraging natural adaptiveness and familiarity with the technology capabilities, these Digital Cowboys bring an acceleration of broader technology adoption to their organizations – organizations that are not "in their way". Pekka and the other leaders sharing their insights here are correct; this is the time to rethink our organizational priorities and how we leverage IT as a key enabler to develop and learn from the next generation of leaders. We have the opportunity, and I think obligation, to move away from a position of "IT" being a separate support function somewhere in a remote data centre to becoming a partner for innovation and the enabler and simplifier of our organizations.

Change is nearly always difficult both for organizations and for individuals. Fundamental change often happens quickly and is accompanied by small indicators that are usually missed by most people. As Birger mentioned earlier, watching the fall of the Berlin Wall was a life-changing experience for him. So it was for me. I was actually there during the days and nights of radical change, experiencing a profound shift that people would have believed impossible only days earlier. With *Agility*, we provided a guide for organizational change through IT. With NO FEAR, we are sharing how leaders can and need to transform themselves and their organizations. I am a strong advocate of developing people and knowledge through collaboration and mentoring; the "NO FEAR community" is a great place to continue the conversation across organizational and functional boundaries.

With the PlayStation generation entering the workforce, we have the opportunity in our organizations to build mentoring relationships between experienced employees who are the repository of institutional and functional knowledge and the Digital Cowboys, unlocking the knowledge of both groups much faster for teams and business priorities. It is critical that we support the Digital Cowboys and our overall organization with state-of-the-art access and collaboration tools that facilitate these relationships.

What is right for the Digital Cowboys is also the path to success for the overall organization. Mårten Mickos, the CEO of Eucalyptus Systems, knows this and has been successfully leading a global contingent of Digital Cowboys for several years. Eucalyptus Systems is the leader in open-source cloud computing platforms for on-premise use. In his previous position as CEO of MySQL AB, Mårten guided the company from its beginnings as a garage start-up to its current status as the second-largest open-source company in the world. A sought-after keynote speaker, Mårten finds his views on technology trends, innovation and leadership noted and discussed worldwide. In the following section, he provides insights on the technology and ideas fuelling the next-generation enterprise.

Mårten Mickos

Building the next-generation enterprise

What do winning enterprises in the 21st century have in common? They build interdependent ecosystems that reach far beyond their own organizational boundaries, and they make use of leading-edge technology to win strategic benefits, positive network effects and economies of scale.

As an example, Netflix has become the juggernaut of movies watched at home thanks to a distribution model that is fantastically convenient for consumers and superbly cost effective for Netflix. Salesforce.com has become a leading platform for those who develop add-on functionality for the CRM system. Facebook combines a technology platform of unprecedented scale and performance with contents generated entirely by its users and with add-on applications and games produced by third-party vendors. And although Apple may do their R&D in a silo, hidden from everyone else behind firewalls, their business edge is vitally founded on ecosystems. They have the widest selection of song titles and the widest selection of smartphone applications. Common to all of these is how they use modern technology to bind everyone in the ecosystem together in such a way that the total is worth more than the sum of the parts.

As the world increasingly moves online, the boundaries between organizations and between people break down. Value creation happens in all these new encounters. Network effects transform small ideas into massive businesses. What matters is not what your own team can produce under a command-and-control regime, but what the whole ecosystem can produce in a self-organizing pattern. The fence around the factory, the physical wall of the office, the firewall of the IT systems – the boundaries that for the last decades and centuries have made us feel safe and protected – are losing out in a world that is increasingly interconnected and interdependent. Singular and hidden production and value creation is still uniquely useful, but it needs to be augmented with the sort of value creation that happens in massive and seemingly spontaneous and ad hoc interactions in the ecosystem.

In order to build for success in this new era, we must organize people and orchestrate IT resources differently. We must fearlessly question the structures of the past. To the extent that such structures in their rigidity limit value creation, we must abandon them and employ new structures. The new structures and models must be employed with rigour, but not with rigidity. We must throw ourselves into the open, engage with the constituents of our business, and create value in massive numbers of relatively modest interactions. In the aggregate, this linkage between our producers, users, customers, partners, influencers and opinion leaders is what creates massive business value and sustainable competitive advantages.

The power of the ecosystem, the total value of network effects and the aggregated wisdom of the crowds today widely exceed the value creation ability of any individual organization. When the world was smaller and less connected, this was not the case. But today, with about two billion people on the Internet and about six billion devices connected to it, there is no singular entity that can measure up against the connected planet in terms of might or productive power.

A real-world example

I experienced this shift first-hand as CEO of MySQL AB from 2001 to 2008. MySQL is the world's most popular open-source database engine. It is used on tens of millions of websites around the world as the place for storing structured data. Google's ad business runs on MySQL, as do Facebook and Twitter. The MySQL product is downloaded over 70,000 times a day by software developers and hobbyists all over the world. The network effects of this massive ecosystem are so vast that our company, with fewer than five hundred employees, could successfully compete against the database giants of the world – IBM, Oracle and Microsoft – which have tens of thousands of salaried employees working on their database products.

We organized our for-profit company in a new way, turning the organization inside out. Most employees worked from their homes in over thirty different countries, and everyone was encouraged to engage and interact with the world around them. By not having a central location or an office to hide in, all our employees became ambassadors of the company, travelling to see users, customers and partners. By empowering the individual employees to build extensive networks in the marketplace, we increased the range of our interaction with the outer world enormously. On paper, we looked like a small organization. But with nearly everyone out there interacting with users, customers, partners and industry opinion leaders, we had a footprint that I would estimate as two orders of magnitude larger.

Having 70 per cent of our employees working from home in 110 major locations across eighteen time zones forced us to come up with smart ways to communicate and cooperate. The main principle was to move everything online. We debated online. We argued online. We did our work online. We were social online. In order to create a genuine company culture across such vast distances, it was important that everyone showed their *real* self online.

Just as you – at the water cooler in an office – exchange comments about your personal life with your colleagues, we did the same online. For each new baby born to an employee, an e-mail would go out to everyone celebrating the YAMB (Yet Another MySQL Baby) and our HR team would send a small gift to the newborn. One team instituted a traditional Christmas party which was held entirely over IRC (Internet relay chat). For Christmas gifts, they sent each other URLs to websites that showed a picture of something they would have liked to give each other.

To keep everyone up to speed with what was happening in the company and its ecosystem, we held monthly conference calls with all employees. During the calls, most participants would be muted, but everyone had access to an IRC channel where they could discuss, debate and ask questions. As the CEO, I first presented the latest about our business and then started answering the questions and comments I saw on the IRC channel. In this way, we shared information broadly and effectively. The conference calls were easy to listen to because just one or a few people would speak on the line. At the same time, they were massively interactive thanks to the backchannel on IRC. To lighten the atmosphere a little, I had a habit of ending every conference call by singing a song for the whole company. I am not really a singer, so I had to resort to what I knew best – simple Scandinavian drinking songs. Our employees all over the world loved it (or so at least they politely told me).

We also built and employed our software and IT tools and systems differently. All employees had PCs or laptops of their own choice. We required only that they had fast and reliable Internet access. The software systems we built and used all had to be available from anywhere on the Internet and usable through an Internet browser. This key principle freed us from the curse of locality. Our employees could be anywhere in the world and still have access to the same resources that were available in our one major office. We were big users of e-mail, IRC and various instant messaging (IM) solutions. We developed our own personnel directory where each employee presented not just their contact information but also an overview of their skills and areas of work. They could also upload pictures and provide information about their family and their hobbies. This made it easier for new employees to get to know their colleagues, despite working at different locations.

We could hold meetings on the fly, irrespective of where the various participants were at the moment. We even had a home-made system of programme scripts that, given the list of attendees for an in-person meeting, would access public travel sites to figure out the cheapest and least time-consuming place for that particular team to meet.

So let us look more closely here at the two main areas that

need modernization: the organization and the underlying IT infrastructure with applications and services to run the business on.

The organization

The main organizational shift from the 20th to the 21st century is that command-and-control as a management principle is largely gone. It matters little that you think you can control your employees when in reality they choose their own tools and work hours, and they channel their passion into topics and projects that excite them. As we keep automating mundane and repetitive tasks, human beings increasingly focus on creative and innovative tasks. In such roles, attention and passion influence the outcome by an order of magnitude of two. You can force an employee to focus on a certain task for ten hours, or you can entice him to do the same job with better results in ten minutes if he or she is passionate about it. Command-and-control is out; vision and culture are in.

With vision and culture as the key management principles, you can build an organization that brings the best out of itself. The leadership of the organization must set a compelling vision. There is a mountain to climb, a cave to descend into, a river to cross or a puzzle to solve. The role of the leader is not to provide answers to the questions that appear on the way, but to paint the vision of the magnificent future that awaits successful completion of the task. With the vision clearly in their minds, employees empowered by self-governance and great tools will come up with the necessary tricks to take them to the goal. The more compelling the vision is, the more it will stimulate creativity and passion among the team members.

No vision, however, is complete without a cultural framework. It is not just about the destination, it is also about the journey. The leadership of the organization must define a clear company culture which the members of the team can respect and subscribe to. Culture is not a set of definitions. Culture is the set of common threads of every action taken in the organization. Defining and setting up the culture means showing a good example, and making sure everyone understands what that example means. Through their body language, actions and routines, leaders become the standard-bearers of the company culture. Human beings are good at reading from each other what's allowed and what's not, and what's expected and what's not. That's at the core of setting up the culture.

With a compelling vision and a strong culture, organizations can reach heights otherwise unattainable. Equipped with the vision and the culture, the individual team members are also ready to successfully engage in the broader world – the online ecosystem that surrounds the corporate enterprise. They know what their mandate is and they know how to go about reaching it. This allows them to work together with individuals from other organizations without losing themselves in a chaos of too many players, each one with their own agenda.

The IT infrastructure

For an organization to be able to fulfil its mandate productively, it needs powerful tools at its disposal. The right plan, the right people, the right data are useless if an organization does not have the right tools. An innovative corporate culture can't reach its full potential without an IT infrastructure that allows the business and its people to be flexible, adaptable, coordinated and cost efficient. In today's world, such tools nearly always comprise software applications accessed through a variety of devices. I've talked about the importance of making all systems available through a browser on the Internet and enabling access from a variety of connected devices – PCs, laptops, smartphones, tablets, and whatever new devices hit the consumer market. Equally important is the way an infrastructure handles data and security.

The world is not going to become less connected, so understanding, managing and protecting data in complex and interconnected systems are the most critical components of your IT infrastructure. Data that can be analysed and distilled becomes business intelligence. This is possible only if it is easy to search through data, determine a way to assign relevance to it and figure out ways to deal with the various degrees of currency of data – real-time, current, older, back-up, archives, etc.

Security surrounding data and access is not a new issue – it's just more multifaceted. Security must be built into the core framework of the IT infrastructure and employ a layered defence. But, at the same time, increased security measures must not complicate how your cowboys access data. Life for users has to be easy – single sign-on, multiple connection options and similar features.

Your infrastructure brings more flexibility to your business when IT stacks are divided into layers. Each layer can be developed and optimized on its own. Yet every layer can support multiple implementations above it and perhaps beneath it. This layered approach enables massive scalability and agility.

You have put an innovative culture and flexible IT infrastructure in place and then disaster, natural or man-made, strikes. If you have not designed for failure, your business may be at risk. You need to build resilience into applications so they can deal with a failure in the underlying infrastructure. On a user level, this may mean having back-up e-mail systems to turn to if the main one fails. On the application level, it typically means having a fresh back-up of the data stored elsewhere so that in the event of a failure, the application can quickly be pointed at the back-up data.

Unfortunately, many of today's current software applications represent visions and frameworks from decades ago. These applications fulfil the very basic function for which they were designed. However, they fail to make use of the enormous advances in computing we have seen in recent times and fail even more to make use of the enormous increase in useful and timely information available to all of us at all times. Every problem is naturally an opportunity. Those who can put modern IT to productive use for their employees will be amply rewarded in higher productivity and more insightful decisions. If you can combine the limitless innovation of the human being with the limitless power of computers to crunch data, you are on a path to success.

Adapting to change

Today, everything is online – people, organizations, information, interactions and so on. Whatever is online is, by definition, connected. For this reason, firm boundaries exist only in specific cases. Most activity is happening in a continuum, in an ecosystem of many players with different mandates. The key to success in such a world is to find a way of letting go of the instinct to strictly control everything. By opening up to useful interactions across boundaries, you invite the world around you to partake in the value creation. Although not all players will have the inclination, desire or ability to do so, many will. The power of that crowd is bigger than the power of your entire cash balance and employee pool. That's the power of the crowd.

When you open up and empower your organization in this way, you will experience a surge in productivity. Your employees will step up to the plate and take more responsibility and more initiative. Superficial tasks that produce no value will be left aside, and people will focus on what really makes a difference for customers. Leadership also becomes easier, because employees take on a bigger share of responsibility themselves.

Organization for this is a new exercise. It is not unlike the models that mankind employed thousands of years ago when villages were formed and people for the first time got together in bigger groups. But in today's world, physical presence and physical distances play little or no role. The entire planet is the village and, hence, the bazaar is much bigger. Navigating that bazaar requires a new approach to organizational culture and to leadership.

Hand in hand with this goes the equipping of the team with

powerful tools. Again, there is nothing new in noting that the tribes or groups with the most powerful tools were the most successful. Today those tools come in the form of online applications with an unprecedented ability to distil insight from vast amounts of ephemeral and permanent data.

To win and to enable others to win, you must fearlessly see the world the way it is today – abandoning old mental models that have become obsolete while maintaining those that carry eternal validity. This requires an open mind, a large portion of curiosity and an ability to quickly try out new models. The best way we know to make things better is to follow the Darwinian model of trial and error. As Darwin said: "It is not the strongest who will survive, nor the most intelligent, but the one most adaptive to change."

To learn more about Mårten and to continue the conversation, go to **www.nofear-community.com**.

8 Fearless means stupidity. NO FEAR can mean success

YEARS AGO, I was in the IMD auditorium listening to Bertrand Piccard tell his incredible story about being the first person to circumnavigate the globe in a hot-air balloon. I was struck by this adventurer's story, especially when he told the audience:

During the nineteen days I sat in the basket of the hot-air balloon, the only ways I had available to impact my own fate were eating right and adjusting the height at which I flew. Airflows in the atmosphere propelled me forward. Once in a while I'd hear on the radio that I was being blown into areas I didn't have permission to overfly. Even then the only thing I could do was hope that the countries below me understood that I was a harmless adventurer and weren't going to shoot me down in a moment. On this trip, I learned how important careful – to the last detail – preparation is. Thinking about all the stuff you didn't do to prepare or what goes through the mind of a Chinese fighter pilot is a waste of time when you sit in a basket six kilometres above the border zone between China and Mongolia. At that moment, you have to concentrate all your strength on the things you can have an effect on. And that's adjusting the height at which I flew.

NO FEAR was born out of a strong belief that a leader today must re-evaluate his or her beliefs and operating methods. The PlayStation generation, plus the freedom provided by technology and globalization, is far too big an opportunity to squander. It's not just a question of dealing with a new generation. It's primarily a question of new opportunities and GROWTH. And by growth, I mean both quality and volume – in a sustainable manner. The discussions and interviews I conducted for this book, as well as the more informal comments I heard from Digital Cowboys, have strengthened my views on this matter.

We are faced with a wave of changes in the coming years and we must prepare for them, both as organizations and as individuals. And the change has already started. Its power and breadth are so huge that passively sitting on the sidelines and observing what happens will be to no one's benefit. Can we as leaders handle this change or will it imprison us?

During this journey of discovery I've noticed that differences in age, gender, education, culture or geography have no impact on our basic conclusion. Not in the least. So I'm not saying that the younger generations will somehow alone make everything great. And don't bother looking for praise for masculine leadership or hosannas for the Western leadership culture. Every discussion we had while developing this book, and all the columns by our guest authors, underlines the importance of diversity. We are talking about a far-reaching equality in many forms – equality between experts, between various organizational levels, and between countries. In companies, all experts must have the same opportunities to do their job and have an impact – including the leaders.

The one thing I'm honestly worried about is our capability here in the West to truly comprehend the speed and depth of the change. We tend to throw around economic growth percentages and increases in consumer demand as gauges of future happiness. It should be clear to everyone that there will be no real growth unless something truly unique is created. Creating something unique requires developing people's know-how, the sensitivity to notice things early enough and the ability to quickly learn what's needed, as well as a bit of luck.

The age of the Digital Cowboy will bring with it a major change in corporate and societal structures. Companies will clearly be either international players or very limited ventures that confine themselves to the local market. The limits of international growth will no longer be tied to the "critical mass" of a certain place. In many industries, we'll see the emergence of players that are the biggest in the world, but whose physical resources and offices do not cover the whole globe. When this inevitably happens, the most important factor in generating international growth will be the birth of a sufficiently international expert cadre and culture.

As far as societal structures go, this same change will divide countries and continents into winners and losers in a more definitive manner. At the moment, countries rich in resources or industrial production reap the rewards of taxes and other incomes. During the next phase, much of the value creation and wealth will be beyond the taxation reach of many countries. Free trade, population mobility and, first and foremost, the ability of the new generation to function and extend trust over national boundaries will change the game. Traditional tax collectors – countries – just don't have the functional mechanisms in place to deal with a situation where a large proportion of production and commodities is distributed digitally. At the same time, those countries that can build the most enticing environments for the best Digital Cowboys can also develop the services their societies offer. Fearlessness and how these changes are faced will irrevocably change not only our companies, but also society at large.

I'm a born optimist, maybe even an idealist, in my belief in positive development. But I also see a dark side to it all. What happens if we fail to really grasp this change? In an organization, the results will manifest first as apathy and then as restlessness as experts start to leave, and then as an unpleasant and uncontrolled downward spiral of negative development.

This applies equally to both societies and companies. The only differences are in timelines and scale.

▶ The burden of doing only the right things

To succeed, a leader must have the ability to combine the hard core of industrial leadership with a front-line sensitivity. Achieving this requires a number of structural and psychological changes that each leader has to undergo in their own unique manner. For many of us, finding the balance between driving through bulky top-down changes and efficiently utilizing the network will be the source of many sleepless nights. Changing what you do in a fundamental way is hard, but the real challenge is in how you help this change make its way through to every expert in your organization.

Planning separates the courageous from the foolhardy. Someone who is foolhardy base-jumps off a cliff after checking that the thing on their back vaguely resembles a parachute. Someone who is courageous will jump too, but he or she will know the direction of the wind and will have observed the performance of a few of the previous jumpers. The courageous one also knows for sure that the thing on his or her back really is a parachute and that it will open automatically if the jumper passes out. If well planned, jumping off a cliff is a sign of courage, but the potential for a fatal error is still always present.

Foolhardiness is *not* a prerequisite for leading Digital Cowboys or taking a business into a new growth market. Just plain courage is enough. You have to realize that visible mistakes will be made, all the time. You have to be able to deal with the fact that your subordinates will be aware of all your snafus through their Twitter accounts even before you've realized you screwed up. If the leader's reaction to all this is more control and steering structures, he or she will probably have started a cultural downward spiral that is impossible to reverse. You have to realize that the greatest insights are found when the strength, challenge and vision of this powerful and rapidly pulsating network is used to develop all individuals, the leader included.

Simplified processes and models are the parachutes of companies and their leaders. They're built with great passion and they provide operations with a basic sense of security. If you jump out of a plane and know that your parachute will open, you can calmly steer yourself towards a good landing spot. In base-jumping, an opening parachute is not sufficient guarantee of survival. In these jumps, the jumper will be travelling at a breathtaking speed very close to a cliff face. The smallest unexpected gust of wind is potentially fatal if it throws the jumper against the cliff. This analogy works well for leading Digital Cowboys. The leader is forced to operate very close to the front line, the customers – the entire ecosystem. You may have the best parachute in the world strapped to your back, but external factors will still mercilessly throw you hither and yon. If you do not react quickly enough, the fact that you did everything by the book will make absolutely no difference at all.

Ten years ago, at a lecture, Professor Bill Fischer gave me a strong push that resulted in me researching the inner life of expert organizations. He got me thinking about which leadership factors really matter. The characteristics related to the inner life of top experts that Bill shared are a very close fit with today's entire PlayStation generation. What was, in 2002, the way a few top experts lived life is now the expectation of a whole generation for themselves.

For me as a leader, learning brings great joy and satisfaction. I've been privileged to have the following bunch as my work group and sparring partners on this NO FEAR voyage of exploration:

The experiences Arkady Dvorkovich, Birger Steen and Victor Orlovski have had in the rapidly developing Russian market strengthen my belief, as a European taxpayer, that shared growth with Russia is vital for our survival. The up-and-coming generation of Digital Cowboys in Russia is very motivated in the areas of international growth and development. This generation has never spent a day within the structures of the Soviet system and the skills of these Digital Cowboys tend to outstrip those in many other European countries. As the old Russian saying goes, "A Russian is slow to saddle a horse, but a very fast rider." It took twenty years and a new generation for modernization and development to really pick up speed.

What Alex Lin has to say about China supports my view that China is already the leading country in the use of social media and technology. No other country outstrips China when it comes to daily use of these technologies. And this is just the beginning. The tiniest fraction of China's economy comes from exploiting intellectual property and service innovations. The situation will be radically different in 2020. And most of the services that are going to change the picture will be created by the PlayStation generation.

Kari Hakola has 40 years of experience and boundless energy for digging deep into the heart of people and organizations. Seeing him in action has been one of the most educational and energizing experiences of my career. It's amazing to see how pitiless analysis and a big heart can be used to both simplify and strengthen what is important. I met Kari at a training session we both attended and thought he was a stubborn old man. I was wrong. Kari has proved to be one of the best Digital Cowboys of all time.

Mark Mueller-Eberstein's and Mårten Mickos's analysis of the impact technology has on leadership and companies steers my thoughts to the importance of strong implementation. I have to ask "when" and "what" can I change in my company in a fast but controlled manner so the tools I offer the Digital Cowboys pulsate and develop at the same rate as my top experts do. It's clear that success involves more than new versions and additional functionalities. Technology exploitation and choices are a change like any other. As a top-down edict, it won't work. Observing and communicating with front-line clusters, as well as a sensitive leadership, are prerequisites for success.

Philipp Rosenthal was once a front-line rider in my Digital Cowboy posse. I will always remember those moments when his piercing questions about new services and corporate development came at me like a hail of bullets. A few years have passed since those days in Munich. After you've read Philipp's text, take note that expectations for leadership and methods of communication that are contrary to many people's ideas of what is "German" have in a very short period of time become mainstream inside and outside that country. Digital Cowboys are no longer the exception. They are the norm for a new way of working. Philipp's text also confirms an observation: Digital Cowboys may seem like *American Idol* contestants who are accustomed to an easy life. The truth is they demand that you deal with them as transparently, honestly and challengingly as they do with you as a leader. This is the true grit of leadership. A leader hiding behind processes, structures and individual metrics does not show true grit. To a Digital Cowboy, this represents an evasion of the responsibility of a leader. You are basically yelling useless instructions from atop a neighbouring hill.

▶ Complete transparency and trust are key

This first leg of the NO FEAR expedition brings me to an intermediate conclusion: the only formula for success is complete transparency in what you do and the guts to push yourself on to the front line. A leader has to be truly interested and engaged, as well as have a deep trust in the people and the experts around him or her. Picking the right people will imbue growth clusters with the kind of energy that will make their development unstoppable. The occasional merciless challenge will do the same even more effectively. When no mistakes are made and no information is shared, nothing is learned.

Leadership in the world of the Digital Cowboys does not require an outsize charisma like Leonard Bernstein's. Instead, you must pick the brightest pearls in the same way the maestro brought together a group of virtuosi to work as a single, productive team. The shiniest pearl is surely the leader's own vision of the direction in which a product, service or innovation is heading. Professional leadership as a phenomenon is dead and buried. I'm fairly sure Bernstein didn't know what the outcome for *West Side Story* would be when he put together his team. I'm equally sure he had his feelers out on the front line throughout the development of the musical. *West Side Story* changed the world of music and carries on Bernstein's name, but it was the result of cooperation by very different people under his strong leadership.

Part of the transparency of leadership involves communicating clearly the things that you want to strengthen and, conversely, simplify. Simplification affects – and should be embraced by – all the things that do not directly create value for your customer. Processes, support functions, leadership structures and metrics are all such things. In the age of the Digital Cowboy, perpetual amplification is concentrated on people's know-how, sharing information and strengthening as well as re-evaluating the company's strategic direction. It's key to understand that the simplification of things can be achieved only as a top-down change. All the areas that require amplification have to be born, raised and evaluated on the front line – close to the customers.

From the earliest stages of the NO FEAR journey, I've been asked one set of questions hundreds of times. Does leading Digital Cowboys inevitably result in those raised in a traditional leadership culture falling off the train? Could it be that despite goodwill and a desire to change, the old ways of doing things and the old comfort zones always get the best of us? If this is the case, does it mean that in order to succeed, a company must get rid of supervisors who are unable to embrace the change?

My answer is: yes, sometimes you must do it. As we saw when we took a look at PSF culture, one bad apple can spoil the whole bunch when it comes to transparency and the sharing of information. I'll also say that I've experienced some incredibly inspiring and positive stories of learning. Learning from the front line has no correlation with age or educational background. How fearlessly and curiously you approach life is what matters. There has been a massive change in which information sources to believe and where to learn – for good. In order to gain access to this information, you have to give more of yourself to others and be hungry. Without fear.

▶ The grand finale: conclusions

We've tried to provide knowledge, our own opinions and a few development suggestions to support our contemplations in this book. The rest is up to you. We encourage you to explore and define your own role in this battle. We strongly believe that, as a leader for today and tomorrow, you must ride at the front of your troops. You need to be an Alexander the Great of the PlayStation generation. As a leader, you have to train your own supervisors to understand that each resource you've got lined up is actually an individual. As an expert, you have to tell your leader your opinions and suggestions for improvement even when he or she thinks there is no time to listen – or when his or her ego gets in the way of learning something new.

We'd like to thank all who accompany us on this journey. We'd especially like to thank all the hundreds of Digital Cowboys with whom we've had the opportunity to ride. They were generous with their scathing whips, as well as knowledge. We learned a lot. Experiencing such energy, fast learning and fearlessness when confronted with something new has been a source of joy.

To support our journey, we hope you add your own opinions, counter-arguments and stories to the journey. It's easy. Just go to **www.nofear-community.com/community** and tell us and other readers what you think.

NO FEAR. Strength and honour.

Beijing, 4 April 2011

Pekka
email: bulldozer@nofear-community.com
twitter: pviljakainen

Mark
email: markme@nofear-community.com
twitter: markMEberstein

Bibliography

Bennett, Drake. "Assessing Wikipedia, Wiki-style, on its 10th anniversary: how the online 'temple of the mind' became the go-to site for looking stuff up: a drama told in the open-source style of Wikipedia", *Bloomberg Business Week*, 6 January 2011.

Bennis, Warren and Robert J. Thomas. *Geeks & Geezers*, Cambridge, MA: Harvard Business School Press, 2002.

Bishop, Todd. *Wikispeed: the future, as viewed from inside a Seattle storage unit*, TechFlash.com, 9 February 2011.

Boorstin, Daniel. *The Americans: The Democratic Experience*, New York, NY: Random House, 1973.

Boynton, Andy and Bill Fischer. *Virtuoso Teams*, London: FT/Prentice-Hall, 2005.

Boynton, Andy and Bill Fischer, with Bill Bole. *The Idea Hunter*, San Francisco, CA: Jossey-Bass, 2011.

Christensen, Clayton. *The Innovator's Dilemma: The Revolutionary Book That Will Change the Way You Do Business*, New York: Harper Paperbacks, 2003.

Clarkin, Larry and Josh Holmes. "Enterprise mash-ups", *Architecture Journal*, Microsoft, 2011.

Collins, Jim. *Good to Great: Why Some Companies Make the Leap ... and Others Don't*, New York: HarperBusiness, 2001.

Collins, Jim. *How The Mighty Fall: And Why Some Companies Never Give In*, New York: Jim Collins, 2009.

Friedrich, Roman, Michael Peterson and Alex Koster. "The rise of Generation C: how to prepare for the connected generation's transformation of the consumer and business landscape", *Strategy + Business*, 62, Spring 2011.

Grossman, Lev. "2045: the year man becomes immortal", *Time*, 10 February 2011.

Guterman, Jimmy. "Do we need leaders?", *Harvard Business Review* blog, 11 February 2011.

Heider, John. *The Tao of Leadership: Lao Tzu's Tao Te Ching Adapted for a New Age*, Lake Worth, FL: Humanics Publishing Group, 2005.

Henderson, John C., Kathleen F. Curley, Stephanie Watts, Andy Corbett, Zachory Halloran and Mark Mueller-Eberstein. *Network Teams: Achieving Exceptional Performance in a Globally Connected World*, Boston, MA: Institute for Global Work, Boston University, January 2010.

Hopkins, Michael S., Steve LaValle and Fred Balboni. "10 insights: a first look at the new Intelligent Enterprise Survey", *MIT Sloan Management Review*, 1 October 2010.

Huy, Quy Nguyen and Henry Mintzberg. "The rhythm of change", *MIT Sloan Management Review*, 44(4), Summer 2003.

Justice, Joe. www.WIKISPEED.com.

Malone, Thomas W. "Is empowerment just a fad? Control, decision making and IT", *Sloan Management Review*, 38(2), Winter 1997.

Maznevski, Martha. *Managing Uncertainty: Simplify and amplify with the three Cs*, Lausanne, Switzerland: IMD, June 2009.

Millar, Michael. "Does your firm need its own mobile app?", *BBC News*, 7 April 2011.

Mueller-Eberstein, Mark. *Agility: Competing and Winning in a Tech-Savvy Marketplace*, New York: Wiley, 2010.

Niccolai, James. "WEB 2.0: so what is an enterprise mashup anyway?", CIO.com, 23 April 2008.

Peters, Tom. *Professional Service Firm,* New York: Alfred A. Knopf, Inc., 1999.

Pink, Daniel. *Drive: The Surprising Truth about What Motivates Us,* New York: Riverhead Books, 2009.

Rudd, Charlie. "Agile innovation, or how to design and build a 100 MPG road car in 3 months", *The Agile CEO,* 2 February 2011.

Salkowitz, Rob. *Young World Rising: How Youth Technology and Entrepreneurship Are Changing the World from the Bottom Up,* New York: Wiley, 2010.

Schwaber, Carey. *The Changing Face of Application Life-Cycle Management,* Cambridge, MA: Forrester Research, Inc., 2006.

Shore, Bill. *The Imaginations of Unreasonable Men,* New York: Public Affairs, 2010.

Silvester, Simon. *Mobile Mania: A Manual for the Second Internet Revolution,* Wundermann, 2010.

Tapscott, Don. *Grown Up Digital: How the Net Generation Is Changing Your World,* New York: McGraw-Hill, 2008.

Willis, David A. *iPad and Beyond: The Media Tablet in Business,* Garnter Research, 15 March 2011, ID no.: G00211735.

Zuboff, Shoshana. *Creating Value in the Age of Distributed Capitalism,* McKinsey & Co., 2010.

Notes

1: Fear equals failure
1. John C. Henderson, Kathleen F. Curley, Stephanie Watts, Andy Corbett, Zachory Halloran and Mark Mueller-Eberstein. *Network Teams: Achieving Exceptional Performance in a Globally Connected World*, Boston, MA: Institute for Global Work, Boston University, January 2010.
2. Mark Mueller-Eberstein, *Agility: Competing and Winning in a Tech-Savvy Marketplace*, Wiley, New York, 2010.
3. Mueller-Eberstein, *Agility*.
4. Thomas W. Malone, "Is empowerment just a fad? Control, decision making and IT", *Sloan Management Review*, 38(2), Winter 1997.

2: Why should I follow you?
1. John Heider, *The Tao of Leadership: Lao Tzu's Tao Te Ching Adapted for a New Age*, Lake Worth, FL: Humanics Publishing Group, 2005.
2. Carey Schwaber, *The Changing Face of Application Life-Cycle Management*, Forrester Research, Inc., Cambridge, MA, 2006.
3. Tom Peters, *Professional Service Firm*, Alfred A. Knopf, Inc., New York, 1999.
4. Jim Collins, *Good to Great: Why Some Companies Make the Leap ... and Others Don't*, HarperBusiness, New York, 2001.
5. Peters, *Professional Service Firm*.
6. Literally, as the term is directly taken from the branding of cattle practised by cowboys protecting the herds that were entrusted to them.
7. Daniel J. Boorstin, *The Americans: The Democratic Experience*, Random House, New York, 1973, p. ix.

8. Not to mention the rising Generation C: connected, communicating, content-centric, computerized, community-oriented, always clicking. Roman Friedrich, Michael Peterson and Alex Koster, "The rise of Generation C", *Strategy + Business*, Spring 2011.
9. Andy Boynton and Bill Fischer, with Bill Bole, *The Idea Hunter*, Jossey-Bass, San Francisco, CA, 2011.
10. Andy Boynton and Bill Fischer, *Virtuoso Teams*, FT/Prentice-Hall, London, 2005.
11. Warren Bennis and Robert J. Thomas, *Geeks & Geezers*, Harvard Business School Press, Cambridge, MA, 2002.
12. Bill Shore, *The Imaginations of Unreasonable Men*, Public Affairs, New York, 2010.

3: Hey, Old Man. What do you know about me? The challenge from the Digital Cowboy's perspective

1. Shoshana Zuboff, *Creating Value in the Age of Distributed Capitalism*, McKinsey & Co., 2010.
2. Daniel Pink, *Drive: The Surprising Truth about What Motivates Us*, Riverhead Books, New York, 2009.
3. Don Tapscott, *Grown Up Digital: How the Net Generation Is Changing Your World*, McGraw-Hill, New York, 2008.

4: How does an industrial-age relic turn into an authentic leader?

1. Martha Maznevski, *Managing Uncertainty: Simplify and amplify with the three Cs*, IMD, Lausanne, Switzerland, June 2009.
2. Jim Collins, *How the Mighty Fall: And Why Some Companies Never Give In*, Jim Collins, New York, 2009.
3. Henry Mintzberg, *The Structuring of Organizations*, http://www.valuebasedmanagement.net/methods_mintzberg_configurations.html.

5: What I as an executive should change in my company, in practice

1. Quy Nguyen Huy and Henry Mintzberg, "The rhythm of change", *MIT Sloan Management Review*, Cambridge, MA, 44(4), Summer 2003.
2. Clayton Christensen, *The Innovator's Dilemma: The Revolutionary Book that Will Change the Way You Do Business*, Harper Paperbacks, New York, 2003.

6: Places of magnificent growth and magnificent failure – emerging markets

1. Heider, John. *The Tao of Leadership: Lao Tzu's Tao Te Ching Adapted for a New Age*, Lake Worth, FL: Humanics Publishing Group, 2005.
2. Jim Collins, *Good to Great: Why Some Companies Make the Leap ... and Others Don't*, HarperBusiness, New York, 2001.
3. Literally "live for centuries, and you will learn for centuries".
4. China Internet Network Information Centre, *The Development Status of Chinese Internet Report*, 27th edn, January 2011.
5. *Gartner Highlights Key Predictions for IT Organizations and Users in 2010 and Beyond*, 13 January 2010.

7: Technology – your saviour or your nemesis

1. Mark Mueller-Eberstein, *Agility: Competing and Winning in a Tech-Savvy Marketplace*, Wiley, New York, 2010.
2. Rob Salkowitz, *Young World Rising: How Youth Technology and Entrepreneurship Are Changing the World from the Bottom Up*, Wiley, New York, 2010.
3. David A. Willis, *iPad and Beyond: The Media Tablet in Business*, Gartner Research, 15 March 2011, ID no.: G00211735.
4. Lev Grossman, "2045: the year man becomes immortal", *Time*, 10 February 2011.
5. Simon Silvester, *Mobile Mania: A Manual for the Second Internet Revolution*, Wundermann, 2010.
6. http://www.oracle.com/technetwork/issue-archive/2011/11-jan/011idm-194107.html?ssSourceSiteId=ocomen.
7. Michael S. Hopkins, Steve LaValle and Fred Balboni, "10 insights: a first look at the new Intelligent Enterprise Survey", *MIT Sloan Management Review*, 1 October 2010.
8. Michael Millar, "Does your firm need its own mobile app?", BBC News, 7 April 2011.
9. Silvester, *Mobile Mania*.
10. http://www.facebook.com/careers, April 2011.
11. Larry Clarkin and Josh Holmes, "Enterprise mash-ups", *The Architecture Journal*, Microsoft, 2011.
12. James Niccolai, "WEB 2.0: so what is an enterprise mashup anyway?", CIO.com, 23 April 2008.
13. This process is fully explained in *Agility*.

14. The single largest factor affecting total cost of ownership (TCO) is staffing cost – 60 per cent, according to the IDC. Recent IDC research has shown that such technology initiatives coupled with organization-wide improvements in IT management processes can reduce IT labour costs by as much as 50 per cent.
15. Jimmy Guterman, "Do we need leaders?", *Harvard Business Review* blog, 11 February 2011.
16. Drake Bennett, "Assessing Wikipedia, Wiki-style, on its 10th anniversary: how the online 'temple of the mind' became the go-to site for looking stuff up: a drama told in the open-source style of Wikipedia", *Bloomberg Business Week*, 6 January 2011.
17. Todd Bishop, "Wikispeed: the future, as viewed from inside a Seattle storage unit", TechFlash.com, 9 February 2011; Charlie Rudd, "Agile innovation, or how to design and build a 100 mpg road car in 3 months", *The Agile CEO*, 2 February 2011; Joe Justice, www.WIKISPEED.com.
18. Thomas W. Malone, "Is empowerment just a fad? Control, decision making and IT", *Sloan Management Review*, 38(2), Winter 1997.